Encyclopedia of Molecular Biology

Encyclopedia of Molecular Biology

Edited by **Gildroy Swan**

R CALLISTO
REFERENCE

New York

Published by Callisto Reference,
106 Park Avenue, Suite 200,
New York, NY 10016, USA
www.callistoreference.com

Encyclopedia of Molecular Biology
Edited by Gildroy Swan

International Standard Book Number: 978-1-63239-274-9 (Hardback)

Printed in the United States of America.

Contents

Preface

There are many different branches in the field of biology, and molecular biology is one of them. It is the study of biology at a molecular level. It majorly concerns itself with understanding the various interactions inside a cell or a system of cells, mainly those between RNA, DNA and protein biosynthesis as well as the method of regulation of these interactions. This is a field that overlaps with other areas of biology and chemistry as well as biochemistry and genetics. A lot of the work done in molecular biology can be said to be quantitative and a lot of work has been done in recent times focusing on the interface of molecular biology and bioinformatics as well as computational biology. There are of course different focus areas in the discipline itself like the study of molecular genetics along with gene structure and function. There are many techniques in molecular biology that can be used for the inference of factors like historical attributes of a species or population. Molecular biology also plays a significant role in helping to understand actions, formation and the regulation of the various parts of cells which are often used efficiently to target new drugs and diagnosis of disease.

This book is an attempt to compile and understand new and ongoing research data in the field of molecular biology. I am thankful to those who put their effort and hard work into this field. I am also grateful to my family and friends who supported me in this endeavour. I especially wish to acknowledge the contributing authors who shared their immense pool of knowledge with me and guided me in the editing process whenever required.

Editor

Molecular Cloning of *phd1* and Comparative Analysis of *phd1, 2,* and *3* Expression in *Xenopus laevis*

Dandan Han,[1,2] **Luan Wen,**[1,2,3] **and Yonglong Chen**[1]

[1] *Key Laboratory of Regenerative Biology, Guangzhou Institutes of Biomedicine and Health, Chinese Academy of Sciences,
510530 Guangzhou, China*
[2] *Graduate University of Chinese Academy of Sciences, 100049 Beijing, China*
[3] *Section on Molecular Morphogenesis, Laboratory of Gene Regulation and Development, Program on Cell Regulation and Metabolism,
Eunice Kennedy Shriver, National Institute Child Health and Human Development (NICHD), National Institutes of Health (NIH),
Building 18T, Room 106, 18 Library DR MSC 5431, Bethesda, MD 20892-5431, USA*

Correspondence should be addressed to Yonglong Chen, chen_yonglong@gibh.ac.cn

Academic Editors: A. Aronheim, M. Gotte, T. K. Kwon, and S. Rodriguez-Enriquez

Intensive gene targeting studies in mice have revealed that prolyl hydroxylase domain proteins (PHDs) play important roles in murine embryonic development; however, the expression patterns and function of these genes during embryogenesis of other vertebrates remain largely unknown. Here we report the molecular cloning of *phd1* and systematic analysis of *phd1, phd2,* and *phd3* expression in embryos as well as adult tissues of *Xenopus laevis*. All three *phds* are maternally provided during *Xenopus* early development. The spatial expression patterns of *phds* genes in *Xenopus* embryos appear to define a distinct synexpression group. Frog *phd2* and *phd3* showed complementary expression in adult tissues with *phd2* transcription levels being high in the eye, brain, and intestine, but low in the liver, pancreas, and kidney. On the contrary, expression levels of *phd3* are high in the liver, pancreas, and kidney, but low in the eye, brain, and intestine. All three *phds* are highly expressed in testes, ovary, gall bladder, and spleen. Among three *phds, phd3* showed strongest expression in heart.

1. Introduction

Aerobic organisms in response to inadequate oxygen availability evolved sophisticated systems to adapt the environment, in which hypoxia-inducible factors (HIFs) play pivotal roles [1–3]. HIF functions as a heterodimer consisting of an unstable alpha subunit, such as HIF1α or HIF2α, and a stable beta subunit, such as HIF1β, also called ARNT1. Under normoxic conditions, the constitutively expressed alpha subunits are hydroxylated by prolyl hydroxylase domain containing proteins, such as PHDs and FIH, whose activity is absolutely dependent on oxygen. The hydroxylation generates binding sites for the von Hippel-Lindau (pVHL) tumor suppressor protein, a component of a ubiquitin ligase complex. Consequently, the alpha subunits are polyubiquitinated and subjected to proteasomal degradation [1, 3]. In contrast, under hypoxic conditions, the activity of PHD proteins is compromised due to low oxygen level and HIF alpha subunits are stabilized, which form active heterodimers with HIF1β to

transcriptionally activate 100–200 genes, including genes involved in erythropoiesis, angiogenesis, autophagy, and energy metabolism [3].

The PHD proteins belong to an Fe(II) and 2-oxoglutarate-dependent oxygenase superfamily. There is only a single PHD family member called Egl9 in worm *Caenorhabditis elegans* and in the fly *Drosophila melanogaster*, while higher metazoans like the vertebrates contain three PHD genes [2, 3]. Although *egl9*-mutant worms are viable [4, 5], inactivation of *egl9* in drosophila and *Phd2* in mice, respectively, both resulted in embryonic lethality [6, 7]. It is intriguing to investigate if deletion of any *phd* genes could cause a lethal phenotype in other vertebrate organisms. $Phd1^{-/-}$ and $Phd3^{-/-}$ mice were normal [7]; however, sophisticated compound and conditional knockout of *Phd1, 2,* and *3* in mice has revealed an important oxygen sensing function of PHDs in angiogenesis [8, 9], erythropoiesis [10–12], and cardiogenesis [7, 13, 14]. The tissue- or cell-type-specific functions

```
H. sapiens     MDSPCQPQPLSQALPQLPGSSSEPLEPEPGRARMGVESYLPCPLLPSYHCPGVPSEASAG 60
M. musculus    MDSPCQPQPALNQALPQLPGSVSESLE--SSRARMGVESYLPCPLLPAYHRPGASGEASAG 58
X. laevis      MDRGHQ-QGKCIRSEVHTANLSEAVP-SVQCVGMGVERFQQYPMKASHPTSDVQSSTNSL 58
               **   *  *           **        ****    *

H. sapiens     SGTPR----------ATATSTTASPLRDG-------------FGGQDGGELRPLQSEGAAA 98
M. musculus    NGTPRTT-------ATATTTTASPLREG-------------FGGQDGGELWPLQSEGAAA 98
X. laevis      DGVHQCHNHLTSVQIKANVEVSSAEQEEQVALLACPVTTLRAMGLASGEMQLVRAAKSGT 118
                *              *   *         *   **

H. sapiens     LVTKG------------CQRLAAQG-ARPE--APKRKWAEDG-------GDAPSPSKRPW 136
M. musculus    LVTKE-----------CQRLAAQG-ARPE--APKRKWAKDG-------GDAPSPSKRPW 136
X. laevis      VVSRERGFIQEIDNNNPCQEVGIQVNGRTDGFATKRKFMENELTEIGVRGIAPESGPSVF 178
                *          **   *   *   * ***        * **

H. sapiens     ARQENQEAEREGGMSCS-----------CSSGSGEASAG-LMEEALPSAPERLALDYIV 183
M. musculus    ARQENQEAKGESGMGCDSGASNSSSSSSNTTSSSGEASAR-LREEVQPSAPERLALDYIV 195
X. laevis      PPMEGKRRKAEVHEDAH-------IHQEYSSTREGQLAQRPVGIKSSAVSPLRMSLDYIV 231
                *      *           *              * *  *****

H. sapiens     PCMRYYGICVKDSFLGAALGGRVLAEVEALKRGGRLRDGQLVSQRAIPPRSIRGDQIAWV 243
M. musculus    PCMRYYGICVKDNFLGAVLGGRVLAEVEALKWGGRLRDGQLVSQRAIPPRSIRGDQIAWV 255
X. laevis      PCITYYGICVKDHFLGEALGSRVLEEVQMLNRSGKFRDGQLVSQRTIPSKNIRGDQIAWV 291
               ** ******** *** ** *** ** *  *  ******** **  ********

H. sapiens     EGHEPGCRSIGALMAHVDAVIRHCAGRLGSYVINGRTKAMVACYPGNGLGYVRHVDNPHG 303
M. musculus    EGHEPGCRSIGALMAHVDAVIRHCAGRLGNYVINGRTKAMVACYPGNGLGYVRHVDNPHG 315
X. laevis      EGKEPGCENIGALMSKIDEVIMHCNGKMENYVINGRTKAMVACYPGNGMGYVRHVDNPNR 351
               ** **** *****   * ** ** *  ****************** *********

H. sapiens     DGRCITCIYYLNQNWDVKVHGGLLQIFPEGRPVVANIEPLFDRLLIFWSDRRNPHEVKPA 363
M. musculus    DGRCITCIYYLNQNWDVKVHGGLLQIFPEGRPVVANIEPLFDRLLIFWSDRRNPHEVKPA 375
X. laevis      DGRCLTCIYYLNQNWDAKVHGGLLQIFPEGRSVVANIEPLFDRLLIFYSDRRNPHEVKPA 411
               **** ********** *************** ***************** ***********

H. sapiens     YATRYAITVWYFDAKERAAAKDKYQLASGQKGVQVPVSQPPTPT 407          51
M. musculus    YATRYAITVWYFDAKERAAARDKYQLASGQKGVQVPVSQPTTPT 419          49
X. laevis      FAMRYAITVWYFDAKERAEAKNKYRLAAGQKGIHVPVSCPGTV- 454        ID (%)
```

FIGURE 1: Predicted primary sequence of *Xenopus phd1* in comparison with human and mouse *Phd1*. Stars indicate identical amino acids in all three species. Hyphens represent gaps introduced for optimizing the alignment. Dashed rectangles demarcate the highly conserved prolyl 4 hydroxylase domain. ID stands for the percentage of amino acid identity of *Xenopus laevis phd1* in comparison with human and mouse *Phd1*.

of Phds defined in mice are well correlated with their abundant expression in corresponding tissues and cells. Except for an early report on the characterization of the temporal mRNA expression profile of *phd2* and *phd3* in *Xenopus* [15], it appears that there are no systematic studies on *phd* genes in other vertebrate organisms. Here, we cloned the open reading frame of *phd1* and examined the temporal and spatial expression profiles of three *phd* genes in developing *Xenopus laevis* embryos as well as in adult tissues. Our data provide a basis for further functional analysis of these genes in the frog system.

2. Materials and Methods

2.1. Cloning of Xenopus laevis phds. As the *Xenopus laevis phd1* (BC159341) in GenBank database is only a partial cDNA lacking the 5′ terminal sequences, we designed

the upstream primer (5′ ACTCTGATCTGCAGTAG-GAGTTGAAT 3′) according to the sequence of *phd1* locus in *Xenopus tropicalis* genome sequence and downstream primer (5′ ATCCCTGTTACACAGTACCAGGGCACGAG 3′) from the partial *phd1* cDNA (BC159341) sequence and successfully amplified the whole open reading frame of *Xenopus laevis phd1* by RT-PCR using *Xenopus laevis* tadpole cDNA as templates. The obtained PCR fragment was cloned into pGEMT-easy (Promega) vector, verified by sequencing, and deposited in GenBank database with accession number (GU108333.1). *Xenopus laevis phd2* and *phd3* cDNAs were also cloned into pGEMT-easy (Promega) by RT-PCR with the following primers designed according to their sequences in GenBank database. *phd2*: forward 5′ AATG-GCTGGTGGAGGAAGCGAGGGTTCTAAC 3′ and reverse 5′ TTCTAGACTTCTTTAACAGCTGGATCAGATG 3′; *phd3*: forward 5′ TATGCCGCCAGGATCTCCCCCATTCGATTTC 3′

```
phd2  MAGGGSEG---------SNQSERDRQYCELCGKMEDLMRCGRCRSSFYCSKEHQRQDWKK   51
phd3  ------------------------------------------------------------
phd1  MDRGHQQGKCIRSEVHTANLSEAVPSVQCVGMGVERFQQYPMKASHPTSDVQSSTNSLDG   60


phd2  HKLFCKIESTIAPASQNTVTVNSKTEQFSSPSDVISSGNTIPNPSCEASGVATESLPSPT   111
phd3  ------------------------------------------MP--------------PGSP    6
phd1  VHQCHNHLTSVQIKANVEVSSAEQEEQVALLACPVTTLRAMGLASGEMQLVRAAKSGTVV   120
                                                              :

phd2  PSDTSQLKKFTDEREELVGGAALRAINE--------------------------------   139
phd3  PLAYSAMPLMQ-------------------------------------------------   17
phd1  SRERGFIQEIDNNNPCQEVGIQVNGRTDGFATKRKFMENELTEIGVRGIAPESGPSVFPP   180
        .   .  :   :

phd2  -------EDYTQASGKAVGSSTSRKVTSGGRPNGQTK--PPLHRIALEYIIPCMNKHGIF!   190
phd3  -----------------------------KPS------LDLEKLALERVVPRLLSSGFC!   41
phd1  MEGKRRKAEVHEDAHIHQEYSSTREGQLAQRPVGIKSSAVSPLRMSLDYIVPCITYYGLC!   240
                               :*         :::*: ::* :   *:

phd2  VLDDFLGQETGDRIECEVKVLHNTGRFTDGQLVSQKS-DSTRDIRGDQITWVEGKELGCK!   249
phd3  YLDNFLGEDIGSRVLDKVRNMHQDGALKDGQLAGHLQGVSKKHLRGDKIAWVSGTEEGCE!   101
phd1  YKDHFLGEALGSRVLEEVQMLNRSGKFRDGQLYSQRT-IPSKNIRGDQIAWVEGKEPGCE!   299
       *.***:  *.*:  :*: ::. * :  ****.:     ..:.:***:*:**.* * **:

phd2  AIGNLMNKMDDLIRHCSGKLGNFTINGRTKAMVACYPGNGTGYVRHVDNPNADGRCVTCI!   309
phd3  PIGLVLSVIDRLVVLCGNRLGQYYVKERSKAMVACYPGNGAGYVRHVDNPTGDGRCITCI!   161
phd1  NIGALMSKIDEVIMHCNGKMENYVINGRTKAMVACYPGNGMGYVRHVDNPNRDGRCLTCI!   359
       ** ::. :* ::  *.::  :: :: :: *:********** ********. ****:***

phd2  YYLNKQWDAKTHGGLLRIFPEGKAQFADIEPKFDRLLLFWSDRRNPHEVQPAFATRYAIT!   369
phd3  YYLNKDWDAKVHGGILRIFPEGSHHVADIEPIFDRLLLFWSDRRNPHEVQPSYSTRYALT!   221
phd1  YYLNQNWDAKVHGGLLQIFPEGRSVVANIEPLFDRLLIFYSDRRNPHEVKPAFAMRYAIT!   419
       ****::****.***:*:*****   .*:*** ***** :*:********** :*::  ***:*

phd2  VWYFDANERARAKEKY-LNTGEKGVRIELNKSSDPAVKEV   408        41
phd3  VWYFDAKERAAARQKF-KRLSES--------REEPPTKES   252        49
phd1  VWYFDAKERAEAKNKYRLAAGQKGIHVP---VSCPGTV--   454       ID (%)
       ******:*** *:.*:         .:.       . *.
```

FIGURE 2: Comparison of the amino acid sequences of three *Xenopus* phds. Stars indicate identical amino acids in all three phds. Hyphens represent gaps introduced for optimizing the alignment. Dashed rectangles demarcate the highly conserved prolyl 4 hydroxylase domain. ID stands for the percentage of amino acid identity of *Xenopus laevis phd1* in comparison with *Xenopus phd2* and *phd3*.

and reverse 5′ TCAGCTTTCTTTAGTGGGAG-GCTCTTCTCTG 3′. These primers were chosen to clone less conserved regions among three *phds* and thus to reduce possible cross signals during whole mount in situ hybridization with antisense probes generated from these plasmids.

2.2. Embryo Manipulation.

Wild-type *Xenopus laevis* eggs were obtained by injecting 1000 IU of human chorionic gonadotrophin (HCG) into the dorsal lymph sacs of adult females 6–8 hours before egg collection. Eggs were fertilized in vitro with minced testes, dejellied with 2% cysteine hydrochloride (pH 7.8–8.0) 30 minutes after fertilization, and cultured in 0.1X MBS (1.76 mM NaCl, 48 μM NaHCO$_3$, 20 μM KCl, 200 μM Hepes, 16 μM Mg$_2$SO$_4$, 8 μM CaCl$_2$, 6 μM Ca(NO$_3$)$_2$, pH 7.4) buffer. Embryos were staged according to Nieuwkoop and Faber [16].

2.3. RNA Extraction and RT-PCR.

Freshly collected tissues were powdered with mortar in liquid nitrogen. Total RNA from embryos and powdered tissues was extracted by using Trizol (Invitrogen) according to the manufacturer's instruction and was digested with DNaseI (Roche). First strand of cDNA was synthesized using superscript I M-MLV reverse transcriptase (Invitrogen). The annealing temperatures and PCR cycle numbers (in parentheses) and the sequences of primers used in the RT-PCR reactions are as follows: *phd1*: (55°C, 28) forward 5′ CAGTCAGAGGACCATACCATC 3′ and reverse 5′ CCTTTGCATCGAAATACCAG 3′; *phd2*: (55°C, 28) forward 5′ CACGGCATCTTTGTGCTTGA 3′ and reverse 5′ GAGTCTTTGCATCCCATTGTTTAT 3′; *phd3*: (55°C, 28) forward 5′ TGCTCTGTGGCAACCGACTT 3′ and reverse 5′ CATGAGGGTTACGCCTATCAG 3′; *ornithine decarboxylase*: (55°C, 23) forward 5′ TGAATTGAT-GAAAGTGGCAAGG 3′ and reverse 5′ CAGGGCTGGGTT-TATCACAGAT 3′.

2.4. Whole Mount In Situ Hybridization.

Embryos were fixed in MEMFA (0.1 M MOPS pH 7.4, 2 mM EDTA, 3.7% Formaldehyde) for 1 hour at room temperature and stored in ethanol at −20°C. Whole-mount *in situ* hybridization was

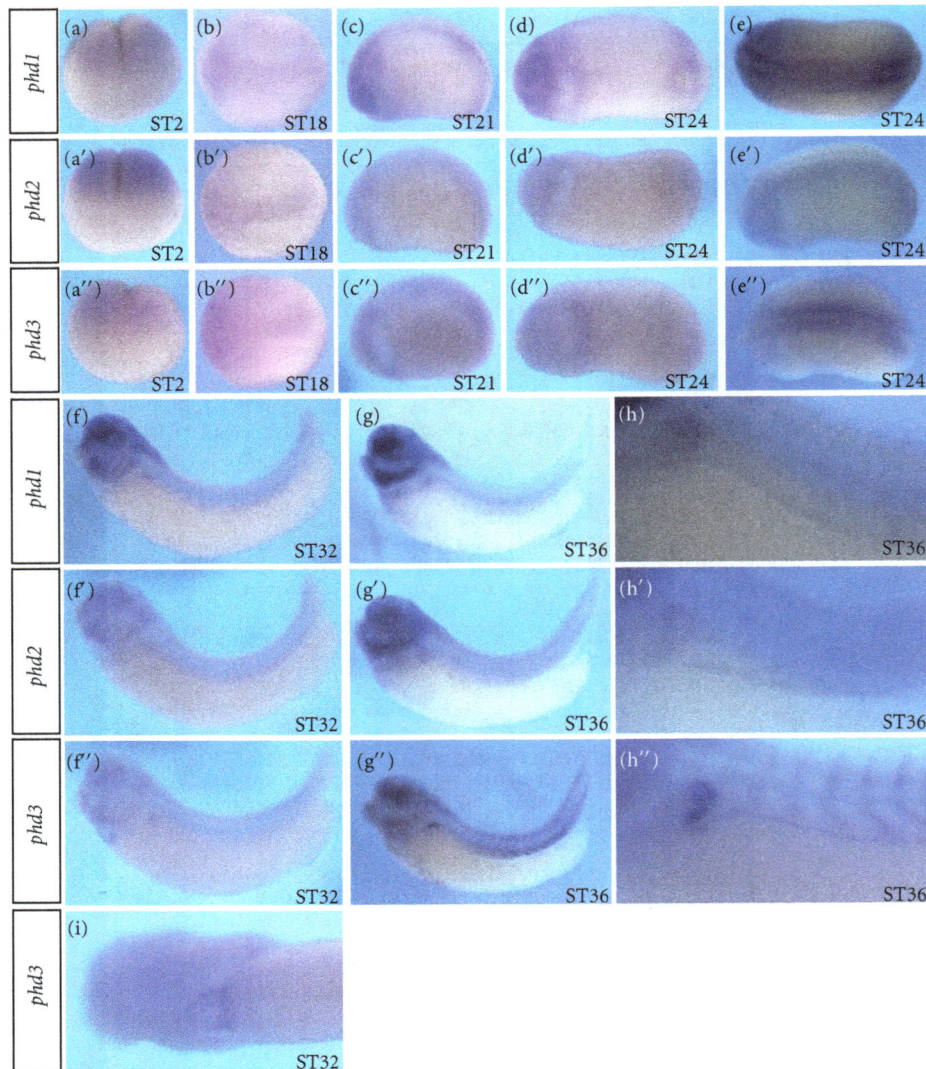

FIGURE 3: Spatial expression of *phd1, 2,* and *3* in *Xenopus* embryos revealed by whole-mount in situ hybridization. (a–a″) Lateral views with animal pole up. (b–b″) Dorsal views with head towards left. (c–c″) Lateral views with head towards left. (d–d″) Ventral views with head towards left. (e–e″) Dorsal views with head towards left. (f–g″) Lateral views with head towards left. (h–h″). Higher magnification views of (g), (g′), and (g″), respectively. (i) Ventral view of (f′) with head towards left.

performed in principle as described by Harland [17], with modifications as reported in Hollemann et al. [18]. To generate digoxigenin-labeled antisense probes, the *phd1*/pGEMT-easy, *phd2*/pGEMT-easy, and *phd3*/pGEMT-easy plasmids were linearized with SalI and transcribed with T7 RNA polymerase.

3. Results

3.1. Isolation of Xenopus laevis phd1. There are three mammalian PHD genes, namely PHD1, PHD2, and PHD3 [3]. Isolation of *Xenopus laevis* homologues of PHD2 and PHD3 has been reported [15]. The amino acid sequence deduced from the whole open reading frame of *Xenopus laevis phd1* shares 51.6% and 49.2% identity with human and mouse PHD1, respectively. Within the highly conserved prolyl 4

hydroxylase domain, the frog sequence shares 80.7% and 80.2% identity with human and mouse prolyl 4 hydroxylase domains, respectively (Figure 1). Among three *Xenopus laevis* phds, the primary amino acid sequence of phd1 shares 41% and 49% identity with those of phd2 and phd3, respectively (Figure 2).

3.2. Spatial and Temporal Expression Profiles of phds. Whole-mount in situ hybridization analyses indicate that at cleavage stages of development, higher levels of maternal transcripts for all three *phd* genes were detected in the animal hemisphere with *phd2* showing the strongest signal (Figure 3(a), 3(a′), and 3(a″)). At neurula stages of development, all three *phds* showed weak and relatively broad expression on the dorsal side (Figure 3(b)–3(c″)). At early tail bud stage of

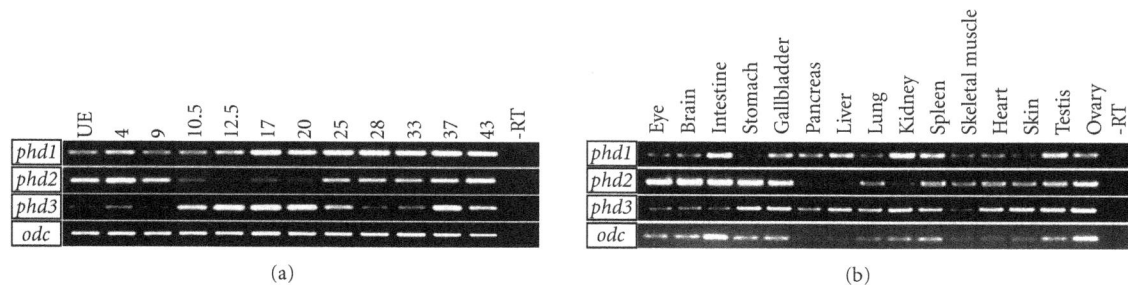

FIGURE 4: Temporal expression profile and adult tissue expression patterns of *phd1, 2,* and *3* revealed by RT-PCR analyses. (a) Temporal expression profile of *phd1, 2,* and *3* in *Xenopus* embryos. (b) Expression of *phds* in *Xenopus* adult tissues. *odc* was employed as a loading control. UE: unfertilized eggs.

development, the dorsal signals became more restricted with *phd1* and *phd3* expression being stronger than phd2 expression (Figures 3(e), 3(e′), and 3(e″)). In addition, a faint signal on the anterior-ventral side of stage 24 embryos was detected for *phd1* and *phd3* (Figures 3(d) and 3(d″)). At tail bud stages of development, more differential expression of all three *phds* was detected in brain, eyes, branchial arches, otic vesicle, and pronephros (Figures 3(f)–3(h″)). A clear signal was detected for *phd3* expression in developing heart (Figure 3(i)).

RT-PCR analysis revealed that, up to stage 33, expression levels of *phd2* and *phd3* just fluctuated in a complementary manner, which has been verified by at least three times of independent experiments (Figure 4(a)). Relatively low level of *phd1* expression maintained till gastrulation and constantly higher expression was detected from neurula stages onwards for *phd1* (Figure 4(a)).

3.3. The Expression of phds in Xenopus Adult Tissues. Overall, transcripts of all three *phds* are detectable in all the adult tissues analyzed (Figure 4(b)). It is of special interest to note that *phd2* and *phd3* showed complementary expression in several tissues. For instance, *phd2* is highly expressed in the eye, brain, and intestine, but low in the liver, pancreas, and kidney. On the contrary, expression levels of *phd3* are high in the liver, pancreas, and kidney, but low in the eye, brain, and intestine. All three *phds* are abundantly expressed in testes, ovary, gall bladder, and spleen. Among three *phds, phd3* showed strongest expression in heart.

4. Discussion

In this study, we report the isolation of the whole open reading frame of *Xenopus laevis phd1* and characterization of the expression profiles of all three *phd3* in *Xenopus* embryos as well as in adult tissues. Consistent with the previous report [15], we detected a complementary fluctuating temporal expression profile of *phd2* and *phd3* during *Xenopus* early embryogenesis. Furthermore, we found complementary expression of *phd2* and *phd3* in several adult tissues. In mice, several lines of evidence have indicated that PHD2 functionally coordinates with PHD3 and *Phd3* is induced upon *Phd2*

loss [13, 14]. The functional link between phd2 and phd3 in *Xenopus* remains to be investigated.

phd3 expression in early *Xenopus* embryos revealed by whole-mount in situ hybridization analysis is reminiscent of zebrafish *phd3* expression [19]. *Xenopus fih* and *hif1α* showed similar spatial expression patterns (data not shown). Thus, in *Xenopus*, it appears that the oxygen homeostasis-related genes, *phd1, 2, 3, fih,* and *hif1α*, may constitute a synexpression group. Consistent with the data in mice [20], *Xenopus phd3* also showed highest levels of expression in adult heart. All three *phds* display expression in the pronephros. It has yet to be defined if *Xenopus* phds play specific roles in the heart and kidney development.

Acknowledgments

This work was supported in part by funds from the National Basic Research Program of China (2009CB941202) and the Key Project of Knowledge Innovation Program of the Chinese Academy of Sciences (KSCX2-YW-R-083).

References

[1] J. Aragonés, P. Fraisl, M. Baes, and P. Carmeliet, "Oxygen sensors at the crossroad of metabolism," *Cell Metabolism*, vol. 9, no. 1, pp. 11–22, 2009.

[2] W. G. Kaelin, "Proline hydroxylation and gene expression," *Annual Review of Biochemistry*, vol. 74, pp. 115–128, 2005.

[3] W. G. Kaelin Jr. and P. J. Ratcliffe, "Oxygen sensing by metazoans: the central role of the HIF hydroxylase pathway," *Molecular Cell*, vol. 30, no. 4, pp. 393–402, 2008.

[4] A. C. R. Epstein, J. M. Gleadle, L. A. McNeill et al., "C. elegans EGL-9 and mammalian homologs define a family of dioxygenases that regulate HIF by prolyl hydroxylation," *Cell*, vol. 107, no. 1, pp. 43–54, 2001.

[5] Z. Shao, Y. Zhang, and J. A. Powell-Coffman, "Two distinct roles for EGL-9 in the regulation of HIF-1-mediated gene expression in Caenorhabditis elegans," *Genetics*, vol. 183, no. 3, pp. 821–829, 2009.

[6] L. Centanin, P. J. Ratcliffe, and P. Wappner, "Reversion of lethality and growth defects in Fatiga oxygen-sensor mutant flies by loss of Hypoxia-Inducible Factor-α/Sima," *EMBO Reports*, vol. 6, no. 11, pp. 1070–1075, 2005.

[7] K. Takeda, V. C. Ho, H. Takeda, L. J. Duan, A. Nagy, and G. H. Fong, "Placental but not heart defects are associated with

elevated hypoxia-inducible factor α levels in mice lacking prolyl hydroxylase domain protein 2," *Molecular and Cellular Biology*, vol. 26, no. 22, pp. 8336–8346, 2006.

[8] L.-J. Duan, K. Takeda, and G.-H. Fong, "Prolyl hydroxylase domain protein 2 (PHD2) mediates oxygen-induced retinopathy in neonatal mice," *American Journal of Pathology*, vol. 178, no. 4, pp. 1881–1890, 2011.

[9] K. Takeda, A. Cowan, and G. H. Fong, "Essential role for prolyl hydroxylase domain protein 2 in oxygen homeostasis of the adult vascular system," *Circulation*, vol. 116, no. 7, pp. 774–781, 2007.

[10] Y. A. Minamishima and W. G. Kaelin Jr., "Reactivation of hepatic EPO synthesis in mice after PHD loss," *Science*, vol. 329, no. 5990, p. 407, 2010.

[11] Y. A. Minamishima, J. Moslehi, N. Bardeesy, D. Cullen, R. T. Bronson, and W. G. Kaelin, "Somatic inactivation of the PHD2 prolyl hydroxylase causes polycythemia and congestive heart failure," *Blood*, vol. 111, no. 6, pp. 3236–3244, 2008.

[12] K. Takeda, H. L. Aguila, N. S. Parikh et al., "Regulation of adult erythropoiesis by prolyl hydroxylase domain proteins," *Blood*, vol. 111, no. 6, pp. 3229–3235, 2008.

[13] Y. A. Minamishima, J. Moslehi, R. F. Padera, R. T. Bronson, R. Liao, and W. G. Kaelin Jr., "A feedback loop involving the Phd3 prolyl hydroxylase tunes the mammalian hypoxic response in vivo," *Molecular and Cellular Biology*, vol. 29, no. 21, pp. 5729–5741, 2009.

[14] J. Moslehi, Y. A. Minamishima, J. Shi et al., "Loss of hypoxia-inducible factor prolyl hydroxylase activity in cardiomyocytes phenocopies ischemic cardiomyopathy," *Circulation*, vol. 122, pp. 1004–1016, 2010.

[15] S. Imaoka, T. Muraguchi, and T. Kinoshita, "Isolation of Xenopus HIF-prolyl 4-hydroxylase and rescue of a small-eye phenotype caused by Siah2 over-expression," *Biochemical and Biophysical Research Communications*, vol. 355, no. 2, pp. 419–425, 2007.

[16] P. D. Nieuwkoop and J. Faber, *Normal Table of Xenopus laevis*, North-Holland Publishing Company, Amsterdam, The Netherlands, 2 edition, 1967.

[17] R. M. Harland, "In situ hybridization: an improved whole-mount method for Xenopus embryos," *Methods in Cell Biology*, vol. 36, pp. 685–695, 1991.

[18] T. Hollemann, F. Panitz, and T. Pieler, "In situ hybridization techniques with Xenopus embryos," in *A Comparative Methods Approach to the Study of Oocytes and Embryos*, pp. 279–290, Oxford University Press, Oxford, UK, 1999.

[19] E. Van Rooijen, E. E. Voest, I. Logister et al., "Zebrafish mutants in the von Hippel-Lindau tumor suppressor display a hypoxic response and recapitulate key aspects of Chuvash polycythemia," *Blood*, vol. 113, no. 25, pp. 6449–6460, 2009.

[20] M. E. Lieb, K. Menzies, M. C. Moschella, R. Ni, and M. B. Taubman, "Mammalian EGLN genes have distinct patterns of mRNA expression and regulation," *Biochemistry and Cell Biology*, vol. 80, no. 4, pp. 421–426, 2002.

2

Study of Enzymatic Hydrolysis of Fructans from *Agave salmiana* Characterization and Kinetic Assessment

Christian Michel-Cuello,[1] **Imelda Ortiz-Cerda,**[2] **Lorena Moreno-Vilet,**[2]
Alicia Grajales-Lagunes,[2] **Mario Moscosa-Santillán,**[2] **Johanne Bonnin,**[3]
Marco Martín González-Chávez,[2] **and Miguel Ruiz-Cabrera**[2]

[1] *Programa Multidisciplinario de Posgrado en Ciencias Ambientales, Universidad Autónoma de San Luis Potosí,
Avenida Dr. Manuel Nava No. 6, Zona Universitaria, 78210 San Luis Potosí, SLP, Mexico*
[2] *Facultad de Ciencias Químicas, Universidad Autónoma de San Luis Potosí, Avenida Dr. Manuel Nava No. 6,
Zona Universitaria, 78210 San Luis Potosí, SLP, Mexico*
[3] *Institut de Chimie Organique et Analytique, Université d'Orléans, Rue d'Issoudun, BP 16729, 45067 Orléans Cedex 02, France*

Correspondence should be addressed to Miguel Ruiz-Cabrera, mruiz@uaslp.mx

Academic Editor: Cesar Mateo

Fructans were extracted from *Agave salmiana* juice, characterized and subjected to hydrolysis process using a commercial inulinase preparation acting freely. To compare the performance of the enzymatic preparation, a batch of experiments were also conducted with chicory inulin (reference). Hydrolysis was performed for 6 h at two temperatures (50, 60°C) and two substrate concentrations (40, 60 mg/ml). Hydrolysis process was monitored by measuring the sugars released and residual substrate by HPLC. A mathematical model which describes the kinetics of substrate degradation as well as fructose production was proposed to analyze the hydrolysis assessment. It was found that kinetics were significantly influenced by temperature, substrate concentration, and type of substrate ($P < 0.01$). The extent of substrate hydrolysis varied from 82 to 99%. Hydrolysis product was mainly constituted of fructose, obtaining from 77 to 96.4% of total reducing sugars.

1. Introduction

The term fructans is a generic name assigned to polymers of fructose linked by fructose-fructose glycosidic bonds. If polymer is composed by 2 to 10 fructose molecules, these are known as fructooligosaccharides (FOSs), whereas a properly named fructan is a polysaccharide with a degree of polymerization (DP) greater than 10 molecules of fructose in the chain [1, 2]. In the literature, references have been made to five groups of fructans, which are classified according to their structure and type of bond such as inulin, levan, graminan, neoseries levan, and neoseries graminan [1–4]. The inulin-type fructan, extracted from chicory (*Cichorium intybus*), artichoke (*Cynara scolymus*), and dahlia plant tubercles (*Dahlia variabilis*) has been the most commonly compound used in the food industry due to their functional properties as well as their health benefits. They are nondigestible polysaccharides, considered as prebiotics, since they stimulate the growth and activity of beneficial colon bacteria such as *Bifidobacteria* and *Lactobacillus* [5–7]. Also, fructans have been associated to a decrease of glucose level in blood, homeostasis of lipids, mineral availability, and immunomodulatory effects [8, 9]. Moreover, fructans possess several properties as texture modification, gel formation, moisture retention, and food stabilization, and consequently they are frequently employed as fat and sugar substitutes [10–13].

Alternatively, fructans have also been considered for the production of high fructose syrups (HFSs) as well as fructooligosaccharides (FOSs) [14–17]. Recently, HFS has gradually replaced refined sucrose and is used as low-caloric sweetener in foods and beverages [18]. In addition, it has been found that some functional properties of foods such as flavor, color, solubility, crystallization inhibition are improved when high-fructose syrups are used [19–21].

At industrial scale, fructose syrups are produced by continuous isomerization of glucose obtained from corn starch. During this industrial process, at least two enzymatic steps with very different reaction conditions are involved. Corn starch is hydrolyzed to maltodextrins by alpha-amylase, then saccharified to glucose by glucoamylase and finally isomerized to fructose by glucose isomerase [22–24]. Due to thermodynamical equilibrium, the product is a syrup containing only about 42% fructose (HFS-42) which is used to produce HFS at different commercial levels in fructose such as HFS-55 and HFS-90 [18, 25, 26]. Among these, HFS-55 has been considered as the standard syrup because it has a similar sweetness to the sucrose obtained from cane sugar [22]. However, the process to enhance the fructose content in syrups is costly and thus makes this method uneconomical [27]. Furthermore, the high corn demand for bioethanol production and the increase in its price stimulate to look for alternative starch sources for the fructose syrups production [26, 28].

Concerning the *Agave* plants, these possess fructans, as their main photosynthetic product, synthesized and stored in the stem. These fructans are used by the plants as a source of energy and as an osmoprotector during drought and cold stress periods [29, 30]. In the particular case of the *Agave tequilana*, over 60% of the soluble carbohydrates represents a complex mixture, mainly composed of highly ramified fructans and neofructans [31–33]. At present, the main use of the fructans from *Agave* is to get fermentable sugars for the manufacturing of alcoholic drinks such as tequila, mescal, and sotol [33–36]. Therefore, some authors such as García-Aguirre et al. [17] have regarded the *Agave* plants as a promising source for fructose syrup production due to its high fructan content. Consequently, in Mexico several distilleries have implemented the same cooking step used in tequila or mezcal elaboration process to obtain fructose syrups. To hydrolyze the fructans, agave heads are cooked in brick ovens for approximately 36 h or cooked in autoclaves for about 12 h. The thermal hydrolysis of fructans in these conditions is not suitable due to undesired degradations (Maillard reaction) and the formation of by-products such as phenolic compounds from lignin which may have a significant impact on the flavor and color of these products [33–35]. Hence, enzymatic hydrolysis based on the use of inulinases constitutes a promissory alternative approach for the production of fructose syrup from agaves. Inulinases are β-fructan fructanohydrolases produced mainly by bacteria, fungi, and yeast. The use of exoinulinase (EC 3.2.1.80) and the endoinulinase (EC 3.2.1.7) acting either alone or combined, have been widely investigated for partial or total inulin hydrolysis [14–16]. In the particular case of agave fructans, a relatively limited investigation had been carried out so far. For example, a commercial enzymatic preparation (Fructozyme L) with endo- and exoinulinase activities was used by Avila-Fernandez et al. [34] to replace the thermal hydrolysis step and by Waleckx et al. [37] to substitute the conventional chemical treatment applied after cooking step, in both cases to hydrolyse the agave fructans during tequila production. On the other hand, García-Aguirre et al. [17] proposed *Kluyveromyces marxianus*, an endogenous strain

isolated from *aguamiel* with capacity for inulinase synthesis which was applied to obtain fructose rich syrups from agave fructans. Generally, the procedure involving enzymatic hydrolysis appears to be very attractive. However, some kinetic aspects of the enzymatic process as well as the assessment of process parameters such as temperature, substrate concentration, and by-product production on the activity and stability of the enzyme need more special attention before their industrial application.

Therefore, the objective of the present study was to investigate the effect of temperature and substrate concentration on the hydrolysis kinetics of fructans extracted from the *Agave salmiana* juice subjected to a commercial inulinase preparation (Fructozyme L) acting freely. Because of the lack concerning data on *Agave salmiana* hydrolysis, it was decided to compare the performance of the commercial enzymatic preparation on a batch of experiments conducted with inulin (standard grade).

2. Materials and Methods

2.1. Obtention of the Agave Fructan Powder. The *Agave salmiana* plants used were from seven to nine years old, with approximately one year of castration, and collected at Ejido de Zaragoza de Solís, a municipality of Villa de Guadalupe, S L P. in the spring of 2010. The Agave juice was extracted using mill and expeller equipment from the local maguey processing plant located on the same Ejido land. This juice was filtered several times to eliminate the fibers and obtain samples with high-fructan content and convenient for spray drying. The first and second filtrations were carried out by means of a stainless steel press filter (Shriv, 405 Type), using bleached cellulose filter paper with a pore diameter of 22 μm and 4 μm and mixed with 1% of diatomaceous material (Celite) as a filtering aid. The third filtration was performed under vacuum conditions, passing the juice through a Whatman no. 42 filter paper with a pore diameter of 2.5 μm. To inactivate the saponins present in the juice, these were treated with heat at 80°C for 30 min in a water bath with continuous agitation [38]. One batch of homogenized juice was obtained which was immediately dried with a Büchi mini spray dryer (Model B-290, Flawil, Switzerland). The atomizer pressure, the feed rate and the inlet air temperature were kept at 1.5 bar, 6.0 mL/min and 170°C, respectively. Carrier agent was not used. The dried samples were subjected to quantitative and qualitative determination of sugars and used to measure the average-degree polymerization (DP) of agave fructan as described later. These powders were also used as substrate during the hydrolysis experiments.

2.2. DP Characterization of Agave Fructan. The DP of the Agave fructan was determined by the technique of Matrix-Assisted Laser Desorption/Ionization Time-Of-Flight Mass Spectrometry (MALDI-TOF-MS), which has been reported as the best choice to establish the DP distribution of these types of carbohydrates. Experiments were performed in positive ion mode using an AutoFlex I mass spectrometer (Bruker Daltonics). The instrument was operated at an accelerating voltage of 20 kV in linear mode and 19 kV in reflectron mode. The pressure was 1.5×10^{-7} mbar. The sample was ionized with a nitrogen laser ($\lambda = 337$ nm). The sample was dissolved in water, and the matrix was 2,5-Dihydroxybenzoic acid (10 mg/mL), prepared in 1:1 methanol:water. Samples (0.5 μL) and matrix solution were spotted and air dried on a stainless-steel plate. Spectra were acquired in linear mode (200–300 laser shots) in the m/z range from 800 to 10,000 and reflectron mode (150–200 laser shots) in the m/z range from 300 to 2,000. A pepmix calibration kit (Bruker Daltonics) and apomyoglobin were used for calibration.

2.3. Carbohydrate Characterization of Agave Fructan and Chicory Inulin.

The dried samples were diluted accordingly for sugars determination (glucose, fructose, sucrose and fructans). This was carried out according to the HPLC-method proposed by Zuleta and Sambucetti [39] with a Waters 600 chromatography equipment (Milford MA. USA), consisting of a degassing device, quaternary pump, column thermocompartment, and a refractive-index detector (Waters 410). An Aminex HPX-87C column ion exchange (7.8 mm d.i. × 300 mm, Bio-Rad Hercules, CA, USA) was used as stationary phase. HPLC grade water with a flow of 0.6 mL/min was used as the mobile phase. The volume of the injected sample was 50 μl (with the injector completely full). The column temperature was kept at 75°C. Environmental temperature was kept constant at 20°C. The samples were filtered through nylon-membrane filters (0.45 μm) coupled to 5 mL polypropylene syringes, both from Waters (Milford, CT), and analyzed immediately.

The Quick Start Empower 5.0 was used for system control and data analysis. References sugars (arabinose, fructose, galactose, glucose, lactose, maltose, mannose, ribose, sucrose, and xylose) as well as the chicory inulin (reference compound) were purchased from Sigma-Aldrich. The chromatographic separation time was 20 min and the carbohydrates present in samples were identified by comparing the retention times with those of references sugars. Quantification of the carbohydrates was performed as a functions of the calibration curves, derived for sugars between 0.1–3.2% w/v ($r = 0.9996$) and for fructans between 0.2–8.0% w/v ($r = 0.9999$). The minimum detection level was 0.068 mg/mL. Correlation coefficients (r) were calculated between refraction values in samples and standards for each carbohydrate to estimate the detector consistency in terms of concentration amplitude. All the determinations were carried out in duplicate, and a mixture of standards sugars was injected daily on order to identify any calibration variations.

2.4. Solid-Phase Microextraction (SPME) of Volatiles Compounds.

For the identification of volatile compounds present in samples, SPME technique was used (SPME device from Supelco Inc. Bellefonte, PA, USA). The fiber used was 60 μm diameter and 1 cm length, polyethylene glycol (PEG) coated. For sampling, the SPME fiber was inserted directly into a 20 mL vial with a valve cap with silicone septa containing 100 mg of fructan sample. After that, the sample was heated to 90°C during 5 min before the extraction. Then, the fiber was maintained 10 min at 90°C. The desorption of analytes

was performed by heating the fiber in the injection port of the GC-MS equipment at 250°C for 5 seconds (split mode, split ratio 10:1).

2.5. Analysis of Volatile Compounds by GC/MS.

GC-MS analysis was performed using an Agilent HP series 6890 N gas chromatograph (Waldbronn, Germany), coupled to a mass selective detector 5973. A Stabilwax capillary column (30 m × 0.25 mm i.d., 0.25 μm film thickness; Restek) was used for separation. The column temperature program was as follows: 50°C hold 3 min, 3°C/min up to 250°C hold 5 min, splitless injection at 250°C. The detector was operated in electron impact mode. The spectra were collected at 71 eV ionization voltage and analyzed mass range was 15–600 m/z. The transfer line temperature was 250°C and the carrier gas helium flow rate was 1 mL/min. The compounds were identified by comparing their mass spectra with mass spectra library NIST 02.L and by comparison of linear retention indices, relative to a mixture of n-alkanes C_{10}–C_{28}. All experiments were repeated two times.

2.6. Enzyme.

The commercial enzyme preparation, Fructozyme L (Novozymes A/S, Denmark), from Aspergillus niger, was purchased from Sigma-Aldrich. This enzyme is a mixture of exoinulinase (EC 3.2.1.80) and endoinulinase (EC 3.2.1.7) with inulinase unit of 2000 INU/g recommended to hydrolyze the linkage of inulin. One inulinase unit is the quantity of enzyme that produces 1 μmol of reducing sugar (calculated as glucose) per minute under the reaction conditions used in Novo Nordisk's standard assay procedure. Protein concentration of Fructozyme L was quantified by the Lowry method [40] and a value of 6.75 ± 0.02 mg/mL was found. The experimental procedure of the inulinase assay with chicory inulin as substrate and described in detail by Ortiz-Cerda [41] was used to select the best reaction parameters such as temperature, pH, enzyme dosage, and substrate concentration.

2.7. Hydrolysis of Agave Fructan and Chicory Inulin.

The enzymatic hydrolysis experiments were carried out in a 125 mL screw-cap Erlenmeyer flask placed in a large water bath with agitation and controlled temperature (Model PB-1400, Boekel Grant Scientific, USA). Preliminary tests allowed to establish a full factorial design considering two types of substrates (chicory inulin and agave fructan), two levels of substrate concentration (40 and 60 mg/mL), and two different temperatures (50 and 60°C). One replicate for each experimental condition was used and a total of 16 runs were conducted (Table 1). 0.0074 mL of fructozyme L was added to 50 mL of substrate solution which indicated that a protein concentration of 0.001 mg/mL was used in all the experiments. The pH of substrate solution was kept at 4.7. The hydrolysis process was monitored by measuring the sugars such as fructose, glucose, and residual fructan by the HPLC method mentioned previously. Therefore, samples of 3 mL of the reaction media were taken periodically and immediately placed in a water bath at 90°C during 1 min and later quenched in ice for 5 min to inactivate the enzymes [20, 37]. Then, samples were filtered in Whatman No. 42 filter

TABLE 1: Values of the rate constants (k) and their respective R^2 determined through regression method for each experimental condition.

Exp No.	Run Order	Temperature (°C)	Substrate concentration (mg/ml)	Substrate type	S_o (mg/ml)	k (min^{-1})	R^2
1	16	50	40	I	41.1	0.0105	0.991
2	2	60	40	I	39.1	0.0165	0.978
3	15	50	60	I	61.3	0.0064	0.996
4	5	60	60	I	57.9	0.0114	0.996
5	12	50	40	Fr	33.7	0.0086	0.997
6	4	60	40	Fr	40.1	0.0100	0.992
7	6	50	60	Fr	50.4	0.0077	0.995
8	9	60	60	Fr	50.7	0.0091	0.997
9	7	50	40	I	39.5	0.0099	0.997
10	11	60	40	I	39.3	0.0148	0.994
11	8	50	60	I	59.1	0.0089	0.998
12	1	60	60	I	57.2	0.0100	0.997
13	10	50	40	Fr	34.5	0.0118	0.998
14	14	60	40	Fr	32.8	0.0164	0.996
15	13	50	60	Fr	47.9	0.0051	0.998
16	3	60	60	Fr	64.8	0.0095	0.997

I: chicory inulin; Fr: agave fructan.

paper with a pore diameter of 2.5 μm and afterward passed through 0.45 μm nylon membrane filter to eliminate the enzymes and obtain samples convenient for sugar analysis.

2.8. Mathematical Modeling of the Hydrolysis Kinetics. The model proposed by Rica et al. [42] which describes the complete hydrolysis of the substrate on a single stage was used to describe the rate of substrate consumption (inulin and agave fructan) and the rate of fructose production:

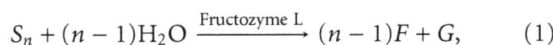

$$S_n + (n-1)\text{H}_2\text{O} \xrightarrow{\text{Fructozyme L}} (n-1)F + G, \qquad (1)$$

where S, F, G, are, respectively, substrate, fructose, glucose; and n the degree of polymerization. By applying a mass balance to the hydrolysis process and expressing the concentration of each species in terms of mass of compound per unit volume, the kinetics of the process can be expressed by (2):

$$\left(-\frac{dS}{dt}\right) + \left(-\frac{dW}{dt}\right) = \frac{dF}{dt} + \frac{dG}{dt}, \qquad (2)$$

where dW/dt expresses the consumption of water involved in the hydrolysis reaction. The model assumes that there exists a stoichiometric relationship between the molar rate of water consumption and the molar rate of fructose production, according to (3):

$$-\frac{dn_W}{dt} = \frac{dn_F}{dt}, \qquad (3)$$

where n_W and n_F express molar concentrations of water and fructose, respectively. Considering the molecular weight of

water (MW_W) and fructose (MW_F), consumption of water in the reaction can be expressed as,

$$-\frac{dW}{dt} = \frac{\text{MW}_W}{\text{MW}_F} \frac{dF}{dt}. \qquad (4)$$

Thus (2) becomes,

$$\left(-\frac{dS}{dt}\right) = \left(1 - \frac{\text{MW}_W}{\text{MW}_F}\right)\frac{dF}{dt} + \frac{dG}{dt}. \qquad (5)$$

The model most often applied to describe the consumption of substrate in enzymatic reactions is the Michaelis-Menten equation (6):

$$-r_s = \left(-\frac{dS}{dt}\right) = \frac{V_{max}S}{K_m + S}. \qquad (6)$$

In this study, preliminary tests suggest that experimental conditions used present a typical unsaturated enzyme behavior. Hence, the initial rate does not depend on enzyme activity but is directly proportional to the concentration of the substrate. Therefore, the Michaelis-Menten model becomes a first order kinetics, as described by (7):

$$-r_s \cong \frac{V_{max}S}{K_m} = kS. \qquad (7)$$

Applying integral method, constant kinetics k can be calculated from experimental results. Moreover, mass balance can be used to estimate product formation.

From the proposed model, the hydrolysis products are fructose and glucose. However, in the case of inulin, the HPLC analysis shows that the formation of glucose is negligible due to its structure (Figure 4(a)). Then, fructose production rate can be directly calculated from substrate consumption rate.

In contrast, for the agave fructan, analysis shows a significant release of glucose concentration as a function of time (Figure 4(b)). Therefore, the relative fractions of fructose and glucose, obtained by HPLC, were used to calculate the evolution of the products concentration.

As mentioned before, for each experiment, the value of constant kinetic k was obtained (Table 1) using the integral method programmed in Scilab 5.2.1. The coefficient of determination (R^2) was used as primary criteria to determine the accuracy of the fit between model proposed and experimental data.

2.9. Statistical Analysis. Analysis of variance (ANOVA) was performed with a confidence level of 99% ($P < 0.01$) with Modde 7.0 (Umetric AB) statistical package. A linear mathematical model with interactions was used to analyze the effect of temperature (T), substrate concentration (C) and substrate type (S) on the rate constants k values:

$$y = \beta_0 + \beta_1 T + \beta_2 C + \beta_3 S + \beta_4 T * C + \beta_5 T * S + \beta_6 C * S, \tag{8}$$

where β_i ($i = 0, 1, \ldots, 6$) represents the regression coefficients of the model.

3. Results and Discussion

3.1. DP Profile of Agave Fructan. The spectra for the low mass fructan content and high-mass fructan content determined as [M + K]+ ions are shown in Figure 1. It can clearly be seen that the molecular mass distribution of Agave fructan comprised oligomers and polymers of molecular weight ranging from 342.8 to 3587.2 Da, which corresponds to a range of DP from 1 to 21. Nevertheless, the DP average was 8, which was calculated based on the intensity of the peaks of the spectra of high mass fructan content (DP from 4 to 21), considering only the number of fructose units. This is the first report of the DP of fructans from *A.salmiana*.

Diverse studies of molecular structure of the fructan from Agave species, more specific *A. tequilana*, have reported that fructan are based on 1-kestose, neokestose, and branched tetrasaccharides, presenting linkages β-(2-1) and β-(2-6) between the fructose and some glucose intermediated forming a branched structure [31, 32]. This suggests that fructan from *A. salmiana* might be a similar structure because is a plant of the same species. Also, it could represent important differences in the process of hydrolysis with respect of fructan type inuline from chicory.

3.2. Carbohydrates Profiles of Substrates. Figure 2 shows the HPLC chromatogram obtained during separation of the different sugars present in the standard chicory inulin and in the fructan extracted from the agave juice. The results showed that fructan and fructose constituted 97% and 3% (dry matter), respectively, of the total sugars present en in the chicory inulin, exhibiting only two peaks with an elution time of 6.6 min and 12.7 min, respectively. Therefore, the

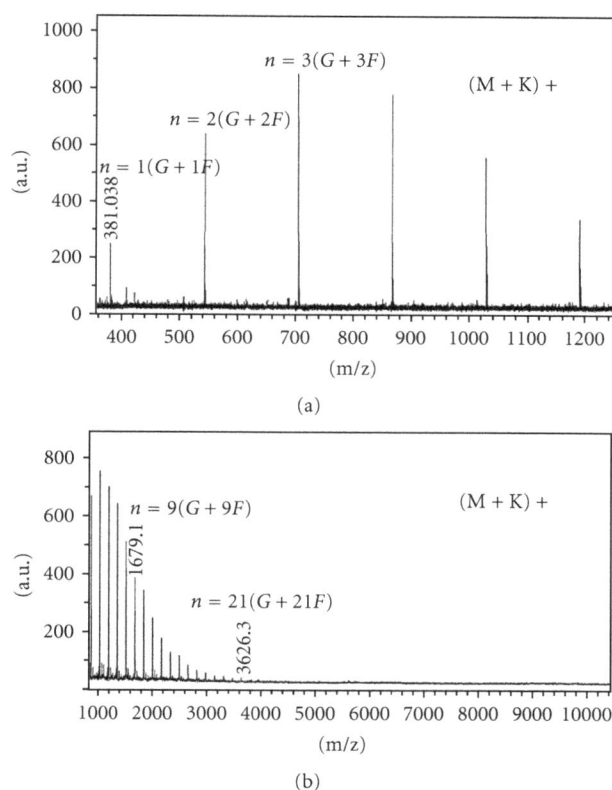

FIGURE 1: MALDI-TOF-MS spectra, in positive-ion mode of *Agave salmiana* fructan. (a) Low-mass spectra (DP from 1 to 6), (b) high-mass spectra (DP from 4 to 21).

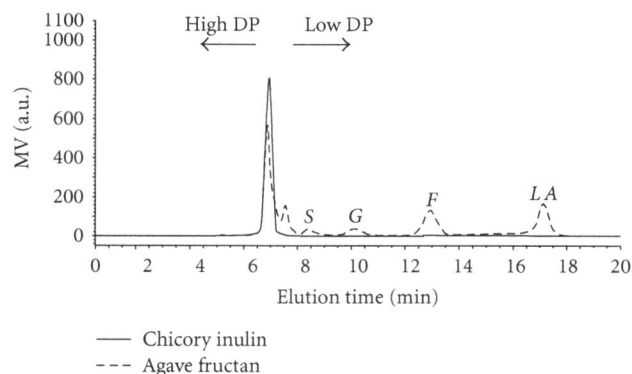

FIGURE 2: HPLC separation of sugars from agave fructan and chicory inulin used as substrates. *S*: sucrose, *G*: glucose, *F*: fructose, *LA*: lactic acid.

chicory inulin from Sigma, used as standard and as substrate in hydrolysis experiments in this work, can be considered as a pure substrate. On the other hand, the average DP of chicory inulin was not indicated by the manufacturer, but it has been reported in literature that commercial inulin are standardized to have an average DP between 12 and 25 [3, 43].

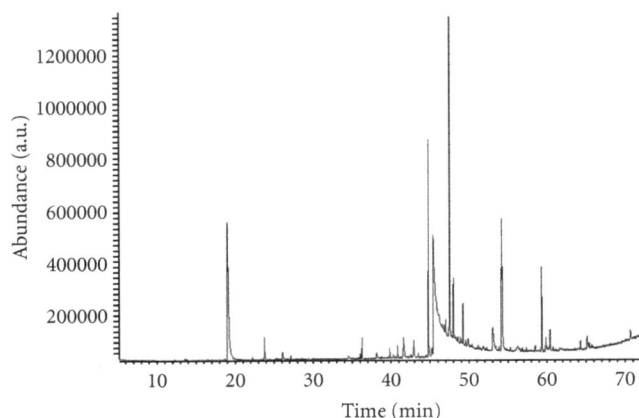

FIGURE 3: Gas chromatogram of polar volatile compounds of powder fructan obtained with PEG fiber and Stabilwax capillary column.

It was also observed that Aminex HPX-87C column separated all the sugars present in the agave fructan powder with a good resolution in a single run of 20 min (Figure 2). Peak separation was good, allowing all peaks to be identified and quantified. Sucrose, glucose, and fructose were identified with elution times of 8.1, 9.8, and 12.7 min, respectively. The main peak of this chromatogram coincides with the peak and elution time (6.7 min) corresponding to the chicory inulin used as a standard. A minor peak with an elution time of 7.5 min was detected just before the sucrose peak which could be considered as a fructan of low DP or fructooligosaccharide. However, for practical considerations regarding this work, these two peaks were quantified as total fructans. It was found in this way that fructan, sucrose, glucose, and fructose constituted $85.6 \pm 2.52\%$, $4.67 \pm 0.22\%$, $3.99 \pm 0.14\%$, and $6.36 \pm 0.54\%$, (dry matter) respectively, of the total sugars present in the *agave* fructan powder. Foregoing, it is evident that agave fructan was partially purified and additional experiments are required using more efficient filtration and purifying process than those used in the present research for treating the *Agave salmiana* juice and obtaining a fructan powder with chemical composition similar to the reference inulin.

The results from the HPLC analysis of samples taken indicate that the kinetics where fructan is used as a substrate, not only are the sugar of interest (Figure 4) but also show the presence of a peak with a retention time (RT) of 17.1 min, which not corresponds with the RT of sugars standards provided. The compound of RT of 17.1 is not a product of the enzymatic hydrolysis of fructan powder, since its concentration remains constant during this process, the tendency of elution in HPLC of fructan goes from fructose polymer to simple sugars (high-to-low-molecular weight, as shown in Figure 3), so we concluded that the peak of 17.1 could be a polar compound and volatile.

The extraction of volatile polar compounds in fructan was performed by SPME and subsequently identified by GC-MS, using both procedures polar fiber and column (Figure 3). The compounds identified by GC-MS are shown

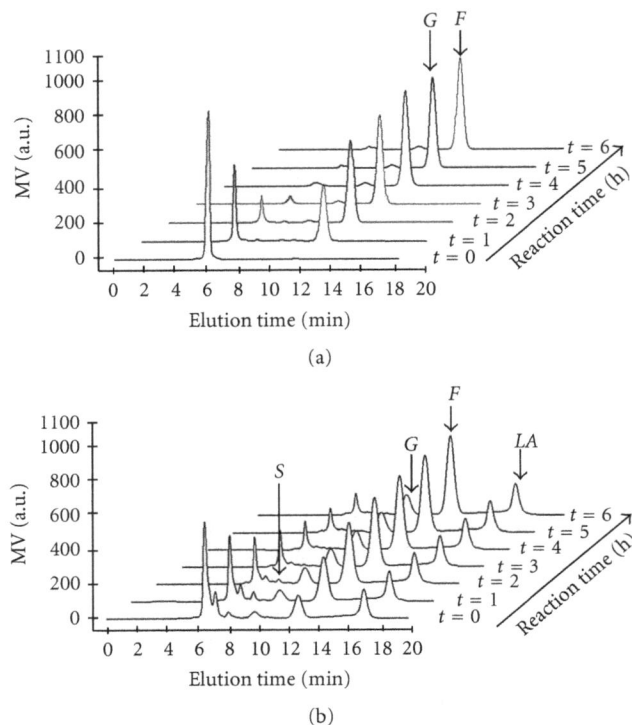

(a)

(b)

FIGURE 4: Chromatographic representation of substrates degradation and release of sugars during the enzymatic hydrolysis. (a) Chicory inulin, (b) agave fructan. S: sucrose, G: glucose, F: fructose, LA: lactic acid.

in Table 2. The compounds correspond to higher percentage of acetic acid and lactic acid, the retention times obtained by HPLC for both acids are acetic acid 13.4 min and 17.1 min for lactic acid. According to this result, we can say that the peak retention time 17.1 min fructan powder corresponds to lactic acid, this was confirmed by powder fructan enriched with lactic acid. The lactic acid is present in samples with high sugar content as a result of fermentation processes [44, 45].

3.3. HPLC Analysis of the Hydrolysis Kinetics. As an example, Figures (4(a) and 4(b)) present the chromatographic evolution of substrate hydrolysis and the corresponding release of sugars for experiments 2 and 6 (Table 1) with chicory inulin and agave fructan as substrates, respectively. In both cases, the reactions were carried out at temperature of 60°C with a substrate concentration of 40 mg/mL. As shown in Figure 4, the degradation of substrates occurred progressively and after 6 h of hydrolysis, fructose was the major product. Figure 4(b) illustrates that when agave fructan was used, the glucose content also increased as a result of fructan and sucrose hydrolysis. The extent of substrate hydrolysis (%) was defined as the ratio of consumed substrate/amount of substrate provided ×100. In this way, at the end of the enzymatic process, values of 92% and 98% were calculated for agave fructan (experiment 6) and for pure inulin (experiment 2), respectively. This disparity in hydrolysis efficiency could be explained by a difference in substrates

TABLE 2: Composition of volatile compounds from solid *Agave salmiana*, obtained from the PEG fiber and Stabilwax column (carbowax).

No	Compound	%
1	Acetic acid	15.83
2	2,3-Butanediol	1.00
3	Benzeneacetaldehyde	0.41
4	2-Furanmethanol	0.16
5	Phenethyl alcohol	0.17
6	2,6-di-tert-butyl-p-cresol	0.79
7	2-Acetylpyrrole	0.26
8	2-Formylpyrrole	0.40
9	Caprylic acid	1.79
10	Pelargic acid	7.46
11	Lactic acid	27.49
12	2,3-dihydro-3,5-dihydroxy-6-methyl-4H-Pyran-4-one	14.86
13	Capric acid	2.97
14	2,4-di-tert-butylphenol	5.52
15	Lauric acid	10.06
16	Butyl phthalate	3.53
17	Unidentified	7.3

materials, as chemical composition, impurities presence, and probably due to average DP. However, the range of the percent hydrolysis achieved in this study are analogous to those reported for thermal hydrolysis of fructan contained in *Agave tequilana* [33].

3.4. Hydrolysis Kinetics. The experimental and simulated kinetics of substrates degradation as well as sugars release as a function of hydrolysis time, when a substrate concentration of 40 mg/mL of chicory inulin and agave fructan was used, are shown in Figures 5(a) and 5(b), respectively. Initially, the rate of substrates hydrolysis and the rate of fructose production were very high, and after 200 min, both processes tend, to remain constant. As a general trend it was observed that as temperature was increased, the % hydrolysis and the amount of fructose increased. Therefore, after 360 min of hydrolysis, the maximum production rate of fructose of 39.5 ± 0.02 mg/mL was observed at temperature of 60°C with inulin as substrate (Figure 5(a)). In this case, standard inulin was hydrolyzed about 99.5% and the results of HPLC revealed that the effluent contained 96.4% fructose and 3.4% constituted of glucose and residual inulin. Although so high values of percent hydrolysis (92–95%) were reached when agave fructan was used as substrate (Figure 5(b)), the maximum amount of fructose obtained in the reaction mixture varied between 27.7 and 30.4 mg/mL at 50 and 60°C, respectively. In this case, fructose constituted between 75 and 77% of total reducing sugars in the reaction product. This is due to the accumulation of considerable amount of glucose during hydrolysis process whose proportion increased from 13.00% to 22.37% of total carbohydrates. A similar increase

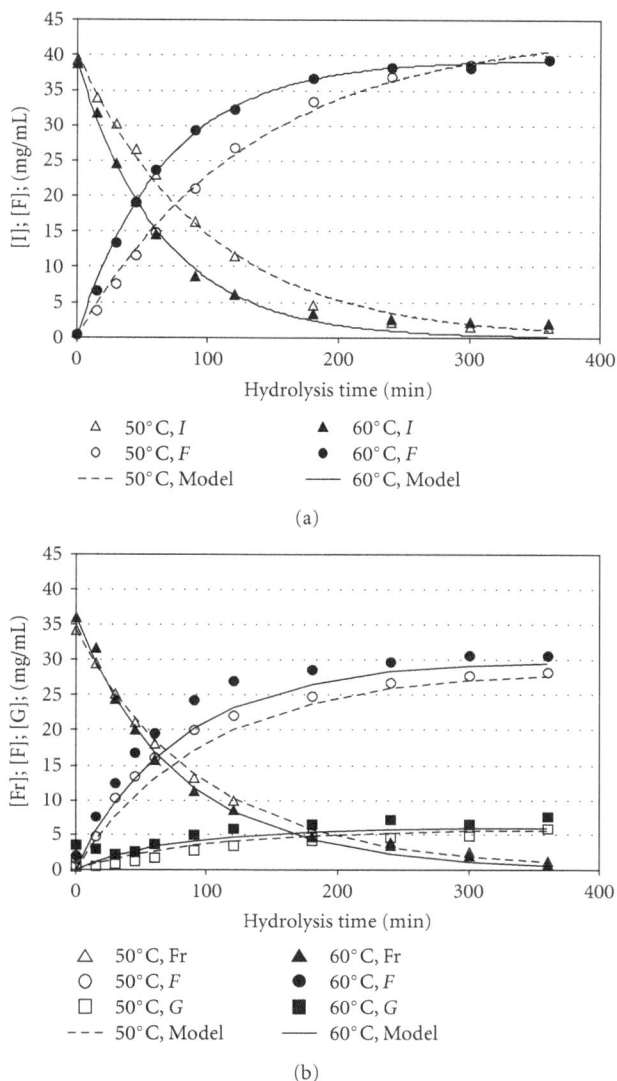

(a)

(b)

FIGURE 5: Kinetics of substrates degradation and sugars released as a function of hydrolysis time at different temperatures. (a) Chicory inulin, (b) Agave fructan. *I*: chicory inulin; Fr: agave fructan; *F*: fructose, *G*: glucose.

of glucose content was also observed by Waleckx et al. [33] with thermal hydrolysis of fructans from *Agave tequilana*.

The simulated curves are also shown in Figure 5. When calculating the values for rate constants (k), the corresponding values for the determination coefficients (R^2) were all greater than 0.978 (Table 1), indicating a good agreement between the experimental and predicted values. An analysis of variance (ANOVA) with a confidence level of 99% ($P < 0.01$) revealed that k were affected by the temperature (T), the substrate concentration (C), and the type of substrate (S). It is important to point out that not significant interaction effect between factors was detected in both cases. The identified values of the rate constants k as a function of temperature can be observed in Figure 6; the corresponding predicted values of k can be also observed.

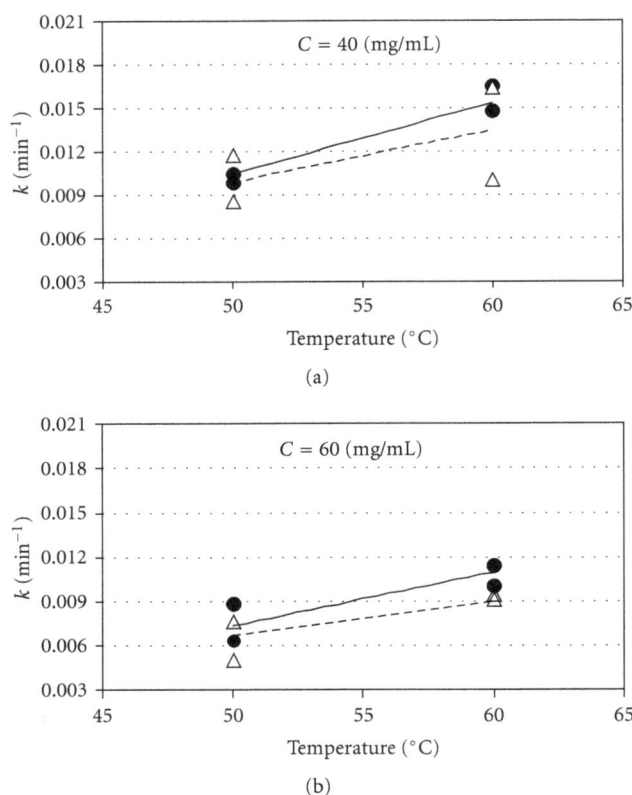

FIGURE 6: Variation of the rate constant k with temperature (a,b).
• Chicory inulin; △ Agave fructan.

An analysis of Figure 6 demonstrates that the higher temperatures and smaller substrate concentration resulted in higher values of k and consequently faster hydrolysis of substrates. These observations are in agreement with those obtained by Catana et al. [46] in their experiments conducted with inulin as substrate where the optimal temperature for Fructozyme L was around 60°C. Although the obtained values of k were smaller when a substrate concentration of 60 mg/mL was used, results concerning % hydrolysis (>92%) in Figure 5 indicated that the amount of enzyme employed (0.001 mg/mL) was enough to hydrolyze almost totally the substrate in 6 h, with no evidence of substrate inhibition. Figure 6 also shows that the experiments conducted with chicory inulin, independent of temperature or substrate concentration, rendered the highest k values, resulting in significantly higher hydrolysis rates for inulin. This could be explained by a difference in the structure of substrates. According to literature, inulin is linear fructans consisting mainly of β-(2-1) fructosyl-fructose links [1–4], while the fructans of agave plants have been considered as a complex mixture of highly branched fructans presenting the two types of linkages β-(2-1) and β-(2-6) between fructose moieties [31, 32].

The Arrhenius equation gives the dependence of the temperature in rate constant k in chemical reactions, but it is important to note that this study did not apply the

Arrhenius equation because only two values were used as the temperature interval between these is relatively short.

It is clear that the fructose production rate increased as the temperature increased and also it falls as the substrate concentration was raised. This last was due to that the same amount of enzyme was used for both substrate concentrations (40 and 60 mg/mL). The experiments conducted with agave fructan, independent of temperature or substrate concentration, resulted in significantly higher fructose production rates from raw agave fructan. This could be explained because agave fructan shows polymers having low DP and branched structure having more external fructose molecules which can be more easily hydrolyzed by exo-inulinases.

According to the results obtained in this work, the enzymatic process with Fructozyme L might thus be of industrial interest for the large-scale production of high fructose syrup from *Agave salmiana* under appropriate conditions of temperature, enzyme dosage, and substrate concentration. The *Agave salmiana* has been a resource of a wide distribution in the Zacatecan-Potosino *altiplano* (high plateau) which has become a promising raw material for the industrial production of fructose syrup, as well as for the production of fructooligosaccharides and/or fructans.

Acknowledgments

The authors wish to acknowledge the financial support from the Secretaría de Desarrollo Agropecuario and Recursos Hidráulicos (SEDARH) and Fundación Produce both of San Luis Potosí, México. The technical support was provided by Plateforme de Spectrométrie de Masse et Protéomique du Centre de Biophysique Moléculaire (Orléans, France), María Estela Nuñez-Pastrana, and Cecilia Rivera-Bautista.

References

[1] M. Wack and W. Blaschek, "Determination of the structure and degree of polymerisation of fructans from *Echinacea purpurea* roots," *Carbohydrate Research*, vol. 341, no. 9, pp. 1147–1153, 2006.

[2] C. Olvera, E. Castillo, and A. López-Munguía, "Fructosiltransferasas, fructanas y fructosa," *Biotecnología*, vol. 14, pp. 327–345, 2007.

[3] M. B. Roberfroid, "Introducing inulin-type fructans," *British Journal of Nutrition*, vol. 93, no. 1, pp. S13–S25, 2005.

[4] J. G. Muir, S. J. Shepherd, O. Rosella, M. Rose, J. S. Barrett, and P. R. Gibson, "Fructan and free fructose content of common Australian vegetables and fruit," *Journal of Agricultural and Food Chemistry*, vol. 55, no. 16, pp. 6619–6627, 2007.

[5] G. R. Gibson and M. B. Roberfroid, "Dietary modulation of the human colonic microbiota: introducing the concept of prebiotics," *Journal of Nutrition*, vol. 125, no. 6, pp. 1401–1412, 1995.

[6] J. O. Lindsay, K. Whelan, A. J. Stagg et al., "Clinical, microbiological, and immunological effects of fructo-oligosaccharide in patients with Crohn's disease," *Gut*, vol. 55, no. 3, pp. 348–355, 2006.

[7] J. Huebner, R. L. Wehling, and R. W. Hutkins, "Functional activity of commercial prebiotics," *International Dairy Journal*, vol. 17, no. 7, pp. 770–775, 2007.

[8] M. H. Davidson and K. C. Maki, "Effects of dietary inulin on serum lipids," *Journal of Nutrition*, vol. 129, no. 7, pp. S49–S55, 1999.

[9] B. Watzl, S. Girrbach, and M. Roller, "Inulin, oligofructose and immunomodulation," *British Journal of Nutrition*, vol. 93, no. 1, pp. S49–S55, 2005.

[10] K. R. Niness, "Inulin and oligofructose: what are they?" *Journal of Nutrition*, vol. 129, no. 7, pp. 1402S–1406S, 1999.

[11] A. Tárrega and E. Costell, "Effect of inulin addition on rheological and sensory properties of fat-free starch-based dairy desserts," *International Dairy Journal*, vol. 16, no. 9, pp. 1104–1112, 2006.

[12] P. J. Hennelly, P. G. Dunne, M. O'Sullivan, and E. D. O'Riordan, "Textural, rheological and microstructural properties of imitation cheese containing inulin," *Journal of Food Engineering*, vol. 75, no. 3, pp. 388–395, 2006.

[13] P. Kip, D. Meyer, and R. H. Jellema, "Inulins improve sensoric and textural properties of low-fat yoghurts," *International Dairy Journal*, vol. 16, no. 9, pp. 1098–1103, 2006.

[14] T. Nakamura, Y. Ogata, A. Shitara, A. Nakamura, and K. Ohta, "Continuous production of fructose syrups from inulin by immobilized inulinase from *Aspergillus niger* mutant 817," *Journal of Fermentation and Bioengineering*, vol. 80, no. 2, pp. 164–169, 1995.

[15] Y. Cho, J. Sinha, J. Park, and J. Yun, "Production of inulooligosaccharides from inulin by a dual endoinulinase system," *Enzyme and Microbial Technology*, vol. 29, no. 6-7, pp. 428–433, 2001.

[16] J. Zhengyu, W. Jing, J. Bo, and X. Xueming, "Production of inulooligosaccharides by endoinulinases from *Aspergillus ficuum*," *Food Research International*, vol. 38, no. 3, pp. 301–308, 2005.

[17] M. García-Aguirre, V. A. Saenz-Alvaro, M. A. Rodríguez-Soto et al., "Strategy for biotechnological process design applied to the enzymatic hydrolysis of agave fructo-oligosaccharides to obtain fructose-rich syrups," *Journal of Agricultural and Food Chemistry*, vol. 57, no. 21, pp. 10205–10210, 2009.

[18] R. A. Forshee, M. L. Storey, D. B. Allison et al., "A critical examination of the evidence relating high fructose corn syrup and weight gain," *Critical Reviews in Food Science and Nutrition*, vol. 47, no. 6, pp. 561–582, 2007.

[19] E. A. Borges da Silva, A. A. Ulson de Souza, S. G. U. de Souza, and A. E. Rodrigues, "Analysis of the high-fructose syrup production using reactive SMB technology," *Chemical Engineering Journal*, vol. 118, no. 3, pp. 167–181, 2006.

[20] E. G. Díaz, R. Catana, B. S. Ferreira, S. Luque, P. Fernandes, and J. M. S. Cabral, "Towards the development of a membrane reactor for enzymatic inulin hydrolysis," *Journal of Membrane Science*, vol. 273, no. 1-2, pp. 152–158, 2006.

[21] E. J. Tomotani and M. Vitolo, "Production of high-fructose syrup using immobilized invertase in a membrane reactor," *Journal of Food Engineering*, vol. 80, no. 2, pp. 662–667, 2007.

[22] W. D. Crabb and J. K. Shetty, "Commodity scale production of sugars from starches," *Current Opinion in Microbiology*, vol. 2, no. 3, pp. 252–256, 1999.

[23] Y. Ge, Y. Wang, H. Zhou, S. Wang, Y. Tong, and W. Li, "Co-immobilization of glucoamylase and glucose isomerase by molecular deposition technique for one-step conversion of dextrin to fructose," *Journal of Biotechnology*, vol. 67, no. 1, pp. 33–40, 1999.

[24] M. E. van der Veen, S. Veelaert, A. J. van der Goot, and R. M. Boom, "Starch hydrolysis under low water conditions: a conceptual process design," *Journal of Food Engineering*, vol. 75, no. 2, pp. 178–186, 2006.

[25] E. A.B. da Silva, A. A. U. de Souza, A. E. Rodrigues, and S. M. A. G. U. de Souza, "Glucose isomerization in simulated moving bed reactor by glucose isomerase," *Brazilian Archives of Biology and Technology*, vol. 49, no. 3, pp. 491–502, 2006.

[26] R. Johnson, G. Padmaja, and S. N. Moorthy, "Comparative production of glucose and high fructose syrup from cassava and sweet potato roots by direct conversion techniques," *Innovative Food Science and Emerging Technologies*, vol. 10, no. 4, pp. 616–620, 2009.

[27] A. Toumi and S. Engell, "Optimization-based control of a reactive simulated moving bed process for glucose isomerization," *Chemical Engineering Science*, vol. 59, no. 18, pp. 3777–3792, 2004.

[28] J. P. Hernández-Uribe, S. L. Rodríguez-Ambriz, and L. A. Bello-Pérez, "Obtención de jarabe fructosado a partir del almidón de plátano (*Musa paradisiaca* L.). Caracterización parcial," *Interciencia*, vol. 33, no. 5, pp. 372–376, 2008.

[29] N. Wang and P. S. Nobel, "Phloem transport of fructans in the crassulacean acid metabolism species *Agave deserti*," *Plant Physiology*, vol. 116, no. 2, pp. 709–714, 1998.

[30] T. Ritsema and S. C. M. Smeekens, "Engineering fructan metabolism in plants," *Journal of Plant Physiology*, vol. 160, no. 7, pp. 811–820, 2003.

[31] M. G. Lopez, N. A. Mancilla-Margalli, and G. Mendoza-Diaz, "Molecular structures of fructans from *Agave tequilana* Weber var. azul," *Journal of Agricultural and Food Chemistry*, vol. 51, no. 27, pp. 7835–7840, 2003.

[32] N. A. Mancilla-Margalli and M. G. Lopez, "Water-soluble carbohydrates and fructan structure patterns from *Agave* and *Dasylirion* species," *Journal of Agricultural and Food Chemistry*, vol. 54, no. 20, pp. 7832–7839, 2006.

[33] E. Waleckx, A. Gschaedler, B. Colonna-Ceccaldi, and P. Monsan, "Hydrolysis of fructans from *Agave tequilana* Weber var. azul during the cooking step in a traditional tequila elaboration process," *Food Chemistry*, vol. 108, no. 1, pp. 40–48, 2008.

[34] A. Avila-Fernandez, X. Rendon-Poujol, C. Olvera et al., "Enzymatic hydrolysis of fructans in the tequila production process," *Journal of Agricultural and Food Chemistry*, vol. 57, no. 12, pp. 5578–5585, 2009.

[35] N. A. Mancilla-Margalli and M. G. Lopez, "Generation of maillard compounds from inulin during the thermal processing of *Agave tequilana* Weber var. azul," *Journal of Agricultural and Food Chemistry*, vol. 50, no. 4, pp. 806–812, 2002.

[36] M. J. Garcia-Soto, H. Jiménez-Islas, J. L. Navarrete-Bolaños, R. Rico-Martinez, R. Miranda-López, and J. E. Botello-Álvarez, "Kinetic study of the thermal hydrolysis of agave salmiana for mezcal production," *Journal of Agricultural and Food Chemistry*, vol. 59, no. 13, pp. 7333–7340, 2011.

[37] E. Waleckx, J. C. Mateos-Diaz, A. Gschaedler et al., "Use of inulinases to improve fermentable carbohydrate recovery," *Food Chemistry*, vol. 124, no. 4, pp. 1533–1542, 2011.

[38] K. M. Tarade, R. S. Singhal, R. V. Jayram, and A. B. Pandit, "Kinetics of degradation of saponins in soybean flour (*Glycine max*) during food processing," *Journal of Food Engineering*, vol. 76, no. 3, pp. 440–445, 2006.

[39] A. Zuleta and M. E. Sambucetti, "Inulin determination for food labeling," *Journal of Agricultural and Food Chemistry*, vol. 49, no. 10, pp. 4570–4572, 2001.

[40] O. H. Lowry, N. J. Rosenbrough, A. L. Farr, and R. J. Randall, "Protein measurement with the Folin phenol reagent," *The Journal of biological chemistry*, vol. 193, no. 1, pp. 265–275, 1951.

[41] I. E. Ortiz-Cerda, *Aprovechamiento del Agave salmiana para la obtención de jarabe fructosado-vía enzimática*, Professional thesis, Facultad de Ciencias Químicas, Universidad Autónoma de San Luis Potosí, San Luis Potosí, México, 2010.

[42] E. Ricca, V. Calabrò, S. Curcio, and G. Iorio, "Fructose production by chicory inulin enzymatic hydrolysis: a kinetic study and reaction mechanism," *Process Biochemistry*, vol. 44, no. 4, pp. 466–470, 2009.

[43] A. Franck and L. de Leenheer, "Inulin," in *Polysaccharides II: Polysaccharides from Eukaryotes*, S. de Baets, E. Vandamme, and A. Steinbüchel, Eds., pp. 439–479, Wiley-VHC, Weinheim, Germany, 2004.

[44] C. J. Bolner de Lima, L. F. Coelho, K. C. Blanco, and J. Contiero, "Response surface optimization of D(-)-lactic acid production by Lactobacillus SMI8 using corn steep liquor and yeast autolysate as an alternative nitrogen source," *African Journal of Biotechnology*, vol. 8, no. 21, pp. 5842–5846, 2009.

[45] G. Chotani, T. Dodge, A. Hsu et al., "The commercial production of chemicals using pathway engineering," *Biochimica et Biophysica Acta*, vol. 1543, no. 2, pp. 434–455, 2000.

[46] R. Catana, M. Eloy, J. R. Rocha, B. S. Ferreira, J. M. S. Cabral, and P. Fernandes, "Stability evaluation of an immobilized enzyme system for inulin hydrolysis," *Food Chemistry*, vol. 101, no. 1, pp. 260–266, 2006.

3

An Optimized Real-Time PCR to Avoid Species-/Tissue-Associated Inhibition for H5N1 Detection in Ferret and Monkey Tissues

LingJun Zhan,[1,2] LinLin Bao,[1,2] FengDi Li,[1,2] Qi Lv,[1,2] LiLi Xu,[1,2] and Chuan Qin[1,2]

[1] *Key Laboratory of Human Diseases Comparative Medicine, Ministry of Health, Institute of Laboratory Animal Science, Chinese Academy of Medical Sciences (CAMS) & Comparative Medicine Centre, Peking Union Medical Collage (PUMC), Pan Jia Yuan Nan Li No. 5, Chao Yang District, Beijing 100021, China*
[2] *Key Laboratory of Human Diseases Animal Model, State Administration of Traditional Chinese Medicine, Pan Jia Yuan Nan Li No. 5, Chao Yang District, Beijing 100021, China*

Correspondence should be addressed to Chuan Qin, chuanqin@vip.sina.com

Academic Editors: G. T. Pharr and I. Valpotic

The real-time PCR diagnostics for avian influenza virus H5N1 in tissue specimens are often suboptimal, since naturally occurring PCR inhibitors present in samples, or unanticipated match of primer to unsequenced species' genome. With the principal aim of optimizing the SYBR Green real-time PCR method for detecting H5N1 in ferret and monkey (Chinese rhesus macaque) tissue specimens, we screened various H5N1 gene-specific primer pairs and tested their ability to sensitively and specifically detect H5N1 transcripts in the infected animal tissues, then we assessed RNA yield and quality by comparing Ct values obtained from the standard Trizol method, and four commonly used RNA isolation kits with small modifications, including Roche High Pure, Ambion RNAqueous, BioMIGA EZgene, and Qiagen RNeasy. The results indicated that a single primer pair exhibited high specificity and sensitivity for H5N1 transcripts in ferret and monkey tissues. Each of the four kits and Trizol reagent produced high-quality RNA and removed all or nearly all PCR inhibitors. No statistically significant differences were found between the Ct values from the isolation methods. So the optimized SYBR Green real-time PCR could avoid species- or tissue-associated PCR inhibition in detecting H5N1 in ferret and monkey tissues, including lung and small intestine.

1. Introduction

Ferrets have emerged as an appropriate and feasible model system of influenza, especially for evaluating the efficacy of antiviral drugs and vaccines. In contrast, the monkey is superior for infection and immunity studies since it is more genetically similar to human [1].

Highly pathogenic avian influenza virus (HPAIV) infection normally targets the mucosal tissues but can rapidly spread to multiple organs, eliciting robust cytokine-mediated systemic inflammation, and possible death. Therefore, in addition to oral and cloacal swabs, tissue biopsies are often used to monitor HPAIV infection in infected birds and animals, especially from lower respiratory tract and digestive tract. Viral transcripts can be effectively detected in infected tissues by qPCR, which has been broadly applied to clinical diagnostics, surveillance, and research [1].

However, qPCR detection in tissues sometimes was not ideal. There were several reasons. First, it could be limited by naturally occurring inhibitory substances that were present in some clinical and environmental samples, including feces, blood, soil, tissues, and urine [2, 3]. Such molecules might be coextracted and copurified with the RNA during isolation from the infected tissues and fecal swabs under examination [4]. But commercially available RNA extraction kits might fail to completely remove such amplification inhibitors [5].

In addition, despite the fact that ferrets have been used in biomedical research for decades, little is known about the ferret genome. Thus, it is difficult to design primer to avoid the potential mismatch between primer sequences of pathogen and the whole genome of animals in PCR. This limitation is, unfortunately, not limited to a single animal type and affects many of the nonhuman primates that have not yet to be sequenced.

TABLE 1: Primer pairs for H5N1 (SZ 406H).

Target gene	Primer name	Sequence (5'-3')
HA	SZHA-F1/R1	CCATTCCACAACATACACCCTC/TTCCCTGCCATCCTCCCT
	SZHA-F2/R2	ACAAGGTCCGACTACAGC/TTCCGTTTCTTACACTTTCC
	SZHA-F3/R3	AGAACAATACATACCCAACA/CACTTTGCCCGTTTACTT
NA	SZNA-F1/R1	CAAAGACAGAAGCCCTC/CTCAGTATGTTGTTCCTCCA
	SZNA-F2/R2	ACAGGGAATCAACACCAA/TACAGCCCATCCTCTAAT
	SZNA-F3/R3	AAGACAGAAGCCCTCACAG/TTTCAATACAGCCACAGCC
NP	SZNP-F1/R1	TCAGCGTTCAGCCCACTT/TCGGGTTCGTTGCCTTTT
	SZNP-F2/R2	CAGCCCACTTTCTCGGTAC/TCGGGTTCGTTGCCTTTT
	SZNP-F3/R3	GCCAGGTCTTTAGTCTCAT/CTTATAGCCCAATATCTACTTC
NS	SZNS-F1/R1	AATGCCGACTTCACGCTAC/TCCCACGATTGCTCCTTC
	SZNS-F2/R2	ATGCCCAAGCAGAAAGTG/TCCGATGAGGACGCCAAT
	SZNS-F3/R3	AAGAAGGAGCAATCGTGG/CGTTTCTGATTTGGAGGG
M	SZM-F1/R1	ATTTGTATTCACGCTCACC/TAGTCACCGTTCCCATCC
	SZM-F2/R2	TTTTGTCCAGAATGCCCTAA/CACCGTTCCCATCCTGTT
	SZM-F3/R3	TACAACAGGATGGGAACG/AGTGGGTTGGTGATGGTT
PA	SZPA-F1/R1	GGAGTGACACGGAGGGAA/TCTCGGATTGACGAAAGG
	SZPA-F2/R2	TGGGATTCCTTTCGTCAA/CTGGAGAAGTTCGGTGGG
	SZPA-F3/R3	TCTATGGGATTCCTTTCG/TCTGGCGTTCACTTCTTT
PB1	SZPB1-F1/R1	CTTGAAGAATCCCACCCA/AAATCTATCAGCCGTCCC
	SZPB1-F2/R2	ACATACCGATGCCACAGA/TCAATTCCCATTTCAAGC
	SZPB1-F3/R3	GCGAGGAGTATCTGTGAG/ATCATTGCCAGAAACATC
PB2	SZPB2-F1/R1	CTGGGTCGGACAGGGTGAT/AACTCGGCGGCGTATTTT
	SZPB2-F2/R2	AAACTGGGAGACCGTGAA/GCTCTGCTTAGGTGGTGC
	SZPB2-F3/R3	AAACTGGGAGACCGTGAA/CTGCTCTGCTTAGGTGGT

In our previous study, we determined that the qPCR based on the SYBR Green reagent was ideal for detecting H5N1 from human nasal swabs, ferret or monkey nasal swabs, respiratory tract lavage and turbinate curettage biopsy, and mouse lung tissue (data not shown), but the results from ferret or monkey tissues were suboptimal. We observed nonspecific amplification products and many instances of complete failure of amplification from confirmed infected tissues. In order to identify an effective qPCR system for these two animal species we had to first identify the most optimal primer pair sequences and technique to extract excellent quality RNA.

2. Materials and Methods

2.1. Reagents. The manufacturer-supplied RNA isolation kits were Qiagen RNeasy mini kit (catalogue #74106; QI), Ambion RNAqueous kit (AM1912; AM), Roche High Pure RNA tissue kit (12033674001; RO), and BioMIGA EZgene tissue RNA miniprep kit (R6311; BI). Invitrogen's Trizol reagent (15596-026; TR) was also used, according to manufacturer's instructions. cDNA was reverse transcribed from isolated total RNA using the SuperScript III first-strand synthesis system (18080-051; Invitrogen). qPCR was carried out using the Power SYBR Green PCR master mix (4367659; Applied Biosystems, Inc.).

2.2. Animals. Ferrets (Mustela Pulourius Furo), 4-5 months of age (Marshall Farms,USA), Monkeys (Chinese rhesus macaques), 3-year-old (the Academy of Military Medical Sciences in Beijing). The animals were serologically negative detected by hemagglutination inhibition (HI) assay for currently circulating influenza viruses including A/California/7/2009 (H1N1), seasonal influenza virus H1N1, H3N2, and avian influenza virus H5N1.

2.3. Tissue Sample Collection and Homogenization. Organ samples were obtained from experimentally infected ferrets and monkeys, which were nasal swab positive for AIV H5N1 (SZ406H) and experiencing obvious clinical symptoms, such as fever, sneezing, and runny nose. Prior to biopsy, the animals were euthanized by injecting Tribromoethanol. The tissues were ground up by a Pro-200 tissue homogenizer (Pro Scientific) to a homogeneous lysate; solid debris was removed by centrifugation, and the remaining liquid was prepared for virus quantitation by qPCR. All the experiment was carried out in ABSL-3 lab [6], and all procedures were approved by the Institute of Animal Use and Care Committee of the Institute of Laboratory Animal Science, Peking Union Medical College (MC-09-7005).

2.4. RNA Isolation from Ferret and Monkey Tissues. The commercial kits and reagents were used applied with the manufacturer's recommendation, including any modifications introduced [6]. The starting material for all procedures was $50\,\mu L$ homogeneous tissue sample (~ 10 mg instead of 30 mg), and the samples were transferred to its starting buffer (e.g., RLT buffer for QI kit) with volume of $500\,\mu L$,

An Optimized Real-Time PCR to Avoid Species-/Tissue-Associated Inhibition for H5N1 Detection in Ferret and Monkey Tissues

19

TABLE 2: Primer pairs recommended by WHO for H5N1.

Target gene	Primer name	Sequence (5′-3′)
HA	H5-1/H5-3	GCCATTCCACAACATACACCC/CTCCCCTGCTCATTGCTATG
M	M30F/M264R	TTCTAACCGAGGTCGAAACG/ACAAAGCGTCTACGCTGCAG
N1	N1-1/N1-2	TTGCTTGGTCGGCAAGTGC/CCAGTCCACCCATTTGGATCC

respectively. At the last step, the isolated RNA was resolved in RNase-free water or Elution buffer with final volume of 50 uL. The manufacturer's protocols were almost completely followed except some small modifications. As for QI kit, after step 7, $50 \mu L$ DNase was added per isolation column and incubated for 1 min at room temperature.

Total RNA was transcribed with SuperScript III First-Strand synthesis system, and the virus-containing supernatant of infected Madin-Darby Canine Kidney (MDCK) epithelial cells was purified and also transcribed as a positive control.

2.5. Primer Design.
Primer pairs were designed for the eight gene fragments of H5N1, three pairs for each gene (Table 1). In addition, primer pairs that recommended by WHO were synthesized and tested (Table 2).

2.6. Primer Screening for SYBR Green Real-Time PCR of Ferret and Monkey Tissues.
The primer pairs listed in Tables 1 and 2 were used to amplify the cDNA of H5N1 virus (SZ406H, Genebank ID: 133711835), and several of the most superior primer pairs were selected for further analysis. The selected primer pairs were then used to amplify cDNAs of infected lung and small intestine, as prepared by the five commercial kits, by SYBR Green qPCR; the primer pair exhibiting the highest specificity and sensitivity was selected from the set. The primer pair was also chosen based upon its suitability in both ferret and monkey tissues. The sensitivity of this primer pair in detection was then evaluated with serial standard samples having known copy numbers of virus (from 10 to 10^8). DNA sequence of some PCR products was confirmed by sequencing (Taihe Biotechnology Co., Ltd.).

2.7. Comparative Analysis of the Commercial RNA Isolation Kits and TR in Ferret and Monkey Tissues.
The quality of different RNA isolation kits and the TR method was evaluated for their ability to remove PCR inhibitors from the complex tissue samples. The Ct value obtained from amplifying the same amount of tissue template was compared among all five methods [6–8].

3. Results

3.1. Identification of an Optimal Primer Pair for SYBR Green Real-Time PCR-Based Detection of H5N1 in Ferret and Monkey Tissues.
A group of candidate primer pairs for eight H5N1 genes was screened for specificity and sensitivity of detection of positive virus template by PCR and agarose gel electrophoresis. The screened primers included SZHA-F1/R1, SZNA-F2/R2, SZM-F2/R2, SZNP-F2/R2, SZNS-F2/R2, SZPA-F1/R1, SZPB1-F3/R3, and SZPB2-F2/R2 (see

(a)

(b)

FIGURE 1: Agarose gel electrophoresis of SYBR Green real-time PCR products of H5N1 transcripts in lung and small intestine tissues. Lane 1: DL2000 Marker; Lane 2: positive control; Lanes 3–5: small intestine samples amplified with primer pairs N1-1/N1-2, M30F/M264R, H5-1/H5-3; Lanes 6–8: lung samples amplified with primer pairs N1-1/N1-2, M30F/M264R, H5-1/H5-3; Lanes 9–13: small intestine samples prepared by TR, QI, BI, AM, and RO, respectively, and amplified by primer pair SZNP-F2/R2; Lanes 14–18: lung samples prepared by TR, QI, BI, AM, and RO, respectively, and amplified with primer pair SZNP-F2/R2; Lane 19: negative control. (a) was for ferret tissues and (b) was for monkey tissues.

Table 1). The primer pair SZNP-F2/R2 among them exhibited the highest specificity and sensitivity and performed better than the WHO recommended primer pairs H5-1/H5-3, M30F/M264R, and N1-1/N1-2 (Figures 1(a) and 1(b)).

3.2. The Specificity and Sensitivity of Primers SZNP-F2/R2 in Ferret and Monkey Tissues.
cDNAs of uninfected ferret and monkey tissues mixed with a range of virus (copy number from 10 to 10^8) were used to evaluate the SZNP-F2/R2 primer pair in SYBR Green qPCR. According to the standard curve, the linear equation for the qRT-PCR was $Y = -3.4779X + 33.317$. The sensitivity of this method, defined as the lowest concentration of cDNA detected in qPCR, was 1 copy/reaction (Figure 2(a)). Moreover, the SZNP-F2/R2 specificity was high, as evidenced by the characteristic monowave profile of the melting curve and the presence of a single band in agarose gel electrophoresis (Figures 2(b) and 2(c)). The DNA sequence of qPCR product was also exactly the same as the fragment 1123–1423 bp of SZNP (Genebank ID: 133711835), as confirmed by DNA sequencing.

3.3. The Ct Values for the Five RNA Isolation Methods Used in Conjunction with SYBR Green Real-Time PCR of Ferret or Monkey Tissues.
cDNAs prepared from RNA isolated by the five different methods were amplified by SYBR-Green

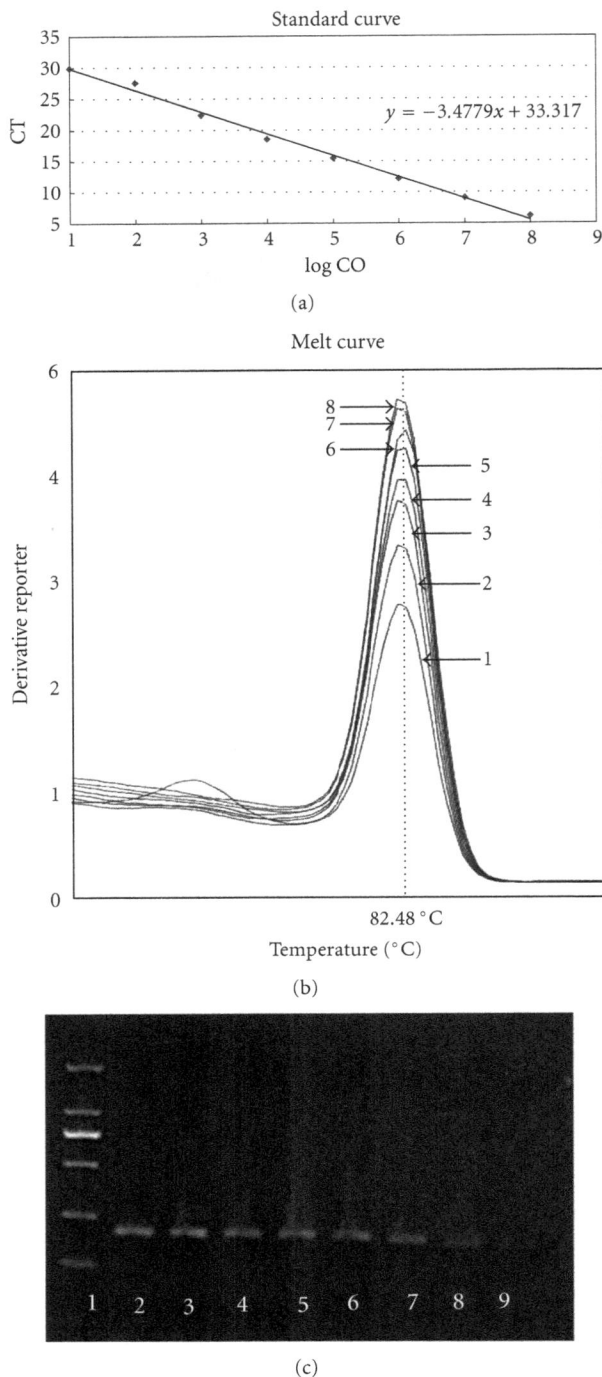

Standard curve

$y = -3.4779x + 33.317$

(a)

Melt curve

82.48 °C

Temperature (°C)

(b)

(c)

FIGURE 2: The sensitivity and specificity achieved with primer pair SZNP-F2/R2 in SYBR Green real-time PCR. (a) Standard curve. LogCO: log10 (copies), slope: -3.4779, R^2: 0.998, eff%: 110%. (b) Melting curve. (c) Agarose electrophoresis of PCR products of H5N1. Lane 1: DL2000 Marker; Lanes 2–9: 10^8–10^1 copies of H5N1 virus cDNA.

qPCR. The respective Ct values from each, using equal amounts of sample material, were listed in Table 3. The melting curves and the agarose gel electrophoresis results were shown in Figures 1(a) and 1(b). The results indicated that the

TABLE 3: Ct values of ferret or monkey cDNAs prepared by five different methods and subjected to SYBR-Green qPCR.

	RO	AM	BI	QI	TR
Ferret lung	20.51	18.54	18.85	17.99	17.31
Ferret small intestine	18.29	18.36	18.31	18.29	18.25
Monkey lung	18.31	18.39	18.58	18.34	18.43
Monkey small intestine	20.11	20.26	19.32	17.05	20.18

Note: The Ct values shown are the average of at least 3 replicates with standard deviations of 6.0–10%. The P values from t-test calculations were >0.05 between the different methods.

four commercially available kits and the Trizol reagent were sufficiently effective to produce high-quality RNA suitable for H5N1 transcripts detection in ferret and monkey tissues. We concluded that the five methods were comparable in removing potentially contaminating PCR inhibitors from tissue samples; our findings in small intestine were particularly insightful since this tissue harbors a complex enzyme profile [6].

4. Discussion

The WHO recommended primer pairs were designed to target Clade 1, 2, and 3 H5N1 viruses from respiratory biopsies, lavage, and swab specimens. However, we found that they were not suitable for detecting H5N1 in ferret and monkey tissues (Figure 1), which may be the result of unanticipated match to sequence of species- and/or tissue-specific cDNAs. The yet undefined whole genomes of ferret and Chinese rhesus macaque (also other macaques) made screening for appropriate and effective H5N1 primers challenging; thus, we designed a group of candidate primers and sought to identify a single pair with the highest sensitivity and specificity for H5N1 virus in these animals. The primer pair SZNP-F2/R2 fit these qualifications and avoided mismatches between primer and tissues cDNA sequences from either ferret or monkey. In addition, SZNP-F2/R2 was able to properly detect H5N1 in nasal swabs and in supernatants from virus-infected cultured cells by using SYBR-Green qPCR and conventional PCR. Therefore, our newly designed SZNP-F2/R2 primers represent the first reported primer pair that was applicable for H5N1 detection in both human swab samples (data not shown) and ferret or monkey samples, including swab, lavage, and tissue specimens.

Numerous commercial kits for total RNA extraction are available, the QI RNeasy mini kit has been reported as sufficient for obtaining RNA from a wide variety of tissue samples and from viruses suspended in culture supernatants [1]. While it has also been reported that the silicon nucleic acid binding column method is not effective when using tissues, cloacal swabs or samples that may contain fecal material; instead, the organic extraction method with Trizol reagent is recommended for use with amnioallantoic fluid samples, oral swabs, tracheal swabs, cloacal swabs, cell culture material, and tissues with complex enzyme profiles [8]. Meanwhile, since the current commercially available column-based RNA isolation kits were developed and optimized using mouse and rat tissue, we needed to determine

An Optimized Real-Time PCR to Avoid Species-/Tissue-Associated Inhibition for H5N1 Detection in Ferret and
Monkey Tissues

21

whether they worked as effectively with ferret and monkey tissues. It is essential to choose a proper RNA isolation method for our own experiment according to different tissues and real-time PCR techniques being used.

In this study, we compared five commonly used RNA isolation methods with some modifications, including QI, AM, RO, BI, and TR. After comparing the Ct value of samples prepared by each, we found that all were capable of sufficiently isolating high quality RNA, without significant statistical differences in yield or quality. Moreover, each method appeared to effectively remove the endogenous PCR inhibitors from the tissues examined; this is an especially insightful finding as the small intestine tissue is presumed to contain a large profile of various enzymes and potential PCR inhibitors [5, 7]. Therefore, the five RNA isolation methods are all effective for viral RNA isolation from tissues of ferret and monkey, and for subsequent detection of H5N1 by SYBR-Green qPCR.

The optimized SYBR-Green qPCR system, though it has many advantages, still has some limitations that must be taken into consideration. The melting curve might sometimes mask nonspecific products; hence, it was recommended to verify the real time PCR products by agarose gel electrophoresis to confirm single- or multiple-band products in preliminary study. Moreover, due to the conservation of NP in influenza virus nucleotide sequence, the primer SZNP-F2/R2 might match with two AIV H9N2 strains by BLAST analysis, which needed further more experimental confirmation. But it was confirmed that the SZNP-F2/R2 was perfect in detecting H5N1 in experimental-infected ferret and monkey tissues.

In conclusion, the findings from this study provide insights into the necessary steps and reagents that will help to resolve the current problems experienced when using SYBR-Green qPCR to detection H5N1 in particular tissues. It is important to consider the possibility of nonspecific amplification from different species and tissues, which may be a result of mismatch or unanticipated extra matching of primer and particular AIV strain or animal genome, respectively. It is also important to determine if failed amplifications are a result of inhibition by contaminating PCR inhibitors in samples. This qPCR-based viral detection is applicable for other viruses and nonviral pathogens detection in a wide array of tissues.

Acknowledgments

The authors thank all the team members of the Emerging Infectious Diseases Center of the Institute of Laboratory Animal Science, Chinese Academy of Medical Sciences (CAMS). The authors are grateful for the basic fund of central non-profit academic institution (DWS201105).

References

[1] Y. Matsuoka, E. W. Lamirande, and K. Subbarao, "The ferret model for influenza," *Current Protocols in Microbiology*, no. 13, pp. 15G.2.1–15G.2.29, 2009.

[2] P. Rådström, R. Knutsson, P. Wolffs, M. Lövenklev, and C. Löfström, "Pre-PCR processing: strategies to generate PCR-compatible samples," *Applied Biochemistry and Biotechnology—Part B*, vol. 26, no. 2, pp. 133–146, 2004.

[3] D. L. Suarez, *Avian Influenza Virus RNA Extraction from Tissue and Swab Material*, Methods in Molecular Biology, Humana Press, Totowa, NJ, USA, 1st edition, 2008.

[4] S. Liu, G. Hou, Q. Zhuang et al., "A SYBR Green I real-time RT-PCR assay for detection and differentiation of influenza A(H1N1) virus in swine populations," *Journal of Virological Methods*, vol. 162, no. 1-2, pp. 184–187, 2009.

[5] H. Tomaso, M. Kattar, M. Eickhoff et al., "Comparison of commercial DNA preparation kits for the detection of Brucellae in tissue using quantitative real-time PCR," *BMC Infectious Diseases*, vol. 10, article 100, 2010.

[6] A. Das, E. Spackman, M. J. Pantin-Jackwood, and D. L. Suarez, "Removal of real-time reverse transcription polymerase chain reaction (RT-PCR) inhibitors associated with cloacal swab samples and tissues for improved diagnosis of Avian influenza virus by RT-PCR," *Journal of Veterinary Diagnostic Investigation*, vol. 21, no. 6, pp. 771–778, 2009.

[7] A. Das, E. Spackman, C. Thomas, D. E. Swayne, and D. L. Suarez, "Detection of H5N1 high-pathogenicity avian influenza virus in meat and tracheal samples from experimentally infected chickens," *Avian Diseases*, vol. 52, no. 1, pp. 40–48, 2008.

[8] A. Das, E. Spackman, D. Senne, J. Pedersen, and D. L. Suarez, "Development of an internal positive control for rapid diagnosis of avian influenza virus infections by real-time reverse transcription-PCR with lyophilized reagents," *Journal of Clinical Microbiology*, vol. 44, no. 9, pp. 3065–3073, 2006.

4

Association of IRGM Polymorphisms and Susceptibility to Pulmonary Tuberculosis in Zahedan, Southeast Iran

Gholamreza Bahari,[1] Mohammad Hashemi,[1,2] Mohsen Taheri,[3] Mohammad Naderi,[4] Ebrahim Eskandari-Nasab,[1] and Mahdi Atabaki[4]

[1] *Department of Clinical Biochemistry, School of Medicine, Zahedan University of Medical Sciences, Zahedan 98167-43463, Iran*
[2] *Cellular and Molecular Research Center, Zahedan University of Medical Sciences, Zahedan 98167-43463, Iran*
[3] *Genetic of Non-Communicable Disease Research Center, Zahedan University of Medical Science, Zahedan 98167-43463, Iran*
[4] *Research Center for Infectious Diseases and Tropical Medicine, Zahedan University of Medical Sciences, Zahedan 98167-43463, Iran*

Correspondence should be addressed to Mohammad Hashemi, mhd.hashemi@gmail.com

Academic Editors: S. K. Behura, H. Schwarzenbach, and A. Toulouse

Tuberculosis (TB) is a major cause of morbidity and mortality worldwide. IRGM1 is an important protein in the innate immune response against intracellular pathogens by regulating autophagy. Polymorphisms in the IRGM genes are known to influence expression levels and may be associated with outcome of infections. This case-control study was done on 150 patients with PTB and 150 healthy subjects to determine whether the IRGM polymorphisms at positions −1208 A/G (rs4958842), −1161 C/T (rs4958843), and −947 C/T (rs4958846) were associated with PTB. The polymorphisms were determined using tetra-amplification refractory mutation system-PCR (T-ARMS-PCR). The results showed that the IRGM −1161 C/T and −947 C/T polymorphisms were associated with decreased susceptibility to PTB (OR = 0.06, 95% CI = 0.03–0.13, $P < 0.001$ and OR = 0.27; 95% CI = 0.013–0.55, $P < 0.001$, resp.). No significant difference was found among the groups regarding −1208 A/G polymorphism. In conclusion we found that the IRGM −1161 C/T and −947 C/T polymorphisms but not −1208 A/G polymorphism provide relative protection against PTB in a sample of Iranian population.

1. Introduction

Tuberculosis is a public health problem especially in developing countries and cause of morbidity and mortality throughout the world. According to the World Health Organization (WHO), it is a global emergency [1] and approximately 2 million peoples annually die due to tuberculosis [2]. Among the one-third of the world infected by TB only 10% develop clinical disease [3]. Increasing evidence indicates that host genetic factors play an important role in susceptibility to TB [3].

Mycobacterium tuberculosis is an intracellular pathogen that can persist within host macrophages. It can reside within phagosomes of macrophages and is able to arrest phagosome maturation [4]. Autophagy is a process in which intracellular components degrade in lysosomes of the cell. It plays a key role against intracellular pathogens such as

mycobacterium [5]. It has been shown that immunity-related GTPase (*IRGM*) induced autophagy in macrophages to control M. tuberculosis [6]. This protein is necessary for immunity against a series of intracellular pathogens in mice, including *Listeria*, *Toxoplasma*, and *Mycobacterium tuberculosis* [2]. There are 3 IRG genes, IRGC, IRGQ, and IRGM, in human genome; only IRGM is functional [7].

The IRGM gene with 5 exons is located on chromosome 5q33.1. The first exon is long and encodes 181 amino acids; the four shorter exons extend more than 50 kb downstream from the first exon [8]. Variations in promoter region of IRGM gene have shown to be associated with an increased risk of TB [2, 6].

There is little and controversial data concerning the impact of IRGM polymorphisms and susceptibility to PTB. Therefore, the present study aimed to evaluate the possible association between −1208 A/G, −1161 C/T, and −947 C/T

IRGM polymorphisms and pulmonary tuberculosis in a sample of Iranian population.

2. Material and Methods

This case-control study was performed from December 2010 to January 2012 in the Research Center for Infectious Diseases and Tropical Medicine, the Bou-Ali Hospital, Zahedan, Iran. A total of 150 PTB patients and 150 healthy subjects were enrolled in the study. Ethics committee of the Zahedan University of Medical Sciences approved the project and informed consent was taken from all patients and healthy subjects. All control subjects were from the same geographical origin and were living in the same region as the patients with PTB. The diagnosis of PTB was based on clinical, radiological, sputum acid-fast bacillus (AFB) smear positivity, culture, and response to antituberculosis chemotherapy as described previously [9, 10]. Two mL of venous blood was drawn from each subject and genomic DNA was extracted from peripheral blood as described previously [11] and stored at $-20°C$.

Tetra-primer amplification refractory mutation system polymerase chain reaction (T-ARMS-PCR) is a simple and rapid method for detection of single nucleotide polymorphism (SNP) [12–14]. We designed T-ARMS-PCR for detection of -1208 A/G (rs4958842), -1161 C/T (rs4958843), and -947 C/T (rs4958846) polymorphisms of IRGM gene. We used two external primers (forward outer and reverse outer) and two inner primers (forward inner and reverse inner) for each position that are shown in Table 1.

Polymerase chain reaction (PCR) was done using commercially available PCR premix (AccuPower PCR PreMix, BIONEER, Daejeon, Republic of Korea) according to the manufacturer protocol. Into a 0.2 mL PCR tube containing the AccuPower PCR PreMix, $1\,\mu$L template DNA ($\sim$ 100 ng/μL), $1\,\mu$L of each primer ($10\,\mu$M), and $15\,\mu$L DNase-free water were added. The PCR cycling conditions were 5 min at 95°C followed by 30 cycles of 30 s at 95°C, 30 s at 63°C for -1208 A/G, 65°C for -1161 C/T, 63°C for -947 C/T, respectively, and 30 s at 72°C, with a final step at 72°C for 10 min to allow for complete extension of all PCR fragments. The PCR products were analyzed by electrophoresis on a 2% agarose gel containing 0.5 μg/mL ethidium bromide and visualized by ultraviolet transilluminator. To ensure genotyping quality, we regenotyped all polymorphisms in random samples and found no genotyping mistake.

For -1208 A/G, the PCR product sizes were 195 bp for A allele, 245 bp for G allele, and 402 bp for two outer primers (control band) (Figure 1). For -1161 C/T, product sizes were 199 bp for C allele, 261 bp for T allele, and 415 bp for control band (Figure 2). Product sizes were 201 bp for C allele, 263 bp for T allele, and 417 bp for control band for -947 C/T (Figure 3).

2.1. Statistical Analysis. The statistical analysis of the data was performed using the SPSS 18.0 software. Demographics and biochemical parameters between the groups were analyses by independent sample t-test for continuous data and

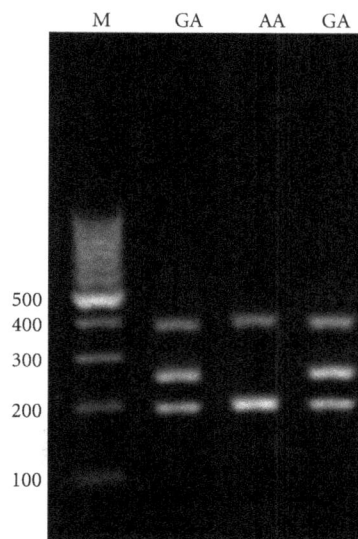

FIGURE 1: Electrophoresis pattern of tetra-amplification refractory mutation system-polymerase chain reaction (T-ARMS-PCR) for detection of SNP in IRGM -1208 A/G. M : DNA marker. Product sizes were 195 bp for A allele, 254 bp for G allele, and 402 bp for two outer primers (control band).

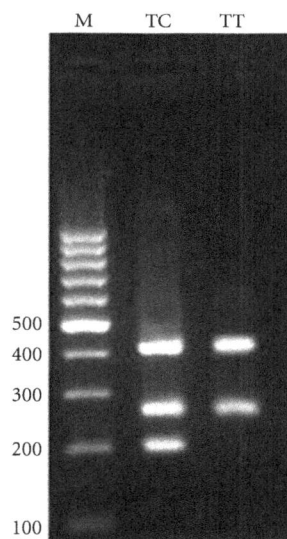

FIGURE 2: Electrophoresis pattern of tetra-amplification refractory mutation system-polymerase chain reaction (T-ARMS-PCR) for detection of SNP in IRGM -1161 C/T. M : DNA marker. Product sizes were 199 bp for C allele, 261 bp for T allele, and 415 bp for control band.

χ^2 test for categorical data. A P value less than 0.05 was considered statistically significant. The associations between genotypes of IRGM gene and PTB were estimated by computing the odds ratio (OR) and 95% confidence intervals (95% CI) from logistic regression analyses.

TABLE 1: Primers used for polymorphism determination.

Primers	−1208 A/G (rs4958842)	−1161 C/T (rs4958843)	−947 C/T (rs4958846)
Forward outer	TGTGAGTATGTGTGGGCCTGTGCACAGA	GGCATGGGTGAGTGTGCACACC	TCCTCAGCCTTGGCGCCCACTCTA
Reverse outer	AGTTGCTGCCCGTGCCTCTCCCTC	CTAAGCCCCTCACTGCCAGGGG	GCTCATAGGGGAGGCTCGGGCTGT
Forward inner	ACAGCATGCTGGCAGCCCTCGAAA	CAGCCTTGGCGCCCACTCTCGT	CAGAGCAGCCATCCGGCCCCTAC
Reverse inner	AGGCTCCGAGAGCCAGCGAGTGC	GCTGAAGGGCTCCTCAAGTGACG	TAAGCCCCTCACTGCCAGGGGACA

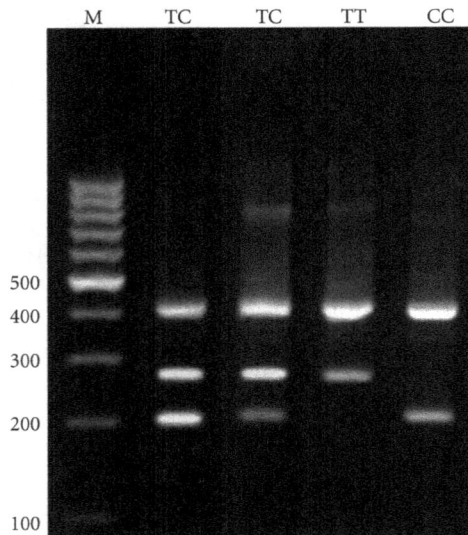

FIGURE 3: Electrophoresis pattern of tetra-amplification refractory mutation system-polymerase chain reaction (tetra ARMS-PCR) for detection of SNP in IRGM −947 C/T. M : DNA marker. Product sizes were 201 bp for C allele, 263 bp for T allele, and 417 bp for control band.

3. Results

For determining of IRGM polymorphisms in PTB patients and comparison of these polymorphisms with healthy individual, a total of 150 pulmonary tuberculosis patients with an average age of 47.5 years (59 male, 91 female; minimum 12 years, maximum 78 years) and 150 healthy subjects with a mean age of 44.13 years (53 male, 97 female; minimum 20, maximum 82) were enrolled in the study. There was no significant difference among the groups regarding sex and age ($P > 0.05$). Allele and genotype frequencies of IRGM −1208 A/G, −1161 C/T, and −947 C/T are given in Table 2. No significant difference was found between the groups concerning −1208 A/G polymorphism ($\chi^2 = 1.19$, $P = 0.274$). A significant difference was observed among case and control groups regarding IRGM −1161 C/T ($\chi^2 = 75.37$, $P < 0.001$) and −947 C/T ($\chi^2 = 15.75$, $P < 0.001$). As shown in Table 2, the −1161 CT genotype as well as −1161 C allele was associated with protection from PTB (OR = 0.06, 95% CI = 0.03–0.13, $P < 0.001$ and

OR = 0.36, 95% CI = 0.26–0.51, $P < 0.001$, resp.). The −947 CT genotype was significantly higher in control group (%23.33) than that in PTB (%8.00) and this polymorphism was negatively associated with susceptibility to PTB (OR = 0.28, 95% CI = 0.14–0.57, $P = 0.002$). The −947 C allele confers a protective role against PTB (OR = 0.27, 95% CI = 0.14–0.54, $P < 0.001$).

4. Discussion

IRGM1 is an important protein in the innate immune system against TB by regulating autophagy in response to intracellular pathogens. In the present study, we examined the impact of IRGM −1208 A/G, −1161 C/T, and −947 C/T polymorphisms on pulmonary tuberculosis (PTB) risk in a sample of Iranian population. Our finding revealed the protective role of −1161 C/T and −947 C/T polymorphisms against PTB in our population. No significant difference was found between control and PTB groups regarding IRGM −1208 A/G polymorphism.

To the best of our knowledge, there is little information regarding the association of IRGM polymorphisms and tuberculosis and this is the first report from Iran. The earliest report about the role of IRGM in autophagy and defense against intracellular pathogens is attributed to Singh et al. study [15]. They found that the murine Irgm1 guanosine triphosphatase induced autophagy and generate large autolysosomal organelles for the elimination of intracellular *Mycobacterium tuberculosis*. They also reported that the human IRGM plays a role in autophagy and in the reduction of intracellular bacillary load [15].

Intemann et al. for the first time investigated the association between IRGM genotypes and tuberculosis. They found that the IRGM genotype −261 TT provides relative protection against Mycobacterium tuberculosis but not by *M. africanum* or *M. bovis* [8]. They predicted that −261 T IRGM variant disrupted several transcription factor-binding sitesed and significantly increased expression of the −261 T IRGM variant compared with the −261 C IRGM variant, suggested that TT genotypeing might enhance expression of IRGM protein. They proposed that IRGM and autophagy have a role in protection against *M. tuberculosis* [8]. Che et al. in a study group of Chinese TB patients sequenced 1.7 kb of IRGM promoter region and identified 29 polymorphisms including 11 novel sites [7]. In contrast to our finding they showed that −1208 A allele and −1208 AA genotype of

TABLE 2: The genotypes and allele distribution of IRGM polymorphisms in case and control groups.

Polymorphism	PTB n (%)	Control n (%)	OR (95%CI)	P	*OR (95%CI)	P
−1208 A/G (rs4958842)						
AA	20 (13.3)	14 (9.3)	Ref.			
AG	130 (86.7)	136 (90.7)	1.49 (0.73–3.08)	0.277	1.53 (0.74–3.18)	0.250
GG	0 (0.0)	0 (0.0)	—	—	—	—
Alleles						
A	170 (56.7)	164 (54.7)				
G	130 (43.3)	136 (45.3)	0.92 (0.67–1.27)	0.681		
−1161 C/T (rs4958843)						
TT	77 (51.3)	9 (6.0)	Ref.			
CT	73 (48.7)	141 (94.0)	0.06 (0.03–0.12)	<0.001	0.06 (0.03–0.13)	<0.001
CC	0 (0.0)	0 (0.0)	—	—	—	—
Alleles						
T	227 (75.7)	159 (53.0)				
C	73 (24.3)	141 (47.0)	0.36 (0.26–0.51)	<0.001		
−947 C/T (rs4958846)						
TT	138 (92.0)	113 (75.3)	Ref.			
CT	12 (8.0)	35 (23.4)	0.28 (0.14–0.57)	0.002	0.27 (0.13–0.55)	<0.001
CC	0 (0.0)	2 (1.3)	—	—	—	—
CT + CC	12	37 (24.66)	0.26 (0.13–0.53)	<0.001	0.26 (0.13–0.53)	<0.001
Alleles						
T	288 (96.0)	261 (87.0)				
C	12 (4.0)	39 (13.0)	0.27 (0.14–0.54)	<0.001		

*Adjusted for age and sex.

IRGM were associated with decreased susceptibility to TB [7].

A study conducted by King et al. on 370 African American and 177 Caucasian tuberculosis (TB) cases and 180 African American and 110 Caucasian controls showed that single nucleotide polymorphism rs10065172 C/T in IRGM is associated with human vulnerability to TB disease among African Americans [2]. Their results showed that there were not differences in IRGM1 expression level based on genotype.

A combination of both innate and adaptive immune responses were involved in the host defense against mycobacterium. Autophagy mediates innate immune responses against mycobacterium by promoting phagolysosomal maturation within macrophages [16], besides, autophagy plays a key role in antigen processing and presentation [5]. IRGM involved in the induction of autophagy in macrophages that infected with mycobacterium. Investigation showed that variations of IRGM gene are associated with an increased risk of several diseases such as Crohn's disease and tuberculosis.

The exact reason as to why only a number of the subjects infected with M. tuberculosis develop clinical disease is clearly unknown. There are some evidences recommending that host genetic factors may be important risk factor for development of tuberculosis [17–20]. The impact of IRGM polymorphism on PTB susceptibility is probably influenced by ethnic background.

In conclusion, our results showed that −1161 C/T and −947 C/T IRGM polymorphisms but not −1208 A/G polymorphism contributes to decreased susceptibility to PTB among an Iranian population. Larger studies with different ethnicitie are required to validate our findings.

Conflict of Interests

There is no conflict of interests to be declared.

Acknowledgments

The authors thankfully acknowledge the Zahedan University of Medical Sciences for grant support. In addition, the authors would like to thank the patients and healthy subjects who willingly participated in the study.

References

[1] W. H. Organization, *Global Tuberculosis Control: Surveillance, Planning, Financing*, World Health Organization, Geneva, Switzerland, 2008.

[2] K. Y. King, J. D. Lew, N. P. Ha et al., "Polymorphic allele of human IRGM1 is associated with susceptibility to tuberculosis in African Americans," *PLoS ONE*, vol. 6, no. 1, article e16317, 2011.

[3] C. M. Stein, "Genetic epidemiology of tuberculosis susceptibility: impact of study design," *PLoS Pathogens*, vol. 7, no. 1, article e1001189, 2011.

[4] D. M. Shin, B. Y. Jeon, H. M. Lee et al., "Mycobacterium tuberculosis eis regulates autophagy, inflammation, and cell

death through redox-dependent signaling," *PLoS Pathogens*, vol. 6, no. 12, article e1001230, 2010.

[5] C. Ní Cheallaigh, J. Keane, E. C. Lavelle, J. C. Hope, and J. Harris, "Autophagy in the immune response to tuberculosis: clinical perspectives," *Clinical and Experimental Immunology*, vol. 164, no. 3, pp. 291–300, 2011.

[6] V. Deretic, "Autophagy in infection," *Current Opinion in Cell Biology*, vol. 22, no. 2, pp. 252–262, 2010.

[7] N. Che, S. Li, T. Gao et al., "Identification of a novel IRGM promoter single nucleotide polymorphism associated with tuberculosis," *Clinica Chimica Acta*, vol. 411, no. 21-22, pp. 1645–1649, 2010.

[8] C. D. Intemann, T. Thye, S. Niemann et al., "Autophagy gene variant IRGM -261T contributes to protection from tuberculosis caused by Mycobacterium tuberculosis but not by M. africanum strains," *PLoS Pathogens*, vol. 5, no. 9, article e1000577, 2009.

[9] M. Naderi, M. Hashemi, H. Kouhpayeh, and R. Ahmadi, "The status of serum procalcitonin in pulmonary tuberculosis and nontuberculosis pulmonary disease," *Journal of the Pakistan Medical Association*, vol. 59, no. 9, pp. 647–648, 2009.

[10] H. R. Kouhpayeh, M. Hashemi, S. A. Hashemi et al., "R620W functional polymorphism of protein tyrosine phosphatase non-receptor type 22 is not associated with pulmonary tuberculosis in Zahedan, southeast Iran," *Genetics and Molecular Research*, vol. 11, pp. 1075–1081, 2012.

[11] M. Hashemi, A. K. Moazeni-Roodi, A. Fazaeli et al., "Lack of association between paraoxonase-1 Q192R polymorphism and rheumatoid arthritis in southeast Iran," *Genetics and Molecular Research*, vol. 9, no. 1, pp. 333–339, 2010.

[12] M. Hashemi, H. Hoseini, P. Yaghmaei et al., "Association of polymorphisms in glutamate-cysteine ligase catalytic subunit and microsomal triglyceride transfer protein genes with nonalcoholic fatty liver disease," *DNA and Cell Biology*, vol. 30, no. 8, pp. 569–575, 2011.

[13] M. Hashemi, A. Moazeni-Roodi, A. Bahari, and M. Taheri, "A tetra-primer amplification refractory mutation system-polymerase chain reaction for the detection of rs8099917 IL28B genotype," *Nucleosides, Nucleotides & Nucleic Acids*, vol. 31, pp. 55–60, 2012.

[14] M. Hashemi, A. K. Moazeni-Roodi, A. Fazaeli et al., "The L55M polymorphism of paraoxonase-1 is a risk factor for rheumatoid arthritis," *Genetics and Molecular Research*, vol. 9, no. 3, pp. 1735–1741, 2010.

[15] S. B. Singh, A. S. Davis, G. A. Taylor, and V. Deretic, "Human IRGM induces autophagy to eliminate intracellular mycobacteria," *Science*, vol. 313, no. 5792, pp. 1438–1441, 2006.

[16] K. Takeda, H. Saiga, and Y. Shimada, "Innate immune effectors in mycobacterial infection," *Clinical and Developmental Immunology*, vol. 2011, Article ID 347594, 8 pages, 2011.

[17] R. Zheng, Y. Zhou, L. Qin et al., "Relationship between polymorphism of DC-SIGN (CD209) gene and the susceptibility to pulmonary tuberculosis in an eastern Chinese population," *Human Immunology*, vol. 72, no. 2, pp. 183–186, 2011.

[18] H. Q. Qu, S. P. Fisher-Hoch, and J. B. McCormick, "Molecular immunity to mycobacteria: knowledge from the mutation and phenotype spectrum analysis of Mendelian susceptibility to mycobacterial diseases," *International Journal of Infectious Diseases*, vol. 15, no. 5, pp. e305–e313, 2011.

[19] M. Hashemi, A. Moazeni-Roodi, A. Bahari et al., "Association of interferon-[gamma]+ 874 T/A polymorphism with nonalcoholic fatty liver disease (NAFLD) in Zahedan, southeast of Iran," *Clinical Biochemistry*, vol. 44, pp. S274–S274, 2011.

[20] M. Naderi, M. Hashemi, H. Karami et al., "Lack of association between rs1024611 (-2581 A/G) polymorphism in CC-chemokine ligand 2 and susceptibility to pulmonary tuberculosis in Zahedan, Southeast Iran," *Prague Medical Report*, vol. 112, pp. 272–278, 2011.

5

Co-Cultures of *Pseudomonas aeruginosa* and *Roseobacter denitrificans* Reveal Shifts in Gene Expression Levels Compared to Solo Cultures

Crystal A. Conway,[1] Nwadiuto Esiobu,[2] and Jose V. Lopez[1]

[1] *Oceanographic Center, Nova Southeastern University, Dania Beach, FL 33004, USA*
[2] *Department of Biological Sciences, Florida Atlantic University, Davie, FL 33314, USA*

Correspondence should be addressed to Jose V. Lopez, joslo@nova.edu

Academic Editors: A. C. Manna and D. Zhou

Consistent biosynthesis of desired secondary metabolites (SMs) from pure microbial cultures is often unreliable. In a proof-of-principle study to induce SM gene expression and production, we describe mixed "co-culturing" conditions and monitoring of messages via quantitative real-time PCR (qPCR). Gene expression of model bacterial strains (*Pseudomonas aeruginosa* PAO1 and *Roseobacter denitrificans* Och114) was analyzed in pure solo and mixed cocultures to infer the effects of interspecies interactions on gene expression *in vitro*, Two *P. aeruginosa* genes (*PhzH* coding for portions of the phenazine antibiotic pathway leading to pyocyanin (PCN) and the *RhdA* gene for thiosulfate: cyanide sulfurtransferase (Rhodanese)) and two *R. denitrificans* genes (*BetaLact* for metallo-beta-lactamase and the *DMSP* gene for dimethylpropiothetin dethiomethylase) were assessed for differential expression. Results showed that *R. denitrificans DMSP* and *BetaLact* gene expression became elevated in a mixed culture. In contrast, *P. aeruginosa* co-cultures with *R. denitrificans* or a third species did not increase target gene expression above control levels. This paper provides insight for better control of target SM gene expression *in vitro* and bypass complex genetic engineering manipulations.

1. Introduction

Interactions among diverse microbial species are dynamic and most likely propel many of the adaptations that allow the occupation of diverse niches that can range from biofilms to host digestive tracts to multiple marine habitats [1, 2]. These interactions among diverse bacteria can be either beneficial such as in symbioses with eukaryotic hosts [3, 4] or antagonistic due to competition within multiple species microcosms [5].

Although necessary for identification and certain microbiological experiments, traditional bacteriological methods which focus on isolating microbes as axenic cultures do not provide much insight into the ecological dynamics of natural habitats, where microbes thrive and interact with different species and within complex communities. Interspecies interactions involve the action of multiple genetic and metabolic pathways which can result in mutualistic or antagonistic

bacterial effects. The molecular basis of some ecological interactions have been linked to secondary metabolites (SMs, also known as "natural products"), which are organic, biosynthesized compounds such as antibiotics and toxins not essential for basic growth or reproduction in organisms. SMs are used for defense, chemical signaling, and host-microbe interactions [6, 7]. SMs may stem from overflow products or evolutionary relics of former physiological functions [2]. Many unique and biologically active SMs continue to increase the interest of academic and industrial researchers [8]. However, pure cultures of microbes often fail to yield reliable or consistent biosynthesis of SMs [9].

Current studies on SM gene expression often rely on expensive, recombinant cloning of target genes into heterologous microbes or large-scale genomic sequencing projects [10]. An alternative experimental strategy would be to induce, measure, and track the expression of microbial SM genes while they grow in mixed culture conditions to better

TABLE 1: Primer sequences used in Quantitative PCR analyses of gene expression of target bacteria in solo and co-cultures.

Target gene	Abbreviation	Source species	Primer sequences, 5' to 3'	Gene product length
DNA directed RNA polymerase, subunit alpha	*HGK*	*P. aeruginosa*	TGATTTCGGTCAGGGACTTC GATGACCTGGAACTGACCGT	139
DNA directed RNA polymerase, subunit alpha	*HGK*	*R. denitrificans*	TCACCTCTGTGCAGATCGAC TGTCACCAGCAGTCACAACA	177
Thiosulfate:cyanide sulfurtransferase (Rhodanese)	*RdhA*	*P. aeruginosa*	AGGAAGTGATCACCCACTGC CTCTACAGGGGTATCGGGGT	140
Biosynthesis of pyocyanin	*PhzH*	*P. aeruginosa*	TGCGCGAGTTCAGCCACCTG TCCGGGACATAGTCGGCGCA	214
Metallo-beta-lactamase	*BetaLact*	*R. denitrificans*	AATACGAATTGCCCAGCATC GCAGGCCATAACAACAACCT	184
Dimethylpropiothetin dethiomethylase	*DMSP*	*R. denitrificans*	GTGCCGCACTGGCTGTGGAT	125

mimic antagonism and interaction in a natural environment. Applying the latter approach to well-studied model bacteria may lead to the elucidation of gene expression patterns from lesser known, nonmodel microbial organisms lacking genomic sequence data.

Based on a primary tenet of bacterial antagonism [5], we now report that targeted SM genes from model bacteria can be reproducibly induced after challenging these microbes *in vitro*. Here, the variable stressor is the "co-culturing" process of marine microbes (defined as growth of >1 bacterial species within one flask). Secondly, levels of specific gene expression were tracked and quantified by quantitative real-time PCR (qPCR) [11]. Model bacteria, *Pseudomonas aeruginosa* and marine *Roseobacter denitrificans*, were chosen for this study due to their available complete genome sequences and possible roles in defense and secondary metabolism [12–14]. Moreover, *Pseudomonas aeruginosa* is a human pathogen, and the particular PAO1 strain has been found in the marine environment playing an important role in biofilms.

This proof-of-principle study now presents specific and reproducible results showing that the selected bacterial genes can be induced by the act of co-culture mixing. Although we did not directly measure each corresponding gene product with chemical methods, the detection of expressed mRNA transcripts serves as a proxy for potential SM production. Moreover, the induced gene expression patterns clearly differ from solo pure cultures.

2. Materials and Methods

The microbial taxa, *Pseudomonas aeruginosa* PAO1 and *Roseobacter denitrificans* Och114, were chosen because they are well-characterized microbial species and can occur in marine habitats, which is a focus of our laboratory. These strains were provided by the Arizona State University (ASU) and the PathoGenesis Corporation, respectively.

All cultures and co-cultures were grown in marine broth before and after mixing and sampled for standard RNA extraction at the different time points indicated. RNA was isolated using the RNeasy Mini Kit (Qiagen, Valencia, CA) following the manufacturer's protocol. Dual co-cultures of *P.*

aeruginosa-R. denitrificans were tracked by quantitative real-time PCR (qPCR) utilizing SYBR green detection [15].

Four genes from the two model bacterial genomes, *Pseudomonas aeruginosa* (GenBank AE004091) and marine *Roseobacter denitrificans* (GenBank CP000362) were retrieved and used for gene-specific primer design: *PhzH, RhdA, BetaLact,* and *DMSP* (Table 1). The qPCR primers were designed using PRIMER BLAST from the National Center for Biotechnology Information website (http://www.ncbi.nlm.nih.gov/tools/primer-blast/).

Expressions of the same genes in solo and co-cultures of *P. aeruginosa* and *R. denitrificans* were compared (first column set to 1.0). In this study, the solo culture with the target genes acted as the control. Then, the Ct values of both the control and the genes in question were normalized to the *P. aeruginosa* housekeeping gene (*RNA polymerase*). All qPCR runs were performed as triplicate reactions with the same DNA template and gene-specific primers, on a single 48-well plate which also included negative (zero DNA) controls.

After qPCR amplification the comparative threshold method ($\Delta\Delta Ct$ analysis) was applied to evaluate the relative changes in gene expression from qPCR experiments [16]. Computer programs GeneX (Bio-Rad) and Excel (Microsoft) were used to calculate the equation: $[delta][delta]Ct = [delta]Ct_{,sample} - [delta]Ct_{,reference}$ (BioRad).

3. Results and Discussion

3.1. Gene Expression Analyses of Solo and Co-Cultures with qPCR. Both *P. aeruginosa* and *R. denitrificans* entered log phase at six hours and were then combined for co-culturing and subsequent gene expression analyses throughout log phase. A housekeeping gene, *DNA directed RNA polymerase (RNA pol), subunit alpha* expression appeared constant throughout all qPCR runs meaning their expression level was unaffected by the experimental conditions (data not shown). A third species, *Salinispora arenicola* (provided by the Joint Genome Institute), was originally intended for SM gene tracking but because of disparate growth patterns was only used for co-culture antagonism.

Figure 1 indicates that the act of co-culturing *P. aeruginosa-R. denitrificans* strains caused a measureable effect, as

Co-Cultures of Pseudomonas aeruginosa and Roseobacter denitrificans Reveal Shifts in Gene Expression Levels Compared to Solo Cultures

29

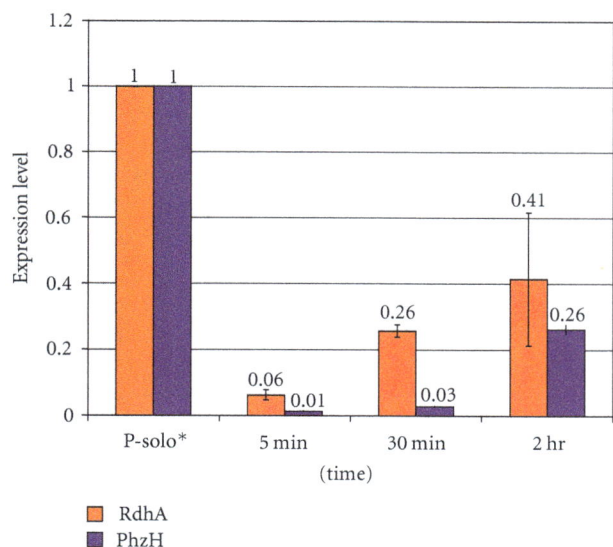

FIGURE 1: *P. aeruginosa* gene expression in the *P. aeruginosa-R. denitrificans R. denitrificans* co-cultures. Relative gene expression levels of *P. aeruginosa RdhA* and *PhzH* genes in dual co-cultures of *P. aeruginosa-R. denitrificans* were determined by quantitative real-time PCR (qPCR) with the SYBR green method [15].

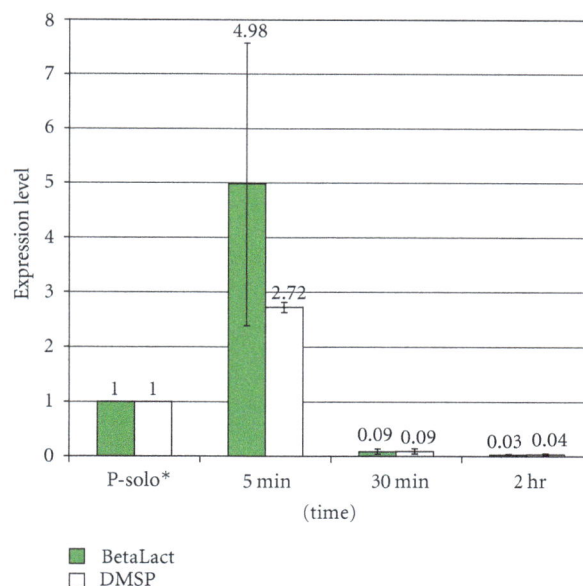

FIGURE 2: *R. denitrificans BetaLact* and *DMSP* gene expression in *R. denitrificans-P. aeruginosa* dual co-culture. Methods were as described in Figure 1 except for the fourth time point added at one hour.

both *P. aeruginosa RdhA* and *PhzH* genes showed overall lower gene expression relative to control solo culture levels (Figure 1). A similar effect of lower *RdhA* gene expression was observed in triplet (*P. aeruginosa-R. denitrificans-S. arenicola*) co-cultures (data not shown).

By contrast, *R. denitrificans BetaLact* and *DMSP* genes showed different patterns including repressed and escalated levels of gene expression across different time points and co-cultures. For example, in *R. denitrificans-P. aeruginosa* dual co-cultures, (Figure 2) both *BetaLact* and *DMSP* gene expression appeared lower than solo levels at initial mixing but then rose by about 2.0 fold after 30 minutes, and then leveled off. At two hours the gene expression levels of both genes decreased below solo culture level.

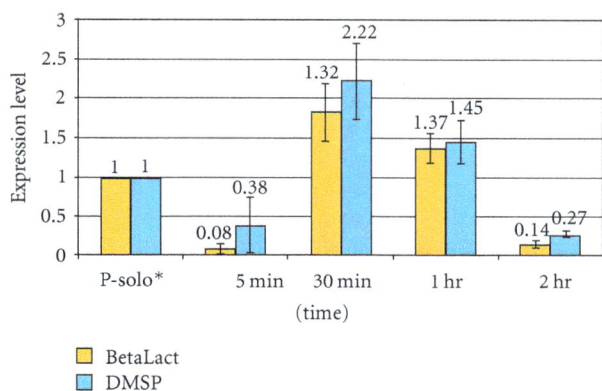

FIGURE 3: *R. denitrificans BetaLact* and *DMSP* relative gene expression in *R. denitrificans-S. arenicola* dual co-culture. Methods were as described in Figure 1.

In a *R. denitrificans-S. arenicola* dual co-culture (Figure 3), the same *R. denitrificans BetaLact* and *DMSP* genes exhibited a large 2.7–4.0-fold increase of gene expression with a more rapid onset after initial mixing of the two bacteria. The *DMSP* gene is expressed about twice as much as the *R. denitrificans* solo culture, but lower than *BetaLact* gene expression which is expressed about three times as much as the *R. denitrificans* solo culture. Similar increases were observed in duplicate experiments. A decrease in the expression of both genes occurred after more than 30 minutes of co-culturing.

4. Discussion

4.1. Gene Expression Patterns Seen in P. aeruginosa and R. denitrificans Co-Cultures. The primary aim of this research was achieved by showing that gene expression of certain targeted genes could be reproducibly induced or affected by systematic co-culturing in multistrain growth conditions. As mentioned above, the choice of genes generally centered on "secondary metabolism" (SM). Because some SMs show therapeutic potential or bioactive effects, large-scale efforts involving more sophisticated biotechnologies have been initiated in recent years to characterize and exploit the rich biochemical and genetic diversity within secondary metabolite producing organisms [17]. For example, with the advances in recombinant DNA technology, efforts have focused on the cloning and sequencing of complete polyketide biosynthetic gene loci (which can be very large) with the expectation of expressing these metabolic pathways in a foreign, heterologous host [6, 7]. The recent creation of a synthetic microbial cell [18] also conforms to an ultimate goal of controlling gene expression through artificial constructs.

Interestingly, with respect to our focus on gene expression, *P. aeruginosa* contains the highest proportion of regulatory genes observed in a microbial genome [13]. Some *P. aeruginosa* strains have diverse antimicrobial activities in different types of marine invertebrates, such as sponges [19]. In *P. aeruginosa,* we focused on the *PhzH* gene, which codes for the production of phenazine-1-carboxamide derived from the common precursor, phenazine-1-carboxylic acid (PCA) [20]. Phenazines are biologically versatile compounds involved in microbial competition, suppressing soil plant pathogens, and virulence in human and animal hosts [21]. A second targeted gene is *RhdA*, a thiosulfate: cyanide sulfurtransferase (Rhodanese) [22]. The *RhdA* gene products in *P. aeruginosa* protect the microbe from cyanide toxicity by converting the cyanide to the less toxic form of thiocyanate.

In co-culture experiments, the *P. aeruginosa RdhA* gene exhibited generally higher and more variable and inducible expression levels compared to *P. aeruginosa PhzH* gene. In contrast, *PhzH* gene expression appeared suppressed throughout most sample time points and never had its gene expression levels higher than its solo culture. However, the gradual increase is consistent with previous studies showing that PCN products appear mostly in late log phase [21].

Roseobacter denitrificans Och114 is a purple marine aerobic anoxygenic phototroph (AAP) [14] that plays an inimitable role in global energy and carbon cycles. A unique trait of this bacteria is that they are able to grow both photoheterotrophically (in the presence of oxygen) and anaerobically (in the dark using nitrate as an electron acceptor) [14]. *R. denitrificans* belongs in a bacterial clade with diverse metabolism, including its designation as one of the first bacteria characterized exhibiting anoxygenic phototrophic features [23].

R. denitrificans genes exhibited much different expression patterns compared to *P. aeruginosa*, with very large gene response spikes to co-culture conditions. Both *R. denitrificans* genes were expressed at higher levels than controls and interestingly behaved in a parallel fashion that tracked each other's rise and fall of expression levels throughout all time periods. *R. denitrificans* metallo-beta-lactamases provide resistance against beta-lactam antibiotics, which account for more than half of the world's antibiotic market. DMSP lyase catalyzes the creation of DMS (dimethyl sulfide) and acrylate from DMSP (dimethylsulfoniopropionate). No exact function of DMSP has been discovered to date, but it has been hypothesized that DMSP provides osmoregulation, some protection from oxidative stress, and herbivory. *R. denitrificans* showed the widest changes in SM gene expression levels throughout the study, possibly because the secondary metabolite genes chosen for this organism are not needed in high levels when in these particular co-cultures. *R. denitrificans BetaLact* gene expression levels rose even though *P. aeruginosa* is not known to produce any beta-lactam antibiotics. This opens the possibility that other SM genes of *R. denitrificans* could be activated upon co-culturing, even though not directly related to defense or antagonism *per se.*

When in a community, microbes will compete with other microbes for both resources and space [24]; for example both *P. aeruginosa* and *R. denitrificans* are competitive microbes, both with strong abilities to outcompete and kill other microorganisms [13, 14, 21]. The spike in gene expression of *R. denitrificans* co-cultures may stem from a defensive reaction to the presence of the second microorganism (Figures 2 and 3). Alternatively since both bacterial species can potentially coexist in diverse environments, lower levels of *P. aeruginosa* SM gene expression observed in the *P. aeruginosa-R. denitrificans* co-culture may be due to relative acclimatization to each other.

4.2. Possible Quorum Sensing in Co-Culture. Although not measured directly *per se,* quorum sensing (QS) factors may have played roles in co-culture gene expression in this study [25]. Bacterial QS compounds change the physiology of conspecific members of the population and represent one other possible explanation for the changes in gene expression during co-culture [26–28]. Throughout the past decade it has become increasingly recognized that bacteria are capable of intercellular communication moderated by QS factors such as autoinducers, or derivatives of homoserine lactone which facilitate adaptations to changing environmental conditions based on the population density of the producing microorganism [29]. This phenomenon probably includes regulating the expression or repression of secondary metabolites [25], which can affect degrees of cooperation or antagonism within and between different species, respectively [5, 26].

In this context, the abrupt decrease in *R. denitrificans* gene expression observed in the *P. aeruginosa-R. denitrificans* dual co-culture (Figure 2) may have stemmed from an interruption in *R. denitrifican* quorum sensing abilities after the initial mixing of the solo bacterial populations. That is, any quorum sensing factors released by the single species became diluted by at least half upon co-culturing. Once the concentration factors fell below minimum threshold levels, they may have lost their ability to affect or maintain the levels of intraspecies signaling present in each solo culture before the mixing. This could represent a switch between intraspecific cooperation and interspecies antagonism.

Secondly, it is quite possible that bacterial interactions in mixed cultures (i.e., in nature) involve the degradation or modification of QS factors secreted by other members of the community. This would result in repression of some gene products, as they are degraded by one of the bacteria. Phenazine PCN participates in a complex pathway of QS regulation [20], and therefore we acknowledge that sufficient explanation for *PhzH* gene expression levels requires further experimentation. Alternatively, none of the target genes may be under QS control.

Although other possible explanations remain, this study shows that varying, yet reproducible, expression levels appear to be gene specific and context dependent. Also, specific gene induction appeared temporary but clearly resulted from the act of mixed species co-culturing. This paper points to future studies and experimental strategies that can focus on factors affecting the structure of artificial or more complex microbial communities and interactions [4]. Finally, finding specific molecules or signals that control unique secondary metabolite pathways and their genes may

Co-Cultures of Pseudomonas aeruginosa and Roseobacter denitrificans Reveal Shifts in Gene Expression
Levels Compared to Solo Cultures

31

have wider ramifications for natural products research, microbial ecology and the pharmaceutical industry.

Acknowledgments

The authors are grateful to the Joint Genome Institute, Arizona State Corporation University and PathoGenesis for the kind gifts of test strains. We also thank Dr. Peter McCarthy, Dr. Maria Hoffman and Amy Doyle for helpful comments on early drafts. This paper is partially a result of funding from the National Oceanic and Atmospheric Administration, Center for Sponsored Coastal Ocean Science, under awards NA07NOS4000200 to Nova Southeastern University for the National Coral Reef Institute (NCRI).

References

[1] E. F. DeLong, "The microbial ocean from genomes to biomes," *Nature*, vol. 459, no. 7244, pp. 200–206, 2009.

[2] D. L. Kirchman, *Microbial Ecology of the Oceans*, Wiley-Less, USA, 2000.

[3] F. E. Dewhirst, T. Chen, J. Izard et al., "The human oral microbiome," *Journal of Bacteriology*, vol. 192, no. 19, pp. 5002–5017, 2010.

[4] N. S. Webster and L. L. Blackall, "What do we really know about sponge-microbial symbioses," *ISME Journal*, vol. 3, no. 1, pp. 1–3, 2009.

[5] K. L. Rypien, J. R. Ward, and F. Azam, "Antagonistic interactions among coral-associated bacteria," *Environmental Microbiology*, vol. 12, no. 1, pp. 28–39, 2010.

[6] J. Piel, D. Hui, N. Fusetani, and S. Matsunaga, "Targeting modular polyketide synthases with iteratively acting acyltransferases from metagenomes of uncultured bacterial consortia," *Environmental Microbiology*, vol. 6, no. 9, pp. 921–927, 2004.

[7] A. Schirmer, R. Gadkari, C. D. Reeves, F. Ibrahim, E. F. DeLong, and C. R. Hutchinson, "Metagenomic analysis reveals diverse polyketide synthase gene clusters in microorganisms associated with the marine sponge *Discodermia dissoluta*," *Applied and Environmental Microbiology*, vol. 71, no. 8, pp. 4840–4849, 2005.

[8] P. R. Jensen and W. Fenical, "Strategies for the discovery of secondary metabolites from marine bacteria: ecological perspectives," *Annual Review of Microbiology*, vol. 48, pp. 559–584, 1994.

[9] A. Muscholl-Silberhorn, V. Thiel, and J. F. Imhoff, "Abundance and bioactivity of cultured sponge-associated bacteria from the Mediterranean Sea," *Microbial Ecology*, vol. 55, no. 1, pp. 94–106, 2008.

[10] S. Dai, Y. Ouyang, G. Wang, and X. Li, "*Streptomyces autolyticus* JX-47 large-insert bacterial artificial chromosome library construction and identification of clones covering geldanamycin biosynthesis gene cluster," *Current Microbiology*, vol. 63, no. 1, pp. 68–74, 2011.

[11] P. Y. Muller, H. Janovjak, A. R. Miserez, and Z. Dobbie, "Processing of gene expression data generated by quantitative real-time RT-PCR," *BioTechniques*, vol. 32, no. 6, pp. 1372–1374, 2002.

[12] A. Buchan, J. M. González, and M. A. Moran, "Overview of the marine *Roseobacter* lineage," *Applied and Environmental Microbiology*, vol. 71, no. 10, pp. 5665–5677, 2005.

[13] C. K. Stover, X. Q. Pham, A. L. Erwin et al., "Complete genome sequence of *Pseudomonas aeruginosa PAO1*, an opportunistic pathogen," *Nature*, vol. 406, no. 6799, pp. 959–964, 2000.

[14] W. D. Swingley, S. Sadekar, S. D. Mastrian et al., "The complete genome sequence of *Roseobacter denitrificans* reveals a mixotrophic rather than photosynthetic metabolism," *Journal of Bacteriology*, vol. 189, no. 3, pp. 683–690, 2007.

[15] D. G. Ginzinger, "Gene quantification using real-time quantitative PCR: an emerging technology hits the mainstream," *Experimental Hematology*, vol. 30, no. 6, pp. 503–512, 2002.

[16] K. J. Livak and T. D. Schmittgen, "Analysis of relative gene expression data using real-time quantitative PCR and the 2-ΔΔCT method," *Methods*, vol. 25, no. 4, pp. 402–408, 2001.

[17] D. J. Faulkner, "Marine natural products," *Natural Product Reports*, vol. 19, no. 1, pp. 1–48, 2002.

[18] D. G. Gibson, J. I. Glass, C. Lartigue et al., "Creation of a bacterial cell controlled by a chemically synthesized genome," *Science*, vol. 329, no. 5987, pp. 52–56, 2010.

[19] J. Kennedy, P. Baker, C. Piper et al., "Isolation and analysis of bacteria with antimicrobial activities from the marine sponge *Haliclona simulans* collected from Irish waters," *Marine Biotechnology*, vol. 11, no. 3, pp. 384–396, 2009.

[20] L. E. P. Dietrich, A. Price-Whelan, A. Petersen, M. Whiteley, and D. K. Newman, "The phenazine pyocyanin is a terminal signalling factor in the quorum sensing network of *Pseudomonas aeruginosa*," *Molecular Microbiology*, vol. 61, no. 5, pp. 1308–1321, 2006.

[21] D. V. Mavrodi, W. Blankenfeldt, and L. S. Thomashow, "Phenazine compounds in fluorescent *Pseudomonas* spp. biosynthesis and regulation," *Annual Review of Phytopathology*, vol. 44, pp. 417–445, 2006.

[22] R. Cipollone, E. Frangipani, F. Tiburzi, F. Imperi, P. Ascenzi, and P. Visca, "Involvement of *P. aeruginosa* rhodanese in protection from cyanide toxicity," *Applied and Environmental Microbiology*, vol. 73, no. 2, pp. 390–398, 2007.

[23] Y. Zhang and N. Jiao, "Roseophage RDJLl, Infecting the aerobic anoxygenic phototrophic bacterium *Roseobacter denitrificans* OChll4," *Applied and Environmental Microbiology*, vol. 75, no. 6, pp. 1745–1749, 2009.

[24] M. E. Hibbing, C. Fuqua, M. R. Parsek, and S. B. Peterson, "Bacterial competition: surviving and thriving in the microbial jungle," *Nature Reviews Microbiology*, vol. 8, no. 1, pp. 15–25, 2010.

[25] B. L. Bassler, "How bacteria talk to each other: regulation of gene expression by quorum sensing," *Current Opinion in Microbiology*, vol. 2, no. 6, pp. 582–587, 1999.

[26] L. Keller and M. G. Surette, "Communication in bacteria: an ecological and evolutionary perspective," *Nature Reviews Microbiology*, vol. 4, no. 4, pp. 249–258, 2006.

[27] M. B. Miller and B. L. Bassler, "Quorum sensing in bacteria," *Annual Review of Microbiology*, vol. 55, pp. 165–199, 2001.

[28] V. Venturi and S. Subramoni, "Future research trends in the major chemical language of bacteria," *HFSP Journal*, vol. 3, no. 2, pp. 105–116, 2009.

[29] N. A. Whitehead, A. M. L. Barnard, H. Slater, N. J. L. Simpson, and G. P. C. Salmond, "Quorum-sensing in Gram-negative bacteria," *FEMS Microbiology Reviews*, vol. 25, no. 4, pp. 365–404, 2001.

6

In Vitro Evaluation of Antiprotozoal and Antiviral Activities of Extracts from Argentinean *Mikania* Species

Laura C. Laurella,[1] **Fernanda M. Frank,**[2] **Andrea Sarquiz,**[3] **María R. Alonso,**[4]
Gustavo Giberti,[4] **Lucia Cavallaro,**[3] **Cesar A. Catalán,**[5] **Silvia I. Cazorla,**[2] **Emilio Malchiodi,**[2]
Virginia S. Martino,[1, 4] **and Valeria P. Sülsen**[1, 4]

[1] *Cátedra de Farmacognosia, Facultad de Farmacia y Bioquímica, Universidad de Buenos Aires, Junín 956 2° P, 1113,
Buenos Aires, Argentina*
[2] *Cátedra de Inmunología, IDEHU UBA-CONICET, Facultad de Farmacia y Bioquímica y Departamento de Microbiología,
Facultad de Medicina, Universidad de Buenos Aires, Junín 956 4° P, 1113, Buenos Aires, Argentina*
[3] *Cátedra de Virología, Facultad de Farmacia y Bioquímica, Universidad de Buenos Aires, Junín 956 4° P, 1113, Buenos Aires, Argentina*
[4] *Instituto de Química y Metabolismo del Fármaco (IQUIMEFA), UBA-CONICET, Junín 956 4° P, 1113, Buenos Aires, Argentina*
[5] *INQUINOA-CONICET y Instituto de Química Orgánica, Facultad de Bioquímica, Química y Farmacia,
Universidad Nacional de Tucumán, Ayacucho 471, T4000INI San Miguel de Tucumán, Argentina*

Correspondence should be addressed to Emilio Malchiodi, emalchio@ffyb.uba.ar and
Virginia S. Martino, vmartino@ffyb.uba.ar

Academic Editors: Y. Hashiguchi and K. E. Kester

The aim of this study was to investigate the antiprotozoal and antiviral activities of four Argentinean *Mikania* species. The organic and aqueous extracts of *Mikania micrantha, M. parodii, M. periplocifolia,* and *M. cordifolia* were tested on *Trypanosoma cruzi* epimastigotes, *Leishmania braziliensis* promastigotes, and dengue virus type 2. The organic extract of *M. micrantha* was the most active against *T. cruzi* and *L. braziliensis* exhibiting a growth inhibition of $77.6 \pm 4.5\%$ and $84.9 \pm 6.1\%$, respectively, at a concentration of $10 \, \mu g/ml$. The bioguided fractionation of *M. micrantha* organic extract led to the identification of two active fractions. The chromatographic profile and infrared analysis of these fractions revealed the presence of sesquiterpene lactones. None of the tested extracts were active against dengue virus type 2.

1. Introduction

Neglected tropical diseases (NTDs) are a group of infectious diseases that cause significant morbidity and mortality in the developing world. American Trypanosomiasis or Chagas' disease, leishmaniasis, and dengue are considered NTDs [1]. According to the World Health Organization (WHO), there are more than 1 billion people that suffer one or more NTD, mostly concentrated in countries of Africa, Asia, and Latin America, where life conditions are linked to poverty. As a consequence, the development of new or better drugs to fight these diseases is not a priority for the pharmaceutical industry.

Chagas' disease, caused by the protozoan parasite *Trypanosoma cruzi*, affects 10 million people worldwide [2], and almost 12 million people are infected with *Leishmania* spp. [3]. Dengue is a viral infection caused by an RNA virus and it is estimated that 50–100 million cases occur annually while approximately half of the world's population is at risk [4].

Chagas' disease, leishmaniasis, and dengue are considered to be, among others, the most common tropical diseases in Argentina [5]. Between 1.5 and 2 million people are reported to be affected by Chagas' disease in this country [6]. American tegumentary leishmaniasis (ATL) is endemic in Northern Argentina where it is frequently associated with Chagas' disease [7] and its incidence, due in particular to *Leishmania braziliensis*, has increased during the last two decades [8]. On the other hand, a dengue epidemic outbreak in 2009 produced more than 25000 cases [9].

The drugs currently available to treat acute Chagas' disease infection are the nitroaromatic compounds, benznidazole and nifurtimox, both of which were discovered in the 70s'. They are effective only in the acute phase of the disease and have serious side effects [2]. The chemotherapy of leishmaniasis is based on pentavalent antimonials, amphotericin B, miltefosine and paromomycin which all have drawbacks [10]. In the case of dengue, however, currently, there are neither licensed vaccines nor any available drug to treat this viral infection [11]. In view of this situation, there is an urgent need to find new drugs to treat these NTDs.

Natural products have played an important role in the drug discovery process, since they are generally small molecules with a wide chemical diversity and more "drug-likeness" than synthetic compounds, so that they are good candidates for lead drug development [12, 13].

In the last decades, the Asteraceae family has been extensively studied due to the great number and variety of active compounds isolated from species belonging to it. Among these, the genus *Mikania*, which comprises nearly 450 species [14], has been reported to contain some interesting chemical substances, mostly terpenoid compounds (sesquiterpene lactones and diterpenes) and flavonoids. These secondary metabolites are known to have important biological activities, including anti-infective properties [15–18]. There are no previous reports concerning the evaluation of *Mikania* spp. on dengue virus, though *M. micrantha* has been reported to be effective against respiratory viruses [19].

The aim of the present study, thus, was to determine the *in vitro* antiprotozoal and antiviral activities of four Argentinean *Mikania* species. Organic and aqueous extracts of *Mikania micrantha, M. periplocifolia, M. parodii,* and *M. cordifolia* were tested against *Trypanosoma cruzi, Leishmania braziliensis,* and dengue virus.

2. Materials and Methods

2.1. Plant Material. The aerial parts of *Mikania micrantha* Kunth (Asteraceae) were collected in Tucumán Province, Argentina in June 2009. Botanical identification was performed by Lic. A. Slanis and Dr. B. Juarez. A voucher specimen (LIL 609699) was deposited at the Herbarium of Instituto Miguel A. Lillo.

Mikania parodii Cabrera (Asteraceae) (BAF 713) and *Mikania cordifolia* (L. f.) Willd. (Asteraceae) (BAF 715) were collected in May 2009 and *Mikania periplocifolia* Hook. & Arn. (Asteraceae) (BAF 732) in November 2011, in all cases in Entre Ríos Province. The plant material was identified by one of the authors and voucher specimens were deposited at the Museo de Farmacobotánica, Facultad de Farmacia y Bioquímica, Universidad de Buenos Aires.

2.2. Microorganisms. *Trypanosoma cruzi* epimastigotes (RA strain) were grown in a biphasic medium and *Leishmania braziliensis* promastigotes (2903 strain) in liver infusion tryptose medium (LIT). Cultures were routinely maintained by weekly passages at 28°C and 26°C, respectively.

The replication of dengue virus type 2 (DENV-2) (16681 strain) was performed using Vero cells (ATCC CCL-81), baby hamster kidney (BHK-21) clon 15.

2.3. Preparation of Plant Extracts. The aerial parts of *Mikania micrantha, M. periplocifolia, M. parodii,* and *M. cordifolia* (20 g each) were air dried and extracted with dichloromethane/methanol (1 : 1) (200 mL) at room temperature for 24 h and then vacuum filtered. The process was repeated twice and the extracts were combined and dried under vacuum. The marc was then extracted with water in the same conditions. Aqueous extracts were freeze-dried.

2.4. Fractionation of Mikania micrantha Organic Extract. The aerial parts of *M. micrantha* (400 g) were extracted with dichloromethane/methanol (1 : 1) as described above. The organic extract was subjected to open column chromatography on silica gel 60 and eluted successively with hexane, hexane: ethyl acetate (1 : 1), ethyl acetate and methanol yielding 48 fractions of 250 mL each. According to their thin-layer chromatography profile, these fractions were combined into 8 final fractions (1–8). These were subsequently tested for trypanocidal activity against *T. cruzi* epimastigotes.

2.5. High Performance Liquid Chromatography Analysis (HPLC). The HPLC analysis of fractions 3 and 4 was performed on a Varian Pro Star instrument equipped with a Rheodyne injection valve (20 μl) and a photodiode array detector set at 210 nm.

A reversed-phase column Phenomenex—C18 (2) Luna (250 mm × 4.6 mm·5 μ.) was used. Samples were eluted with a gradient of water (A) and acetonitrile (B) from 0% B to 75% B in 30 min and 75% B to 100% B in 2 min. The flow rate was 1.0 mL/min and the separation was done at room temperature. Chromatograms were recorded and processed using the Varian Star Chromatography Workstation version 6.x.

Fractions 3 and 4 were dissolved in methanol and water (90 : 10) to a final concentration of 5 mg/mL.

Water employed to prepare the mobile phase was of ultrapure quality (Milliq). Acetonitrile (HPLC) J. T. Baker and methanol (HPLC) J. T. Baker were used.

2.6. Infrared Spectroscopic Analysis. The IR-spectra of the active fractions 3 and 4 were recorded on FT-IR spectrophotometer (Bruker IFS-25) in chloroform solution.

2.7. Trypanocidal and Leishmanicidal Activity Assay. Growth inhibition of *T. cruzi* epimastigotes and *L. braziliensis* promastigotes was evaluated using the previously described [³H] thymidine uptake assay [15]. Parasites were adjusted to a cell density of 1.5×10^6/mL and cultured in the presence of each extract or fraction for 72 h. Benznidazole (5 to 20 μM; Roche) and Amphotericin B (0.27–1.6 μM, ICN) were used as positive controls. The percentage of inhibition was calculated as 100 − [(cpm of treated parasites)/(cpm of untreated parasites)] × 100 [16]. Organic and aqueous extracts of *Mikania* species were tested on both parasites at concentrations of 100, 10 and 1 μg/mL for 72 h. Extracts which showed an inhibition below 30% at a concentration of 10 μg/mL were no further tested. Fractions 1–8 were assayed on *T. cruzi* at concentrations of 100 and 10 μg/mL.

2.8. Antiviral Activity Assay. Vero cells were seeded in Minimal Essential Medium (MEM), supplemented with 10% fetal bovine serum (FBS, PAA) in microwell plates (96 wells) at a density of 2.2×10^4 cells per well. After 24 h in a 5% CO_2 incubator at 37°C, the cells were infected with DENV-2 in MEM supplemented with 2% FBS that induced an 80–90% cytopathic effect (CPE) on the sixth day postinoculation in absence of the drug.

The cytotoxic concentration 50% (CC_{50}), defined as the concentration that inhibits the proliferation of exponentially growing cells by 50%, was calculated for organic and aqueous extracts.

Two-fold serial dilutions of organic (12–0.75 μg/mL) and aqueous extracts (500–31.25 μg/mL) of *Mikania* species were tested in quadruplicate. Mock-infected cells with and without extracts and infected cells without extract were included as controls. Ribavirin (100–1 μg/mL; Sigma-Aldrich) was used as a positive control. The CPE was determined by the measurement of cell viability using the MTS/PMS method (CellTiter 96 Aqueous) (Promega, Madison, WI) as previously described [20]. All extracts were tested at concentrations below their CC_{50}.

2.9. Statistical Analysis. The results are expressed as mean ± SEM. The level of statistical significance was determined submitting the data to one-way analysis of variance (ANOVA) using GraphPad Prism 3.0 software (GraphPad Software Inc., San Diego, CA). All data were referred to the control group. *P* values of <0.05 were considered significant.

3. Results

3.1. Trypanocidal Activity. The trypanocidal activity of organic and aqueous extracts of *Mikania* species was evaluated *in vitro* on *T. cruzi* epimastigotes. Results are presented in Table 1.

The organic extracts of the four *Mikania* species were found to be active against *T. cruzi* epimastigotes with an inhibition above 85% at 100 μg/mL. The organic extract of *M. micrantha* proved to be the most active of all tested species, showing an inhibition of 77.6 ± 4.5% and 14.2 ± 4.6%, when applied at concentrations of 10 and 1 μg/mL, respectively. The aqueous extracts of the four *Mikania* species showed inhibitions ranged between 13 and 40% at 100 μg/mL (Table 1).

According to these results, the organic extract of *M. micrantha* was selected for further fractionation.

3.2. Leishmanicidal Activity. The leishmanicidal effect of extracts of *Mikania* species was evaluated on *L. braziliensis* promastigotes. The results are shown in Table 2.

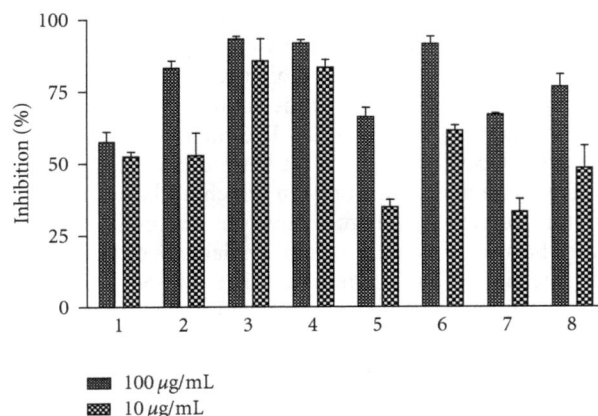

FIGURE 1: Trypanocidal activity of fractions 1–8 of *Mikania micrantha* on *T. cruzi* epimastigotes.

At a concentration of 100 μg/mL, organic extracts of *M. micrantha* and *M. parodii* displayed leishmanicidal activity with growth inhibition rates above 85%. At the lowest concentration tested (1 μg/mL), *M. micrantha* was the most active extract with an inhibition of 77.8 ± 1.1%.

Aqueous extracts displayed inhibition rates below 30% at a concentration of 10 μg/mL.

3.3. Antiviral Activity. None of the organic and aqueous extracts from the four tested *Mikania* species was able to inhibit the replication of DENV-2. Approximately 30% reduction of the CPE effect was observed when infected cells were treated with 500 μg/mL aqueous extracts.

3.4. Bioassay-Guided Fractionation of Mikania micrantha Organic Extract. The fractionation of the organic extract of *M. micrantha* by column chromatography yielded eight final fractions (1–8), which were tested *in vitro* against *T. cruzi* epimastigotes. Fractions 3, 4, and 6, at a concentration of 100 μg/mL, showed trypanocidal activity with percentages of growth inhibition of 93.2 ± 1.0%, 91.8 ± 1.2%, and 91.4 ± 2.6%, respectively (Figure 1).

At the lowest tested concentration (10 μg/mL), fractions 3 and 4 were the most active against *T. cruzi* with inhibition rates of 85.7 ± 7.6% and 83.4 ± 2.8%, respectively (Figure 1). The analysis of the HPLC profile of these two fractions showed the presence of the same three major peaks with retention times of 16.5, 19.2, and 20.6 min and UV maximum at 219, 217, and 223 nm, respectively (Figure 2).The infrared spectroscopic analysis of these fractions showed the presence of bands between 1750–1790 cm^{-1}, corresponding to γ-lactone carbonyl group (data not shown).

4. Discussion

In the present study, the antiprotozoal and antiviral effects of extracts of four Argentinean *Mikania* species against *Trypanosoma cruzi, Leishmania braziliensis,* and dengue virus type 2 were evaluated.

TABLE 1: Trypanocidal activity of organic and aqueous extracts of *Mikania micrantha*, *M. periplocifolia*, *M. parodii*, and *M. cordifolia*.

Species	Extract	% of growth inhibition ± SEM		
		100 μg/mL	10 μg/mL	1 μg/mL
Mikania micrantha	Organic	91.1 ± 3.7	77.6 ± 4.5	14.2 ± 4.6
	Aqueous	40.4 ± 1.4	23.2 ± 2.9	n.d.
Mikania periplocifolia	Organic	95.5 ± 0.4	56.7 ± 5.0	7.0 ± 4.2
	Aqueous	40.2 ± 2.5	25.2 ± 1.0	n.d.
Mikania parodii	Organic	94.9 ± 0.5	33.0 ± 1.3	2.3 ± 1.5
	Aqueous	30.2 ± 2.0	19.0 ± 1.8	n.d.
Mikania cordifolia	Organic	86.2 ± 1.8	10.5 ± 2.5	n.d.
	Aqueous	13.6 ± 2.8	12.2 ± 5.4	n.d.

Results are expressed as mean ± SEM. n.d.: not determined.

TABLE 2: Leishmanicidal activity of organic and aqueous extracts of *Mikania micrantha*, *M. periplocifolia*, *M. parodii*, and *M. cordifolia*.

Species	Extract	% of growth inhibition ± SEM		
		100 μg/mL	10 μg/mL	1 μg/mL
Mikania micrantha	Organic	90.9 ± 0.8	84.9 ± 6.1	77.8 ± 1.1
	Aqueous	41.6 ± 4.1	29.9 ± 1.5	n.d.
Mikania periplocifolia	Organic	73.4 ± 5.3	69.2 ± 2.0	53.5 ± 4.3
	Aqueous	78.4 ± 7.2	11.4 ± 4.0	n.d.
Mikania parodii	Organic	87.3 ± 1.7	73.0 ± 0.6	58.7 ± 3.9
	Aqueous	48.7 ± 8.4	7.7 ± 1.5	n.d.
Mikania cordifolia	Organic	69.7 ± 6.6	55.7 ± 7.4	35.3 ± 7.5
	Aqueous	38.9 ± 3.5	5.0 ± 1.1	n.d.

Results are expressed as mean ± SEM. n.d.: not determined.

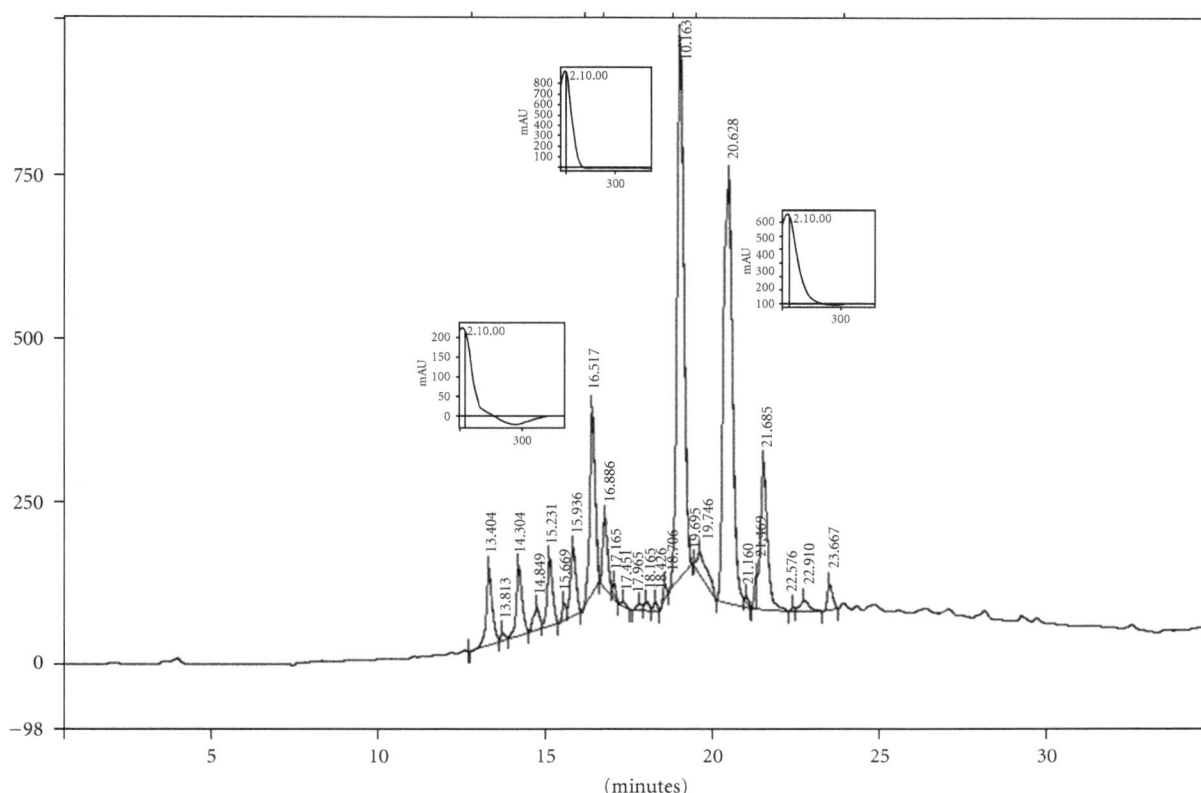

FIGURE 2: HPLC chromatographic profile of fraction 4 from *Mikania micrantha*.

The organic extracts of *M. micrantha, M. periplocifolia, M. parodii,* and *M. cordifolia* showed significant *in vitro* antiprotozoal activity against *T. cruzi* epimastigotes and *L. braziliensis* promastigotes. The *M. micrantha* organic extract was the most active against the two protozoans. All aqueous extracts displayed moderate to low activity against *T. cruzi* and *L. braziliensis.* This is the first time that trypanocidal and leishmanicidal activities of *M. parodii* are reported.

In the case of antiviral activity, neither the organic nor the aqueous extracts were able to inhibit the replication of dengue virus type 2 under the described experimental conditions.

Previous reports on the chemical composition of species of the genus *Mikania* (Asteraceae) describe terpenoid compounds and flavonoids as the main constituents [21–24]. There are some references about the antiprotozoal and antiviral activities of *Mikania* spp. [19, 25, 26] and particularly, in the case of the four studied species, there are some reports of studies performed on different strains, stages, and parasite species than the ones used herein [27–29].

The bioguided fractionation of *M. micrantha* organic extract resulted in the identification of the most active fractions (3, 4) against *T. cruzi.* The chromatographic profile of these fractions revealed the presence of three major peaks with UV spectra that could be attributed to terpenoid compounds. Besides, in the infrared spectrum of these fractions, characteristic bands of sesquiterpene lactones could be observed. Thus, the active fractions 3 and 4 contain sesquiterpene lactones.

Terpenoids, mainly sesquiterpene lactones and diterpenes, are characteristic constituents of the genus *Mikania* and some of these metabolites have shown trypanocidal activity [24]. Thus, the trypanocidal activity of fractions 3 and 4 could be due to the presence of sesquiterpene lactones, since some of these compounds have been previously reported in *M. micrantha* [30–32].

These findings reveal the importance of the *Mikania* genus as a rich source of antiprotozoal molecules. Isolation and purification of the bioactive compounds from the active fractions of *M. micrantha* and bioassay-guided fractionation of the other active extracts are under way.

Authors' Contribution

These two authors contributed equally to this work L. C. Laurella, F. M. Frank.

Acknowledgments

The authors wish to thank Dr. Monica Esteva and Estela Lammel for kindly providing *Leishmania* promastigotes and *T. cruzi* epimastigotes, respectively, and Mrs. Cristina Aguilera and Mrs. Teresa Fogal for their valuable technical assistance. Dengue virus strain was kindly provided by Dr. Andrea Gamarnik from Instituto Leloir, Argentina. This research was supported in part by PIP 01540 (Consejo Nacional de Investigaciones Científicas y Técnicas) and UBACYT 20020090300115, 20020090200478 and 20020100100201.

References

[1] World Health Organization, "First WHO report on neglected tropical diseases," Tech. Rep., 2010, http://www.who.int/neglected_diseases/2010report/en/.

[2] World Health Organization, "Chagas disease (American trypanosomiasis)," Fact Sheet 340, 2010, http://www.who.int/mediacentre/factsheets/fs340/en/index.html.

[3] World Health Organization, *Leishmania*sis, http://www.who.int/Leishmaniasis/en/.

[4] World Health Organization, Dengue control, http://www.who.int/denguecontrol/en/index.html.

[5] Instituto Nacional de Medicina Tropical, "Enfermedades más frecuentes en Argentina," http://www.msal.gov.ar/inmet/frecuentes.php.

[6] Drugs for Neglected Diseases Initiative, "Argentina: more action needed," *Newsletter*, vol. 19, article 5, 2010.

[7] F. M. Frank, M. M. Fernández, N. J. Taranto et al., "Characterization of human infection by *Leishmania* spp. in the Northwest of Argentina: immune response, double infection with *Trypanosoma cruzi* and species of *Leishmania* involveds," *Parasitology*, vol. 126, no. 1, pp. 31–39, 2003.

[8] O. D. Salomón, S. A. Acardi, D. J. Liotta et al., "Epidemiological aspects of cutaneous *Leishmania*sis in the Iguazú falls area of Argentina," *Acta Tropica*, vol. 109, no. 1, pp. 5–11, 2009.

[9] Ministerio de Salud, Plan Nacional de control del dengue y la fiebre amarilla, http://www.msal.gov.ar/inmet/documentos.php, 2009.

[10] Drugs for Neglected Diseases Initiative, *Leishmania*sis. Current treatments, http://www.dndi.org/diseases/vl/current-treatment.html.

[11] J. Whitehorn and J. Farrar, "Dengue," *British Medical Bulletin*, vol. 95, no. 1, pp. 161–173, 2010.

[12] B. Wang, J. Deng, Y. Gao, L. Zhu, R. He, and Y. Xu, "The screening toolbox of bioactive substances from natural products: a review," *Fitoterapia*, vol. 82, no. 8, pp. 1141–1151, 2011.

[13] A. L. Harvey, "Natural products in drug discovery," *Drug Discovery Today*, vol. 13, no. 19-20, pp. 894–901, 2008.

[14] W. Herz, "Terpenoid chemistry of *Mikania* species," *Journal of the Indian Chemical Society*, vol. 75, no. 10–12, pp. 559–564, 1998.

[15] V. P. Sülsen, F. M. Frank, S. I. Cazorla et al., "Trypanocidal and leishmanicidal activities of sesquiterpene lactones from *Ambrosia tenuifolia* Sprengel (Asteraceae)," *Antimicrobial Agents and Chemotherapy*, vol. 52, no. 7, pp. 2415–2419, 2008.

[16] V. P. Sülsen, F. M. Frank, S. I. Cazorla et al., "Psilostachyin C: a natural compound with trypanocidal activity," *International Journal of Antimicrobial Agents*, vol. 37, no. 6, pp. 536–543, 2011.

[17] D. Chaturvedi, "Sesquiterpene lactones: structural diversity and their biological activities," in *Opportunity, Challenge and Scope of Natural Products in Medicinal Chemistry*, V. Tiwari and B. Mishra, Eds., pp. 313–334, Research Signpost, Kerala, India, 2011.

[18] T. P. T. Cushnie and A. J. Lamb, "Antimicrobial activity of flavonoids," *International Journal of Antimicrobial Agents*, vol. 26, no. 5, pp. 343–356, 2005.

[19] P. P. H. But, Z. D. He, S. C. Ma et al., "Antiviral constituents against respiratory viruses from *Mikania micrantha*," *Journal of Natural Products*, vol. 72, no. 5, pp. 925–928, 2009.

[20] L. M. Finkielsztein, E. F. Castro, L. E. Fabián et al., "New 1-indanone thiosemicarbazone derivatives active against BVDV," *European Journal of Medicinal Chemistry*, vol. 43, no. 8, pp. 1767–1773, 2008.

[21] M. Krautmann, E. C. de Riscala, E. Burgueño-Tapia, Y. Mora-Pérez, C. A. N. Catalán, and P. Joseph-Nathan, "C-15-functionalized eudesmanolides from *Mikania campanulata*," *Journal of Natural Products*, vol. 70, no. 7, pp. 1173–1179, 2007.

[22] C. A. N. Catalán, M. D. R. Cuenca, L. R. Hernández, and P. Joseph-Nathan, "cis,cis-Germacranolides and melampolides from *Mikania thapsoides*," *Journal of Natural Products*, vol. 66, no. 7, pp. 949–953, 2003.

[23] E. Ohkoshi, S. Kamo, M. Makino, and Y. Fujimoto, "Ent-Kaurenoic acids from *Mikania hirsutissima* (Compositae)," *Phytochemistry*, vol. 65, no. 7, pp. 885–890, 2004.

[24] X. A. Dominguez, "Eupatorieae—chemicals review," in *The Biology and Chemistry of the Compositae Vol I*, V. Heywood, J. Harbone, and B. Turner, Eds., pp. 487–502, Academic Press, London, UK, 1977.

[25] A. M. Do Nascimento, J. S. Chaves, S. Albuquerque, and D. C. R. de Oliveira, "Trypanocidal properties of *Mikania stipulacea* and *Mikania hoehnei*," *Fitoterapia*, vol. 75, no. 3-4, pp. 381–384, 2004.

[26] P. S. Luize, T. S. Tiuman, L. G. Morello et al., "Effects of medicinal plant extracts on growth of *Leishmania (L.) amazonensis* and *Trypanosoma cruzi*," *Brazilian Journal of Pharmaceutical Sciences*, vol. 41, no. 1, pp. 85–94, 2005.

[27] J. S. Chaves, A. M. D. Nascimento, A. P. Soares et al., "Screening of Southeastern Brazilian *Mikania* species on *Trypanosoma cruzi*," *Pharmaceutical Biology*, vol. 45, no. 10, pp. 749–752, 2007.

[28] S. Muelas-Serrano, J. J. Nogal, R. A. Martínez-Díaz, J. A. Escario, A. R. Martínez-Fernández, and A. Gómez-Barrio, "*In vitro* screening of American plant extracts on *Trypanosoma cruzi* and *Trichomonas vaginalis*," *Journal of Ethnopharmacology*, vol. 71, no. 1-2, pp. 101–107, 2000.

[29] Á. I. Calderón, L. I. Romero, E. Ortega-Barría et al., "Screening of Latin American plants for antiparasitic activities against malaria, Chagas disease, and Leishmaniasis," *Pharmaceutical Biology*, vol. 48, no. 5, pp. 545–553, 2010.

[30] M. D. R. Cuenca, A. Bardon, C. A. N. Catalan, and W. C. M. C. Kokke, "Sesquiterpene lactones from *Mikania micrantha*," *Journal of Natural Products*, vol. 51, no. 3, pp. 625–626, 1988.

[31] G. Nicollier and A. C. Thompson, "Essential oil and terpenoids of *Mikania micrantha*," *Phytochemistry*, vol. 20, no. 11, pp. 2587–2588, 1981.

[32] H. Huang, W. Ye, P. Wu, L. Lin, and X. Wei, "New sesquiterpene dilactones from *Mikania micrantha*," *Journal of Natural Products*, vol. 67, no. 4, pp. 734–736, 2004.

7

Changes in Bacterial Composition of Zucchini Flowers Exposed to Refrigeration Temperatures

F. Baruzzi, M. Cefola, A. Carito, S. Vanadia, and N. Calabrese

Institute of Sciences of Food Production, National Research Council of Italy (CNR-ISPA), Via. G. Amendola 122/o, 70126 Bari, Italy

Correspondence should be addressed to F. Baruzzi, federico.baruzzi@ispa.cnr.it

Academic Editor: Mario Vaneechoutte

Microbial spoilage is one of the main factors affecting the quality of fresh fruits and vegetables, leading to off-flavor, fermented aroma, and tissue decay. The knowledge of microbial growth kinetics is essential for estimating a correct risk assessment associated with consuming raw vegetables and better managing the development of spoilage microorganisms. This study shows, for the first time, that only a part of total microbial community, originally present on fresh harvested female zucchini flowers, was able to adapt itself to refrigerated conditions. Through the study of microbial growth kinetics it was possibleto isolate forty-four strains belonging to twenty-two species of the genera *Acinetobacter, Arthrobacter, Bacillus, Enterobacter, Erwinia, Klebsiella, Pantoea, Pseudoclavibacter, Pseudomonas, Serratia, Staphylococcus,* and *Weissella,* suggesting *Enterobacteriaceae* as potentially responsible for pistil spoilage.

1. Introduction

The fresh-like and healthy traits of ready-to-use (RTU) products such as leafy vegetables (lettuce, rocket, spinach, etc.) and fruits (melon, pineapple, apricot, peach, apple, etc.), with a minimal time of preparation before consumption, are characteristics sought by consumers [1–3]. Microflora causing release of off-flavors and tissue decay is well known in both RTU vegetables and fruits [4–9]. The dominating bacterial population during low temperature storage mainly consists of species belonging to the *Pseudomonadaceae* and *Enterobacteriaceae;* besides some species belonging to the lactic acid bacteria and moulds may be present at relatively lower numbers [4, 10–13].

The consumption of zucchini (*Cucurbita pepo* L.) flowers with the immature fruit attached is spread in many countries, specially in the Mediterranean area, but, due to their high perishability, they are in general marketed only locally.

Different authors [14, 15] studied the effects of cold storage on zucchini flowers processed as a new RTU vegetable; these papers showed that the shelf life of flowers was affected either by low storage temperatures and by cultivars. However, no study is reported in literature on the evolution under refrigerated conditions of natural microbial population occurring on female zucchini flowers that, composed by petals and pistil, presents different tissues and nutrient availabilities.

Aim of this work was to gain a better knowledge about the survival of natural microorganisms occurring on female zucchini flowers, potentially involved in vegetable spoilage, in order to lay the basis for the development of a new RTU vegetable.

2. Materials and Methods

2.1. Microbiological Analyses of Zucchini Flowers. Microbial evaluation on growth kinetics in zucchini female flowers (cv. Aquilone) was carried out in two consecutive years (trial A and B). Clean flowers, with their immature fruit attached, were harvested and immediately subjected to microbiological analyses.

At the same time, zucchini flowers were stored, as harvested, in gas permeable polypropylene open bags at 4°C ± 1°C for 11 days. Three replicate bags of eight zucchini flowers were prepared for each sampling time (4, 8, and 11 days).

Microbial analysis was carried out on about 25 g of petals and pistils transferred aseptically and separately to a stomacher bag, containing 9 parts (w/w) of sterile saline solution (0.9% NaCl) and homogenized for 1 min using a stomacher (BagMixer, Interscience, St Nom, France). Petal and pistil dilutions were plated in triplicate on different agar media. Total mesophilic bacteria (Plate Count Agar, Oxoid S.p.A. Garbagnate, Milano, Italy) and *Enterobacteriaceae* (Violet Red Bile Glucose Agar) were counted after 24 h of growth at 30°C, yeasts and moulds (Potato Dextrose Agar) at 25°C for four days, whereas presumptive lactic acid bacteria (Rogosa agar under anaerobiosis) were enumerated after 48 h of growth at 30°C. In order to prevent yeast growth, 150 mg L^{-1} of cycloheximide were added to Rogosa and PCA, while to prevent bacterial growth 150 mg L^{-1} of chloramphenicol were added to PDA. Microbial biotyping and species identification were carried out on isolates from the dominant bacterial population.

For each storage time (0, 4, 8, and 11 days), about 10–15 well-isolated colonies were picked up from PCA plates with the highest number of colonies, inoculated in PC broth (Tryptone 5.0 g L^{-1}, Yeast extract 2.5, Glucose 1.0 g L^{-1}, Oxoid S.p.A. Garbagnate, Milano, Italy), incubated at 30°C for 24 h and frozen at −80°C for further analyses.

2.2. Microbial Biotyping and Species Identification. DNA extraction was carried out using one mL of viable cell suspension by using Wizard Genomic DNA Purification kit (Promega Italia Srl, Milano, Italy) following manufacture's instructions. DNA quantity and quality were determined by electrophoresis with known amounts of molecular weight marker (Marker IV, Roche Diagnostics, Milan, Italy) as a standard.

Isolates were submitted to the two-step RAPD analysis for biotyping [16] using the M13 primer (5′-GAGGGTGGC-GGTTCT-3′) capable of characterizing bacterial isolates from apples [17]. After electrophoresis run (2% agarose in TBE buffer, QIAGEN, 70 V for 3–5 h), different isolates showing the same electrophoretic pattern were grouped. Isolates were clustered, sequenced, and identified as recently defined by Baruzzi et al. [18].

2.3. Evaluation of Spoilage Symptoms. Fruits and flowers were individually scored for spoilage symptoms (petal necrosis or water soaking, wilting, pistil acid rot, browning, and mycelia growth, etc.) by a group of six trained people using a 5 to 1 subjective scale, with 5 = excellent, no spoilage; 4 = very good, minor spoilage; 3 = fair, moderate spoilage; 2 = poor, major spoilage; 1 = inedible. A score of 3 was considered as the limit of marketability and a score of 2 as the limit of edibility [19].

3. Results and Discussion

3.1. Enumeration of Different Microbial Populations Occurring on Zucchini Flowers. With the aim to identify the microorganisms, which could influence the spoilage, different microbial populations occurring on pistils and petals were evaluated.

Separating microbial counts by pistils or petals, pistil microbial load was always higher than the corresponding petal microbial population by about 1 or 2 logarithmic unit. The greatest differences in microbial populations were found for total mesophilic aerobic bacteria; in particular, for each sampling time and for both experiments, total mesophilic bacteria on pistils were found to be from 10 to 1000 times more than on petals. The differences between pistils and petals for *Enterobacteriaceae* and presumptive lactobacilli were far less marked. The nutrient availability on pistils can explain the higher microbial growth on pistils rather than on petals.

The growth kinetics of total aerobic bacteria, yeasts and moulds, presumptive lactobacilli, and *Enterobacteriaceae* on pistils during the storage at 4°C are shown in Figure 1. The viable cell counts of total aerobic bacteria, yeasts and moulds, and *Enterobacteriaceae* showed a similar behavior for both trials; after an initial reduction, microbial counts remained stable for some days and then increased after eight days in storage, reaching the highest value recorded for flowers in refrigerated conditions at the end of the experiment when spoilage symptoms were detected. In particular, viable cell counts of total aerobic bacteria (Figure 1) decreased from 10 (trial A) to 100 (trial B) times reaching about 6 log cfu g^{-1} in both trials after eleven days of cold storage. The viable *Enterobacteriaceae* counts decreased from 1.0×10^5 to 2.4×10^4 cfu g^{-1} in trial A and from 1.6×10^6 to 8.2×10^3 cfu g^{-1} in trial B, increasing to about 5 log cfu g^{-1} at the end of experiment for both trials. Total presumptive lactobacilli, usually known to be related to vegetable spoilage [10], showed counts (Figure 1) that remained relatively low, reaching 2.8×10^4 cfu g^{-1} only after eleven days of cold storage in trial B. Yeasts and moulds decreased 100 times in both trials with final values of 3.6×10^4 cfu g^{-1} and 5.6×10^2 cfu g^{-1} in trial A and B, respectively.

We can suppose that, after a selective effect of low temperatures on microflora counted on fresh female zucchini flowers, surviving microorganisms fitted well with the new cold environment, starting to grow. Microbial growth kinetics resulted similarly in both experiments; in particular, total aerobic counts resulted in the dominant microflora, followed by *Enterobacteriaceae*, whereas yeasts and moulds and presumptive lactobacilli remained stable or increased slowly.

Since total mesophilic aerobic bacterial population occurring in pistils was numerically dominant, it was considered, according to Ragaert et al. [2], responsible for female zucchini flower spoilage and then deeper investigated.

3.2. Microbial Biotyping and Taxonomic Identification. Biodiversity of aerobic mesophilic microflora from eight different pistil samples was studied analyzing RAPD-PCR profiles of 118 colonies from PCA plates. In general, every sample showed many different electrophoretic fingerprints, helping to identify the different strains present: clustering 118 colonies gave 44 strains. 16SrDNA analysis showed that in some cases different electrophoretic patterns belonged to the same species, confirming the ability of two-step RAPD analysis to differentiate isolates belonging to the same

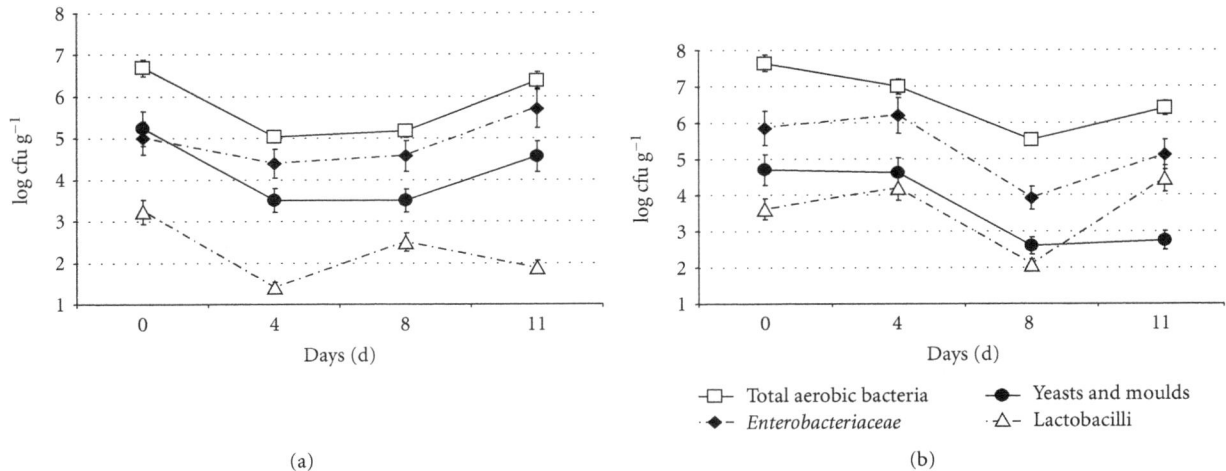

FIGURE 1: Growth dynamics of microbial populations from zucchini female flower pistils during cold storage at 4°C in trial A and B. Mean value ± standard deviation.

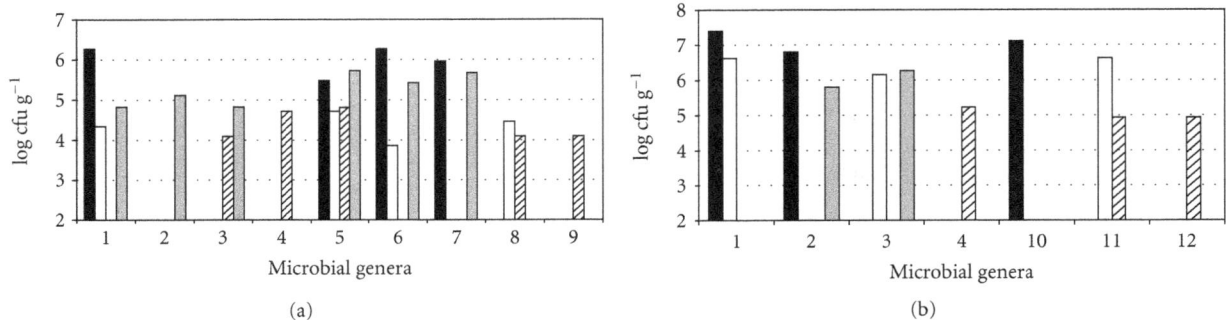

FIGURE 2: Distribution of bacterial genera and relative total load from zucchini female flower pistils in trial A and trial B after 0 (black bars), 4 (white bars), 8 (diagonal stripes bars), and 11 (grey bars) days of cold storage at 4°C. Bacterial genera are defined as: 1, *Acinetobacter* spp.; 2, *Serratia* spp.; 3, *Klebsiella* spp.; 4, *Pantoea* spp.; 5, *Erwinia* spp.; 6, *Arthrobacter* spp.; 7, *Staphylococcus* spp.; 8, *Weissella* spp.; 9, *Pseudoclavibacter* spp.; 10, *Enterobacter* spp.; 11, *Pseudomonas* spp.; 12, *Bacillus* spp.

species. The list of all strains isolated from pistils is presented in Table 1, whereas the distribution and the viable cell load of bacterial genera from pistils during trials A and B are shown in Figure 2. The cluster analysis carried out after RAPD-PCR and 16SrDNA analysis showed wide differences between trial A and B microflora. The trial A showed eight different fingerprints in naturally occurring bacteria on flowers, whereas isolates of trial B were clustered in four groups. During cold storage microbial isolates from trial A were grouped in six, eight, and fourteen RAPD-PCR cluster after four, eight, and eleven days of storage, respectively. The trial B was characterized by a lower level of microbial biodiversity showing six groups of isolates at any sampling time. The unstored flowers microflora was dominated by *Acinetobacter* sp. (Figure 2) in both trials, but the relative microbial load was higher in trial B. Together with this genus, after four days of storage other strains belonging to *Weissella* spp., *Erwinia* spp. and *Arthrobacter* spp. for trial A and *Pseudomonas* spp. and *Klebsiella* spp. for trial B appeared. After eight days, when the early damage symptoms became evident in zucchini flowers in the two experiments, other

microbial genera were identified: *Pantoea* spp. in both trials, *Erwinia* spp., *Pseudoclavibacter* spp. and *Weissella* spp. in trial A or *Pseudomonas* spp. and *Bacillus* spp. isolated from trial B.

At the end of zucchini flowers cold storage, the dominant microflora in trial B were mainly *Klebsiella* and *Serratia* genera. In trial A, in addition to these genera, *Staphylococcus* spp., *Acinetobacter* spp., *Arthrobacter* spp., and *Erwinia* spp. were also detected. The microflora of both trials showed isolates belonging to *Acinetobacter*, *Serratia*, *Klebsiella*, and *Pantoea* genera. Isolates belonging to *Erwinia* spp. were always found in trial A and never in trial B, whereas, differently from trial A, no *Staphylococcus* strain was isolated from trial B samples.

Differences in microflora composition between trials could be also due to different environmental conditions during the growth of the zucchini flowers.

The strains isolated from both trials belonged to the following species: *Acinetobacter calcoaceticus*, *Staphylococcus succinus* and *S. xylosus*, *Arthrobacter nicotianae*, *Serratia marcescens*, *Enterobacter aerogenes*, *Pantoea agglomerans*, and *Klebsiella oxytoca*. The most frequently isolated species was

TABLE 1: List of strains identified from zucchini pistils harvested in Trial A and B at 0 (sample T0), 4 (sample T4), 8 (sample T8), and 11 (sample T11) days of cold storage.

Trial A			Trial B		
Strain	Taxonomic identification	Isolated in sample	Strain	Taxonomic identification	Isolated in sample
A5	*Acinetobacter calcoaceticus*	T0	C35	*Acinetobacter calcoaceticus*	T4
A12	*Acinetobacter calcoaceticus*	T0, T4	C36	*Acinetobacter calcoaceticus*	T4
A13	*Acinetobacter calcoaceticus*	T0, T8, T11	C37	*Acinetobacter calcoaceticus*	T4
A83	*Acinetobacter sp.*	T11	C5	*Acinetobacter calcoaceticus*	T0, T4, T8
A63	*Arthrobacter arilaiti*	T11	C3	*Acinetobacter sp.*	T0
A16	*Arthrobacter nicotianae*	T0	C86	*Bacillus Megaterium*	T8
A35	*Arthrobacter sp.*	T4	C107	*Enterobacter aerogenes*	T11
A80	*Arthrobacter sp.*	T0, T11	C20	*Enterobacter aerogenes*	T0
A81	*Arthrobacter sp.*	T11	C74	*Enterobacter sp.*	T8
A9	*Erwinia persicina*	T0, T4	C110	*Klebsiella oxytoca*	T11
A46	*Erwinia persicina*	T8, T11	C38	*Klebsiella oxytoca*	T4, T11
A38	*Erwinia sp.*	T4	C71	*Pantoea agglomerans*	T8
A64	*Erwinia sp.*	T8, T11	C73	*Pantoea agglomerans*	T8
A56	*Klebsiella oxytoca*	T8	C51	*Pseudomonas sp.*	T4, T8
A72	*Klebsiella sp.*	T11	C122	*Serratia marcescens*	T11
A42	*Pantoea agglomerans*	T4	C19	*Serratia marcescens*	T0
A57	*Pantoea agglomerans*	T8			
A41	*Pantoea sp.*	T8			
A48	*Pseudoclavibacter helvolus*	T8			
A77	*Serratia sp.*	T11			
A82	*Serratia sp.*	T11			
A65	*Staphylococcus saprophyticus*	T11			
A17	*Staphylococcus succinus*	T0, T11			
A20	*Staphylococcus succinus*	T0			
A67	*Staphylococcus succinus*	T11			
A62	*Staphylococcus xylosus*	T11			
A26	*Weissella sp.*	T4			
A53	*Weissella viridescens*	T8			

Acinetobacter calcoaceticus, present at each storing day in trial A and initially and after four and eight days in trial B.

Acinetobacter and *Staphylococcus* presented five strains; the remaining species showed one or two strains in both trials. As concerns strain growth kinetics, some strains (*A. calcoaceticus, E. persicina, S. marcescens, K. oxytoca,* and *P. agglomerans*) were recovered in more than one sample. The species present at the highest concentrations were *A. calcoaceticus* and *Staphylococcus succinus* (trial A) and *A. calcoaceticus, S. marcescens,* and *K. oxytoca* (trial B) isolated from fresh flowers in which total aerobic load showed the highest values. The strains with the lowest microbial load were *A. calcoaceticus* and *K. oxytoca* found at the end of storage in trial A and *Pseudomonas* sp., *Pantoea* sp., *Bacillus megaterium,* and *S. marcescens* detected in trial B. Even when it was possible to identify the same strain in more than one sample, no single strain (or single species) was ever identified as being entirely responsible for the spoilage symptoms showed. The microbial isolates identified during

the two trials were quite similar, having many common species. The identification of *Pantoea agglomerans, Erwinia persicina, Serratia marcescens,* and *Enterobacter aerogenes* in all samples confirms the importance of *Enterobacteriaceae* in microbial spoilage of cold stored vegetables, according to previous indications [1, 10, 20].

A. baumannii and *A. calcoaceticus* form a phylogenetically complex of which only *A. calcoaceticus* is usually considered a nonpathogen soil-borne species [21]. We retained reliable the identification of *Acinetobacter* strains as belonging to *A. calcoaceticus* species, when the max score of alignment, calculated against sequences present in the Reference Sequence (RefSeq) database of other *Acinetobacter* species, was no higher than the 95% of that calculated for the top sequence match.

The microbial composition of zucchini flowers here reported partially resembles data from Janisiewicz and Buyer [22] relative to microflora associated with nectarine brown rot, that, in addition with strains belonging to other bacterial genus, reported the isolation of *Enterobacter, Erwinia, Pantoea, Pseudomonas, Serratia,* and *Staphylococcus* strains associated over a period of 7 weeks of nectarine development.

The microbial data showed many species and sometimes different strains of a single species (Table 1); since some strains were found both initially and at the end of cold storage, whereas other grew only during storage, it is possible to assume that only a part of total microbial community, originally present on flowers, was adaptable to refrigerated conditions.

The strain kinetic during cold storage highlighted the microbial biodiversity of flowers, underestimated in fresh flowers that usually are not subjected to severe changes in environmental conditions.

The *Acinetobacter* strains are often associated with food spoilage as previously demonstrated for meat, milk, and cheeses [18, 23, 24].

The finding of some potentially pathogenic bacteria—such as *Acinetobacter spp.* [25], *K. oxytoca, Enterobacter aerogenes* [26], and *Serratia* spp. [27, 28]–commonly found on vegetables, does not represent a risk for consumer health, since these vegetables are usually cooked before eating. Notably, in some cases (*Erwinia persicina, Enterobacter aerogenes, Klebsiella oxytoca,* and *Serratia marcescens*) the counts for single strains or species, from PCA plates, was significantly higher (more than 10 times) than *Enterobacteriaceae* counted on VRBGA plates. In addition, in both trials the shape of the *Enterobacteriaceae* growth curve appears the same, but with lower values, of total aerobic bacteria curve.

The results of this work seem to indicate that VRBGA is excessively selective for *Enterobacteriaceae* present on the flowers, leading us to underestimate this microbial load. The presence of $1.5 \, \mathrm{g} \, \mathrm{L}^{-1}$ of bile salts, needed to evaluate the coliforms present in milk and dairy products, could be selective against *Enterobacteriaceae* naturally occurring on vegetables where bile salts are absent. If similar results will be obtained for other vegetables, then the use of VRBGA for evaluating *Enterobacteriaceae* in these foods should be avoided.

3.3. Evaluation of Spoilage Symptoms. During both experiments (trial A and trial B), petals and pistils showed different types of spoilage. Generally, petals remained intact but their orange-yellow color lost natural brightness. In addition, petals turgidity reduction (score 3 or less) was observed after eleven days of refrigerated storage. Pistil spoilage occurred quickly just after eight days of cold storage, when about 80% of pistils appeared brown with gelatinous materials on the surface. Even though we did not perform assays direct to differentiate physiological from microbial spoilage, symptoms on petals remember physiological and biochemical tissue disorders caused by storage at low temperature for long time [29, 30], whereas pistil spoilage resembled soft rot caused by pectolytic and macerating enzymes of many plant pathogenic bacteria on different organs and succulent leaves of vegetables [31, 32]. The evaluation of visual symptoms occurring on petals, pistils, and fruits of female zucchini flowers leads us to identify the pistils as the main perishable part limiting the marketability during the cold storage.

4. Conclusions

In this work we evaluated, for the first time, the microbial kinetics of natural occurring microflora of zucchini flowers developing under cold storage conditions; the most perishable part of female zucchini flower was represented by pistils, well known to have high amount of free simple sugars, in comparison with petals, with limited nutrient availability. Molecular tools enabled us to isolate forty-four strains from complex microflora developing on pistils suggesting *Enterobacteriaceae* as potentially responsible for pistil spoilage. However, due to the high microbial biodiversity found in all samples of both trials, the joint responsibility of other bacterial genera cannot be excluded.

A parallel result coming from data comparison from strain biotyping, taxonomic identification, and microbial isolation of different media is that the VRBGA media is excessively selective to evaluate *Enterobacteriaceae* from vegetables.

Microbial data from this work could be useful to set tools and technologies able to preserve the most delicate part of zucchini flower, the pistil, against spoiler microorganisms.

References

[1] I. Babic, S. Roy, A. E. Watada, and W. P. Wergin, "Changes in microbial populations on fresh cut spinach," *International Journal of Food Microbiology*, vol. 31, no. 1–3, pp. 107–119, 1996.

[2] P. Ragaert, F. Devlieghere, and J. Debevere, "Role of microbiological and physiological spoilage mechanisms during storage of minimally processed vegetables," *Postharvest Biology and Technology*, vol. 44, no. 3, pp. 185–194, 2007.

[3] D. Rico, A. B. Martín-Diana, J. M. Barat, and C. Barry-Ryan, "Extending and measuring the quality of fresh-cut fruit and vegetables: a review," *Trends in Food Science & Technology*, vol. 18, no. 7, pp. 373–386, 2007.

[4] M. H. J. Bennik, W. Vorstman, E. J. Smid, and L. G. M. Gorris, "The influence of oxygen and carbon dioxide on the growth of prevalent *Enterobacteriaceae* and *Pseudomonas* species isolated from fresh and controlled-atmosphere-stored vegetables," *Food Microbiology*, vol. 15, no. 5, pp. 459–469, 1998.

[5] M. Giménez, C. Olarte, S. Sanz, C. Lomas, J. F. Echávarri, and F. Ayala, "Relation between spoilage and microbiological quality in minimally processed artichoke packaged with different films," *Food Microbiology*, vol. 20, no. 2, pp. 231–242, 2003.

[6] L. Jacxsens, F. Devlieghere, P. Ragaert, E. Vanneste, and J. Debevere, "Relation between microbiological quality, metabolite production and sensory quality of equilibrium modified atmosphere packaged fresh-cut produce," *International Journal of Food Microbiology*, vol. 83, no. 3, pp. 263–280, 2003.

[7] O. Lamikanra, K. L. Bett-Garber, D. A. Ingram, and M. A. Watson, "Use of mild heat pre-treatment for quality retention of fresh-cut cantaloupe melon," *Journal of Food Science*, vol. 70, no. 1, pp. C53–C57, 2005.

[8] M. A. Rojas-Graü, G. Oms-Oliu, R. Soliva-Fortuny, and O. Martín-Belloso, "The use of packaging techniques to maintain freshness in fresh-cut fruits and vegetables: a review," *International Journal of Food Science & Technology*, vol. 44, no. 5, pp. 875–889, 2009.

[9] I. Vandekinderen, J. Van Camp, F. Devlieghere et al., "Effect of decontamination on the microbial load, the sensory quality and the nutrient retention of ready-to-eat white cabbage,"

European Food Research and Technology, vol. 229, no. 3, pp. 443–455, 2009.

[10] C. Nguyen-The and F. Carlin, "The microbiology of minimally processed fresh fruits and vegetables," *Critical Reviews in Food Science and Nutrition*, vol. 34, no. 4, pp. 371–401, 1994.

[11] K. Vankerschaver, F. Willocx, C. Smout, M. Hendrickx, and P. Tobback, "The influence of temperature and gas mixtures on the growth of the intrinsic micro-organisms on cut endive: predictive versus actual growth," *Food Microbiology*, vol. 13, no. 6, pp. 427–440, 1996.

[12] K. T. Rajkowski and X. Fan, "Microbial quality of fresh-cut iceberg lettuce washed in warm or cold water and irradiated in a modified atmosphere package," *Journal of Food Safety*, vol. 28, no. 2, pp. 248–260, 2008.

[13] M. Martínez-Ferrer and C. Harper, "Reduction in microbial growth and improvement of storage quality in fresh-cut pineapple after methyl jasmonate treatment," *Journal of Food Quality*, vol. 28, no. 1, pp. 3–12, 2005.

[14] A. M. Villalta, M. Ergun, A. D. Berry, N. Shaw, and S. A. Sargent, "Quality changes of yellow summer squash blossoms (*Cucurbita pepo* L.) during storage," *Acta Horticulturae*, vol. 659, pp. 831–834, 2004.

[15] N. Calabrese, F. Baruzzi, G. Signorella, and G. Damato, "Yield and microbial evaluation of summer squash pistillate flowers for "ready to use product". First results," *Acta Horticulturae*, vol. 741, pp. 221–227, 2007.

[16] F. Baruzzi, M. Morea, A. Matarante, and P. S. Cocconcelli, "Changes in the *Lactobacillus* community during Ricotta forte cheese natural fermentation," *Journal of Applied Microbiology*, vol. 89, no. 5, pp. 807–814, 2000.

[17] A. Ricelli, F. Baruzzi, M. Solfrizzo, M. Morea, and F. P. Fanizzi, "Biotransformation of patulin by *Gluconobacter oxydans*," *Applied and Environmental Microbiology*, vol. 73, no. 3, pp. 785–792, 2007.

[18] F. Baruzzi, R. Lagonigro, L. Quintieri, L. Caputo, and M. Morea, "Occurrence of non-lactic acid microflora in high moisture cold stored Mozzarella cheese," *Food Microbiology*. In press.

[19] B. Pace, M. Cefola, F. Renna, and G. Attolico, "Relationship between visual appearance and browning as evaluated by image analysis and chemical traits in fresh-cut nectarines," *Postharvest Biology and Technology*, vol. 61, no. 2-3, pp. 178–183, 2011.

[20] E. Wevers, P. Moons, R. Van Houdt, I. Lurquin, A. Aertsen, and C. W. Michiels, "Quorum sensing and butanediol fermentation affect colonization and spoilage of carrot slices by *Serratia plymuthica*," *International Journal of Food Microbiology*, vol. 134, no. 1-2, pp. 63–69, 2009.

[21] H. C. Chang, Y. F. Wei, L. Dijkshoorn, M. Vaneechoutte, C. T. Tang, and T. C. Chang, "Species-level identification of isolates of the *Acinetobacter calcoaceticus-Acinetobacter baumannii* complex by sequence analysis of the 16S-23S rRNA gene spacer region," *Journal of Clinical Microbiology*, vol. 43, no. 4, pp. 1632–1639, 2005.

[22] W. J. Janisiewicz and J. S. Buyer, "Culturable bacterial microflora associated with nectarine fruit and their potential for control of brown rot," *Canadian Journal of Microbiology*, vol. 56, no. 6, pp. 480–486, 2010.

[23] M. Gennari, M. Parini, D. Volpon, and M. Serio, "Isolation and characterization by conventional methods and genetic transformation of *Psychrobacter* and *Acinetobacter* from fresh and spoiled meat, milk and cheese," *International Journal of Food Microbiology*, vol. 15, no. 1-2, pp. 61–75, 1992.

[24] P. Munsch-Alatossava and T. Alatossava, "Phenotypic characterization of raw milk-associated psychrotrophic bacteria," *Microbiological Research*, vol. 161, no. 4, pp. 334–346, 2006.

[25] L. Dortet, P. Legrand, C. J. Soussy, and V. Cattoir, "Bacterial identification, clinical significance, and antimicrobial susceptibilities of *Acinetobacter ursingii* and *Acinetobacter schindleri*, two frequently misidentified opportunistic pathogens," *Journal of Clinical Microbiology*, vol. 44, no. 12, pp. 4471–4478, 2006.

[26] Y. A. Markova, A. S. Romanenko, and A. V. Dukhanina, "Isolation of bacteria of the family *Enterobacteriaceae* from plant tissues," *Microbiology*, vol. 74, no. 5, pp. 575–578, 2005.

[27] V. Livrelli, C. De Champs, P. Di Martino, A. Darfeuille-Michaud, C. Forestier, and B. Joly, "Adhesive properties and antibiotic resistance of *Klebsiella*, *Enterobacter*, and *Serratia* clinical isolates involved in nosocomial infections," *Journal of Clinical Microbiology*, vol. 34, no. 8, pp. 1963–1969, 1996.

[28] A. B. Christensen, K. Riedel, L. Eberl et al., "Quorum-sensing-directed protein expression in *Serratia proteamaculans* B5a," *Microbiology*, vol. 149, no. 2, pp. 471–483, 2003.

[29] S. Chung, J. E. Staub, and G. Fazio, "Inheritance of chilling injury: a maternally inherited trait in cucumber," *Journal of the American Society for Horticultural Science*, vol. 128, no. 4, pp. 526–530, 2003.

[30] C. B. Watkins and J. H. Ekman, "Storage technologies: temperature interactions and effects on quality of horticultural products," *Acta Horticulturae*, vol. 682, pp. 1527–1533, 2005.

[31] C. B. Wegener, "Induction of defence responses against *Erwinia* soft rot by an endogenous pectate lyase in potatoes," *Physiological and Molecular Plant Pathology*, vol. 60, no. 2, pp. 91–100, 2002.

[32] Y. Aysan, A. Karatas, and O. Cinar, "Biological control of bacterial stem rot caused by *Erwinia chrysanthemi* on tomato," *Crop Protection*, vol. 22, no. 6, pp. 807–811, 2003.

8

Association of *Acinetobacter baumannii* EF-Tu with Cell Surface, Outer Membrane Vesicles, and Fibronectin

Shatha F. Dallo,[1] **Bailin Zhang,**[2] **James Denno,**[3] **Soonbae Hong,**[1] **Anyu Tsai,**[1] **Williams Haskins,**[1,4,5,6,7,8] **Jing Yong Ye,**[2] **and Tao Weitao**[1,9]

[1] *Department of Biology, The University of Texas at San Antonio, One UTSA Circle, San Antonio, TX 78249, USA*
[2] *Department of Biomedical Engineering, The University of Texas at San Antonio, One UTSA Circle, San Antonio, TX 78249, USA*
[3] *Department of Biology, The University of Texas at Austin, 1 University Station, Austin, TX 78712, USA*
[4] *Pediatric Biochemistry Laboratory, The University of Texas at San Antonio, San Antonio, TX 78249, USA*
[5] *Department of Chemistry, The University of Texas at San Antonio, San Antonio, TX 78249, USA*
[6] *RCMI Proteomics and Protein Biomar Feers Cores, The University of Texas at San Antonio, San Antonio, TX 78249, USA*
[7] *Center for Research & Training in The Sciences, The University of Texas at San Antonio, San Antonio, TX 78249, USA*
[8] *Division of Hematology/Oncology, Department of Medicine, Cancer Therapy & Research Center, The University of Texas Health Science Center at San Antonio, San Antonio, TX 78229, USA*
[9] *Department of Biology, College of Science and Mathematics, Southwest Baptist University, 1600 University Avenue, Bolivar, MO 65613, USA*

Correspondence should be addressed to Tao Weitao, weitaosjobs@yahoo.com

Academic Editors: G. Bruant, T. Darribere, S. F. Porcella, R. Rivas, and M. Vaneechoutte

A conundrum has long lingered over association of cytosol elongation factor Tu (EF-Tu) with bacterial surface. Here we investigated it with *Acinetobacter baumannii*, an emerging opportunistic pathogen associated with a wide spectrum of infectious diseases. The gene for *A. baumannii* EF-Tu was sequenced, and recombinant EF-Tu was purified for antibody development. EF-Tu on the bacterial surface and the outer membrane vesicles (OMVs) was revealed by immune electron microscopy, and its presence in the outer membrane (OM) and the OMV subproteomes was verified by Western blotting with the EF-Tu antibodies and confirmed by proteomic analyses. EF-Tu in the OM and the OMV subproteomes bound to fibronectin as detected by Western blot and confirmed by a label-free real-time optical sensor. The sensor that originates from photonic crystal structure in a total-Internal-reflection (PC-TIR) configuration was functionalized with fibronectin for characterizing EF-Tu binding. Altogether, with a novel combination of immunological, proteomical, and biophysical assays, these results suggest association of *A. baumannii* EF-Tu with the bacterial cell surface, OMVs, and fibronectin.

1. Introduction

A Gram-negative and obligate aerobic bacterial species, *Acinetobacter baumannii,* has emerged as one of the most important nosocomial pathogens [1–4], raising risks not only regional but also global in the aftermath of war and natural disasters. *A. baumannii* was identified in the US military personnel deployed to Iraq and Afghanistan [5]. Interestingly, more than 60% of the isolates were related to three pan-European clones that, in fact, had been disseminated in geographically distinct areas [6]. Besides, *Acinetobacter* infections are associated with natural disasters, such as the 1999 earthquake in Turkey [7] and the 2008 earthquake in China [8].

For such austerity of *A. baumannii* infections, little is known about the pathogenesis. To explore the fundamental mechanisms, we tested whether extracellular proteins of *A. baumannii* mediate the bacterial attachment [9]. Proteins were extracted from whole-cell lysate, outer membrane (OM) fractions, and cell-free spent cultures (CFCs) of the wild-type and the biofilm mutants of *A. baumannii* we isolated [9]. With a proteomic approach, translation elongation factor (EF-Tu) of *A. baumannii* was detected in cell-free cultures, the data suggesting release of EF-Tu from the bacterial cells. The release appeared unlikely to result from cell death and lysis but rather likely to be regulated, because the mutants, as viable as the wild type, exhibited deficiency in the release and cell adhesion [9]. The EF-Tu release seemed to be a puzzle to us as the primary function of EF-Tu, while remaining to be characterized for *A. baumannii*, is translation elongation as deduced from the *E. coli* EF-Tu, because EF-Tu and translation are highly conserved throughout the bacterial domain [10–12]. Specifically, in the first step of peptide chain elongation on ribosomes, EF-Tu·GTP serves as a carrier of codon-specified aminoacyl-tRNA to the ribosomal aminoacyl site. Eubacterial EF-Tus belong to the superfamily of GTP-binding proteins. It is not a membrane protein, since EF-Tu lacks a signal sequence and transmembrane domains that mediate protein translocation across cell membrane.

This has led to a conundrum concerning EF-Tu release. The original clue to this question may come from a study with the sucrose-dependent spectinomycin-resistant mutants of *Escherichia coli* grown in the absence of sucrose [13]. EF-Tu was detected in the OM fractions; its presence in OM did not result from artificial binding during membrane preparation. It was also found in the periplasm of *Neisseria gonorrhoeae* [14]. Two decades after the initial finding, *E. coli* EF-Tu was detected again in the OM fractions of the cells adherent to abiotic surface [15]. The bacterial surface association of EF-Tu has been further evidenced by EF-Tu involvement in *Staphylococcus aureus* biofilm development [16], in mediating attachment to human cells by *Lactobacillus johnsonii* [17] or *P. aeruginosa* [18]. The EF-Tu surface association has been attested by its acting as a part of pathogen-associated molecular patterns recognized by receptors on eukaryote hosts [19], as a target for a serine-threonine phosphatase involved in virulence and survival of *Listeria monocytogenes* in the infected host [20], and as an active protein eliciting innate [21] and acquired immunity [16, 22].

How the surface-associated EF-Tu is released still seems to be an enigma. Our hypothesis was that *A. baumannii* EF-Tu is associated with outer membrane vesicles (OMVs). The rationale is based on the proteomic analyses that have implicated EF-Tu association with OMVs in multiple bacterial species [23] and with OM in *A. baumannii* [24], and *A. baumannii* actually produces OMVs [25]. To test it, we cloned and sequenced the EF-Tu encoding gene, purified the recombinant EF-Tu (rEF-Tu), and produced EF-Tu antibodies. Then we employed a combination of transmission electron microcopy (TEM), proteomics, Western blot, and an optical sensor to show that EF-Tu is associated with OMVs and OM and binds to the host extracellular matrix protein fibronectin.

2. Results

2.1. A. baumannii EF-Tu. The EF-Tu encoding gene of *A. baumannii* ATCC19606 strain was sequenced and the protein was purified for antibody development. The ATCC 19606 strain was chosen for novelty because its genome was not completely sequenced and the EF-Tu encoding gene was

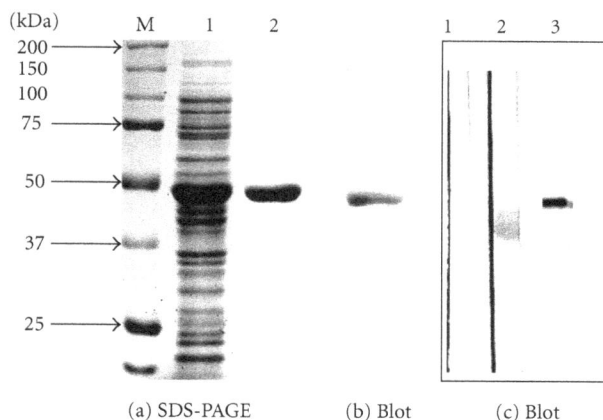

FIGURE 1: Purification of *A. baumannii* EF-Tu. Purification of *A. baumannii* rEF-Tu. (a) Overexpressed (lane 1) and column-purified rEF-Tu (lane 2). (b) Immunoblot of column-purified rEF-Tu with anti-His-tag monoclonal antibody. (c) Immunoblot of *A. baumannii* cell lysate with rabbit prebleed (lane 1) and anti-rEF-Tu antibodies (2). Immunoblot of rEF-Tu with the anti-rEF-Tu (3).

not studied at the time we started our investigation. The availability of genome sequencing data for the ATCC 17978 strain greatly facilitated our study. Based on the genome data, there are two genes for EF-Tu, namely *tufAa* and *tufBa*, both identical [26], with reference to *tufAe* and *tufBe* of *E. coli*. The *tufAe* deletion caused *E. coli* growth defect in rich media, while the *tufBe* deletion did not [27], the observations suggesting that *tufAe* is functional. These data led us to clone and sequence *tufAa* of the *A. baumannii* 19606 strain. Comparison of the *tufAa* sequences from 17978 and 19606 strains showed 99.8% identity; the small difference resulted from two nucleotide changes located in 1,032 and 1,137 (Figure S1 in Supplementary Material available online at doi: 10.1100/2012/128705)—GCA of the 19606 strain but GCG of the 17978 strain—a silent mutation in the codon for alanine. The gene of the 19606 strain was cloned and His-tagged; rEF-Tu (48 kDa) was expressed and purified to homogeneity (Figure 1(a) lane 2). Immunoblots of the His-tagged rEF-Tu showed that the tagged rEF-Tu reacted with anti-His monoclonal antibodies (b), verifying that the purified protein was His-tagged. The identity of rEF-Tu was confirmed with proteomic analysis as we described before [9]. Furthermore, the antiserum specific to rEF-Tu was produced. Immunoblots with the sera indicate that the antiserum recognized both 43 kDa EF-Tu in cell lysate (Figure 1(c) lane 2) and 48 kDa rEF-Tu in the purified fraction (lane 3), but the preimmune serum did not (lane 1). The band of EF-Tu from the whole-cell extract appeared wider (lane 2) than that from the purified fraction (lane 3), suggesting that EF-Tu undergoes slight degradation in the cell extract, in line with the previous data about cleavage of *E. coli* EF-Tu by a phage-exclusion system [28].

2.2. EF-Tu Associated with OMVs and Cell Surface of A. baumannii. Immune TEM of *A. baumannii* OMVs and the cells was conducted with the antibodies specific for rEF-Tu

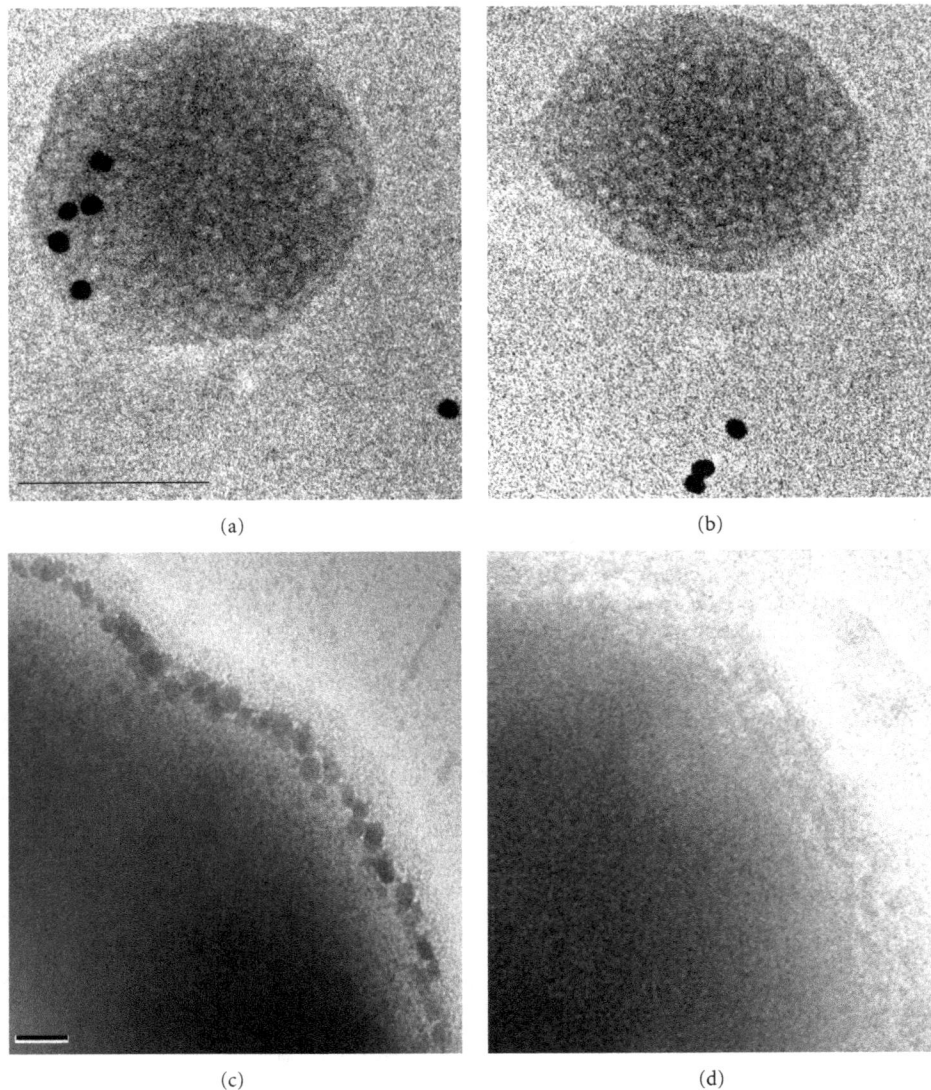

(a)   (b)

(c)   (d)

FIGURE 2: . EF-Tu visualized on *A. baumannii* OMVs and cells by immune TEM with EF-Tu antibodies. Immunogold TEM with anti-EF-Tu antibodies showing localization of (a) EF-Tu on the isolated OMV and (c) on cell surface of *A. baumannii*. Immunogold with preimmune serum shows no localization on OMVs (b) and cells (d). $n = 40$ OMVs. Bar: 100 nm.

in order to examine whether EF-Tu is physically associated with *A. baumannii* OMVs and cell surface. OMVs or cells were incubated with the anti-rEF-Tu antibodies or the pre-immune serum as a control. After washes, the samples were probed with the gold-labeled anti-IgG antibodies and examined under TEM (Figure 2). When OMVs were probed with the primary EF-Tu antibodies and the secondary gold-labeled antibodies, high-density dots of gold particles were often observed associated with OMVs (a). When the cells were probed, the gold particles appeared to deposit on the cell surface (c), the result consistent with a previous finding with thin sectioning of bacterial cells [17]. In contrast, when the pre-immune serum was used, the gold dots were mostly washed off (b, d). The rEF-Tu antibodies appeared highly specific not only for rEF-Tu (Figure 1(c) lane 3) but also for EF-Tu in the cell lysate (Figure 1(c) lane 2), the OM, and

the OMV fractions (Figure 3). Evidently, deposition of the gold-labeled antibodies specific for the EF-Tu antibodies on OMVs and cells appeared reflective of EF-Tu on the surfaces of OMVs and the cells. The results provide the physical evidence for association of EF-Tu with OMVs and cell surface of *A. baumannii*.

2.3. EF-Tu Detected in OM and OMV Fractions. The physical association of EF-Tu with OMVs and cell surface prompted us to verify the presence of EF-Tu in the OM and the OMV subproteomes. We performed 1D and 2D gel-based Western blotting analyses. First, we resolved the proteins of the OM (lane 1) and the OMV fractions (lane 2) in SDS-PAGE (Figure 3(A)(a)). After the proteins in the gel were transferred onto a membrane and probed with the anti-rEF-Tu antibodies (Figure 3(A)(b)), a protein band of 43 kDa

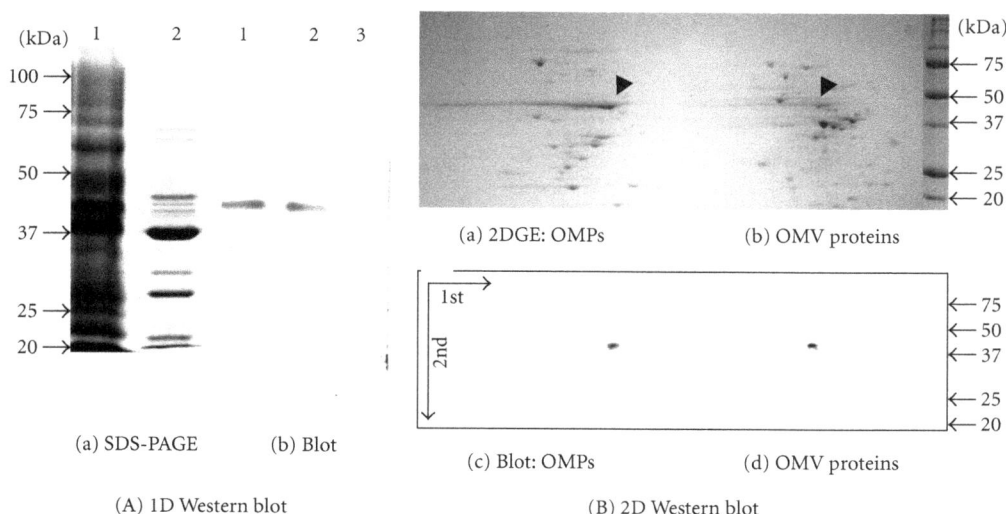

(a) 2DGE: OMPs   (b) OMV proteins

(a) SDS-PAGE   (b) Blot

(c) Blot: OMPs   (d) OMV proteins

(A) 1D Western blot   (B) 2D Western blot

FIGURE 3: EF-Tu detected in OMV and OM fractions. Panel (A): (a) SDS-PAGE of proteins from OM (lane 1) and OMV fractions (2). (b) Western blot of proteins from OM (lane 1) and OMV fractions (2) reacted with the EF-Tu antibody diluted at 1 : 3000. Control blot with pre-immune serum (lane 3). Panel (B): 2D-based Western blot. Proteins from OM (a and c) and OMV fractions (b and d) were resolved by isoelectric focusing (1st D) and then separated on a second dimension SDS-PAGE (2nd D). Proteins from the gel were blotted onto PVDF membranes and probed with the anti-EF-Tu antibodies (c and d). Arrows: EF-Tu. Blot with pre-immune serum not shown.

the same as the EF-Tu mass was detected in the OM (lanes 1) and the OMV fractions (lane 2) [10]. In contrast, when the membrane was probed with the pre-immune serum, this protein band was not detected (lane 3). Second, to address the limitation of 1D resolution, we conducted the 2D gel electrophoresis and probed the proteins for EF-Tu (Figure 3(B)). EF-Tu was still detected by the antibodies in the OM (a, c) and the OMV fractions (b, d), but not by the pre-immune serum control.

While the evidence of Western blot for presence of EF-Tu in the OM and the OMV fractions appeared to be convincing, there were some drawbacks of Western blotting analyses, such as lack of scope due to limitation on utilizing costly arrays of antibodies to target multiple proteins. Proteomic analyses of the OMV and the OM subproteomes were carried out to address them. Briefly, total proteins in the lane of SDS-PAGE as shown in Figure 3(A)(a) were subjected to the *in vitro* trypsin proteolysis and capillary LC/MS/MS. The degraded peptide masses were determined and searched across the bacterial protein databases with the $P < 0.05$-based MOWSE scoring algorithm [29]. By this significance threshold and the cut-off score of 50, 144 proteins were identified in the OMV and the OM fractions (Table 1). EF-Tu was detected in both OMV and OM subproteomes (no. 57 in Table S2). The consistent results of EF-Tu obtained by the immunological and the proteomic analyses attested the validity of both methodologies in detection of EF-Tu. The proteomic analyses also detected OmpA (no. 51) in the OM and the OMV fractions, the results consistent with the former finding [25]. The consistent results of the immune TEM with the Western blotting and proteomic analyses demonstrate that EF-Tu is indeed associated with OMVs, unlikely due to protein contamination during protein sample preparation.

TABLE 1: Summary of OMV and OM subproteomes.

Subproteomes	SP*-TM** domains (n)	TM (n)	None (n)	Total (n)
OMV	12% (6)	59% (29)	29% (14)	34% (49)
OM	46% (32)	29% (20)	25% (17)	48% (69)
Common	39% (10)	27% (7)	34% (9)	18% (26)
	33% (48)	39% (56)	28% (40)	100% (144)

*Signal peptide; **transmembrane domains. Common: proteins detected in both OMV and OM fractions.

2.4. The OMV- or OM-Associated EF-Tu Binding to Fibronectin.

As *A. baumannii* EF-Tu was detected in OMVs and on the bacterial cell surface (Figures 2 and 3) and *Mycoplasma pneumoniae* EF-Tu was found to bind the host extracellular matrix protein fibronectin [30], it could be hypothesized that the OMV- and the OM-associated EF-Tus of *A. baumannii* bind to fibronectin. This hypothesis was tested with the Western-based binding assays. The proteins extracted from the OM (lane 1 of Figure 4(a)) and the OMV fractions (lane 2) were fractioned together with rEF-Tu (lane 3) by SDS-PAGE and transferred onto PVDF membranes. The membrane strips (1, 2, and 3) were incubated with fibronectin and then blotted against the fibronectin antibodies. A band of 43 kDa corresponding to the EF-Tu mass was detected in the OM (lane 1) and OMV fractions (lane 2). A 48 kDa band known to be rEF-Tu (lane 3) was detected but not by the pre-immune serum (lane 4). Identity of EF-Tu in each of the bands was confirmed by proteomic analysis as we described previously [9]. To deal with the limited power of 1D resolution, we performed the 2-D gel electrophoresis of the proteins from OM (Figure 4(b)) and probed the proteins for fibronectin binding as above.

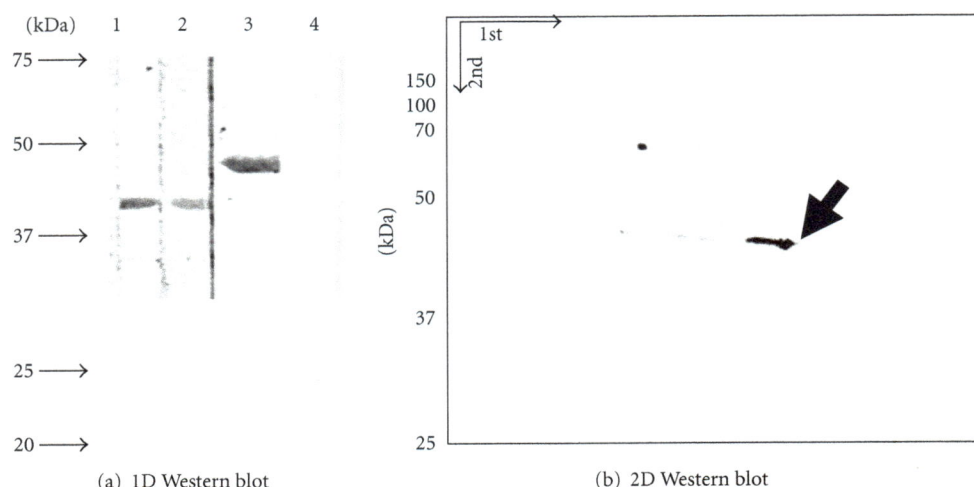

(a) 1D Western blot

(b) 2D Western blot

FIGURE 4: Binding of EF-Tu to fibronectin examined by Western-blot-based binding assays. (a) Proteins were resolved by SDS-PAGE and blotted onto PVDF membranes. Proteins from OM (lane 1) and OMVs (2) and the purified rEF-Tu (3) were probed with FN and anti-FN. rEF-Tu (4) and proteins from OM (not shown) probed with anti-FN alone. (b) Immunoblot of 2-DE of proteins from OM probed with FN followed by anti-FN. Arrow indicates EF-Tu confirmed by protein sequencing.

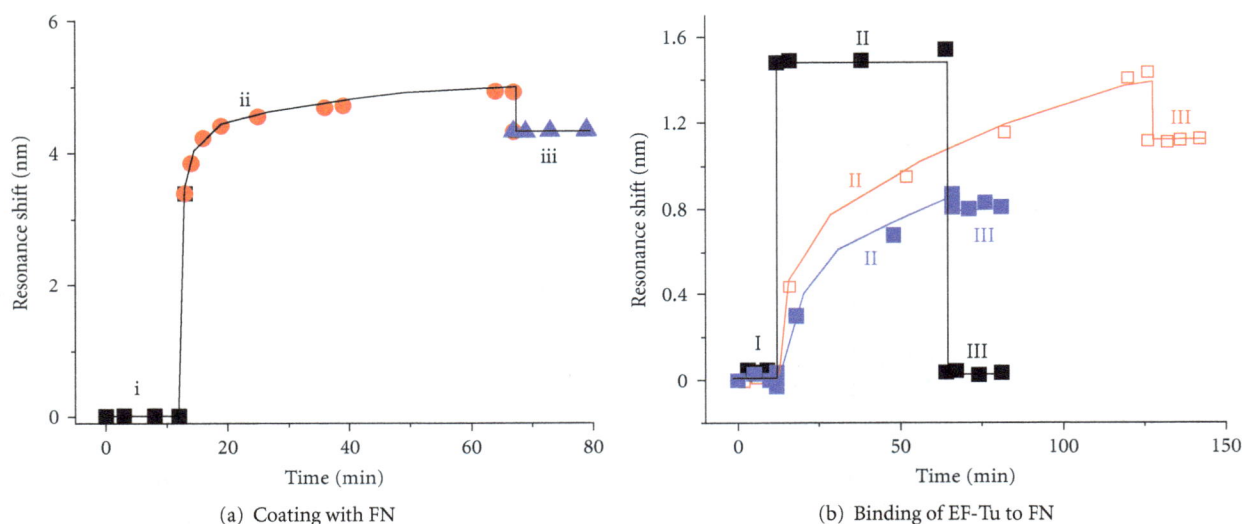

(a) Coating with FN

(b) Binding of EF-Tu to FN

FIGURE 5: Interaction of EF-Tu and fibronectin characterized by a PC-TIR sensor. Binding of an analyte to a reactant causes resonant wavelength shift that was measured throughout the process as a function of time. (a) Coating the sensor with fibronectin. (i) The sensor surface was calibrated with PBS and (ii) coated with fibronectin (FN at 200 μg/mL in a 200 μL volume of PBS at 25°C). Immobilization of FN onto the sensor caused resonant wavelength shift. (iii) The subsequent washes. (b) Binding of EF-Tu to FN. (I) Detection baseline of the FN functionalized sensor. (II) EF-Tu at 20 μg/mL (blue curve) or 50 μg/mL (red) and a negative control (a portion of proteoglycan 4) at 20 μg/mL (black) were incubated with FN at 25°C for the indicated times. (III) The sensor was washed.

One conspicuous spot was seen, having a size of 43 kDa; the protein identity was determined to be EF-Tu by proteomic analysis as above. Evidently, the fibronectin-EF-Tu complexes were recognized by the fibronectin antibodies.

Binding of rEF-Tu to fibronectin was further characterized by a novel label-free optical sensor (Figure 5). The sensor is based on a photonic-crystal structure in a total-internal-reflection (PC-TIR) configuration. The unique working principle and high sensitivity of the PC-TIR sensor

have been demonstrated by Ye and his colleagues [31–34]. The assays with the PC-TIR sensor encompass two steps: sensor coating and protein binding, each including three phases. For coating as indicated in Figure 5(a), (i) the baseline was calibrated with PBS; (ii) the sensor was coated with fibronectin (200 μg/mL) and the coating was detected by measuring the resonant wavelength shift; (iii) the subsequent washes removed the unbound protein molecules, leading to a minor reduction in the resonance shift, but the substantial

(a) TEM of vesiculation

(b) Model for OMV protein distribution

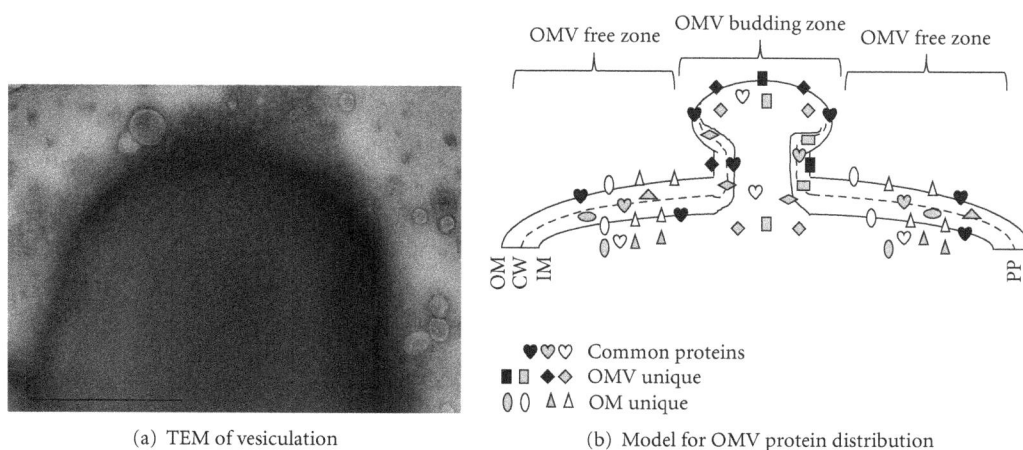

FIGURE 6: EF-Tu delivery model. (a) A cell with budding and released OMVs revealed by TEM. Bar: 250 nm. (b) A model derived from the OM and the OMV subproteomic data as well as the immune TEM observations, depicting an OMV budding upwards (see Discussion for details). OM: outer membrane; CW: cell wall; IM: inner membrane; PP: periplasm. Proteins in three subproteomes (common, OMV unique and OM unique) are indicated therein.

changes still remained indicating the effective coating of the sensor by fibronectin. The binding thickness of fibronectin on the sensor was determined to be 4.8 nm corresponding to the 4.2 nm resonance shift calculated with a transfer matrix method [31]. For protein binding as demonstrated in Figure 5(b) (blue), after (I) the coating baseline was calibrated, (II) EF-Tu was added onto the fibronectin-coated sensor surface. A gradual increase in resonance shift was observed, demonstrating EF-Tu binding to fibronectin. (III) The unbound or weakly bound protein molecules were washed off subsequently. The resonance shift level remained high after wash, indicating strong binding of EF-Tu to fibronectin. When the EF-Tu concentration increased from 20 to 50 μg/mL, the resonant shift level changed from 0.8 nm to 1.1 nm (blue and red in Figure 5(b)). The binding thicknesses of EF-Tu are 0.91 nm and 1.26 nm, respectively. The reason that the binding thickness did not increase as much as the increase of the protein concentrations may be attributed to binding saturations. In a sharp contrast, after addition of the proteoglycan 4 control, resonant shift was not observed; the resonant wavelength returned to the baseline after wash (Black).

3. Discussion

A combination of biological, immunological, biophysical, and proteomic methods was employed to test the hypothesis stating that EF-Tu is associated with OMVs and binds to fibronectin. The results revealed by immune TEM show that EF-Tu was physically associated with the OMVs and the cell surface. Its presence in the OMV and the OM subproteomes was verified by Western blotting and proteomic analyses. EF-Tu carried by OMVs and OM was found to bind to fibronectin as detected with the Western blot- and the PC-TIR-based sensor assays.

3.1. OMV-OM Subproteomes and Possible Mechanisms for EF-Tu Delivery into OMVs.
Our subproteomic analyses provide clues to the mechanisms of EF-Tu delivery into OMVs. We categorized the proteins in the OMV and the OM subproteomes (Table 1). The first consisted of the proteins detected only in the OMV subproteome (Table S1). The second was the common category of the proteins shared by the two subproteomes (Table S2). The third comprised the proteins present only in the OM subproteome (Table S3). While the biological meaning of these subproteomes remains to be deciphered, the current finding implies that there may be the OMV-budding zones from which OMVs bud from OM. The budding might be random as indicated in Figure 6(a). If so, OMVs produced via random budding should have displayed irregular protein distributions; yet this presumption does not appear reconciled with the subproteomic observations. Rather, our data let us suggest the OMV budding zones for protein delivery into OMVs (Figure 6(b)). First, the proteins of the OMV subproteome seem to be located in the OMV budding zones from which OMVs bud out and carry these proteins with OMVs (Table S1). Second, the common proteins seem likely to scatter over the two zones (Table S2). Third, the proteins of the OM subproteome seem to be distributed in the OMV-free zones and so not to be seen in OMVs (Table S3). Based on the premise, EF-Tu that belongs to the OM-OMV common subproteomes seems to scatter over the OMV-budding and the OMV-free zones; it seems to be delivered into OMVs through OMV budding from OM (Figure 6(b)). This model can be used to explain the presence of EF-Tu in the OM, the OMVs, and the CFC fractions. Understandable is the absence of the OM-only proteins (e.g., #78, 43 kDa glucose-sensitive porin; #91, 46 kDa urocanase; #96, putative aromatic compound porin; #109, 43 kDa l-sorbosone dehydrogenase in Table S3) from the OMVs (Table S2) and the CFC fractions. In *Bacillus subtilis*, EF-Tu localizes underneath the cell membrane, colocalizing and

interacting with MreB, an actin-like cytoskeletal element that plays a role in cell shape maintenance [35]. These predictions may stimulate future studies for their verification.

3.2. EF-Tu and Other Cytosolic Proteins in OMV and OM Subproteomes. One of the intriguing observations is the presence of cytosolic proteins in the OMV and the OM subproteomes, for example, DNA binding proteins (#18, 26, 45) in OMV and EF-Tu (#57) in the common subproteome. Considering OM's hydrophobic nature, we were tempted to suspect cytosolic protein contamination. Nevertheless, detection of DNA-binding proteins in OMVs seems unlikely to be attributed to contamination, as DNA-binding proteins and DNA were detected with in *N. gonorrhoeae* OMVs [36, 37]. Since DNA was detected in *A. baumannii* OMVs (Figure S2), it seems possible that the proteins are hitched by DNA into OMVs or vice versa. Moreover, the presence of EF-Tu in both OMV and OM subproteomes is not just coincident but consistently documented [23]. Its presence in OM did not result from artificial binding during membrane preparation [13]. EF-Tu was detected in both subproteomes of multiple species [23]. EF-Tu was found in OM fractions of *A. baumannii* [24]. It also was present in OMVs of *N. meningitides* [38, 39] and *E. coli* [40]. However, a former proteomic analysis did not detect EF-Tu in the *A. baumannii* OMV fraction [25]. This discrepancy with ours may be due to different strains and growth conditions used in their and our studies. Kwon et al. used *A. baumannii* from clinical isolates, but we employed the standard ATCC19606 strain. They grew the culture under shaking condition while we used plates. While a combination of physical, immunological, and biochemical evidence appears to be convincing, we plan to compare data acquired from the standard and the clinical strains concerning the discrepancy.

3.3. Implications for Binding of A. baumannii EF-Tu to Fibronectin: A. baumannii EF-Tu Bound to Fibronectin (Figures 4 and 5). The role of the binding seems to be intriguing. EF-Tus of *L. johnsonii* [17] and *P. aeruginosa* [18] are involved in bacterial attachment to human cells. Particularly, EF-Tu is involved in bacterial infection of human monocyte-like cells via binding to the cell-surface-associated nucleolin [41]. Given that fibronectin was found to bind to macrophage as documented [42] and that *A. baumannii* EF-Tu was detected in OMVs and on the bacterial cell surface, these data seem to support a notion that *A. baumannii* EF-Tu contributes to mediating adhesion of the bacterial cells and OMVs to macrophages through binding to fibronectin on the host cells, a hypothesis to be tested in the future.

4. Experimental Procedures

4.1. Expression and Purification of EF-Tu. The gene *tufAa* encoding EF-Tu of *A. baumannii* was cloned by following the manufacturer instruction (Novagen, San Diego, CA, USA). Since the genome of the *A. baumannii* 19606 strain was not available when we conducted this study, *tufAa* of *A. baumannii* 17987 was used for designing specific primers

and PCR was performed with the 19606 DNA template. Forward primer: 5′-GACGACGACAAGATGATGGCTAAA-GCCAAG-3′ and the reverse primer: 5′-GAGGAGAAG-CCCGGTCCGTCACTATATTATGCTTATGC-3′. The PCR product (1,191 bp) was cloned into the pTriEX-4 Ek/LIC expression vector (Novagen, San Diego, CA, USA), which added an N-terminal hexa-histidine (6xHis) for purification by affinity chromatography and S-tag. The positive recombinant EF-Tu (rEF-Tu) clone DNA was verified by DNA sequencing analysis. The expressed rEF-Tu was fused with N-terminus His-S-tags containing 48 amino acids and purified by nickel affinity chromatography under native conditions by following the manufacturer protocol (Qiagen). The purified protein was analyzed with SDS-PAGE to determine the presence of rEF-Tu and was confirmed further by immunoblot by using monoclonal antibody to His-tag. The identity of rEF-Tu was further confirmed by N-terminal microsequencing as we described before [9].

4.2. Preparation of Antibody against rEF-Tu. The antiserum specific to rEF-Tu was developed by ProSci Incorporated. Briefly, rabbits were injected subcutaneously with 100–200 μg rEF-Tu with complete Freund's adjuvant. Individual rabbits were boosted 3 times with the same amount of antigen in incomplete Freund's adjuvant at intervals of 21 days. Serum samples were collected and used in immunological studies. ProSci Incorporated provided us with sera prior to immunization and 3 bleed collected serums.

4.3. Isolation of OMV and OMPs. *A. baumannii* (ATCC19606) OMVs were isolated from LB agar plates as described previously [43] with modifications. Colonies grown on LB agar plates were scraped off and suspended in PBS with gentle agitation to $OD_{600\,nm}$ of 5. Then, bacteria were collected by low-speed centrifugation (6,000 g) for 5 min, and the recovered supernatant was centrifuged at 12,000 g for 10 min and further passed through the 0.2-μm pore size filters (Millipore). OMVs in the supernatant were then collected by ultra centrifugation at 100,000 g for 12 hours at 4°C and resuspended in PBS. OMPs were extracted according to Caldwell et al. [44].

4.4. TEM and Immunogold TEM. TEM was conducted according to a standard protocol [25, 45]. For immunogold TEM, *A. baumannii* cells and purified OMVs were immunogold-labeled with the anti-EF-Tu antibodies as described previously with some modifications [30]. Briefly, the cells and OMVs were incubated with 100 mM Tris-HCl buffer (pH 7.5) containing 1% bovine serum albumin (BSA) supplemented with 1% heat inactivated goat serum (buffer A) to reduce nonspecific binding. They were incubated with anti-EF-Tu diluted 1 : 100 in buffer A at 37°C for 2 hrs, washed with buffer A, and incubated with goat anti-rabbit immunoglobulin-G- (IgG-) gold complex (average size particle, 10 nm, 1 : 20 dilution) suspended in PBS containing 1% BSA (buffer B). After sequential washing with buffer B, PBS, and deionized water, bacterial cells and

OMVs were mounted onto Holey carbon film nickel grids by fixing with 1% glutaraldehyde-4% formaldehyde for 20 min at room temperature. Grids were stained with 7% uranyl acetate followed by Reynolds lead citrate for TEM.

4.5. Western Blot. Proteins from OMV and OM fractions were separated on 10% SDS-PAGE [46] and stained by Coomassie blue. The proteins were transferred electrophoretically onto nitrocellulose membranes [47]. The membrane strips were incubated with the *A. baumannii* anti-rEF-Tu antibodies at a dilution of 1 : 3000 in 1% (w/v) blotto for 2 hrs at 25°C. The membrane strips then were washed, incubated in alkaline phosphatase-conjugated goat anti-rabbit antibodies (Santa Cruz Biotechnologies, Santa Cruz, CA), at a dilution of 1 : 5000 in 1% (w/v) blotto for 1 hr at 25°C, washed, and developed with 5-bromo-4-chloro-3-indolyl phosphate/Nitroblue Tetrazolium (BCIP/NBT, Sigma). For 2D gel electrophoresis, proteins from OMV and OM fractions were solubilized for iso-electric focusing (IEF) in 8 M urea, 2% CHAPS, 2% IPG buffer (Amersham Biosciences), 20 mM DTT, and traces of bromophenol blue. Sample was loaded on a 13 cm Immobiline DryStrip pH 3–10 (Amersham Biosciences) by rehydration and left under oil overnight. IEF was conducted under 17,000 Vh by using the Multiphor II (Amersham Biosciences) at 20°C. The IEF strips were equilibrated in Equilibration solution (0.05 M Tris-HCl, 0.4 M urea, 30% glycerol, 1% SDS (w/v), and 0.02 M DTT) for 10 minutes. SDS-PAGE (10% w/v) was conducted as previously [46]. After separation of the supernatant proteins by electrophoresis, the proteins in the gels were transferred electrophoretically onto nitrocellulose membranes [47]. Membranes were blocked and incubated with the anti-rEF-Tu at a dilution of 1 : 5000 in 1% blotto for 1 hr, washed, and developed with Sigma FAST BCIP/NBT solution.

4.6. OM-OMV Subproteomic Analyses. The analysis was performed as described previously [45, 48]. Briefly, OMPs were isolated as above, and OMV proteins were extracted by resuspending in 2% (w/v) SDS and 100 mmol l^{-1} DTT and incubating at 25°C for 5 min. The proteins (30 μg) were fractionated by SDS-PAGE (10%, w/v). After staining, the proteins-containing lane in replicate was sliced into pieces (1 × 1 mm) for *in vitro* trypsin proteolysis. Capillary liquid chromatography-tandem mass spectrometry (LC/MS/MS) was performed, and the peptides derived from the proteins in the gel slices were determined with a linear ion trap tandem mass spectrometer in which the top 7 eluting ions were fragmented by collision-induced dissociation. Proteins were identified by following a standard protocol [49], in which MS/MS spectra were searched against the NCBI nonredundant protein database (version 20100306; 10551781 sequences and 3596151245 residues) with a probability-based database searching algorithm (Mascot, Matrix Science). A score of each peptide entry was calculated by the molecular weight search (MOWSE) peptide-mass database developed previously [29] and the scoring algorithm. The significance threshold was set for $P \leq 0.05$ in a search for random matches, and the proteins consistently detected in the replicates were counted.

4.7. Binding of Released EF-Tu to Fibronectin. Nitrocellulose membranes transferred with proteins from OMV and OM fractions were blocked, washed, and incubated with pure fibronectin (Sigma, human plasma fibronectin) at 10 μg/mL for 24 hr at 4°C. Then, the membranes were washed and incubated with the rabbit antifibronectin antibodies at 1 : 10,000 dilution in 1% blotto for 2 hr at room temperature. Subsequently, the membranes were washed and incubated with the alkaline-phosphatase-conjugated goat anti-rabbit antibodies at a dilution of 1 : 20,000 in 1% blotto for 1 hr, washed, and developed as above.

4.8. Functionalization of PC-TIR Sensor. The sensor originates from a prototype of a photonic crystal structure in a total-internal-reflection (PC-TIR) configuration [33]. The design and fabrication of the PC-TIR sensor was reported by Ye and his coworkers [31, 32, 34]. Briefly, a photonic crystal (PC) structure with five alternating layers of silica and titania was fabricated on a transparent BK7 glass substrate by electron beam physical vapor deposition. A thin film of poly(methyl methacrylate) (PMMA) (A6, MicroChem) was spin-coated on the top of the structure at 500 rpm for 10 sec, followed by 4,200 rpm for 45 sec. The sensor chip was baked at 60°C for 30 minutes. Two sample wells were fabricated with polydimethylsiloxane and placed in a tight contact with the top surface of the sensor chip. One is used as the reference channel and the other as the signal channel. Both wells were filled with PBS, and the resonant wavelengths of the two channels were recorded to establish the detection baseline. Fibronectin was immobilized on the chip surface [50] through physical absorption by directly adding fibronectin (200 μL, 200 μg/mL) on the sensor chip followed by incubation at 25°C for the indicated time as in Figure 5. Shift in resonant wavelength was measured and recorded. The sensor surface was washed with PBS twice and refilled with 200 μL PBS for measuring the thickness of bound protein. Finally, PBS was replaced with the analyte solution of EF-Tu or the proteoglycan 4 control in PBS (100 nM, 200 μL). For the preparation of the control, the DNA (1.2 kb) encoding a portion of proteoglycan 4 was cloned; the polypeptide (44 kDa) was expressed and purified from baboon temporomandibular joint cells; it was tested negative in fibronectin binding by the same antibody-based assay as shown in Figure 5 (unpublished data by Jennifer McDaniel Schulze et al.).

Acknowledgments

The authors thank Dr. JM. Schulze for assistance with preparation of recombinant EF-Tu and the control protein; Dr. RA. Esparza-Munoz for TEM; Mr. V. Pericherla and the RCMI Proteomics and Protein Biomarkers Cores at UTSA (NIH 2G12RR013646-11) for assistance with the proteomic experiment design, the sample preparation, the data collection; the Computational Biology Initiative (UTSA/UTHSCSA) for providing access and training to the analysis software used; Mr. E. Rodriguez for protein data search; Dr. R. LeBaron for kind support. This work was supported by San Antonio Area Foundation and UTSA Collaborative Research Seed Grant Program and partly supported by San Antonio Life Sciences Institute (SALSI) funding.

References

[1] S. F. Dallo and T. Weitao, "Insights into acinetobacter war-wound infections, biofilms, and control," *Advances in skin & Wound Care*, vol. 23, no. 4, pp. 169–174, 2010.

[2] L. Dijkshoorn, A. Nemec, and H. Seifert, "An increasing threat in hospitals: multidrug-resistant Acinetobacter baumannii," *Nature Reviews Microbiology*, vol. 5, no. 12, pp. 939–951, 2007.

[3] J. Garnacho-Montero and R. Amaya-Villar, "Multiresistant Acinetobacter baumannii infections: epidemiology and management," *Current Opinion in Infectious Diseases*, vol. 23, no. 4, pp. 332–339, 2010.

[4] D. E. Karageorgopoulos and M. E. Falagas, "Current control and treatment of multidrug-resistant Acinetobacter baumannii infections," *The Lancet Infectious Diseases*, vol. 8, no. 12, pp. 751–762, 2008.

[5] K. M. Hujer, A. M. Hujer, E. A. Hulten et al., "Analysis of antibiotic resistance genes in multidrug-resistant Acinetobacter sp. isolates from military and civilian patients treated at the Walter Reed Army Medical Center," *Antimicrobial Agents and Chemotherapy*, vol. 50, no. 12, pp. 4114–4123, 2006.

[6] H. van Dessel, L. Dijkshoorn, T. Van Der Reijden et al., "Identification of a new geographically widespread multiresistant Acinetobacter baumannii clone from European hospitals," *Research in Microbiology*, vol. 155, no. 2, pp. 105–112, 2004.

[7] O. Öncül, Ö. Keskin, H. V. Acar et al., "Hospital-acquired infections following the 1999 Marmara earthquake," *Journal of Hospital Infection*, vol. 51, no. 1, pp. 47–51, 2002.

[8] Y. Wang, P. Hao, B. Lu et al., "Causes of infection after earthquake, China, 2008," *Emerging Infectious Diseases*, vol. 16, no. 6, pp. 974–975, 2010.

[9] S. F. Dallo, J. Denno, S. Hong, and T. Weitao, "Adhesion of acinetobacter baumannii to extracellular proteins detected by a live cell-protein binding assay," *Ethnicity & Disease*, vol. 20, no. 1, supplement 1, pp. S1–S7, 2010.

[10] M. Baensch, R. Frank, and J. Köhl, "Conservation of the amino-terminal epitope of elongation factor Tu in eubacteria and archaea," *Microbiology*, vol. 144, no. 8, pp. 2241–2246, 1998.

[11] J. M. Ogle and V. Ramakrishnan, "Structural insights into translational fidelity," *Annual Review of Biochemistry*, vol. 74, pp. 129–177, 2005.

[12] S. Weber, F. Lottspeich, and J. Kohl, "An epitope of elongation factor Tu is widely distributed within the bacterial and archaeal domains," *Journal of Bacteriology*, vol. 177, no. 1, pp. 11–19, 1995.

[13] M. Dombou, S. V. Bhide, and S. Mizushima, "Appearance of elongation factor Tu in the outer membrane of sucrose-dependent spectinomycin-resistant mutants of *Escherichia coli*," *European Journal of Biochemistry*, vol. 113, no. 2, pp. 397–403, 1981.

[14] R. C. Judd and S. F. Porcella, "Isolation of the periplasm of Neisseria gonorrhoeae," *Molecular Microbiology*, vol. 10, no. 3, pp. 567–574, 1993.

[15] K. Otto, J. Norbeck, T. Larsson, K. A. Karlsson, and M. Hermansson, "Adhesion of type 1-fimbriated *Escherichia coli* to

[16] R. A. Brady, J. G. Leid, A. K. Camper, J. W. Costerton, and M. E. Shirtliff, "Identification of Staphylococcus aureus proteins recognized by the antibody-mediated immune response to a biofilm infection," *Infection and Immunity*, vol. 74, no. 6, pp. 3415–3426, 2006.

[17] D. Granato, G. E. Bergonzelli, R. D. Pridmore, L. Marvin, M. Rouvet, and I. E. Corthésy-Theulaz, "Cell surface-associated elongation factor Tu mediates the attachment of lactobacillus johnsonii NCC533 (La1) to human intestinal cells and mucins," *Infection and Immunity*, vol. 72, no. 4, pp. 2160–2169, 2004.

[18] A. Kunert, J. Losse, C. Gruszin et al., "Immune evasion of the human pathogen *Pseudomonas aeruginosa*: elongation factor Tuf is a factor H and plasminogen binding protein," *Journal of Immunology*, vol. 179, no. 5, pp. 2979–2988, 2007.

[19] C. Zipfel, G. Kunze, D. Chinchilla et al., "Perception of the bacterial PAMP EF-Tu by the receptor EFR restricts agrobacterium-mediated transformation," *Cell*, vol. 125, no. 4, pp. 749–760, 2006.

[20] C. Archambaud, E. Gouin, J. Pizarro-Cerda, P. Cossart, and O. Dussurget, "Translation elongation factor EF-Tu is a target for Stp, a serine-threonine phosphatase involved in virulence of Listeria monocytogenes," *Molecular Microbiology*, vol. 56, no. 2, pp. 383–396, 2005.

[21] G. Kunze, C. Zipfel, S. Robatzek, K. Niehaus, T. Boller, and G. Felix, "The N terminus of bacterial elongation factor Tu elicits innate immunity in Arabidopsis plants," *Plant Cell*, vol. 16, no. 12, pp. 3496–3507, 2004.

[22] N. I. A. Carlin, S. Lofdahl, and M. Magnusson, "Monoclonal antibodies specific for elongation factor Tu and complete nucleotide sequence of the tuf gene in Mycobacterium tuberculosis," *Infection and Immunity*, vol. 60, no. 8, pp. 3136–3142, 1992.

[23] E. Y. Lee, D. S. Choi, K. P. Kim, and Y. S. Gho, "Proteomics in Gram-negative bacterial outer membrane vesicles," *Mass Spectrometry Reviews*, vol. 27, no. 6, pp. 535–555, 2008.

[24] S. Martí, J. Sánchez-Céspedes, E. Oliveira, D. Bellido, E. Giralt, and J. Vila, "Proteomic analysis of a fraction enriched in cell envelope proteins of Acinetobacter baumannii," *Proteomics*, vol. 6, pp. S82–S87, 2006.

[25] S. O. Kwon, Y. S. Gho, J. C. Lee, and S. I. Kim, "Proteome analysis of outer membrane vesicles from a clinical Acinetobacter baumannii isolate," *FEMS Microbiology Letters*, vol. 297, no. 2, pp. 150–156, 2009.

[26] M. G. Smith, T. A. Gianoulis, S. Pukatzki et al., "New insights into Acinetobacter baumannii pathogenesis revealed by high-density pyrosequencing and transposon mutagenesis," *Genes and Development*, vol. 21, no. 5, pp. 601–614, 2007.

[27] A. M. Zuurmond, A. K. Rundlöf, and B. Kraal, "Either of the chromosomal tuf genes of E. coli K-12 can be deleted without loss of cell viability," *Molecular and General Genetics*, vol. 260, no. 6, pp. 603–607, 1999.

[28] Y. T. N. Yu and L. Snyder, "Translation elongation factor Tu cleaved by a phage-exclusion system," *Proceedings of the National Academy of Sciences of the United States of America*, vol. 91, no. 2, pp. 802–806, 1994.

[29] D. J. C. Pappin, P. Hojrup, and A. J. Bleasby, "Rapid identification of proteins by peptide-mass fingerprinting," *Current Biology*, vol. 3, no. 6, pp. 327–332, 1993.

[30] S. F. Dallo, T. R. Kannan, M. W. Blaylock, and J. B. Baseman, "Elongation factor Tu and E1 β subunit of pyruvate

dehydrogenase complex act as fibronectin binding proteins in Mycoplasma pneumoniae," *Molecular Microbiology*, vol. 46, no. 4, pp. 1041–1051, 2002.

[31] Y. Guo, C. Divin, A. Myc et al., "Sensitive molecular binding assay using a photonic crystal structure in total internal reflection," *Optics Express*, vol. 16, no. 16, pp. 11741–11749, 2008.

[32] Y. Guo, J. Y. Ye, C. Divin et al., "Real-time biomolecular binding detection using a sensitive photonic crystal biosensor," *Analytical Chemistry*, vol. 82, no. 12, pp. 5211–5218, 2010.

[33] J. Y. Ye, Y. Guo, T. B. Norris, and J. R. Baker Jr., Novel Photonic Crystal Sensor, patent No 7,639,362 patent 7,639,362, 2009.

[34] J. Y. Ye and M. Ishikawa, "Enhancing fluorescence detection with a photonic crystal structure in a total-internal-reflection configuration," *Optics Letters*, vol. 33, no. 15, pp. 1729–1731, 2008.

[35] H. J. D. Soufo, C. Reimold, U. Linne, T. Knust, J. Gescher, and P. L. Graumann, "Bacterial translation elongation factor EF-Tu interacts and colocalizes with actin-like MreB protein," *Proceedings of the National Academy of Sciences of the United States of America*, vol. 107, no. 7, pp. 3163–3168, 2010.

[36] D. W. Dorward and C. F. Garon, "DNA-binding proteins in cells and membrane blebs of Neisseria gonorrhoeae," *Journal of Bacteriology*, vol. 171, no. 8, pp. 4196–4201, 1989.

[37] D. W. Dorward, C. F. Garon, and R. C. Judd, "Export and intercellular transfer of DNA via membrane blebs of Neisseria gonorrhoeae," *Journal of Bacteriology*, vol. 171, no. 5, pp. 2499–2505, 1989.

[38] D. M. B. Post, D. Zhang, J. S. Eastvold, A. Teghanemt, B. W. Gibson, and J. P. Weiss, "Biochemical and functional characterization of membrane blebs purified from Neisseria meningitidis serogroup B," *The Journal of Biological Chemistry*, vol. 280, no. 46, pp. 38383–38394, 2005.

[39] C. Vipond, J. Suker, C. Jones, C. Tang, I. M. Feavers, and J. X. Wheeler, "Proteomic analysis of a meningococcal outer membrane vesicle vaccine prepared from the group B strain NZ98/254," *Proteomics*, vol. 6, no. 11, pp. 3400–3413, 2006.

[40] E. Y. Lee, Y. B. Joo, W. P. Gun et al., "Global proteomic profiling of native outer membrane vesicles derived from *Escherichia coli*," *Proteomics*, vol. 7, no. 17, pp. 3143–3153, 2007.

[41] M. Barel, A. G. Hovanessian, K. Meibom, J. P. Briand, M. Dupuis, and A. Charbit, "A novel receptor—ligand pathway for entry of Francisella tularensis in monocyte-like THP-1 cells: interaction between surface nucleolin and bacterial elongation factor Tu," *BMC Microbiology*, vol. 8, article 145, 2008.

[42] D. R. Schmidt and W. J. Kao, "The interrelated role of fibronectin and interleukin-1 in biomaterial-modulated macrophage function," *Biomaterials*, vol. 28, no. 3, pp. 371–382, 2007.

[43] T. Henry, S. Pommier, L. Journet, A. Bernadac, J. P. Gorvel, and R. Lloubès, "Improved methods for producing outer membrane vesicles in Gram-negative bacteria," *Research in Microbiology*, vol. 155, no. 6, pp. 437–446, 2004.

[44] H. D. Caldwell, J. Kromhout, and J. Schachter, "Purification and partial characterization of the major outer membrane protein of Chlamydia trachomatis," *Infection and Immunity*, vol. 31, no. 3, pp. 1161–1176, 1981.

[45] R. Maredia, N. Devineni, P. Lentz et al., "Vesiculation from *Pseudomonas aeruginosa* under SOS," *The Scientific World Journal*, vol. 2012, Article ID 402919, 18 pages, 2012.

[46] U. K. Laemmli, "Cleavage of structural proteins during the assembly of the head of bacteriophage T4," *Nature*, vol. 227, no. 5259, pp. 680–685, 1970.

[47] H. Towbin, T. Staehelin, and J. Gordon, "Electrophoretic transfer of proteins from polyacrylamide gels to nitrocellulose sheets: procedure and some applications," *Proceedings of the National Academy of Sciences of the United States of America*, vol. 76, no. 9, pp. 4350–4354, 1979.

[48] M. Wilm, A. Shevchenko, T. Houthaeve et al., "Femtomole sequencing of proteins from polyacrylamide gels by nano-electrospray mass spectrometry," *Nature*, vol. 379, no. 6564, pp. 466–469, 1996.

[49] J. N. Williams, P. J. Skipp, H. E. Humphries, M. Christodoulides, C. D. O, and J. E. Heckels, "Proteomic analysis of outer membranes and vesicles from wild-type serogroup B Neisseria meningitidis and a lipopolysaccharide-deficient mutant," *Infection and Immunity*, vol. 75, no. 3, pp. 1364–1372, 2007.

[50] P. E. Vaudaux, F. A. Waldvogel, J. J. Morgenthaler, and U. E. Nydegger, "Adsorption of fibronectin onto polymethylmethacrylate and promotion of Staphylococcus aureus adherence," *Infection and Immunity*, vol. 45, no. 3, pp. 768–774, 1984.

9

Antibiotic Resistant *Salmonella* and *Vibrio* Associated with Farmed *Litopenaeus vannamei*

Sanjoy Banerjee,[1] Mei Chen Ooi,[2] Mohamed Shariff,[1, 2] and Helena Khatoon[1]

[1] *Institute of Bioscience, Universiti Putra Malaysia, Selangor, 43400 Serdang, Malaysia*
[2] *Faculty of Veterinary Medicine, Universiti Putra Malaysia, Selangor, 43400 Serdang, Malaysia*

Correspondence should be addressed to Mohamed Shariff, pshariff@gmail.com

Academic Editors: J. Qiu and G. Salvat

Salmonella and *Vibrio* species were isolated and identified from *Litopenaeus vannamei* cultured in shrimp farms. Shrimp samples showed occurrence of 3.3% of *Salmonella* and 48.3% of *Vibrio*. The isolates were also screened for antibiotic resistance to oxolinic acid, sulphonamides, tetracycline, sulfamethoxazole/trimethoprim, norfloxacin, ampicillin, doxycycline hydrochloride, erythromycin, chloramphenicol, and nitrofurantoin. *Salmonella enterica* serovar Corvallis isolated from shrimp showed individual and multiple antibiotic resistance patterns. Five *Vibrio* species having individual and multiple antibiotic resistance were also identified. They were *Vibrio cholerae* (18.3%), *V. mimicus* (16.7%), *V. parahaemolyticus* (10%), *V. vulnificus* (6.7%), and *V. alginolyticus* (1.7%). Farm owners should be concerned about the presence of these pathogenic bacteria which also contributes to human health risk and should adopt best management practices for responsible aquaculture to ensure the quality of shrimp.

1. Introduction

Shrimp is an important commodity in the global fishery trade due to its increasing demand and competitive international price [1]. As a result of rising shrimp exports, traditional shrimp farming which began in Malaysia in the 1930s has given way to intensive farming system. At present, the bulk of the production consists of *Litopenaeus vannamei* having a production of 52,926 tonnes [2]. It has been introduced in the early 2000 due to the advantages in terms of disease management and most widely cultured in intensive system throughout Malaysia for local consumption as well as for export.

With the change to intensive culture system having high stocking density, disease problems appear frequently causing heavy economic losses to the industry. Antibiotics are normally used to prevent or treat disease outbreaks in shrimp farming [3]. However, extensive use of antibiotics in shrimp farming can cause the development of antibiotic-resistant pathogens which can infect both cultured animals as well as humans [4, 5].

Shrimp intended for export have to meet the bacteriological standards of the importing countries. *Salmonella* and *Vibrio* species are important foodborne pathogens and most importing countries do not accept them in raw frozen shrimp. Contamination of tropical shrimp with *Salmonella* due to growth in polluted waters has been a problem in many parts of the world and is reported to be a part of the natural population of the brackishwater cultured shrimp [6]. In the United States, most *Salmonella* contamination problems in seafood were in shrimp [7]. In addition, opportunistic *Vibrio* spp. is the most common bacterial pathogen found in shrimp which can cause lethal infections following primary infections with other pathogens, environmental stress, nutritional imbalance, and/or predisposing lesions [8]. In humans, *Vibrio* spp. are known to cause gastroenteritis, cholera, and septicemia [9].

Therefore, the present study was carried out in major shrimp producing areas to identify the incidence of *Salmonella* and *Vibrio* in *Litopenaeus vannamei* cultured in commercial shrimp farms and their resistance to some of the commonly used antibiotics in aquaculture.

TABLE 1: Characteristics of the three farms from where shrimp and water were sampled.

	Farm 1	Farm 2	Farm 3
Average pond size (ha)	0.4	0.4	0.4
Stocking density (PL/m²)	80	80	80–100
Age of farm (years)	13	16	10
Number of ponds	20	14	52
Source of water	Brackish water river	Brackish water river	Brackish water sea
Use of antibiotics	No	No	No
Use of probiotics	No	Yes	Yes
Source of feed	Direct-feed miller	Direct-feed miller	Direct-feed miller
Nearby farms	No	Chicken farm	Shrimp farm
Market	Local	Local, export	Local, export

2. Materials and Methods

2.1. Sampling. Clinically healthy with no external lesions or clinical signs and alive *L. vannamei* were collected from three farms from growout ponds (80–120 days; total 6 ponds) situated in Carey Island (2°52′0″ N, 101°22′0″ E) and Kuala Selangor (3°21′0″ N, 101°15′0″ E), Malaysia. The live shrimp were transported in oxygenated pond water filled plastic bags and put into another plastic bag filled with ice flakes and placed in styrofoam boxes. Water samples from ponds were collected in sterile bottles and transported in ice. Samples were transported to the Aquatic Animal Health Unit in Universiti Putra Malaysia and processed immediately for examination. The three farms had well laid out ponds and had characteristics as shown in Table 1.

2.2. Bacterial Isolation

2.2.1. Shrimp Sample. For isolation of *Salmonella*, the head and exoskeleton of shrimp were removed aseptically. Muscle and intestine (10 g) were separately taken aseptically from the shrimp, placed in individual sterile test tubes, and homogenized in 3 mL autoclaved seawater using stomacher for 1 m. The homogenized sample was incubated in buffered peptone water at 37°C for 24 h for preenrichment. Selective enrichment was done using 10 mL of Rappaport and Vassiliadis (RVS) broth (Merck, Germany) inoculated with 1 mL culture from buffered peptone water and incubated at 42°C for 24–48 h. A loopful of sample from RVS was then streaked on selective media xylose-lysine-tergitol 4 (XLT-4) (Merck, Germany) and brilliant-green phenol-red lactose sucrose (BPLS) agar (Merck, Germany) and incubated at 37°C for 24–48 h. Subculture on XLT-4 and BPLS was done to obtain pure culture. *Salmonella* spp. colonies appear black or black-centered on XLT-4 and red centered pink on BPLS.

For isolation of *Vibrio* from shrimp, one loopful of haemolymph was taken from hepatopancreas and cultured in thiosulphate citrate bile salt sucrose (TCBS) agar (Merck, Germany) with 3% sodium chloride (NaCl) (Merck, Germany). The culture was incubated at 25°C for 18 to 24 h. Yellow and green single colonies were subcultured on TCBS and tryptic soy agar (TSA) (Merck, Germany) with 3% NaCl at 25°C for 24 h to obtain pure culture.

2.2.2. Water Sample. Pond water (10 mL) was diluted with 90 mL buffered peptone water and processed according to shrimp samples for isolation of *Salmonella*. In the case of *Vibrios*, pond water was plated on TCBS agar using the spread plate technique and processed according to shrimp samples.

2.3. Bacterial Identification. Different biochemical tests such as triple sugar iron (TSI), urease test, lysine iron agar (LIA), sulfide indole motility (SIM), slide agglutination test, and *Salmonella* serotyping were performed for identification of *Salmonella* spp.

Vibrio species identification was done using API 20E (bioMérieux, France) identification system. *Escherichia coli* (ATCC 25922) was used as control. Gram staining was also done from single pure culture colony.

2.4. Antibiotic Sensitivity Test. Antibiotic susceptibility test was conducted using disc diffusion method on Mueller Hinton agar at 37°C for 24 h. Procedure was based on the standardized disc agar diffusion method of the National Committee for Clinical Laboratory Standards for antimicrobial susceptibility tests [10]. After incubation, the diameter of the zone of inhibition was measured and compared with BBL zone interpretative chart to determine the sensitivity of the isolates to the antibiotics. The BBL zone interpretative chart was used in the absence of standard interpretative scheme for environmental isolates or for shrimp pathogens. The antimicrobials oxolinic acid 2 μg (OA 2), compound sulphonamides 300 μg (S3 300), tetracycline 30 μg (TE 30), sulfamethoxazole 23.75 μg/trimethoprim 1.25 μg (SXT 25), norfloxacin 10 μg (NOR 10), ampicillin 10 μg (AMP 10), doxycycline hydrochloride 30 μg (DO 30), and erythromycin 15 μg (E 15) were selected as they are veterinary important antimicrobials. Chloramphenicol 30 μg (C 30) and nitrofurantoin 300 μg (F 300) were also included since they were in use few years ago. The antimicrobials were from Oxoid, UK.

3. Results

3.1. Bacterial Isolation and Identification

3.1.1. Shrimp and Water Samples. One *Salmonella* and five *Vibrio* spp. were isolated and identified from shrimp and

TABLE 2: Percentage (%) of *Salmonella* and *Vibrio* spp. in three farms isolated from shrimp and water.

Isolated bacteria	Farm 1		Farm 2		Farm 3	
	Shrimp (*n* = 60)	Water	Shrimp (*n* = 60)	Water	Shrimp (*n* = 60)	Water
Salmonella enterica Serovar Corvallis	0	Yes	10		3.3	
Vibrio mimicus	15	Yes	40	Yes	10	
Vibrio vulnificus	5		20		3.3	Yes
Vibrio cholera	0	Yes	60	Yes	16.67	Yes
Vibrio parahaemolyticus	0		0		20	Yes
Vibrio alginolyticus	0		0		3.3	

Isolates from water did not state percentage as only one sample was taken per pond to determine if the bacteria isolated came from water.

water samples. *Salmonella enterica* serovar Corvallis was isolated from water sample in Farm 1 and shrimp samples from Farm 2 and 3. However, five *Vibrio* spp., namely, *V. alginolyticus*, *V. cholera*, *V. mimicus*, *V. parahaemolyticus*, and *V. vulnificus* were isolated from both shrimp and water samples from all the three farms (Table 2). Farm 1 and 3 had low occurrence of *Vibrio* spp., while Farm 2 had moderate to high occurrence. The overall occurrence of *Salmonella* and *Vibrio* in shrimp was 3.33% and 48.3%, respectively.

3.2. Antibiotic Sensitivity Test. The *Salmonella enterica* serovar Corvallis isolates from water in Farm 1 and shrimp in Farm 3 were found to be resistant to erythromycin only. However, the two *S. enterica* serovar Corvallis isolates from shrimp in Farm 2 were found to be resistant to all antibiotics except for nitrofurantoin and norfloxacin (Table 3).

In Farm 1, *V. vulnificus* and *V. mimicus* (S1, S2) isolated from shrimp were resistant to AMP 10. On the other hand, *V. mimicus* (S1, S2) isolated from water was resistant to SXT 25, S 300, and AMP 10, whereas *V. cholerae* (S1, S2) was resistant to AMP 10 only.

In case of Farm 2, *V. cholerae* (S1, S3) isolated from shrimp were resistant to DO 30, AMP 10, and TE 30, *V. cholerae* (S2) to DO 30 and TE 30, whereas *V. cholerae* (S4, S5, S6) were resistant to AMP 10. *V. cholerae* isolated from water was resistant to AMP 10 only. Besides, *V. mimicus* isolated from shrimp and water, and *V. vulnificus* isolated from shrimp were found to be resistant to AMP 10 also. In Farm 3, all the vibrios isolated from shrimp and water were resistant to AMP 10 (Table 3).

All *Vibrio* isolates except for two were resistant to ampicillin. Tetracycline was the second highest and doxycycline was the third highest that the *Vibrio* spp. were resistant to (Table 3). Farm 1 and 2 were found to have more antibiotic-resistant patterns (one to four antibiotics) than Farm 3 (one to two antibiotics). Out of the five *Vibrio* species, *V. cholerae* showed the most antibiotic-resistant pattern.

4. Discussion

Results of this survey showed that *Salmonella* and *Vibrio* isolated and identified from the three shrimp farms are a serious cause for concern since they are of public health significance.

Salmonella is facultative anaerobes and belongs to the family Enterobacteriaceae, and more than 2500 serovars of *Salmonella* are considered potential pathogens in animal and human. Many *Vibrio* spp. are pathogenic to humans and have been implicated in foodborne disease.

Several studies have been done on prevalence of *Salmonella* in the tropics [11–13]. In the present study, *Salmonella* was found in water samples from Farm 1 and shrimp samples from Farm 2 and 3. This is in accordance with the studies where *Salmonella* have been reported from shrimp pond water [1, 12, 14] and shrimp [11, 12, 15]. Studies by Iyer and Varma [16], Bhaskar et al. [11], and Wan Norhana et al. [12] emphasized that *Salmonella* is natural part of the microflora of the shrimp culture practice. However, the absence of *Salmonella* from water and shrimp samples in some farms leading to a low occurrence in the present study could mean that *Salmonella* is not a common normal flora in shrimp culture environment. This is in accordance with a study done in Thailand by Dalsgaard et al. [17] who reported the absence of *Salmonella* from shrimp, sediment, water, and pelleted feed. Study by Koonse et al. [14] also showed that *Salmonella* is not part of the natural flora of the shrimp culture environment or naturally present in shrimp growout ponds. It is related to the concentration of fecal bacteria in the source of water supply to the growout pond water. In the present survey, the water source for two of the shrimp farms were from Langat River. The Langat River is one of the principal rivers draining a densely populated and developed area of Selangor. The major pollution sources affecting Langat River are sewage treatment plants, manufacturing industries not equipped with proper effluent treatment facilities, livestock, and pig farms [18]. Therefore, farmers should treat the water properly before introducing into the culture ponds.

In the present study, all *S. enterica* serovar Corvallis isolated from shrimp and water showed resistance to erythromycin. This is in agreement with the study done by Wan Norhana et al. [12] where *S. enterica* serovar Weltevreden, *S. enterica* serovar Hvittingfoss, *S. enterica* serovar Litchfield, *S. enterica* serovar Agona, *S. enterica* serovar Paratyphi, *S. enterica* serovar Benin, and *S. enterica* serovar Java isolated from shrimp were resistant to erythromycin. However, *Salmonella* isolated from Farm 2 showed multiple antibiotic resistances (eight antibiotics) compared to Farm 1 and 3 which were resistant to one antibiotic only. The occurrence of multiple

TABLE 3: Susceptibility (zone of inhibition in mm) of *Salmonella* and *Vibrio* species isolated from different shrimp farms to antibiotics.

	Isolates	NOR 10	DO 30	E 15	SXT 25	F 300	S3 300	AMP 10	OA 2	C 30	TE 30
					Farm 1						
Shrimp	*V. mimicus* (S1)	S-31	S-24	MS-18	S-26	S-25	S-20	R-6.5	S-29.5	S-31	S-23.5
	V. mimicus (S2)	S-31	S-23	MS-19	S-24.5	S-23	S-30	R-6	S-28	S-32	S-22
	V. mimicus (S3)	S-28	S-17	MS-20	S-22	S-22	S-28	MS-16	S-22.5	S-29	MS-18
	V. vulnificus	S-18	S-21	MS-18.5	S-23.5	S-23	S-30	R-7	S-28	S-30	S-24
Water	*Salmonella*	S-40	S-21	R-12	S-18	S-24	S-30	S-24	S-26	S-26	MS-17
	V. cholerae (S1)	S-31	S-18.5	MS-20	S-24	S-23	S-24	R-6	S-30	S-31	S-20
	V. cholerae (S2)	S-26	S-22	MS-16	S-25	S-27.5	S-26	R-6	S-29	S-32	S-23
	V. mimicus	S-29.5	S-17	MS-20.5	R-6	S-23	R-6	R-7.5	S-25.5	S-29.5	R-13
					Farm 2						
Shrimp	*Salmonella* (S1)	S-26	R-6	R-6	R-6	S-20	R-6	R-6	R-6	R-6	R-6
	Salmonella (S2)	S-28.5	R-8	R-6	R-6	S-21	R-6	R-6	R-6	R-6	R-6
	V. cholerae (S1)	S-29	R-9	MS-18	S-24	S-23.5	S-28	R-6	S-27	S-30	R-14
	V. cholerae (S2)	S-29	R-10	MS-16	S-23.5	S-21	S-22	MS-13.5	S-28.5	S-28	R-11.5
	V. cholerae (S3)	S-32	R-10.5	MS-18	S-27	S-24	S-25	R-6	S-30.5	S-31.5	R-10.5
	V. cholerae (S4)	S-29	S-20	MS-19.5	S-22.5	S-22	MS-15	R-7.5	S-27.5	S-31	S-19
	V. cholerae (S5)	S-26.5	MS-14	MS-17.5	S-22.5	S-21.5	S-27	R-7	S-24.5	S-32	MS-16
	V. cholerae (S6)	S-24	S-18	MS-16.5	S-20.5	S-20	S-27	R-6	S-24	S-29	MS-16
	V. mimicus (S1)	S-29	S-23	MS-18	S-24	S-24	S-22	R-6	S-29.5	S-35	S-25
	V. mimicus (S2)	S-28	S-22	MS-20	S-22	S-24.5	S-28	R-6	S-27	S-32	S-24
	V. mimicus (S3)	S-27.5	S-18.5	MS-20	S-25	S-24	S-31	R-13	S-29	S-31	S-21.5
	V. mimicus (S4)	S-26	S-21	MS-18.5	S-23	S-23.5	S-30	R-6	S-25	S-30	S-22
	V. vulnificus (S1)	S-28	MS-14	MS-19	S-24	S-22	S-28	R-8	S-25	S-30	MS-18
	V. vulnificus (S2)	S-26	S-20	MS-18.5	S-23	S-21.5	S-26	R-7.5	S-29	S-31	S-23
Water	*V. cholerae*	S-30	S-21	MS-20	S-26	S-23	S-22	R-8	S-28.5	S-30	S-19.5
	V. mimicus	S-27	S-16.5	MS-19.5	S-22.5	S-22.5	MS-14	R-6	S-24	S-29.5	S-20.5
					Farm 3						
Shrimp	*Salmonella* (S1)	S-42	MS-15	R-10	S-25	S-23	S-20	S-26	S-28	S-25	S-20
	Salmonella (S2)	S-41	MS-13	R-10	S-27	S-21.5	S-20	S-26	S-27.5	S-25	MS-17
	V. alginolyticus	S-23	S-24	MS-17	S-20.5	S-23	S-17	R-6	S-22	S-32	S-24
	V. cholerae (S1)	S-25	S-24	MS-20	S-24	S-25	S-26	R-6	S-22	S-31	S-25
	V. cholerae (S2)	S-24	S-22.5	MS-19	S-22.5	S-24	S-25	R-6	S-19	S-30.5	S-25
	V. cholerae (S3)	S-21	MS-13	MS-16	S-22	S-22	S-27	R-7.5	S-27	S-26.5	R-14
	V. cholerae (S4)	S-24	S-19.5	MS-17.5	S-24	S-21	S-25	R-6	S-27.5	S-27	S-20.5
	V. cholerae (S5)	S-25	S-18	MS-15.5	S-23.5	S-19	S-24	R-6	S-27	S-26.5	S-21
	V. mimicus (S1)	S-27	S-22	MS-17	S-25	S-21.5	S-21	R-9	S-26.5	S-30	S-23
	V. mimicus (S2)	S-26	S-20	MS-17.5	MS-15	S-22	S-19.5	R-8	S-26.5	S-25	S-20
	V. mimicus (S3)	S-25	S-20	MS-16	S-18	S-18	S-25	R-10	S-24	S-28	S-21.5
Water	*V. parahaemolyticus* (S1)	S-28	S-23.5	MS-19	S-25	S-22	S-23.5	R-6	S-28	S-31	S-24.5
	V. parahaemolyticus (S2)	S-23	S-20.5	MS-17	S-21.5	S-20	S-25	R-8	S-25	S-30	S-21
	V. parahaemolyticus (S3)	S-24	S-22	MS-17	S-24	S-21	S-20	R-7.5	S-25	S-29	S-21.5
	V. parahaemolyticus (S4)	S-24	S-20.5	MS-16.5	S-21	S-20	S-19	R-9	S-24.5	S-30	S-20
	V. parahaemolyticus (S5)	S-22.5	S-18	MS-16.5	S-23	S-20	S-26	R-8.5	S-22	S-26.5	S-20
	V. parahaemolyticus (S6)	S-23	S-17.5	MS-18	S-20.5	S-19.5	S-25	R-12.5	S-23	S-28.5	S-20
	V. vulnificus (S1)	S-22.5	S-21	MS-17.5	S-22	S-24	S-22.5	R-6	S-22.5	S-28	S-21.5
	V. vulnificus (S2)	S-26	S-18.5	MS-16.5	S-16	S-22	S-26.5	R-9	S-27	S-28	S-20
	V. cholerae	S-30	S-26	S-24	S-27.5	S-30	S-28	R-7	S-25.5	S-33	S-28

TABLE 3: Continued.

Isolates	NOR 10	DO 30	E 15	SXT 25	F 300	S3 300	AMP 10	OA 2	C 30	TE 30
V. parahaemolyticus (S1)	S-29	S-23	MS-19.5	S-25	S-24	S-23.5	R-6	S-28	S-30	S-24
V. parahaemolyticus (S2)	S-26	S-20	S-27.5	S-22.5	S-21	S-17	R-9	S-25	S-30	S-20.5
V. vulnificus	S-34	S-30	S-24	S-36	S-32	S-40	R-10	S-30	S-40	S-34

NOR 10: norfloxacin 10 μg; DO 30: doxycycline hydrochloride 30 μg; E 15: erythromycin 15 μg; SXT 25: sulfamethoxazole 23.75 μg/trimethoprim 1.25 μg; F 300: nitrofurantoin 300 μg; S3 300: compound sulphonamides 300 μg; AMP 10: ampicillin 10 μg; OA 2: oxolinic acid 2 μg; C 30: chloramphenicol 30 μg; TE 30: tetracycline 30 μg; S: susceptible; MS: moderately susceptible; R: resistant.

antibiotic resistances could be due to the presence of chicken farm nearby that maybe using different types of antibiotics. Antibiotic is used in poultry as therapeutic as well as growth promotant. According to Singer and Hofacre [19], antibiotics and their metabolites as well as bacteria can spread from poultry farms into waterways. In addition, poultry litter can also help in their dissemination onto open field. Petersen et al. [20] have reported that integrated broiler chicken-fish farm contributed to antimicrobial-resistant bacteria in a pond environment. The antibiotic residues from the nearby chicken farm could have led to multiple antibiotic resistance observed in the present survey.

The natural occurrence of vibrios in marine and estuarine environment has been reported by Varnam and Evans [21]. Incidence of vibrios in marine-caught seafoods including shrimp has been reported by Adeleye et al. [22], while Boinapally and Jiang [23] showed that vibrios are also found in pond-reared shrimp.

In general, the incidence of bacteria resistance to shrimp samples was higher than those in water samples from the same location. In the present survey, differences in antibiotic resistance patterns in a species of *Vibrio* could be due to presence of different strains. Bacteria resistance to AMP10 was the highest followed by TE30. Ampicillin is not commonly used in shrimp culture. So there is a possibility that these vibrios could have acquired resistance from other places. The widespread use of tetracycline because of its low toxicity and broad-spectrum antibiotic activity against a wide range of Gram-positive and Gram-negative bacteria [24] and also as a successful prophylaxis and therapy against *Vibrio* [25, 26] could have led to high resistance.

In the present survey, although the managers of the farms stated that they did not use any antibiotics, the possibility of the presence of antibiotic-resistant bacteria in shrimp could be from postlarvae. Yasuda and Kitao [27] reported that *Vibrio* spp. were the dominant genera in the digestive tract of the zoea of *Penaeus japonicus*. According to Baticados and Paclibare [28], a variety of drugs are used in shrimp hatcheries. The use of these drugs leads to resistance to certain antimicrobials during the rearing of postlarvae in the hatchery which remain in the shrimp gut when transferred to the growout ponds [29].

The other possibility of the presence of antibiotic resistant bacteria could be the use of probiotics. According to Mathur and Singh [30], there are reports that commensal bacteria including lactic acid bacteria may act as reservoirs of antibiotic resistant genes that can be transferred to pathogenic bacteria. Therefore, the use of probiotics in the

surveyed farms may have led to the incidence of multidrug resistant bacteria.

Four out of five *Vibrio* spp. isolated in the present study were similar to the findings of Bhaskar and Setty [31] who reported the presence of *V. alginolyticus* as the most common followed by *V. cholerae*, *V. parahemolyticus*, and *V. vulnificus* in *P. monodon* culture system. Farm 3 had two *Vibrio* species more than Farm 1 and 2 that had three species each. This could be because of the different source of postlarvae or different source of water. Farm 1 and 2 obtained postlarvae from the same hatchery and had the same source of water.

The result of this survey reveals the presence of multidrug resistant *Salmonella* and *Vibrio* in shrimp farms. Antibiotic resistance is a legitimate concern which may affect future therapy of shrimp and human disease. Farm owners should be concerned about the presence of these pathogenic bacteria which also contributes to human health risk and should adopt best management practices for responsible aquaculture to ensure the quality of shrimp.

Acknowledgment

The authors would like to thank the managers of the shrimp farms who have made the sampling possible.

References

[1] N. Bhaskar, T. M. Setty, G. V. Reddy et al., "Incidence of *Salmonella* in cultured shrimp *Penaeus monodon*," *Aquaculture*, vol. 138, no. 1–4, pp. 257–266, 1995.

[2] Department of Fisheries (DoF), *Annual Fisheries Statistics*, Department of Fisheries, Putrajaya, Malaysia, 2009.

[3] K. Holmström, S. Gräslund, A. Wahlström, S. Poungshompoo, B. E. Bengtsson, and N. Kautsky, "Antibiotic use in shrimp farming and implications for environmental impacts and human health," *International Journal of Food Science and Technology*, vol. 38, no. 3, pp. 255–266, 2003.

[4] G. G. Khachatourians, "Agricultural use of antibiotics and the evolution and transfer of antibiotic-resistant bacteria," *Canadian Medical Association Journal*, vol. 159, no. 9, pp. 1129–1136, 1998.

[5] C. Willis, "Antibiotics in the food chain: their impact on the consumer," *Reviews in Medical Microbiology*, vol. 11, no. 3, pp. 153–160, 2000.

[6] P. J. A. Reilly and D. R. Twiddy, "*Salmonella* and *Vibrio cholerae* in brackishwater cultured tropical prawns," *International Journal of Food Microbiology*, vol. 16, no. 4, pp. 293–301, 1992.

[7] J. Allshouse, J. Buzby, D. Harvey, and D. Zorn, "Seafood safety and trade. Issues in diet, safety and health," Agricultural

Information Bulletin 789-7, United States Department of Agriculture, Washington, DC, USA, 2004.

[8] D. V. Lightner, "Diseases of cultured penaeid shrimp and prawns," in *Disease Diagnosis and Control in North American Aquaculture*, C. J. Sindermann and D. V. Lightner, Eds., pp. 42–47, Elsevier, New York, NY, USA, 1998.

[9] J. G. Morris and R. E. Black, "Cholera and other vibrioses in the United States," *New England Journal of Medicine*, vol. 312, no. 6, pp. 343–350, 1985.

[10] S. M. Finegold and W. J. Martin, "Antimicrobial susceptibility tests and assays," in *Bailey and Scott's Diagnostic Microbiology*, W. R. Bailey, E. G. Scott, S. M. Finegold, and W. J. Martin, Eds., pp. 385–404, The CV Mosby Company, St. Louis, Mo, USA, 1982.

[11] N. Bhaskar, T. M. R. Setty, S. Mondal et al., "Prevalence of bacteria of public health significance in the cultured shrimp (*Penaeus monodon*)," *Food Microbiology*, vol. 15, no. 5, pp. 511–519, 1998.

[12] M. N. Wan Norhana, M. Y. Johara, and A. M. Ramlah, "Occurrence of pathogens from major shrimp and oyster production areas in Peninsular Malaysia," *Malaysian Fisheries Journal*, vol. 2, pp. 176–184, 2001.

[13] G. Jeyasekaran and S. Ayyappan, "Postharvest microbiology of farm-reared, tropical freshwater prawn (Macrobrachium rosenbergii)," *Journal of Food Science*, vol. 67, no. 5, pp. 1859–1861, 2002.

[14] B. Koonse, W. Burkhardt, S. Chirtel, and G. P. Hoskin, "*Salmonella* and the sanitary quality of aquacultured shrimp," *Journal of Food Protection*, vol. 68, no. 12, pp. 2527–2532, 2005.

[15] Murachman and Darius, "Study of handling, sanitation and hygiene of shrimp from brackishwater ponds in East Java," 1991, Paper Presented at the Eighth Session of the Indo-Pacific Fishery Commission Working Party of Fish Technology and Marketing, Yogjakarta, Indonesia, 24-27 September 1991.

[16] T. S. G. Iyer and P. R. G. Varma, "Sources of contamination with *Salmonella* during processing of frozen shrimp," *Fishery Technology*, vol. 27, pp. 60–63, 1990.

[17] A. Dalsgaard, H. H. Huss, A. H-Kittikun, and J. L. Larsen, "Prevalence of *Vibrio cholerae* and *Salmonella* in a major shrimp production area in Thailand," *International Journal of Food Microbiology*, vol. 28, no. 1, pp. 101–113, 1995.

[18] I. Rosnani, "River water quality status in Malaysia," in *Proceedings of the National Conference on Sustainable River Basin Management in Malaysia*, Kuala Lumpur, Malaysia, November 2001.

[19] R. S. Singer and C. L. Hofacre, "Potential impacts of antibiotic use in poultry production," *Avian Diseases*, vol. 50, no. 2, pp. 161–172, 2006.

[20] A. Petersen, J. S. Andersen, T. Kaewmak, T. Somsiri, and A. Dalsgaard, "Impact of integrated fish farming on antimicrobial resistance in a pond environment," *Applied and Environmental Microbiology*, vol. 68, no. 12, pp. 6036–6042, 2002.

[21] A. H. Varnam and M. G. Evans, *Food Borne Pathogens*, Wolfe Publishing, London, UK, 1991.

[22] I. A. Adeleye, F. V. Daniels, and V. A. Enyinnia, "Characterization and pathogenicity of *Vibrio* spp. contaminating seafoods in Lagos, Nigeria," *Internet Journal of Food Safety*, vol. 12, pp. 1–9, 2010.

[23] K. Boinapally and X. Jiang, "Comparing antibiotic resistance in commensal and pathogenic bacteria isolated from wild-caught South Carolina shrimps vs. farm-raised imported shrimps," *Canadian Journal of Microbiology*, vol. 53, no. 7, pp. 919–924, 2007.

[24] V. M. Moretti, G. L. Maggi, A. Albertini et al., "High-performance liquid chromatographic determination of oxytetracycline in channel catfish (*Ictalurus punctatus*) muscle tissue," *Analyst*, vol. 119, no. 12, pp. 2749–2751, 1994.

[25] R. Williams and D. Lightner, "Regulatory status of therapeutants for penaeid shrimp culture in the United States," *Journal of the World Aquaculture Society*, vol. 19, pp. 188–196, 1998.

[26] G. Carignan, K. Carrier, and S. Sued, "Assay of oxytetracycline residues in salmon muscle by liquid chromatography with ultraviolet detection," *Journal of Association of Official Agricultural Chemists International*, vol. 76, pp. 325–328, 1993.

[27] K. Yasuda and T. Kitao, "Bacterial flora in the digestive tract of prawns, *Penaeus japonicus* Bate," *Aquaculture*, vol. 19, no. 3, pp. 229–234, 1980.

[28] M. C. L. Baticados and J. O. Paclibare, "The use of chemotherapeutic agents in aquaculture in the Philippines," in *Diseases in Asian Aquaculture I. Fish Health Section*, M. Shariff, R. P. Subasinghe, and J. R. Arthur, Eds., pp. 531–546, Asian Fisheries Society, Manila, Philippines, 1992.

[29] E. A. Tendencia and L. D. de la Peña, "Antibiotic resistance of bacteria from shrimp ponds," *Aquaculture*, vol. 195, no. 3-4, pp. 193–204, 2001.

[30] S. Mathur and R. Singh, "Antibiotic resistance in food lactic acid bacteria—a review," *International Journal of Food Microbiology*, vol. 105, no. 3, pp. 281–295, 2005.

[31] N. Bhaskar and T. M. R. Setty, "Incidence of vibrios of public health significance in the farming phase of tiger shrimp (*Penaeus monodon*)," *Journal of the Science of Food and Agriculture*, vol. 66, no. 2, pp. 225–231, 1994.

10

Bactericidal Effects of 405 nm Light Exposure Demonstrated by Inactivation of *Escherichia, Salmonella, Shigella, Listeria, and Mycobacterium* Species in Liquid Suspensions and on Exposed Surfaces

Lynne E. Murdoch, Michelle Maclean, Endarko Endarko, Scott J. MacGregor, and John G. Anderson

The Robertson Trust Laboratory for Electronic Sterilisation Technologies, Department of Electronic and Electrical Engineering, University of Strathclyde-Glasgow, Glasgow G1, 1XW, UK

Correspondence should be addressed to Lynne E. Murdoch, lynne.murdoch@strath.ac.uk

Academic Editor: Kent E. Kester

The bactericidal effect of 405 nm light was investigated on taxonomically diverse bacterial pathogens from the genera *Salmonella, Shigella, Escherichia, Listeria, and Mycobacterium*. High-intensity 405 nm light, generated from an array of 405-nm light-emitting diodes (LEDs), was used to inactivate bacteria in liquid suspension and on exposed surfaces. *L. monocytogenes* was most readily inactivated in suspension, whereas *S. enterica* was most resistant. In surface exposure tests, *L. monocytogenes* was more susceptible than Gram-negative enteric bacteria to 405 nm light when exposed on an agar surface but interestingly less susceptible than *S. enterica* after drying onto PVC and acrylic surfaces. The study findings, that 405 nm light inactivates diverse types of bacteria in liquids and on surfaces, in addition to the safety advantages of this visible (non-UV wavelength) light, indicate the potential of this technology for a range of decontamination applications.

1. Introduction

Despite enormous investments in public health research, bacterial pathogens transmitted in food, water, and from other environmental sources remain a major cause of illness in both the developed and developing world. Examples of such ubiquitous pathogens include enteric Gram-negative bacteria such as *Salmonella, Escherichia,* and *Shigella* which continue to cause significant diarrhoeal infections worldwide [1]. The foodborne pathogen *Listeria monocytogenes* also, has significant impact on health statistics through its propensity for causing serious illness in the immunocompromised [2]. Actinobacteria from the *Mycobacterium* genus are also a major cause of human morbidity and mortality and pathogens from the *Mycobacterium tuberculosis* complex such as *M. bovis* and *M. tuberculosis* remain amongst the most serious causes of infective disease worldwide [3].

Whilst many traditional decontamination methods give sterling service, they are not without limitations. Problems associated with product/material damage related to the use of physical methods and development of microbial resistance as well as the formation and persistence of potentially harmful residues are longstanding issues related to the use of chemical disinfectants such as sodium hypochlorite, ozone, and H_2O_2 [4–7]. Mycobacteria have innate resistance to chemical decontamination as they possess an unusual cell envelope containing peptidoglycan, arabinoglycan, and mycolic acid. This restricts the passage of many chemicals/drugs across the cell envelope, contributing to the hardiness of this microorganism [8, 9].

In response to the ongoing challenges presented by established and emerging pathogens and to provide additional or alternative approaches to microbial control, considerable interest has developed concerning novel methods of disinfection and decontamination.

Such alternative methods of decontamination include continuous and pulsed UV light, which has been shown

Bactericidal Effects of 405 nm Light Exposure Demonstrated by Inactivation of Escherichia, Salmonella, Shigella, Listeria, and Mycobacterium Species in Liquid Suspensions and on Exposed Surfaces

61

to be highly germicidal. However, limitations such as poor transmissibility, degradative effects on materials, and potential carcinogenic effects in humans mean that there are restrictions in its use [10–13]. A safe, non-UV, light-based decontamination technology termed high-intensity narrow-spectrum (HINS) light has been recently described. HINS light of 405 nm stimulates endogenous microbial porphyrin molecules to produce oxidising reactive oxygen species (ROS), predominantly singlet oxygen ($^1O_2$) that damages cells leading to microbial death [14–16]. Specifically 405 nm light has been shown to be capable of inactivating a range of predominantly nosocomial pathogens and also Gram negative food-related pathogens [17–22].

This study further investigates the inactivating effect of high-intensity 405 nm light exposure on taxonomically diverse bacterial pathogens. Inactivation data on three species of enteric facultatively anaerobic Gram negative bacilli (*Salmonella enterica*, *Shigella sonnei*, and *Escherichia coli*) were compared with the facultatively anaerobic Gram-positive coccobacillus *Listeria monocytogenes* and with the aerobic, acid fast Gram positive bacillus *Mycobacterium terrae*. The study also aimed to determine the effectiveness of 405 nm light for inactivating bacteria in both liquid suspensions and on exposed surfaces.

2. Materials and Methods

2.1. Microorganisms. The bacteria used in this study were *Salmonella enterica* serovar *enteritidis* NCTC 4444 (formerly known as *Salmonella enteritidis*), *Shigella sonnei* NCTC 12984, *Escherichia coli* serotype O157:H7 NCTC 12900, *Listeria monocytogenes* NCTC 11994, and *Mycobacterium terrae* LMG 10394. All cultures were obtained from the National Collection of Type Cultures, Colindale, UK, except *M. terrae*, which was obtained from the Laboratorium voor Microbiologie, Universiteit Gent, Belgium. *M. terrae* was chosen as a safe, comparative, surrogate microorganism for the highly pathogenic *M. tuberculosis*. *S. sonnei*, *S. enterica*, and *E. coli* were inoculated into 100 mL Nutrient Broth, *L. monocytogenes* into 100 mL Tryptone Soya Broth (all Oxoid, Basingstoke, UK), and *M. terrae* into 100 mL of Middlebrook 7H9 Broth containing ADC enrichment media (Becton Dickinson and Company, NJ, USA). Broths were cultivated at 37°C for 18 hours under rotary conditions (120 rpm), with the exception of *M. terrae* which was cultivated at 37°C for 14 days, after which the purity was checked using Ziehl-Neelsen stain (Kinyoun method). After cultivation, broths were centrifuged at 3939× g for 10 minutes, and the resultant pellet resuspended in 100 mL phosphate-buffered saline (PBS) and serially diluted to give the appropriate starting population of 1-2 × 10^5 CFU mL^{-1} for experimental use.

2.2. 405 nm Light Exposure of Bacterial Suspensions

2.2.1. Light Source. An indium-gallium-nitride 99-DIE light-emitting diode (LED) array (Opto Diode Corp, CA, USA) was used for exposure of bacterial suspensions. This array was 20 mm × 16 mm in size and had an emission at 405 nm

with a bandwidth of 14 nm at full-width half maximum (FWHM). A cooling fan and heat sink were attached to the array to dissipate heat from the source and this also served to minimise any heat transfer to the sample. The LED array was mounted in a polyvinyl chloride (PVC) housing designed to fit a 12-well micro plate (NUNC, Roskilde, Denmark).

2.2.2. Treatment of Bacterial Suspensions. For exposure of bacterial suspensions, the LED array was set in a fixed position 2 cm directly above a micro plate well, which held a 2 mL volume of bacterial suspension. To ensure that all bacteria in the suspension were uniformly suspended and exposed to the same dose over the exposure period, a small magnetic follower was incorporated in the well and the sample dish and LED array were positioned on a magnetic stirrer which permitted continuous stirring of the sample during light exposure. The LED array was powered by a DC power supply (0.5 A ± 0.05 and 11.2 V ± 0.2), giving an approximate irradiance (or power density) of 10 mW cm^{-2} at the surface of the bacterial suspension.

In order to quantitatively examine the inactivation process, it was necessary to account for any attenuation of the irradiance of the 405 nm light as it passed through a bacterial sample; attenuation is a result of light absorption and scattering. Attenuation by the samples used in the study was examined by measuring the irradiance as the light entered the surface of the sample and comparing that value with the irradiance immediately below the sample depth of 7 mm, after allowing for the transmission loss through the base of the sample dish. These measurements showed that, for samples containing bacterial populations of 10^7 CFU mL^{-1} and less, no measurable attenuation, occurred, therefore; to ensure that no attenuation effects occurred, bacterial suspensions of 1-2 × 10^5 CFU mL^{-1} were used in the current study.

2.2.3. Dose Delivery Experiments. In order to establish if inactivation of bacteria was dose dependent, suspensions of *L. monocytogenes* of a population density of 1-2×10^5 CFU mL^{-1} were exposed to 108 J cm^{-2} of 405 nm light applied using three different regimes. Three different light irradiances were used (10 mW cm^{-2}, 20 mW cm^{-2}, and 30 mW cm^{-2}), and in order to keep the total dose constant in each case (i.e., 108 J cm^{-2}), the sample exposure time was adjusted according to the equation:

$$E' = P't, \tag{1}$$

where E' is the energy density (dose) in J cm^{-2}, P' is the irradiance (power density) in W cm^{-2}, and t is the time in seconds.

2.3. 405 nm Light Exposure of Bacteria Seeded onto Surfaces

2.3.1. Light Source. An ENFIS QUATTRO Mini Air Cooled Light Engine (ENFIS Ltd, Swansea, UK) was used for exposure of bacteria seeded onto agar and inert surfaces as this source allowed more effective treatment of larger surface areas. This source was an array of 144 LEDs (40 mm × 40 mm in size) with emission at 405 nm (16 nm FWHM) and was

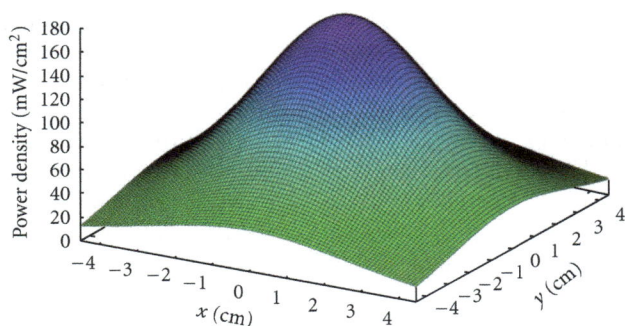

FIGURE 1: Three-dimensional model demonstrating the power density (mW cm^{-2}) distribution of the emission from the 405 nm ENFIS QUATTRO LED array across a 9 cm agar plate.

powered by a 48 V power supply. The light engine had an integrated heat sink and cooling fan to minimise any heat buildup during experimentation. The optical distribution of the 405 nm LED array emission was measured across the length and breadth of (i) a 9 cm diameter agar plate and (ii) 6 × 4 cm coupons of inert materials (polyvinyl chloride and acrylic), in 0.5 cm increments, using a radiant power density meter calibrated at 405 nm. These measurements were then used to generate a three-dimensional model using Maxima Software release 5.2.2.1 (Figure 1). The average irradiance emitted by the light source was found to be 71 mW cm^{-2} across the agar plate and 110 mW cm^{-2} across the coupons, as calculated with Maxima Software release 5.2.2.1 using the Romberg Method:

$$P = \frac{\iint_{-4.5}^{4.5} 5^{(3+m)} \times E_0 \times (25 + x^2 + y^2)^{-(3+m)/2} dx\, dy}{\pi \times r^2}, \quad (2)$$

where P is the average irradiance/power density, E_0 is the irradiance/power density at the plate centre, m is the Lambertian mode number and x and y are the cartesian coordinates.

2.3.2. Quantitative Agar Surface Exposure Experiments. Bacterial suspensions of *S. enterica*, *E. coli*, *L. monocytogenes*, and *S. sonnei*, containing approximately 10^2 CFU mL^{-1}, were pipetted and spread onto the surface of 9 cm Tryptone Soya Agar (TSA) (Oxoid, Basingstoke, UK) plates, giving approximately 200–250 CFU/per plate, equivalent to 2.3-2.4 log$_{10}$ CFU/plate. Seeded agar plates were then exposed to increasing durations of high-intensity 405 nm light. Nonexposed control plates were prepared for each 405 nm light-exposed sample. This test was not carried out for *M. terrae* as the organism does not grow on TSA and an accurate comparison could not have been drawn with different growth media.

2.3.3. Qualitative Agar Surface Exposure Experiments. For a qualitative indication of the bactericidal effects of 405 nm light on surfaces, loopfuls of 10^6 CFU mL^{-1} bacterial suspensions of *S. enterica*, *E. coli*, *L. monocytogenes*, and *S. sonnei* were streaked in individual lines of length 6 cm (1 cm apart)

onto the surface of a TSA plate. Half of each inoculum line (3 cm) was exposed to high-intensity 405 nm light, with the other half covered with aluminium foil to prevent light exposure. Plates were exposed to 15, 30, and 45 minutes of high-intensity 405 nm light, with an average irradiance of 71 mW cm^{-2} across the plate. After exposure, plates were incubated at 37°C for 16 hours and photographs of each plate were taken. *M. terrae* was not included in this test due to its unique growth requirements.

2.3.4. Inert Surface Exposure Experiments. Bacterial suspensions of *S. enterica* (as a representative Gram negative bacterium) and *L. monocytogenes* (as a representative Gram positive bacterium) containing approximately 10^5 CFU mL^{-1} were loaded into a 6-jet Collison nebuliser (BGI Inc, MA, USA). Coupons of acrylic (6 cm × 4 cm) and polyvinyl chloride (PVC) (6 cm × 4 cm), pre-sterilised with 80% ethanol, were held at a distance of 3 cm from the nebuliser for 15 seconds allowing the aerosolised bacteria to deposit and immediately dry onto the surface of the coupon. Coupons after seeding were immediately exposed to increasing durations of high-intensity 405 nm light with an average irradiance of 110 mW cm^{-2} across the coupon surface. Following exposure, coupons were pressed onto a TSA surface for 5 seconds to recover the surviving bacteria from the seeded surface. This seeding and recovery process was also carried out for both the initial seeded population and the nonexposed control samples. These experiments were repeated in triplicate.

2.4. Plating and Enumeration. For the suspension exposure experiments, test and control samples were plated onto Nutrient Agar (*E. coli*, *S. enterica*, *S. sonnei*) (Oxoid, Basingstoke, UK), TSA (*L. monocytogenes*) (Oxoid, Basingstoke, UK), or Middlebrook 7H10 Agar with OADC enrichment media (*M. terrae*) (Becton Dickinson and Company, NJ, USA) plates. A WASP 2 spiral plater (Don Whitely Scientific Ltd, Shipley, UK) was used to plate out the samples, either as 50 μL spiral plate or 100 μL spread plate samples, with each sample being plated in at least triplicate. One millilitre pour plates were prepared if low counts were expected. Plates were incubated at 37°C for 24 hours, except *M. terrae* plates which were incubated at 37°C for 7 days, after which the plates were enumerated and results reported as CFU mL^{-1}.

For the agar and inert surface exposure experiments, plates were incubated directly following exposure at 37°C for 24 hours with results reported as CFU per plate. Qualitative experiment plates were photographed using a Sony Cybershot DSC-T2 digital camera.

2.5. Statistical Analysis. Data points on each figure represent the mean results of two or more independent experiments, with each individual experimental data point being sampled in triplicate at least. Data points include the standard deviation and significant differences obtained from results. Significant differences were calculated at the 95% confidence interval using ANOVA (one way) with MINITAB software release 15.

Bactericidal Effects of 405 nm Light Exposure Demonstrated by Inactivation of Escherichia, Salmonella, Shigella, Listeria, and Mycobacterium Species in Liquid Suspensions and on Exposed Surfaces

63

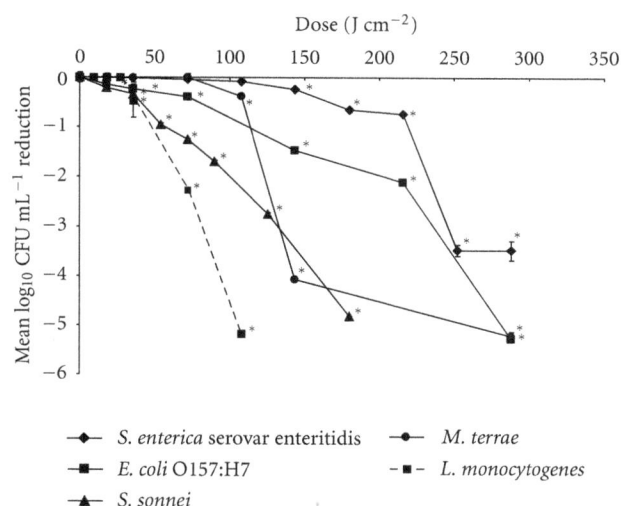

FIGURE 2: Inactivation of *S. sonnei*, *E. coli*, *S. enterica*, *L. monocytogenes*, and *M. terrae* in liquid suspension, by exposure to high-intensity 405 nm light of an irradiance of approximately 10 mW cm^{-2}. Control samples remained constant throughout experimentation in all cases (data not shown). *Indicates where a light-exposed bacterial count was significantly different from the non-exposed control count ($P \leq 0.05$ calculated at the 95% confidence interval).

3. Results

Liquid bacterial suspensions were exposed to 405 nm light at an irradiance of 10 mW cm^{-2} for increasing time periods. From these values, the absolute dose applied in each experiment can be calculated using (2). Figure 2 demonstrates the effect of 405 nm light exposure on suspensions of Gram negative bacteria—*E. coli*,—*S. enterica*, *S. sonnei*—and Gram positive bacteria—*L. monocytogenes* and *M. terrae*. *S. enteritidis* and *E. coli* were inactivated by 3.5-log$_{10}$ CFU mL^{-1} and 5-log$_{10}$ CFU mL^{-1}, respectively, at 288 J cm^{-2}, and *S. sonnei* was inactivated by 5-log$_{10}$ CFU mL^{-1} at a lower dose of 180 J cm^{-2}. *M. terrae* was inactivated by 4-5 log$_{10}$ CFU mL^{-1} between a dose of 144 and 288 J cm^{-2}, and *L. monocytogenes* was inactivated by 5-log$_{10}$ CFU mL^{-1} at 108 J cm^{-2}. It should be noted that the population densities of control samples for all five bacterial species remained constant throughout this series of experiments (data not shown).

Table 1 contains the results of exposing suspensions of *L. monocytogenes* to 108 J cm^{-2} of 405 nm light, with the 108 J cm^{-2} dose being applied in varying regimes. The results demonstrate that application of the dose, regardless of how it is applied (i.e., lower irradiance for longer exposure time or higher irradiance for shorter exposure time), yields very similar final populations (no statistical significant difference), in this case approximately a 5-log$_{10}$ CFU mL^{-1} reduction in bacterial numbers.

In order to eliminate the possibility of inactivation being the result of heat transfer from the LED array during sample exposure, the temperatures of the bacterial suspensions were monitored during experimentation. Temperature readings (taken every 15 minutes) showed that the bacterial suspen-

sions experienced minimal temperature changes during light exposure. Temperature was monitored over a 480-minute period (longest exposure time). The initial temperature of the suspensions was 26°C, which fluctuated ±1°C throughout the exposure period.

The inactivation achieved when bacterial-seeded agar plates were exposed to high-intensity 405 nm light (with an average irradiance of approximately 71 mW cm^{-2} across the plate) is shown in Table 2. With all four bacterial species tested, almost complete (100%) inactivation was achieved (<1 CFU/plate). *L. monocytogenes* was again inactivated at the fastest rate, with 100% reduction at an average dose of 128 J cm^{-2}. *S. enterica*, *S. sonnei*, and *E. coli* were inactivated by 2.28 (100%), 2.10 (99.3%), and 2.18 (99.8%) log$_{10}$ CFU/plate, respectively, at an average dose of 192 J cm^{-2}.

Figure 3 shows a qualitative representation of the bactericidal effect of high-intensity 405 nm light on the four bacteria, *E. coli*, *S. sonnei*, *S. enterica*, and *L. monocytogenes*. The streaks of bacteria were light-exposed for 15, 30, and 45 minutes (Figures 3(a), 3(b), and 3(c), resp.). It can be seen that *L. monocytogenes* was the most susceptible and that by 45 minutes of 405 nm light exposure all of the tested species of bacteria were effectively inactivated.

The results of the 405 nm light inactivation of *S. enterica* and *L. monocytogenes* seeded onto acrylic and PVC surfaces are reported in Table 3. Both *S. enterica* and *L. monocytogenes* were inactivated more readily on the PVC surface using high-intensity 405 nm light than on acrylic surfaces. Interestingly, the Gram negative *S. enterica* was more rapidly inactivated than the Gram positive *L. monocytogenes* on both surfaces. *S. enterica* was inactivated by 2.19 log$_{10}$ CFU/plate (100%) on PVC with an average dose of 50 J cm^{-2} and 1.63 log$_{10}$ CFU/plate (98%) on acrylic with an average dose of 66 J cm^{-2}. *L. monocytogenes* was inactivated by 0.90 log$_{10}$ CFU/plate (90%) on PVC with an average dose of 50 J cm^{-2}, and only 0.42 log$_{10}$ CFU/plate (61%) on acrylic with an average dose of 66 J cm^{-2}. When non-exposed control counts were compared to the counts of the initial seeded population it was found that between 70 and 80% of the seeded bacteria were nonrecoverable from acrylic and PVC surfaces after 7.5–10 minutes, likely as a result of inactivation by desiccation. Therefore, inactivation results were calculated as the reduction of light-exposed bacteria when compared to the respective non-exposed control and are reported as both reduction in log$_{10}$ CFU/plate and % inactivation.

4. Discussion

This study has demonstrated that 405 nm light has a significant bactericidal effect on a number of important and taxonomically diverse bacterial pathogens. The current study set out with two aims: (1) to test the relative susceptibility of several diverse types of bacterial pathogens, including mycobacteria, which are among the most difficult of bacteria to inactivate using conventional decontamination techniques and (2) to determine the effectiveness of 405 nm light for inactivation of selected types of pathogens in both liquid suspensions and on surfaces.

TABLE 1: The effect of applying a dose of $108\,\text{J}\,\text{cm}^{-2}$ using three different power density/exposure time regimes on the 405 nm light inactivation of *L. monocytogenes* in suspension.

Power density ($\text{mW}\,\text{cm}^{-2}$)	Exposure time (min)	Dose ($\text{J}\,\text{cm}^{-2}$)	Log_{10} CFU mL^{-1} reduction
10	180	108	5.18* (SD ±0.2)
20	90	108	5.05* (SD ±0.0)
30	60	108	4.90* (SD ±0.4)

*Indicates where a light-exposed bacterial count was significantly different from the non-exposed control count ($P \leq 0.05$ calculated at the 95% confidence interval).
SD: standard deviation of averaged results.

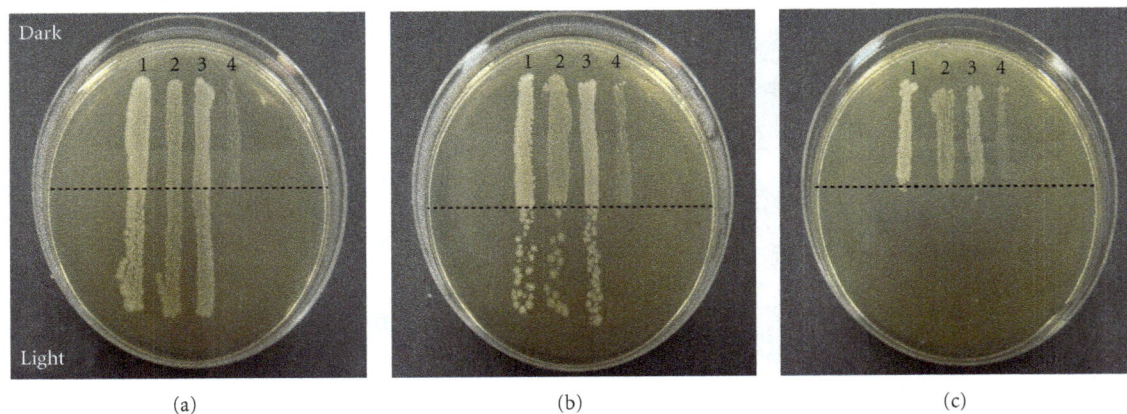

(a) (b) (c)

FIGURE 3: Qualitative representation of the bactericidal effect of high-intensity 405 nm light on (1) *E. coli*, (2) *S. sonnei*, (3) *S. enterica*, and (4) *L. monocytogenes*. Plates (a), (b), and (c) demonstrate the inactivating effect of 15, 30, and 45 minutes light exposure, respectively, on streaks of the foodborne pathogenic bacteria, with half of each plate being light exposed and the other half being kept in darkness to provide the equivalent non-light-exposed control.

The most resistant bacterium in liquid suspension was *S. enterica*, which was inactivated by 3.5-log_{10} CFU mL^{-1} at a dose of $288\,\text{J}\,\text{cm}^{-2}$, around 2.5 times the dose required for 5-log_{10} CFU mL^{-1} inactivation of the least susceptible bacterium *L. monocytogenes* ($108\,\text{J}\,\text{cm}^{-2}$). *M. terrae* was inactivated by 4-5 log_{10} CFU mL^{-1} between 144 and $288\,\text{J}\,\text{cm}^{-2}$, and *E. coli* and *S. sonnei* were inactivated by 5-log_{10} CFU mL^{-1} at $288\,\text{J}\,\text{cm}^{-2}$ and $180\,\text{J}\,\text{cm}^{-2}$, respectively.

L. monocytogenes also proved to be the most readily inactivated organism when seeded onto agar surfaces, with 100% inactivation achieved with an average dose of $128\,\text{J}\,\text{cm}^{-2}$. The least susceptible microorganism, of those tested, in the agar surface exposure experiments appeared to be *S. sonnei* with a $2.10\,\text{log}_{10}$ CFU/plate (99.3%) reduction in bacterial numbers achieved at an average dose of $192\,\text{J}\,\text{cm}^{-2}$, however; in statistical tests this level of inactivation was not significantly different from the percentage inactivation rates achieved for *S. enterica* and *E. coli* at the same dose. It would be interesting to directly compare the susceptibility of the test bacteria exposed in liquid suspension to those seeded onto surfaces; however, the inactivation doses cannot be directly compared due to differences in the experimental arrangements and exposure conditions. Under suspension test conditions, uniform exposure of a well-mixed bacterial suspension was achieved, whereas with surface exposure tests, light irradiance varied over the surface, therefore requiring that an average power density/irradiance value be calculated in order to determine the corresponding dose values.

Reasons for the variable susceptibility of different bacteria to 405 nm light are as yet undetermined. Studies have reported that Gram positive species, in general, were more susceptible to 405 nm light inactivation than Gram negative species, which is generally consistent with the results obtained in the current study [21]. It is also theorised that the difference in inactivation kinetics may be due to organism-specific differences in porphyrin levels, porphyrin types, porphyrin wavelength absorption maxima, or as a consequence of the relatively short distance that singlet oxygen molecules can diffuse ($\sim$20 nm^3 in solution) within cell structures [13, 21, 23]. In addition it has been speculated that less oxygen-tolerant bacterial species may be particularly susceptible to the effects of ROS as microorganisms such as some microaerophilic species have been found to possess fewer key oxidative regulators than most aerobes [24, 25]. Studies have shown blue light to be capable of inactivating the anaerobic oral pathogens *Prevotella*, *Porphyromonas*, and *Fusobacterium* as well as microaerophilic pathogens such as *Propionibacterium acnes* and *Helicobacter pylori* [15, 26–29]. However, inactivation results achieved with anaerobic/microaerophilic bacteria have not provided conclusive evidence as to whether oxygen-sensitive bacteria are any more susceptible than aerobes as many of these bacteria are also known to accumulate high levels of porphyrins.

The inactivation data for bacteria exposed in suspension in the present study can be compared with the 405 nm inactivation data obtained in other studies [21, 22].

Bactericidal Effects of 405 nm Light Exposure Demonstrated by Inactivation of Escherichia, Salmonella, Shigella, Listeria, and Mycobacterium Species in Liquid Suspensions and on Exposed Surfaces

65

TABLE 2: Inactivation of bacteria seeded onto agar surfaces upon exposure to high-intensity 405 nm light for different times. Agar plates (9 cm diameter) were exposed to an average irradiance of 71 mW cm^{-2}.

Exposure time (min)	Dose (J cm^2)	S. enterica (Log$_{10}$ CFU/plate ± SD)			S. sonnei (Log$_{10}$ CFU/plate ± SD)			E. coli O157:H7 (Log$_{10}$ CFU/plate ± SD)			L. monocytogenes (Log$_{10}$ CFU/plate ± SD)		
		Non-light exposed	Light exposed	Log$_{10}$ reduction (% reduction)	Non-light exposed	Light exposed	Log$_{10}$ reduction (% reduction)	Non-light exposed	Light exposed	Log$_{10}$ reduction (% reduction)	Non-light exposed	Light exposed	Log$_{10}$ reduction (% reduction)
10	60	—	—	—	—	—	—	—	—	—	2.44 (±0.02)	2.37 (±0.03)	0.07* (15.3%)
15	90	2.38 (±0.03)	2.33 (±0.02)	0.05 (11.3%)	2.28 (±0.02)	1.82 (±0.12)	0.46* (64.4%)	2.32 (±0.02)	2.09 (±0.13)	0.23* (39.4%)	—	—	—
20	120	—	—	—	—	—	—	—	—	—	2.33 (±0.10)	1.65 (±0.30)	0.68* (76.4%)
30	180	2.33 (±0.01)	1.19 (±0.30)	1.14* (91.4%)	2.21 (±0.09)	0.93 (±0.40)	1.28* (93.3%)	2.27 (±0.03)	0.90 (±0.10)	1.37* (95.7%)	2.25 (±0.06)	0 (±0.00)	2.25* (100%)
45	270	2.28 (±0.02)	0 (±0.00)	2.28* (100%)	2.26 (±0.04)	0.16 (±0.30)	2.10* (99.3%)	2.18 (±0.02)	0 (±0.00)	2.18* (99.8%)	—	—	—

* Indicates where a light-exposed sample value was statistically significant from a non-exposed control value ($P \leq 0.05$ calculated at the 95% confidence interval).

SD: standard deviation of averaged results.

TABLE 3: Inactivation of bacteria aerosolised onto PVC and acrylic surfaces upon exposure to $110\,\mathrm{mW\,cm^{-2}}$ of high-intensity 405 nm light.

Surface material	Exposure time (min)	Dose $(\mathrm{J\,cm})^{-2}$	S. enterica ($\mathrm{Log_{10}}$ CFU/plate ± SD)			L. monocytogenes ($\mathrm{Log_{10}}$ CFU/plate ± SD)		
			Non-exposed	Light exposed	$\mathrm{Log_{10}}$ reduction (% reduction)	Non-exposed	Light exposed	$\mathrm{Log_{10}}$ reduction (% reduction)
PVC	2.5	15	2.54 (±0.10)	0.53 (±0.21)	2.01* (98%)	2.08 (±0.45)	1.41 (±0.18)	0.68* (78%)
	5	30	2.62 (±0.15)	0.72 (±0.40)	1.90* (99%)	1.93 (±0.28)	1.0 (±0.42)	0.93* (86%)
	7.5	45	2.19 (±0.03)	0 (±0.0)	2.19* (100%)	1.69 (±0.40)	0.79 (±0.1)	0.90* (90%)
Acrylic	2.5	15	2.76 (±0.13)	1.59 (±0.05)	1.18* (93%)	2.45 (±0.06)	2.21 (±0.22)	0.24 (36%)
	5	30	2.42 (±0.16)	1.18 (±0.31)	1.24* (93%)	2.37 (±0.06)	2.15 (±0.12)	0.22 (39%)
	7.5	45	2.66 (±0.1)	1.20 (±0.26)	1.46* (96%)	2.24 (±0.10)	2.02 (±0.17)	0.21 (36%)
	10	60	2.09 (±0.16)	0.46 (±0.15)	1.63* (98%)	2.18 (±0.16)	1.75 (±0.21)	0.42* (61%)

*Indicates where a light-exposed sample value was statistically significant from a non-exposed control value ($P \le 0.05$ calculated at the 95% confidence interval).
SD: standard deviation of averaged results.

L. monocytogenes, although much more readily inactivated than the microorganisms *S. sonnei, M. terrae, S. enterica,* and *E. coli,* was less susceptible than the majority of the medically significant Gram positive organisms (*Staphylococcus, Streptococcus, Clostridium* species) investigated by Maclean et al. [21]. All Gram negative organisms (*Acinetobacter, Proteus, Pseudomonas, Klebsiella,* and *Escherichia* species) tested by Maclean et al. [21] had inactivation rates comparable with the values found in the present study for *S. sonnei, E. coli,* and *S. enterica.* A notable exception to this is the Gram negative microaerophilic organism *Campylobacter jejuni,* investigated in a study by Murdoch et al. [22], which exhibited a particularly high sensitivity to 405 nm light. Interestingly, *M. terrae* showed similar inactivation kinetics to the Gram negative pathogens, particularly *S. sonnei.* Possible explanations for this difference in inactivation susceptibility could be that the unusual cell envelope, characteristic of mycobacteria, confers some resistance to 405 nm light penetration and/or that mycobacterial cell envelopes provide greater innate resistance to ROS given that mycobacteria species are capable of evading phagocytic oxidative burst damage [30].

A study by Bohrerova and Linden [31] into UV light inactivation showed *M. terrae* to be more resistant to the effects of the UV light than other representative Gram positive and Gram negative bacterial pathogens. The study demonstrated UV-light susceptibility in *M. terrae* and *M. tuberculosis* to be almost identical which corresponds well with studies by Maclean et al. [21], where they achieved similar inactivation kinetics between bacteria of the same genus when exposed to 405 nm light. The results of these studies also indicate the suitability of *M. terrae* as a reliable *M. tuberculosis* surrogate organism in light inactivation studies.

This study has demonstrated that exposure to high-intensity 405 nm light is capable of inactivating a variety of taxonomically diverse bacterial pathogens without the requirement for exogenous photosensitiser molecules. The bactericidal effect was demonstrated quantitatively in liquid suspension and both quantitatively and qualitatively on agar plates and inert surfaces. The inactivation process has been shown to be dose dependent; therefore, higher-intensity

light-sources could achieve lethal doses in shorter time periods. The findings that the bacteria can be inactivated using high-intensity 405 nm light whilst seeded on both nutritious and inert surfaces is particularly significant for potential practical application within the food and healthcare industry, where cross-contamination from environmental contact surfaces and equipment is a problem. The fact that 405 nm light falls within the visible light range and does not require the containment conditions of harmful UV light potentially permits the continuous treatment of food contact areas in the presence of operator personnel, a uniquely advantageous feature. In order to evaluate the effectiveness of 405 nm LED arrays for practical applications, such as the continuous treatment of large surface areas, custom-designed light sources are required that can achieve a more uniform power density distribution, perhaps through the use of lenses. Future work to assess the treatment potential of 405 nm light, as well as further studies on the inactivation of bacterial biofilms, will also be important in order to fully assess the potential of this inactivation technology for applications within the food industry alongside other safety control methods.

Conflict of Interests

The authors declare that they have no conflict of interests.

Acknowledgments

The first author would like to thank the Engineering and Physical Sciences Research Council (EPSRC) for support through a Doctoral Training Grant (Awarded 2007–2010). All authors would like to thank The Robertson Trust for their funding support.

References

[1] World Health Organisation (WHO), "Initiative for vaccine research: diarrhoeal diseases," 2009, http://www.who.int/vaccine_research/diseases/diarrhoeal/en/index.html.

[2] N. E. Freitag, G. C. Port, and M. D. Miner, "*Listeria mono-cytogenes*—from saprophyte to intracellular pathogen," *Nature Reviews Microbiology*, vol. 7, no. 9, pp. 623–628, 2009.

[3] K. O'Riordan, D. S. Sharlin, J. Gross et al., "Photoinactivation of mycobacteria in vitro and in a new murine model of localized *Mycobacterium bovis* BCG-induced granulomatous infection," *Antimicrobial Agents and Chemotherapy*, vol. 50, no. 5, pp. 1828–1834, 2006.

[4] J. G. Kim, A. E. Yousef, and S. Dave, "Application of ozone for enhancing the microbiological safety and quality of foods: a review," *Journal of Food Protection*, vol. 62, no. 9, pp. 1071–1087, 1999.

[5] A. D. Russell, "Bacterial resistance to disinfectants: present knowledge and future problems," *Journal of Hospital Infection*, vol. 43, supplement 1, pp. S57–S68, 1999.

[6] S. Springthorpe, "Disinfection of surfaces and equipment," *Journal of the Canadian Dental Association*, vol. 66, no. 10, pp. 558–560, 2000.

[7] F. Barbut, D. Menuet, M. Verachten, and E. Girou, "Comparison of the efficacy of a hydrogen peroxide dry-mist disinfection system and sodium hypochlorite solution for eradication of *Clostridium difficile* spores," *Infection Control and Hospital Epidemiology*, vol. 30, no. 6, pp. 507–514, 2009.

[8] V. Jarlier and H. Nikaido, "Mycobacterial cell wall: structure and role in natural resistance to antibiotics," *FEMS Microbiology Letters*, vol. 123, no. 1-2, pp. 11–18, 1994.

[9] L. J. Alderwick, H. L. Birch, A. K. Mishra, L. Eggeling, and G. S. Besra, "Structure, function and biosynthesis of the *Mycobacterium tuberculosis* cell wall: arabinogalactan and lipoarabinomannan assembly with a view to discovering new drug targets," *Biochemical Society Transactions*, vol. 35, no. 5, pp. 1325–1328, 2007.

[10] N. J. Rowan, S. J. MacGregor, J. G. Anderson, R. A. Fouracre, L. McIlvaney, and O. Farish, "Pulsed-light inactivation of food-related microorganisms," *Applied and Environmental Microbiology*, vol. 65, no. 3, pp. 1312–1315, 1999.

[11] A. L. Andrady, S. H. Hamid, X. Hu, and A. Torikai, "Effects of increased solar ultraviolet radiation on materials," *Journal of Photochemistry and Photobiology B*, vol. 46, no. 1–3, pp. 96–103, 1998.

[12] K. L. Bialka and A. Demirci, "Decontamination of *Escherichia coli* O157:H7 and *Salmonella enterica* on blueberries using ozone and pulsed UV-light," *Journal of Food Science*, vol. 72, no. 9, pp. M391–M396, 2007.

[13] M. R. Hamblin and T. Hasan, "Photodynamic therapy: a new antimicrobial approach to infectious disease?" *Photochemical and Photobiological Sciences*, vol. 3, no. 5, pp. 436–450, 2004.

[14] M. Elman and J. Lebzelter, "Light therapy in the treatment of *Acne vulgaris*," *Dermatologic Surgery*, vol. 30, no. 2, pp. 139–146, 2004.

[15] O. Feuerstein, I. Ginsburg, E. Dayan, D. Veler, and E. I. Weiss, "Mechanism of visible light phototoxicity on *Porphyromonas gingivalis* and *Fusobacterium nucleatum*," *Photochemistry and Photobiology*, vol. 81, no. 5, pp. 1186–1189, 2005.

[16] M. Maclean, S. J. MacGregor, J. G. Anderson, and G. A. Woolsey, "The role of oxygen in the visible-light inactivation of *Staphylococcus aureus*," *Journal of Photochemistry and Photobiology B*, vol. 92, no. 3, pp. 180–184, 2008.

[17] M. Maclean, *An investigation into the light inactivation of medically important microorganisms*, Ph.D. thesis, University of Strathclyde, Glasgow, UK, 2006.

[18] J. S. Guffey and J. Wilborn, "*In vitro* bactericidal effects of 405 nm and 470 nm blue light," *Photomedicine and Laser Surgery*, vol. 24, no. 6, pp. 684–688, 2006.

[19] C. S. Enwemeka, D. Williams, S. Hollosi, D. Yens, and S. K. Enwemeka, "Visible 405 nm SLD light photo-destroys methicillin-resistant *Staphylococcus aureus* (MRSA) *In vitro*," *Lasers in Surgery and Medicine*, vol. 40, no. 10, pp. 734–737, 2008.

[20] M. Maclean, S. J. MacGregor, J. G. Anderson, and G. Woolsey, "High-intensity narrow-spectrum light inactivation and wavelength sensitivity of *Staphylococcus aureus*," *FEMS Microbiology Letters*, vol. 285, no. 2, pp. 227–232, 2008.

[21] M. Maclean, S. J. MacGregor, J. G. Anderson, and G. A. Woolsey, "Inactivation of bacterial pathogens following exposure to light from a 405 nm LED array," *Applied and Environmental Microbiology*, vol. 75, no. 7, pp. 1932–1937, 2009.

[22] L. E. Murdoch, M. MacLean, S. J. MacGregor, and J. G. Anderson, "Inactivation of *Campylobacter jejuni* by exposure to high-intensity visible light," *Foodborne Pathogens and Disease*, vol. 7, no. 10, pp. 1211–1216, 2010.

[23] Z. Malik, H. Ladan, and Y. Nitzan, "Photodynamic inactivation of Gram-negative bacteria: problems and possible solutions," *Journal of Photochemistry and Photobiology B*, vol. 14, no. 3, pp. 262–266, 1992.

[24] D. Jean, V. Briolat, and G. Reysset, "Oxidative stress response in *Clostridium perfringens*," *Microbiology*, vol. 150, no. 6, pp. 1649–1659, 2004.

[25] C. Murphy, C. Carroll, and K. N. Jordan, "Environmental survival mechanisms of the foodborne pathogen *Campylobacter jejuni*," *Journal of Applied Microbiology*, vol. 100, no. 4, pp. 623–632, 2006.

[26] C. A. Henry, B. Dyer, M. Wagner, M. Judy, and J. L. Matthews, "Phototoxicity of argon laser irradiation on biofilms of *Porphyromonas* and *Prevotella* species," *Journal of Photochemistry and Photobiology B*, vol. 34, no. 2-3, pp. 123–128, 1996.

[27] H. Ashkenazi, Z. Malik, Y. Harth, and Y. Nitzan, "Eradication of *Propionibacterium acnes* by its endogenic porphyrins after illumination with high intensity blue light," *FEMS Immunology and Medical Microbiology*, vol. 35, no. 1, pp. 17–24, 2003.

[28] M. R. Hamblin and T. Hasan, "Photodynamic therapy: a new antimicrobial approach to infectious disease?" *Photochemical and Photobiological Sciences*, vol. 3, no. 5, pp. 436–450, 2004.

[29] N. S. Soukos, S. Som, A. D. Abernethy et al., "Phototargeting oral black-pigmented bacteria," *Antimicrobial Agents and Chemotherapy*, vol. 49, no. 4, pp. 1391–1396, 2005.

[30] J. Chan, T. Fujiwara, P. Brennan et al., "Microbial glycolipids: possible virulence factors that scavenge oxygen radicals," *Proceedings of the National Academy of Sciences of the United States of America*, vol. 86, no. 7, pp. 2453–2457, 1989.

[31] Z. Bohrerova and K. G. Linden, "Assessment of DNA damage and repair in Mycobacterium terrae after exposure to UV irradiation," *Journal of Applied Microbiology*, vol. 101, no. 5, pp. 995–1001, 2006.

11

Roquefort Cheese Proteins Inhibit *Chlamydia pneumoniae* Propagation and LPS-Induced Leukocyte Migration

Ivan M. Petyaev,[1] **Naylia A. Zigangirova,**[2] **Natalie V. Kobets,**[2] **Valery Tsibezov,**[2] **Lydia N. Kapotina,**[2] **Elena D. Fedina,**[2] **and Yuriy K. Bashmakov**[1]

[1] *Lycotec Ltd. Granta Park Campus, Cambridge CB21 6GP, UK*
[2] *Gamaleya Institute of Epidemiology and Microbiology, Ministry of Health, 18 Gamaleya Street, Moscow 123098, Russia*

Correspondence should be addressed to Ivan M. Petyaev; ykb75035@aol.com

Academic Editors: J. Sóki and D. Zhou

Inflammation in atherosclerosis, which could be associated with some subclinical infections such as *C. pneumoniae*, is one of the key factors responsible for the development of clinical complications of this disease. We report that a proprietary protein extract isolated from Roquefort cheese inhibits the propagation of *C. pneumoniae* in a human HL cell line in a dose-dependent manner, as revealed by the immunofluorescence analysis. These changes were accompanied by a significant reduction in the infective progeny formation over the protein extract range of 0.12–0.5 μg/mL. Moreover, short term feeding of mice with Roquefort cheese (twice, 10 mg per mouse with an interval of 24 hours) led to the inhibition of the migration of peritoneal leukocytes caused by intraperitoneal injection of *E. coli* lipopolysaccharide. These changes were complemented by a reduction in neutrophil count and a relative increase in peritoneal macrophages, suggesting that ingestion of Roquefort could promote regenerative processes at the site of inflammation. The ability of this protein to inhibit propagation of *Chlamydia* infection, as well as the anti-inflammatory and proregenerative effects of Roquefort itself, may contribute to the low prevalence of cardiovascular mortality in France where consumption of fungal fermented cheeses is the highest in the world.

1. Introduction

An epidemiological link between the reduction of cardiovascular mortality and moderate wine consumption in French and Mediterranean adults, known as the "French paradox" [1], has attracted the steady interest of medicobiological researchers over the past two decades. It has been proven in numerous clinical and experimental studies that common constituents of red wine—resveratrol and other polyphenols—have multiple health benefits resulting from their anti-inflammatory [2], anticarcinogenic [3], and antiatherogenic [4] properties and thereby contribute to the occurrence of the "French paradox." The molecular mechanism behind the biological activity of dietary polyphenols has been extensively reviewed [5] and is believed to be mediated by sirtuins, a family of NAD^+-dependent histone deacetylases [6].

Thorough ongoing research into the matter has led to the gradual realization that there are other dietary and nondietary determinants contributing to reduced cardiovascular mortality in the French and Mediterranean populations. In particular, higher consumption of omega-3 fatty acid [7], flavonoids [8], and dietary fiber [9] may also contribute to the mechanisms of the "French paradox."

Cheese and cheese-containing products are prominent ingredients of the Mediterranean diet. It has been assumed that high cheese consumption may somehow contribute to favorable changes in lipid profile and reduced risk of cardiovascular disease [10]. Indeed, the varieties of blue-veined and other fungal fermented cheeses, which are a trademark of French culinary culture, may possess some measurable health benefits due to the presence of numerous functional substances in their core. Leading commercial brands of blue cheese have been shown to contain short-chain fatty acids, methyl ketones, and secondary alcohols [11]. Assessment of the microbial population in blue cheese reveals that

Penicillium roqueforti, *Penicillium glaucum*, and *Geotrichum candidum* are three major distinguishable fungi, while *Lactococcus lactis*, *Lactococcus garvieae*, and *Lactococcus raffinolactis* can be identified in blue cheese specimens during different stages of ripening [12]. *P. roqueforti* metabolites in particular show a wide range of pharmacological activity. Andrastins A, B, C, and D are consistently produced in blue-veined cheese and are potent inhibitors of farnesyltransferase, a major enzyme of cholesterol biosynthesis [13]. Andrastin A is also known to display strong antitumor properties [13]. Other substances, including roquefortine, a compound with some neurotoxic properties, constrain Gram-positive bacterial growth by inhibiting cytochrome P-450 [14]. The biological activity of metabolites produced by other fungi has yet to be studied.

In the present paper we report that Roquefort cheese extract inhibits propagation of *C. pneumoniae* in cultured cell line, while Roquefort feeding attenuates the LPS-induced migratory response of peritoneal leukocytes and causes significant changes in immune cell subpopulations.

2. Materials and Methods

2.1. Reagents and Organisms.
All reagents were from Sigma-Aldrich unless specified otherwise. HL cells (Washington Research Foundation, Seattle, USA) as well as *C. pneumoniae* (strain Kajaani6, *K6*) were kindly provided by Dr. P. Saikku (University of Oulu, Finland). Roquefort Societe (Société) was purchased from a general grocery supplier in Cambridge, United Kingdom. Cheese specimens were homogenized and processed for protein extraction before expiration dates. A/JSnYCit (A/Sn)/c mice, males aged from 2 to 4 months, were bred and kept under conventional conditions at the Animal Facilities of the Institute of Epidemiology and Microbiology (Moscow, Russia) in accordance with guidelines from the Russian Ministry of Health (number 755). Food and water were provided ad libitum. All experimental procedures were performed under a protocol approved by the Institutional Animal Care Committee.

2.2. Roquefort Fractionation.
To obtain protein extracts a 10–15 g specimen of Roquefort cheese was placed in 10–15 mL of PBS and the samples were homogenized using an Omni TH-115. The resulting suspensions were kept for 1 hour at 4°C and centrifuged for 15 min at 10 000 g using an Eppendorf 5810R centrifuge. The obtained supernatant was centrifuged again for another 15 min at 10 000 g on an Eppendorf 5115D centrifuge. The resulting supernatant was used for further fractionation.

The protein extract was fractionated by gelfiltration on a 1.5 × 9.0 cm column with Sephadex G-25 Medium equilibrated with PBS. The column was precalibrated to determine free and total volume using Dextran Blue and DNP-L-Ala. For each experiment 3 mL of the cheese extract was introduced to the column and the elution fractions were collected as the following volumes: fraction number "1" as 6 mL, intermediate fraction "2" as 4 mL, and the last fraction "3" as 10 mL. Protein concentrations were determined by absorption at 280 nm on

a Shimadzu UV-1.800 spectrophotometer. All three protein fractions were combined, dialyzed, lyophilized, and kept at −80°C for further studies.

2.3. In Vitro Studies.
C. pneumoniae was initially propagated in HL cells and elementary bodies (EB) were purified by Renografin gradient centrifugation [15]. Chlamydial titers were determined by infecting HL cells with 10-fold dilutions of thawed stock suspension. Purified elementary bodies (EB) of known titer were suspended in sucrose-phosphate-glutamic acid buffer and used as inoculums for HL cells. Cells were grown in 24 well plates until a confluence rate of 80% was reached. HL plates were infected with *C. pneumoniae* at multiplicities of infection (MOI) of 1 in DMEM with FBS without cycloheximide and centrifuged for 1 hour at 1500 g. Addition of the cheese protein exract dissolved in PBS at concentrations of 0.12, 0.25, and 0.5 mg/mL was done simultaneously with the inoculation of *C. pneumoniae*.

2.3.1. Immunofluorescence Staining.
Infected HL monolayers grown on coverslips in 24-well plates for 72 hours were fixed with methanol. Permeabilized cells were stained by direct immunofluoresence (IF) using FITC—conjugated monoclonal antibody against chlamydial lipopolysaccharide (Nearmedic Plus, RF). Inclusion-containing cells were visualized using a Nikon Eclipse 50i fluorescence microscope at ×1350 magnification.

2.3.2. Assessment of Infective Progeny.
In order to assess the infective progeny accumulation in HL cells after a 72-hour cultivation period, HL cells were harvested, frozen, and thawed, as described elsewhere. Serial dilutions of lysates were inoculated onto HL cells and centrifuged for 1 hour at 1500 g. The infected cells were visualized with chlamydial LPS-specific monoclonal antibody after 72 hours of the postinfection period.

2.4. In Vivo LPS Stimulation.
Mice were gavaged with 0.2 mL of PBS containing 10.0 mg of Roquefort cheese (once daily) or PBS alone. 24 hours after the last gavage procedure, control and cheese-fed mice were injected with 0.5 mg/kg LPS (*E. coli*, strain O111:B4) dissolved in PBS or PBS alone. Peritoneal washes were obtained 48 hours after LPS injection. Resulting cells were counted and recruited cell populations were analyzed by flow cytometry.

Peritoneal exudate cells were washed, resuspended in Fc block (Biolegend, USA) for 10 min and incubated with monoclonal antibodies (mAb) in FACS buffer (PBS, 2% FCS, and 0.05% sodium azide) for 15 min. Cells were then washed, resuspended, and analyzed using a FACS Calibur flow cytometer (BD Biosciences), CELLQUEST acquisition software (BD Biosciences), and FCS EXPRESS V.3 analysis software (DeNovo software, USA). The mAbs used were fluorescein-isothiocyanate- (FITC-) conjugated anti-Ly-6G (Myltenyi Biotech, Germany), RPE-conjugated anti-F4/80 (Invitrogen, USA), RPE-conjugated anti-CD11c (Invitrogen,

TABLE 1: HL cells were plated, grown, and infected with *C. pneumoniae* at MOI 1. Additions of the Roquefort cheese extract fraction with final concentrations of 0.12, 0.25, and 0.55 mg/mL were performed simultaneously with the bacterial pathogen inoculation. The cell monolayers were harvested after 72 hours incubation at 35°C in 5% CO_2 and infective progeny number was determined as described in Section 2 . Infective progeny formation in U-937 cells infected with *C. pneumoniae* in the presence of cheese protein extract.

Cheese extract (mg/mL)	Infective progeny IFU/mL
0	$2.4 \times 10^5 \pm 1.3$
0.12	$2.8 \times 10^4 \pm 0.9^*$
0.25	$7.2 \times 10^2 \pm 2.1^*$
0.50	$5.8 \times 10^1 \pm 2.7^*$

*Significant changes as compared with control, $P \leq 0.05$.

USA)- fluorescein-isothiocyanate- (FITC-) conjugated anti-CD86 (Invitrogen, USA), RPE-conjugated anti-B220 (Invitrogen, USA), phycoerythrin- (PE-) conjugated anti-TCR (Biolegend, USA), fluorescein-isothiocyanate- (FITC-) conjugated anti-CD4 (Biolegend, USA), and PerCP-conjugated anti-CD8 (Biolegend, USA).

2.5. Statistical Analysis. All values are expressed as the mean ± SD. Variation between data sets was evaluated using Student's *t*-test. Changes with *P* values less than 0.05 were considered statistically significant. All experiments were repeated at least 3 times. Most representative IF images were chosen for publishing.

3. Results

As can be seen from Figure 1, infecting HL cells with *C. pneumoniae* strain leads to the formation of typical, densely stained, round-shaped inclusions of different sizes inside the host cells. Addition of the Roquefort cheese protein extract led to dose-dependent inhibition of inclusion body formation with their complete disappearance at a concentration of 0.5 mg/mL. Lower concentrations of cheese extract (0.12 and 0.25 mg/mL) induced the formation of atypical pleomorphic inclusion bodies, which were generally smaller in size and poorly stained and seen in a smaller number of cells as compared to the control. Cytotoxicity assessment showed that incubation of the host cells with cheese extract alone was not accompanied by changes in cell growth and their viability (results not shown). The results obtained were in good agreement with the data revealing the effect of Roquefort cheese extract on *C. pneumoniae* infection progeny formation. Table 1 shows that Roquefort protein extract caused stepwise reduction in the infective progeny number. Interestingly, on multiple repeats, the highest concentration of the cheese extract used in our experiments did not cause complete eradication of infective progeny of *C. pneumoniae*.

The development and outcomes of chlamydial infections are intrinsically predetermined by both the virulence of the pathogen and the ability of the host organism to develop an adequate physiological response to bacterial insult. Therefore,

next, we tried to evaluate the ability of the Roquefort protein extract to affect an intraperitoneal cell migratory response induced by injection of LPS *in vivo*. As it can be seen from Figure 2, IP injection of LPS induces an exuberant cellular migratory response into the peritoneal cavity which was revealed by ~3.5-fold increase in the number of cells detectable in peritoneal washes. Although Roquefort feeding by itself did not affect the peritoneal cell count, IP injection of LPS into cheese-fed mice was accompanied by a significant attenuation of the peritoneal migratory response manifested by a smaller increase in the number of peritoneal cells.

Next we decided to analyze the subpopulations of the peritoneal cells recruited into the LPS-induced migratory response. First, we compared Gr-1$^+$ (neutrophil) and F4/80$^+$ (macrophage) populations as two major and crucial players in early inflammatory events. As shown in Figure 3, the number of neutrophils (Gr-1$^+$) was extremely low in peritoneal washes obtained from intact mice. However, it was dramatically increased in the LPS stimulated mice. The neutrophil count after LPS injection was noticeably lower in the cheese-fed mice as compared to the group of mice with no cheese feeding. On the other hand, Roquefort feeding promoted accumulation of macrophages in the peritoneal washes upon LPS stimulation, whereas it had no effect on macrophage counts in the PBS-injected mice.

In addition we analyzed CD11c$^+$ dendritic cell subpopulations in the peritoneal washes obtained from LPS-stimulated mice. Depending on CD8 expression two major DC populations are traditionally referred to as myeloid (CD11c$^+$CD8$^-$) and plasmocytoid (CD11c$^+$CD86$^+$) dendritic cells [16]. As shown in Figure 4, LPS stimulation in the absence of Roquefort feeding results in a pronounced increase in myeloid dendritic cells expressing CD86. The level of this cell population in the cheese-fed mice was unchanged in comparison with control values. In contrast, the level of plasmocytoid dendritic cells expressing CD86 was higher in the Roquefort-fed LPS-treated mice as compared to LPS-treated mice and control group.

4. Discussion

C. pneumoniae is an airborne obligate bacterial pathogen responsible for a significant number of respiratory infections around the world and implemented in the pathogenesis of atherosclerosis. Unlike other Gram-negative bacteria, *C. pneumoniae* has a unique biology and is capable of completing its life cycle inside host cells. The intracellular location of the pathogen as well as its ability to infect cells with migratory potential (mononuclear leukocytes and lymphocytes) promotes the generalization of *C. pneumoniae* infection in the human body and creates a significant challenge for pharmacotherapy [17]. There are multiple reports regarding identification of *C. pneumoniae* in the tissues of the cardiovascular system, joints, brain, and meninges [18]. These observations raise a valid question about the etiological significance of *C. pneumoniae* in atherosclerosis, arthritis, and some neurological diseases.

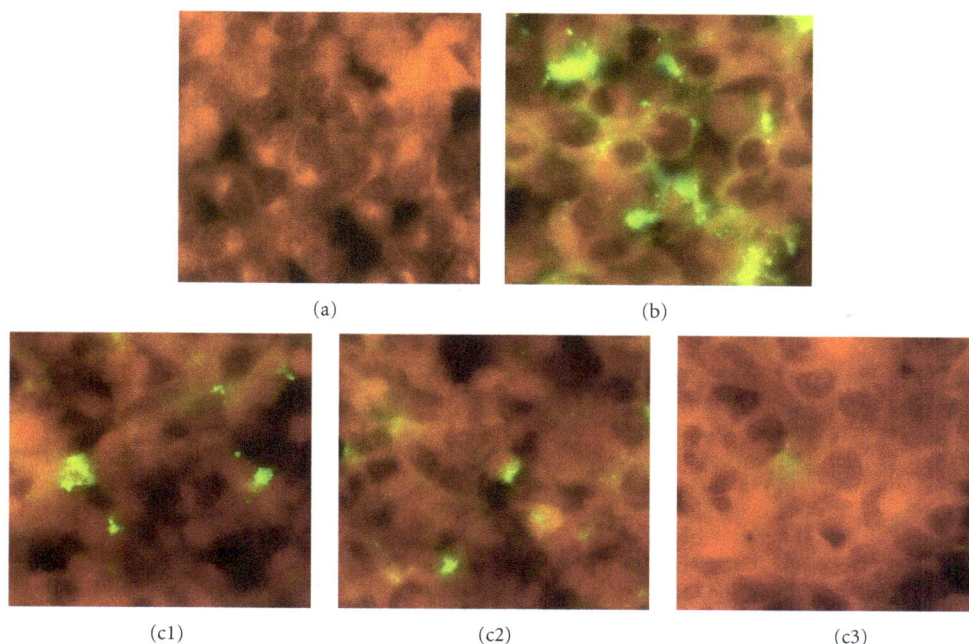

(a) (b)

(c1) (c2) (c3)

FIGURE 1: HL cells were plated, grown, and infected with *C. pneumoniae* at MOI 1. Additions of the Roquefort cheese protein extract with final concentrations of 0.12, 0.25, and 0.50 mg/mL were performed simultaneously with the bacterial pathogen inoculation. The cell monolayers were harvested after 72 h incubation at 35°C in 5% CO_2 and immunofluorescent staining conducted as described in Section 2. Inhibition of *C. pneumonia* growth by roquefort protein extract ((a) uninfected cells; (b) infected cells; (c1), (c2), and (c3) infected cells grown in the presence of 0.12, 0.25, and 0.50 mg/mL Roquefort protein extract, resp.).

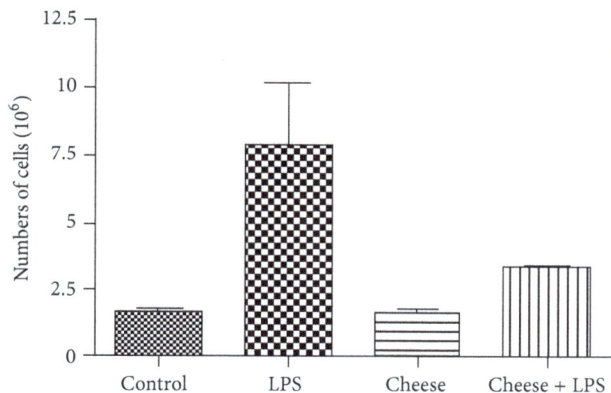

FIGURE 2: Peritoneal cell count in LPS-stimulated mice fed with Roquefort cheese. Mice were gavaged twice (once daily) with PBS containing 10.0 mg of blue-veined cheese or PBS alone (control). 24 hours later, the animals were injected intraperitoneally with LPS or its solvent (PBS). Peritoneal washes were obtained 48 hours after LPS injection and the resulting cell suspensions were counted as described in Section 2.

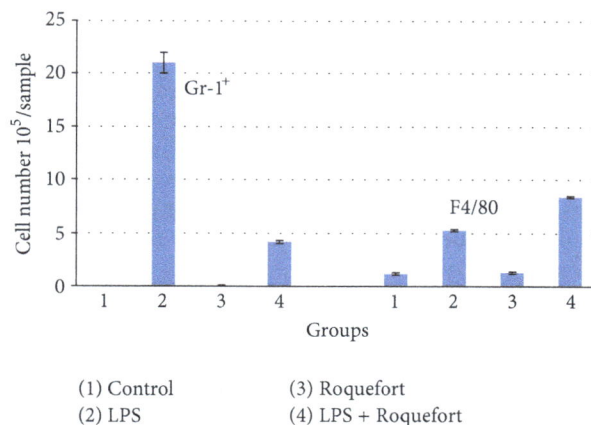

(1) Control (3) Roquefort
(2) LPS (4) LPS + Roquefort

FIGURE 3: Neutrophil (Gr-1$^+$) and macrophage (F4/80) counts in peritoneal washes obtained from LPS-stimulated mice fed with Roquefort cheese. Mice were gavaged twice (once daily) with PBS containing 10.0 mg of the Roquefort protein extract or PBS alone (control). 24 hours later, the animals were injected intraperitoneally with LPS or its solvent (PBS). Peritoneal washes were obtained 48 hours after LPS injection. Neutrophil (Gr-1$^+$) and macrophage (F4/80) populations were quantified as described in Section 2.

Our major finding reported in this paper is that a protein fraction obtained from Roquefort cheese exhibits significant inhibitory activity on *C. pneumoniae* growth in host cells. This observation has been made and repeated multiple times in our experiments using a classical immunofluorescence assay in U-cells as well as measurement of the infective progeny formation in HL cells. Despite our confidence in the observed phenomenon, it has to be yet determined which particular component/s is/are responsible for the antichlamydial activity of the Roquefort cheese. The cheese core of ripened moulded cheeses contains a unique variety of substances of mammalian, bacterial and fungal origin

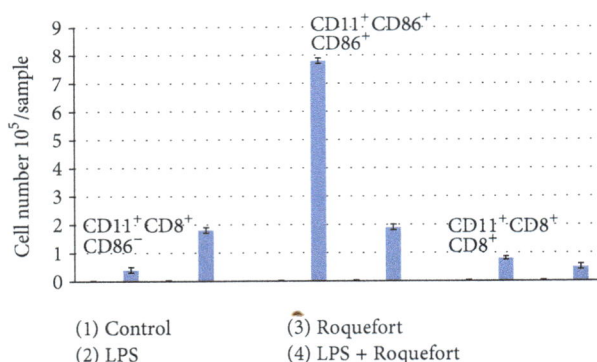

FIGURE 4: Dendritic cell subpopulations in peritoneal washes obtained from LPS-stimulated mice fed with Roquefort cheese. Mice were gavaged twice (once daily) with PBS containing 10.0 mg of Roquefort cheese or PBS alone (control). 24 hours later, the animals were injected intraperitoneally with LPS or its solvent (PBS). Peritoneal washes were obtained 48 hours after LPS injection. Dendritic cell subpopulations (CD11$^+$CD8$^+$CD86$^-$, CD11$^+$CD8$^+$CD86$^+$, and CD11$^+$CD8$^+$CD$^+$8$^+$) were quantified as described in Section 2.

[19]. Interestingly, the genome of *Penicillium roqueforti*, a major fungus used for Roquefort production, does not contain genes encoding penicillin biosynthesis [20]. Some other substances, in particular roquefortine, are believed to have antibacterial activity [14]. It is also important to mention that in our preliminary studies a crude Roquefort cheese homogenate had noticeable cytotoxicity which was attributable, in our opinion, to the presence of organic substances of low molecular weight. This activity became redundant after size-exclusion chromatography and subsequent dialysis of the resulting fractions. Indeed, Roquefort cheese has been shown to contain a wide range of organic compounds with potential cytotoxicity—mycophenolic acid, putrescine, tyramine, and others [11]. Only purified protein fraction of Roquefort cheese were found in our experiments to have reproducible and dose-dependent antichlamydial activity. It has to be determined in future whether our *in vitro* observation has any possible implication for *Chlamydia* infection *in vivo* conditions. Animal studies, and possibly clinical trials, will be required to assess this potential.

Besides antichlamydial activity, Roquefort cheese proteins have a noticeable impact on the mechanisms of innate immunity. Using an *in vivo* LPS stimulation model we have demonstrated that oral supplementation with Roquefort attenuated the overall peritoneal migratory response. The attenuation of the cell migratory response has clearly occurred as a result of the reduced neutrophil influx, while the macrophage number in the peritoneal washes of LPS-stimulated mice fed with the cheese-derived proteins was consistently raised. It is worthwhile mentioning that attenuated cell migratory response of neutrophils is generally believed to lessen tissue damage during subsequent release of inflammatory mediators, whereas macrophage influx promotes reparative events at the site of inflammation [21].

In addition we observed that the Roquefort feeding can modulate the dendritic cell subsets *in vivo* conditions.

A possible role of these cells in the development of atherosclerosis has been of a particular interest. For example, it has been shown that statins can exert their ability to inhibit atherosclerosis by affecting the dendritic cell subpopulations [16]. In our study Roquefort cheese feeding decreased the recruitment of "myeloid" dendritic cells, while some increase in recruitment was observed for plasmacytoid dendritic cells. The exact beneficial or pathogenic consequences of these differential activities of Roquefort cheese on dendritic cells need further elucidation.

Although the identification of proteins responsible for Roquefort antichlamydial activity and the exact mechanisms of their actions remain unclear, our results represent the first evidence demonstrating an antichlamydia activity associated with Roquefort cheese. It is tempting to assume that systematic dietary intake of fungal fermented cheeses, as a key element of French gastronomic tradition, may have some positive impact on the level of the control of this infection and associated inflammatory conditions, and consequently on the prevalence of cardiovascular disease in this country. Although this assumption is highly speculative, there is epidemiological evidence suggesting that Southern France, a region historically associated with high Roquefort cheese production and consumption, has remarkably low rates of cardiovascular mortality.

Acknowledgment

This work is supported by Lycotec Ltd. United Kingdom.

References

[1] A. Evans, "The French paradox and other ecological fallacies," *International Journal of Epidemiology*, vol. 40, no. 6, pp. 1486–1489, 2011.

[2] K. Rahal, P. Schmiedlin-Ren, J. Adler et al., "Resveratrol has antiinflammatory and antifibrotic effects in the peptidoglycan-polysaccharide rat model of Crohn's disease," *Inflammatory Bowel Diseases*, vol. 18, no. 4, pp. 613–623, 2012.

[3] M. E. Juan, I. Alfaras, and J. M. Planas, "Colorectal cancer chemoprevention by trans-resveratrol," *Pharmacological Research*, vol. 65, no. 6, pp. 584–591, 2012.

[4] R. S. Matos, L. A. Baroncini, L. B. Précoma et al., "Resveratrol causes antiatherogenic effects in an animal model of atherosclerosis," *Arquivos Brasileiros de Cardiologia*, vol. 98, no. 2, pp. 136–142, 2012.

[5] L. H. Opie and S. Lecour, "The red wine hypothesis: from concepts to protective signalling molecules," *European Heart Journal*, vol. 28, no. 14, pp. 1683–1693, 2007.

[6] A. F. Fernández and M. F. Fraga, "The effects of the dietary polyphenol resveratrol on human healthy aging and lifespan," *Epigenetics*, vol. 6, no. 7, pp. 870–874, 2011.

[7] C. R. Harper and T. A. Jacobson, "Beyond the mediterranean diet: the role of omega-3 fatty acids in the prevention of coronary heart disease," *Preventive Cardiology*, vol. 6, no. 3, pp. 136–146, 2003.

[8] S. M. Nadtochiy and E. K. Redman, "Mediterranean diet and cardioprotection: the role of nitrite, polyunsaturated fatty acids, and polyphenols," *Nutrition*, vol. 27, no. 7-8, pp. 733–744, 2011.

[9] M. Tabernero, K. Venema, A. J. Maathuis, and F. D. Saura-Calixto, "Metabolite production during in vitro colonic fermentation of dietary fiber: analysis and comparison of two European diets," *Journal of Agricultural and Food Chemistry*, vol. 59, no. 16, pp. 8968–8975, 2011.

[10] D. M. Colquhoun, S. Somerset, K. Irish, and L. M. Leontjew, "Cheese added to a low fat diet does not affect serum lipids," *Asia Pacific Journal of Clinical Nutrition*, vol. 12, supplement, S65, 2003.

[11] I. V. Wolf, M. C. Perotti, and C. A. Zalazar, "Composition and volatile profiles of commercial Argentinean blue cheeses," *Journal of the Science of Food and Agriculture*, vol. 91, no. 2, pp. 385–393, 2011.

[12] A. B. Flórez and B. Mayo, "Microbial diversity and succession during the manufacture and ripening of traditional, Spanish, blue-veined Cabrales cheese, as determined by PCR-DGGE," *International Journal of Food Microbiology*, vol. 110, no. 2, pp. 165–171, 2006.

[13] K. F. Nielsen, P. W. Dalsgaard, J. Smedsgaard, and T. O. Larsen, "Andrastins A-D, Penicillium roqueforti Metabolites consistently produced in blue-mold-ripened cheese," *Journal of Agricultural and Food Chemistry*, vol. 53, no. 8, pp. 2908–2913, 2005.

[14] C. Aninat, Y. Hayashi, F. André, and M. Delaforge, "Molecular requirements for inhibition of cytochrome P450 activities by roquefortine," *Chemical Research in Toxicology*, vol. 14, no. 9, pp. 1259–1265, 2001.

[15] H. D. Caldwell, J. Kromhout, and J. Schachter, "Purification and partial characterization of the major outer membrane protein of Chlamydia trachomatis," *Infection and Immunity*, vol. 31, no. 3, pp. 1161–1176, 1981.

[16] P. Puddu, G. M. Puddu, E. Cravero, S. Muscari, and A. Muscari, "The functional role of dendritic cells in atherogenesis," *Molecular Medicine Reports*, vol. 3, no. 4, pp. 551–554, 2010.

[17] M. R. Hammerschlag and S. A. Kohlhoff, "Treatment of chlamydial infections," *Expert Opinion on Pharmacotherapy*, vol. 13, no. 4, pp. 545–552, 2012.

[18] H. W. Doerr and J. Cinatl, "Recent publications in medical microbiology and immunology: a retrospective," *Medical Microbiology and Immunology*, vol. 201, no. 1, pp. 1–5, 2012.

[19] M. M. EL-Sheikh, M. H. EL-Senaity, Y. B. Youssef, N. M. Shahein, and N. S. Abd Rabou, "Effect of ripening conditions on the properties of Blue cheese produced from cow's and goat's milk," *Journal of American Science*, vol. 7, no. 1, pp. 485–490, 2011.

[20] S. Gente, N. Durand-Poussereau, and M. Fevre, "Controls of the expression of aspA, the aspartyl protease gene from *Penicillium roqueforti*," *Molecular and General Genetics*, vol. 256, no. 5, pp. 557–565, 1997.

[21] A. Alber, S. E. Howie, W. A. Wallace, and N. Hirani, "The role of macrophages in healing the wounded lung," *International Journal of Experimental Pathology*, vol. 93, no. 4, pp. 243–251, 2012.

12

Defluorination of Sodium Fluoroacetate by Bacteria from Soil and Plants in Brazil

Expedito K. A. Camboim,[1] **Michelle Z. Tadra-Sfeir,**[2]
Emanuel M. de Souza,[2] **Fabio de O. Pedrosa,**[2] **Paulo P. Andrade,**[3] **Chris S. McSweeney,**[4]
Franklin Riet-Correa,[1] **and Marcia A. Melo**[1]

[1] Unidade Acadêmica de Medicina Veterinária, Universidade Federal de Campina Grande, Avenida Universitária, s/n, Bairro Sta. Cecília, Patos, PB, CEP: 58700-970, Brazil
[2] Laboratório de Fixação Biológica de Nitrogênio, Departamento de Bioquímica e Biologia Molecular, Universidade Federal do Paraná, Curitiba, PR, CEP: 81531-980, Brazil
[3] Departamento de Genética, Universidade Federal de Pernambuco, Recife, PE, CEP: 50670-901, Brazil
[4] CSIRO Livestock Industries, Queensland Bioscience Precinct, Carmody Road, 306, St Lucia, 4067, QLD, Australia

Correspondence should be addressed to Marcia A. Melo, marcia.melo@pq.cnpq.br

Academic Editor: Fumihiko Takeuchi

The aim of this work was to isolate and identify bacteria able to degrade sodium fluoroacetate from soil and plant samples collected in areas where the fluoroacetate-containing plants *Mascagnia rigida* and *Palicourea aenofusca* are found. The samples were cultivated in mineral medium added with $20\,mmol\,L^{-1}$ sodium fluoroacetate. Seven isolates were identified by 16S rRNA gene sequencing as *Paenibacillus* sp. (ECPB01), *Burkholderia* sp. (ECPB02), *Cupriavidus* sp. (ECPB03), *Staphylococcus* sp. (ECPB04), *Ancylobacter* sp. (ECPB05), *Ralstonia* sp. (ECPB06), and *Stenotrophomonas* sp. (ECPB07). All seven isolates degraded sodium-fluoroacetate-containing in the medium, reaching defluorination rate of fluoride ion of $20\,mmol\,L^{-1}$. Six of them are reported for the first time as able to degrade sodium fluoroacetate (SF). In the future, some of these microorganisms can be used to establish in the rumen an engineered bacterial population able to degrade sodium fluoroacetate and protect ruminants from the poisoning by this compound.

1. Introduction

In Brazil, thirteen species of plants causing sudden death associated with physical effort are responsible for nearly 500.000 cattle deaths each year: *Palicourea marcgravii, P. aeneofusca, P. juruana, P. grandiflora, Pseudocalymma elegans, Arrabidaea bilabiata, A. japurensis, Mascagnia rigida, M. elegans, M. pubiflora, M. aff. rigida, M. exotropia,* and *M. sepium* [1, 2]. Sodium fluoroacetate was identified as the active principle of *P. maracgravii* [3] and *A. bilabiata* [4] and is probably present in other plants of these genera. It disrupts the tricarboxylic acid (TCA) cycle, being first converted to fluorocitrate which in turn inhibits the enzymes aconitase and succinate dehydrogenase resulting in citrate accumulation in tissues and plasma and ultimately causing energy deprivation and death of the animal [5].

The use of sodium fluoroacetate is prohibited in Brazil, occurring exclusively as a natural product in plants. In Australia the compound is used in impregnated baits for the control of rabbit, fox, dingo, and other mammal populations; however, when introduced in the environment it can select fluoroacetate-degrading microorganisms [6].

Microbial degradation of sodium fluoroacetate is catalyzed by a haloacetate halidohydrolase, which is able to cleave the strong carbon-fluorine bond [7]. Twenty-four fluoroacetate-degrading microorganisms were isolated from soil in Central Australia, from seven bacterial genera (*Acinetobacter, Arthrobacter, Aureobacterium, Bacillus, Pseudomonas, Weeksella,* and *Streptomyces*) and four genera of fungi (*Aspergillus, Fusarium, Cryptococcus,* and *Penicillium*) [8].

The possibility to prevent fluoroacetate poisoning in ruminants by the ruminal inoculation of genetically modified bacteria, containing a gene encoding fluoroacetate dehalogenase has been investigated [9, 10]. A thorough search among samples taken from environment, such as soil, leaf, and digestive tract contents of herbivores that come in contact with fluoroacetate, would be important to ascertain the diversity and prevalence of microorganisms that can degrade these toxins. In the future, genes encoding enzymes from such microorganisms may be used to engineer new bacterial strains able to colonize the rumen and degrade fluoroacetate from toxic plants, thereby protecting the animal from poisoning by this compound.

This study aimed to isolate and identify bacteria able to degrade sodium fluoroacetate from soil and plant samples collected in the State of Paraíba, Brazil.

2. Materials and Methods

2.1. Samples Collection. Plant and soil samples were collected in the State of Paraíba, Brazil, in areas where *Mascagnia rigida* and *Palicourea aenofusca* were present. Soil samples were collected at the plant base, 1 to 8 cm depth. Leaves and flowers were also collected. All samples were placed in individual 50 mL Falcon type tubes and sent to the laboratory under refrigerated conditions for the immediate cultivation of associated bacteria.

2.2. Bacterial Isolation. Bacterial isolation was performed in 50 mL Falcon type tubes in mineral medium (Brunner) added with vitamins (http://www.dsmz.de/microorganisms/medium/pdf/DSMZ_Medium457.pdf) and 20 mmol L^{-1} sodium fluoroacetate (SF) (Sigma-Fluka) as single-carbon source. This medium will be here designated as Brunner medium. Samples were incubated at 28°C in an orbital shaker. After 48 hours, one mL of the first growth was transferred to test tubes containing nine mL of Brunner medium and incubated under the same conditions described above.

The SF defluorination was measured with an F$^-$ selective electrode (Thermo Electron Corporation) in 24-well plates containing 500 μL of culture and 500 μL of Total Ionic Strenght Adjustment Buffer-TISAB (diaminocyclohexane, sodium chloride, and glacial acetic acid, pH 5.5). The fluoride ion released from the microbial degradation of the sodium fluoroacetate was expressed in millimoles (mmol), the defluorination rate of 20 mmol L^{-1} corresponding to the release of 20 mmol L^{-1} F$^-$.

Samples showing SF defluorination were cultivated in serial dilutions from 10^{-1} to 10^{-9}. To obtain pure colonies the highest dilution that presented SF defluorination was plated on Brunner agar (Brunner medium added with agar 1%) and incubated at 28°C for 72 hours. Subsequently, each colony was used to inoculate three test tubes containing 9 mL of Brunner medium, which were monitored for SF defluorination. *Pseudomonas fluorescens* (strain DSM 8341) was used as positive control for fluoroacetate dehalogenase activity. Nine mL of Brunner medium without bacterial inoculum were incubated under the same conditions to evaluate the sodium fluoroacetate degradation background.

The standard sample (strain DSM 8341) and the bacteria isolated from soil and plants were grown into Brunner medium with increased concentrations of SF (20 mmol L^{-1}, 40 mmol L^{-1}, 60 mmol L^{-1}, and 80 mmol L^{-1} to 200 mmol L^{-1}) to evaluate the highest defluorination rate. Additionally, to evaluate defluorination in the presence of other carbon sources, strain DSM 8341 and the bacteria isolated from soil and plants were also grown in Brunner medium enriched with yeast extract and glucose, on the following conditions: (1) Brunner medium alone; (2) medium supplemented with yeast extract 0.01% and glucose 2%; (3) medium with yeast extract 0.01%; (4) medium with glucose 2%.

2.3. 16S rRNA Gene Sequence Identification. Bacteria displaying defluorination activity were identified by polymerase chain reaction (PCR) amplification and sequencing of the 16S rRNA gene. DNA extraction was performed with Brazol (LGC Biotechnology) according to the manufacturer's specifications. 16S rRNA gene was amplified in buffer containing 0.5 μM of 27f and 1492r universal primers [11], 2U of Taq DNA polymerase, 0.2 mM of dNTP and 100 ng of DNA and ultrapure water to a final volume of 20 μL. In the negative control, the DNA volume was substituted by ultrapure water. The amplified products were applied into agarose gel 1% and submitted to electrophoresis. DNA was stained with ethidium bromide and bands visualized with an imaging system (UVP-Bioimaging Systems).

The sequencing reaction was performed with BigDye kit according to manufacturer's recommendations (Applied Biosystems) and the product sequenced in the Genetic Analyzer 3500 XL sequencer (Applied Biosystems).

2.4. Sequence Analysis and Phylogram. 16S rRNA gene sequences were assembled with the CAP3 Sequence Assembly Program (http://pbil.univ-lyon1.fr/cap3.php). DNA sequences were analyzed by Basic Local Alignment Search Tool (BLAST) available on the website of the National Center for Biotechnology Information (NCBI—http://www.ncbi.nlm.nih.gov/BLAST). Species identification was based on maximum score, identity, and coverage values. The Greengenes database and workbench were used to corroborate species identification (http://greengenes.lbl.gov/). The phylogram tree was generated with MEGA 5 software using the default parameters (http://www.megasoftware.net/mega.php) [12, 13].

3. Results

Following the analysis of all 16S rRNA genes, seven isolates were identified as *Paenibacillus* sp. (ECPB01), *Burkholderia* sp. (ECPB02), *Cupriavidus* sp. (ECPB03), *Staphylococcus* sp. (ECPB04), *Ancylobacter* sp. (ECPB05), *Ralstonia* sp. (ECPB06), and *Stenotrophomonas* sp. (ECPB07) (Table 1). *Cupriavidus* sp. and *Ralstonia* sp. were isolated from both soil and plants samples, whereas *Paenibacillus* sp., *Burkholderia* sp. and *Ancylobacter* sp. were found from soil samples and, *Staphylococcus* sp. and *Stenotrophomonas* sp. were isolated from plant samples.

TABLE 1: Results of BLAST for the 16S rRNA sequences obtained from bacteria isolated from soil and plants samples.

Isolate code no.	Most similar species*	Isolated from	16S rRNA sequence length	Coverage (%)	Max score	E-value	Identity (%)
ECPB01	*Paenibacillus* sp.	Soil	1425	99	2619	0.0	99
ECPB02	*Burkholderia* sp.	Soil	1398	99	2553	0.0	99
ECPB03	*Cupriavidus* sp.	Soil and plant	1398	99	2540	0.0	99
ECPB04	*Staphylococcus* sp.	Plant	1402	100	2569	0.0	99
ECPB05	*Ancylobacter* sp.	Soil	1368	99	2435	0.0	99
ECPB06	*Ralstonia* sp.	Soil and plant	1407	99	2551	0.0	99
ECPB07	*Stenotrophomonas* sp.	Plant	1417	99	2606	0.0	99

*Genera were identified based on maximum score, identity, and coverage.

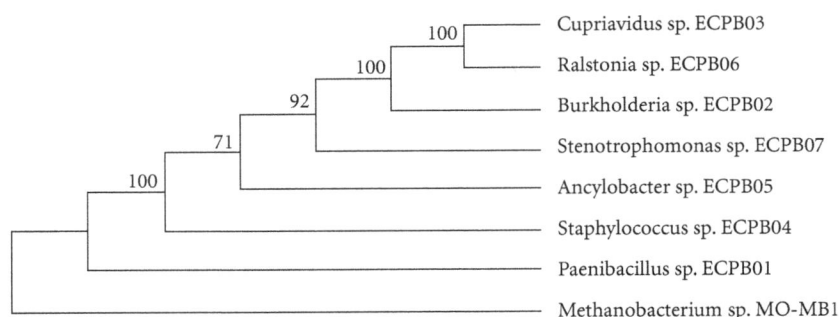

FIGURE 1: Phylogenetic tree based on 16S rRNA sequences by Maximum Parsimony analysis. ECPB01 to ECPB07 represent the isolate code and *Methanobacterium* sp. MO-MB1 (gi|311141366|dbj|AB598270.1|) the outgroup. The evolutionary history was inferred using the Maximum Parsimony method. The bootstrap consensus tree inferred from 1000 replicates is taken to represent the evolutionary history of the taxa analyzed. Branches corresponding to partitions reproduced in less than 50% bootstrap replicates are collapsed. The percentage of replicate trees in which the associated taxa clustered together in the bootstrap test (1000 replicates) is shown next to the branches. The MP tree was obtained using the Close-Neighbor-Interchange algorithm with search level 1 in which the initial trees were obtained with the random addition of sequences (10 replicates). The analysis involved 8 nucleotide sequences. All positions containing gaps and missing data were eliminated. There were a total of 1235 positions in the final dataset. Evolutionary analyses were conducted in MEGA5.

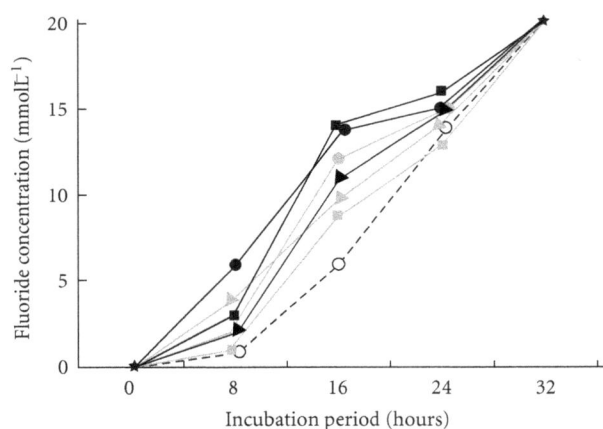

FIGURE 2: Sodium fluoroacetate degradation rate by bacteria isolated from both soil and plants in the State of Paraíba, Brazil. Symbols are as follows. The white circle: ECPB01; the black triangle: ECPB02; the black square: ECPB03; the grey circle: ECPB04; the grey square: ECPB05; the grey triangle: ECPB06; the black circle: ECPB07.

In order to further support the attribution of genera based on BLAST maximum scores, identity and coverage values,

a phylogram tree was built from the sequences previously trimmed to have the same length and indicated that five isolates were phylogenetically closely related (Figure 1): *Ancylobacter* sp., three Burkholderiaceae (*Cupriavidus* sp., *Ralstonia* sp., *Burkholderia* sp.) and *Stenotrophomonas* sp. Two other species, *Paenibacillus* sp. and *Staphylococcus* sp. formed a heterogeneous group.

All bacterial isolates displayed SF degradation activity, reaching a 20 mmol L^{-1} level of released fluoride ion 32 hours after incubation in Brunner medium containing 20 mmol L^{-1} of SF (Figure 2). The same result was observed with the *Pseudomonas fluorescens* control strain (DSM 8341). There was no release of fluoride ions when the Brunner medium was incubated in the absence of bacteria.

When samples were grown in different concentrations of SF the maximum rate of defluorination was 140 mmol L^{-1} F$^-$ in 140 mmol L^{-1} of concentration. At higher SF concentrations (160 mmol L^{-1}, 180 mmol L^{-1} and 200 mmol L^{-1}) the levels of degradation were still 140 mmol L^{-1}. When the isolates were cultured into Brunner medium added with others carbon sources (yeast extract 0.01% and glucose 2%), an intense bacterial growth was observed after 24 hours, but SF degradation achieved only 20 mmol L^{-1} F$^-$ between 48 to 64 hours with 20 mmol L^{-1} SF initial concentration.

4. Discussion

Ancylobacter sp., the Burkholderiaceae (*Cupriavidus* sp., *Ralstonia* sp. and *Burkholderia* sp.) and *Stenotrophomonas* sp. belong to the Alpha, Beta and Gamma-proteobacteria classes, respectively. The two other species, *Paenibacillus* sp. and *Staphylococcus* sp. are from *Paenibacillaceae* and *Staphylococcaceae* families, respectively, forming a heterogeneous group; the genus assignment based on BLAST values was coherent with the phylogram, which is in agreement with the phylogeny of these bacteria. The existence of a similar haloacid dehalogenase activity among distantly related microorganisms isolated over a restricted area points towards a common and effective selective pressure acting in a broad set of soil microorganisms.

Annotated genomic sequences are available at the NCBI genome database for the species *Burkholderia* sp., *Cupriavidus taiwanensis*, *Staphylococcus saprophyticus*, *Paenibacillus* sp., *Ralstonia* sp., and *Stenotrophomonas maltophilia*. The *Ancylobacter dichloromethanicus* genome has not yet been sequenced, but there is 16S rRNA sequence for this bacterial species deposited in Genbank. For the genera listed above only once was the detoxifying enzyme designated as fluoroacetate dehalogenase (Fac-Dex FA1) for *Burkholderia* sp. [14]. For *Cupriavidus* sp., *Staphylococcus* sp., *Paenibacillus* sp., *Ralstonia* sp., *Stenotrophomonas* sp., and *Ancylobacter* sp. the dehalogenases were annotated as haloacid dehalogenase hydrolase domain-containing protein [15, 16], haloacid dehalogenase-like hydrolase [17, 18], 2-haloalkanoic acid dehalogenase [17, 18], haloacetate dehalogenase [18, 19], and dichloromethane dehalogenase [20]. However, when the seven isolated were grown in the presence of SF the substrate was defluorinated. This finding may be explained by the unspecificity of the dehalogenases, which can use as substrates other compounds structurally similar to SF, which is known as cross-adaptation [21]. Liu and colleagues [22] verified that fluoroacetate dehalogenase degrades other halogenated compounds such as chloroacetate, bromoacetate, iodoacetate, and dichloroacetate. Similar results were obtained by Donnelly and Murphy [23] that found fluoroacetate dehalogenase activity catalyzing chloroacetate, bromoacetate, and ethyl fluoroacetate. Also sodium fluoroacetate can be defluorinated by L-2 haloacid dehalogenase [24]. Another possibility is the lateral transfer of fluoroacetate dehalogenase genes.

Although a couple of other environmental bacteria such as *Moraxella* sp. [25], *Acinetobacter*, *Arthrobacter*, *Aureobacterium*, *Bacillus*, *Pseudomonas*, *Weeksella*, and *Streptomyces* [8] also present fluoroacetate dehalogenase activity, there are no previous reports on such an activity from microorganisms belonging to any of the six species described here: *Cupriavidus* sp., *Staphylococcus* sp., *Paenibacillus* sp., *Ralstonia* sp., *Stenotrophomonas* sp., and *Ancylobacter* sp., since genes for haloacid dehalogenases were annotated in the genomes of the first five species. It is therefore reasonable to assume that all seven isolates have the fluoroacetate genes in their chromosomes and not in plasmids, which further support the existence of a strong, durable and common environmental selective pressure over these organisms.

In all isolates defluorination occurred up to 32 hours of cultivation, with the release of 20 mmol L^{-1} fluoride ion in the presence of 20 mmol L^{-1} sodium fluoroacetate. These results were similar to those reported by Davis and colleagues [26] with *Burkholderia* sp., which also degraded SF in 32 hours, releasing 20 mmol L^{-1} F^- with 20 mmol L^{-1} of SF.

Growing the strains in different concentrations of SF indicated that the highest degradation rate was 140 mmol L^{-1} F^- with 140 mmol L^{-1} of substrate. For higher fluoride concentrations the defluorination rate did not increase, which may be due to exhaustion of other nutrients.

When the bacteria were cultivated in Brunner medium added with yeast extract and glucose, the dehalogenase activity was delayed, probably due to the availability of the others energy sources and to the difficult cleavage of the strong carbon-fluorine bond in fluoroacetate [7]. In conclusion, seven fluoroacetate degrading bacteria were isolated from both soil and plants, six of which were not previously reported as able to degrade SF. The possibility to use some of these bacteria to establish in the rumen an engineered bacterial population able to degrade fluoroacetate and protect ruminants from the poisoning by this compound should be explored in the future.

Acknowledgment

This work was supported by the INCT for the control of plant poisonings: Grant no. 573534/2008-0.

References

[1] C. H. Tokarnia, P. V. Paixoto, and J. Dobereiner, "Toxic plants affecting heart function of cattle in Brazil," *Pesquisa Veterinária Brasileira*, vol. 10, pp. 1–10, 1990.

[2] S.V. Schons, T. L. Mello, F. Riet-Correa, and A. L. Schild, "Poisoning by amorimia *(Mascagnia) sepium* in sheep in Northern Brazil," *Toxicon*, vol. 57, pp. 781–786, 2011.

[3] M. M. Oliveira, "Chromatographic isolation of monofluoroacetic acid from *Palicourea marcgravii St. Hil*," *Experientia*, vol. 19, no. 11, pp. 586–587, 1963.

[4] H. C. Krebs, W. Kemmerling, and G. Habermehl, "Qualitative and quantitative determination of fluoroacetic acid in *Arrabidea bilabiata* and *Palicourea marcgravii* by 19F-NMR spectroscopy," *Toxicon*, vol. 32, no. 8, pp. 909–913, 1994.

[5] P. A. Mayes, "O ciclo do ácido cítrico: o catabolismo da acetil-coA," in *Harper: Bioquímica*, K. R. Murray et al., Ed., pp. 182–189, Atheneu, São Paulo, Brazil, 9th edition, 2002.

[6] L. E. Twigg and D. R. King, "The impact of fluoroacetate-bearing vegetation on native Australian fauna: a review," *Oikos*, vol. 61, no. 3, pp. 412–430, 1991.

[7] S. Fetzner and F. Lingens, "Bacterial dehalogenases: biochemistry, genetics, and biotechnological applications," *Microbiological Reviews*, vol. 58, no. 4, pp. 641–685, 1994.

[8] L. E. Twigg and L. V. Socha, "Defluorination of sodium monofluoroacetate by soil microorganisms from central Australia," *Soil Biology and Biochemistry*, vol. 33, no. 2, pp. 227–234, 2001.

[9] K. Gregg, C. L. Cooper, D. J. Schafer et al., "Detoxification of the plant toxin fluoroacetate by a genetically modified rumen bacterium," *Nature*, vol. 12, no. 13, pp. 1361–1365, 1994.

[10] K. Gregg, B. Hamdorf, K. Henderson, J. Kopecny, and C. Wong, "Genetically modified ruminal bacteria protect sheep

from fluoroacetate poisoning," *Applied and Environmental Microbiology*, vol. 64, no. 9, pp. 3496–3498, 1998.

[11] W. G. Weisburg, S. M. Barns, D. A. Pelletier, and D. J. Lane, "16S ribosomal DNA amplification for phylogenetic study," *Journal of Bacteriology*, vol. 173, no. 2, pp. 697–703, 1991.

[12] J. Felsenstein, "Confidence limits on phylogenies: an approach using the bootstrap," *Evolution*, vol. 39, pp. 783–791, 1985.

[13] K. Tamura, D. Peterson, N. Peterson, G. Stecher, M. Nei, and S. Kumar, "MEGA5: molecular evolutionary genetics analysis using maximum likelihood, evolutionary distance, and maximum parsimony methods," *Molecular Biology and Evolution*, vol. 28, no. 10, pp. 2731–2739, 2011.

[14] T. Kurihara, T. Yamauchi, S. Ichiyama, H. Takahata, and N. Esaki, "Purification, characterization, and gene cloning of a novel fluoroacetate dehalogenase from *Burkholderia* sp. FA1," *Journal of Molecular Catalysis B*, vol. 23, pp. 347–355, 2003.

[15] S. Lucas, A. Copeland, A. Lapidus et al., "*Paenibacillus* sp. Y412MC10, complete genome," 2011, http://www.ncbi.nlm.nih.gov/.

[16] S. Lucas, A. Copeland, A. Lapidus et al., "*Stenotrophomonas maltophilia* R551-3, complete genome," 2011, http://www.ncbi.nlm.nih.gov/.

[17] M. Kuroda, A. Yamashita, H. Hirakawa et al., "Whole genome sequence of *Staphylococcus saprophyticus* reveals the pathogenesis of uncomplicated urinary tract infection," *Proceedings of the National Academy of Sciences of the United States of America*, vol. 102, no. 37, pp. 13272–13277, 2005.

[18] C. Amadou, G. Pascal, S. Mangenot et al., "Genome sequence of the beta-rhizobium *Cupriavidus taiwanensis* and comparative genomics of rhizobia," *Genome Research*, vol. 18, pp. 1472–1483, 2008.

[19] D. Ward, A. Earl, M. Feldgarden et al., "The genome sequence of *Ralstonia* sp. strain 5_7_47FAA," 2011, http://www.ncbi.nlm.nih.gov/.

[20] J. E. Firsova, N. V. Doronina, E. Lang, C. Spröer, S. Vuilleumier, and Y. A. Trotsenko, "Ancylobacter dichloromethanicus sp. nov.—a new aerobic facultatively methylotrophic bacterium utilizing dichloromethane," *Systematic and Applied Microbiology*, vol. 32, no. 4, pp. 227–232, 2009.

[21] I. S. Melo and J. L. Azevedo, "Como isolar microrganismos degradadores de moléculas xenobióticas," in *Microbiologia Ambiental*, I. S. Melo and J. L. Azevedo, Eds., pp. 167–183, EMBRAPA-CNPMA, Jaguariuna, Brazil, 1997.

[22] J. Q. Liu, T. Kurihara, S. Ichiyama et al., "Reaction mechanism of fluoroacetate dehalogenase from *Moraxella* sp. B," *Journal of Biological Chemistry*, vol. 273, no. 47, pp. 30897–30902, 1998.

[23] C. Donnelly and C.D. Murphy, "Purification and properties of fluoroacetate dehalogenase from Pseudomonas fluorescens DSM 8341," *Biotechnology Letters*, vol. 31, pp. 245–250, 2009.

[24] W. Y. Chan, M. Wong, J. Guthrie et al., "Sequence- and activity-based screening of microbial genomes for novel dehalogenases," *Microbial Biotechnology*, vol. 3, no. 1, pp. 107–120, 2010.

[25] H. Kawasaki, K. Tsuda, I. Matsushita, and K. Tonomura, "Lack of homology between two haloacetate dehalogenase genes encoded on a plasmid from *Moraxella* sp. strain B," *Journal of General Microbiology*, vol. 138, no. 7, pp. 1317–1323, 1992.

[26] C. K. Davis, S. E. Denman, L. I. Sly, and C. S. Mcsweeney, "Development of a colorimetric colony-screening assay for detection of defluorination by micro-organisms," *Letters in Applied Microbiology*, vol. 53, no. 4, pp. 417–423, 2011.

13

In Vitro Synergistic Effect of *Psidium guineense* (Swartz) in Combination with Antimicrobial Agents against Methicillin-Resistant *Staphylococcus aureus* Strains

Tiago Gomes Fernandes,[1] **Amanda Rafaela Carneiro de Mesquita,**[1] **Karina Perrelli Randau,**[2] **Adelisa Alves Franchitti,**[3] **and Eulália Azevedo Ximenes**[1]

[1] *Laboratório de Fisiologia e Bioquímica de Microorganismos, Departamento de Antibióticos, Centro de Ciências Biológicas, Universidade Federal de Pernambuco, 50670-901 Recife, PE, Brazil*
[2] *Laboratório de Farmacognosia, Departamento de Farmácia Universidade Federal de Pernambuco, Centro de Ciências da Saúde, 50670-901 Recife, PE, Brazil*
[3] *Department of Biochemistry, Kansas State University, 141 Chalmers Hall, Manhattan, KS 66506, USA*

Correspondence should be addressed to Eulália Azevedo Ximenes, eulaliaximenes@yahoo.com.br

Academic Editors: W. A. Gebreyes and T. Hatano

The aim of this study was to evaluate the antimicrobial activity of aqueous extract of *Psidium guineense* Swartz (Araçá-do-campo) and five antimicrobials (ampicillin, amoxicillin/clavulanic acid, cefoxitin, ciprofloxacin, and meropenem) against twelve strains of *Staphylococcus aureus* with a resistant phenotype previously determined by the disk diffusion method. Four *S. aureus* strains showed resistance to all antimicrobial agents tested and were selected for the study of the interaction between aqueous extract of *P. guineense* and antimicrobial agents, by the checkerboard method. The criteria used to evaluate the synergistic activity were defined by the fractional inhibitory concentration index (FICI). All *S. aureus* strains were susceptible to *P. guineense* as determined by the microdilution method. The combination of the *P. guineense* extract with the antimicrobial agents resulted in an eight-fold reduction in the MIC of these agents, which showed a FICI ranging from 0.125 to 0.5, suggesting a synergistic interaction against methicillin-resistant *Staphylococcus aureus* (MRSA) strains. The combination of the aqueous extract of *P. guineense* with cefoxitin showed the lowest FICI values. This study demonstrated that the aqueous extract of *P. guineense* combined with beta lactamics antimicrobials, fluoroquinolones, and carbapenems, acts synergistically by inhibiting MRSA strains.

1. Introduction

The widespread emergence of bacteria resistance to a large number of antimicrobial agents poses major health problems because of difficulties in treatment [1]. *Staphylococcus aureus* is an important pathogen both in community acquired and nosocomial infections and a common etiological agent of infections of many different tissues and organs (e.g., furuncle, carbuncle, abscess, myocarditis, endocarditis, pneumonia, meningitis, bacterial arthritis, and osteomyelitis) [2]. The multidrug resistant *Staphylococcus aureus* strains now pose serious problems to hospitalized patients and their care providers. The organism has successfully evolved numerous strategies for resisting, the action to practically all antimicrobial agents [3]. Thus, it is extremely important to find new antimicrobial agents or new ways for the treatment of infectious diseases caused by multidrug resistant microorganisms [4].

The screening of plant extracts and phytochemicals for antimicrobial activity has shown that higher plants, especially their secondary metabolites, are a source to provide structurally diverse bioactive compounds with different pharmacological activities, including antimicrobials [5].

According to Hemaiswarya et al. [6], plants-derived antimicrobials are less potent, which enhances the need to adopt a synergistic interaction between its bioactive compounds to combat infections. Based on this knowledge, several studies have been published using crude plant extracts or phytochemicals combined with antimicrobial agents. These combinations can enhance the efficacy of the antimicrobial agents and are an alternative to treat infections caused by multidrug resistant microorganisms, which do not have any effective therapy available [4, 7, 8].

Psidium guineense Swartz (Myrtaceae), popularly known as wild guava, is a shrub native and widely dispersed throughout Tropical America. In Brazil, it mainly occurs across the coastline [9, 10].

Few studies have been focused on the phytochemical profile and pharmacological activity of this plant; however, it is widely used in folk medicine in regions of South America to treat infections of the gastrointestinal and genitourinary tract [11]. These authors associated the presence of mainly flavonoids and tannins in the fruits and leaves of *P. guineense* to be effective against *Streptococcus mutans* strains.

The aim of the present study was to evaluate the antimicrobial activity of the aqueous extract of *P. guineense* and to determine the synergistic potential of the combination of this extract with five antimicrobial agents against MRSA strains.

2. Materials and Methods

2.1. Plant Material and Extract Preparation. The plant materials used in this study consisted of *Psidium guineense* Swartz (leaves) that were collected in November/2009 from Moreno city, Pernambuco, Brazil, (08°07′07″S-35°05′32″W). This plant was identified by the biologist Olivia Cano, and the voucher specimen was deposited in the Herbarium Dárdano de Andrade Lima, Instituto Agronômico de Pernambuco and indexed under registration 83.564.

The leaves were air-dried for two weeks and then grinded into fine powder using an electric dry mill. A total of 50 g of the powder was soaked in 1000 mL of boiled distilled water and extracted exhaustively at room temperature for 48 hours. The extracts were filtered through Whatman No. 2 filter paper. The aqueous extract was lyophilized to obtain a dry powder extract.

2.2. Phytochemical Screening. The preliminary phytochemical screening was carried out by various secondary metabolites present on aqueous extract of *P. guineense* using standard procedures [12–14]. The chromatographic analysis was made by TLC on Si gel (MERCK, Germany, 105553) developed by different solvent systems: (EtOAc-HCOOH–AcOH-H_2O (100 : 11 : 11 : 26, v/v)) to investigate the presence of saponins, flavonoids, tannins, phenylpropanoids, alkaloids, condensed proanthocyanidins, leucoanthocyanidins, and iridoids and (n-BuOH–Me_2CO-buffer phosphate pH = 5.0 (40 : 50 : 10 v/v)) to evaluate the presence of sugars.

2.3. Total Polyphenol Content (TPC). The total polyphenol content (TPC) was determined colorimetrically by using Folin-Ciocalteu's reagent and expressed as pyrogallol, according to the Brazilian Pharmacopoeia with modifications [15]. Thus, samples of aqueous extract of *P. guineense*

(1000 μg/mL) were prepared in distilled water. Concentrations (10 to 30 μg/mL) were measured. The standard pyrogallol (1000 μg/mL) was prepared in distilled water. 80 microliters of each solution was transferred to a 25 mL flask containing distilled water (10 mL) and Folin-Ciocalteu's reagent (1 mL). The volume was completed with anhydrous sodium carbonate solution (10.75% w/v), resulting in final concentrations of 3.2 μg/mL of pyrogallol. A standard curve was determined with concentrations between 1.6 and 4.8 μg/mL (r^2 = 0.999). Final results were given as pyrogallol equivalents. The samples were scanned in a UV/VIS spectrophotometer (Evolution 60S, Thermo Scientific), and measurements were obtained 15 min after addition of the sodium-carbonate solution and scanning ranging from 500 to 900 nm wavelength. Distilled water was used as a blank. All measurements were performed in triplicate.

2.4. Bacterial Strains and Inoculum Standardization. S. aureus strains (n = 11) were isolated from clinical specimens and food. The standard strain used was *S. aureus* ATCC 25923. Strains were isolated in sheep blood agar and after identification they were stored in brain heart infusion (BHI) plus glycerol 20% v/v [16]. The *S. aureus* strains used in this study showed a resistant phenotype, by the diffusion method, to several antimicrobial agents such as beta-lactams, aminoglycosides, macrolides, fluoroquinolones, tetracycline, chloramphenicol, and lincosamides. These strains were cultured onto Mueller Hinton Agar (MHA) (Acumedia Manufacturers, Baltimore, USA) and incubated at 37°C for 18 hours. Single colonies were selected and inoculated into Mueller-Hinton broth (Acumedia Manufacturers, Baltimore, USA) to turbidity comparable to that of 0.5 McFarland standard, which is equivalent to a bacterial count of approximately 10^8 CFU/mL.

After that, the bacterial suspension was diluted in saline (1 : 10) to obtain a final inoculum of 10^7 CFU/mL.

2.5. Antimicrobial Agents. The standard reference powders of ampicillin; amoxicillin/clavulanic acid; cefoxitin; ciprofloxacin; gentamicin; meropenem were provided by Eurofarma Laboratório LTDA, Brazil. Resistance was defined for each case: ampicillin (AMP, MIC ≥ 0.25 μg/mL); amoxicillin/clavulanic acid (AMC, MIC ≥ 8 μg/mL); cefoxitin (CFO, MIC ≥ 4 μg/mL); ciprofloxacin (CIP, MIC ≥ 4 μg/mL); gentamicin (GEN, MIC ≥ 8 μg/mL; meropenem (MER, MIC ≥ 16 μg/mL).

2.6. Antimicrobial Activity. The Minimal Inhibitory Concentration (MIC) test was performed by the microdilution broth method, following the recommendations established by Clinical Laboratory Standards Institute, [17], with some modifications. Serial two-fold dilutions of aqueous extract of *P. guineense* and antimicrobial agents were prepared in sterile 96-well microplates containing Mueller Hinton broth (MHB). Five microliters of bacterial suspension were inoculated in each well to give a final concentration of 10^4 CFU/mL. *P. guineense* extract and antimicrobial agents concentrations ranged from 7.25 to 1000 μg/mL and 3.12 to

In Vitro Synergistic Effect of Psidium guineense (Swartz) in Combination with Antimicrobial Agents against
Methicillin-Resistant Staphylococcus aureus Strains

81

$400 \,\mu g/mL$, respectively. The growth inhibition was demonstrated by optical density at 630 nm using a microplate reader (Thermo plate—TP Reader). Considering the total growth (100%) in the control well (MHB + bacteria), the percentage of growth reduction was attributed to the remaining wells. The MIC was reported as the lowest concentration of *P. guineense* extract or antimicrobial agents that inhibited the bacterial growth after 24 h of incubation at 37°C. In order to determine the Minimal Bactericidal Concentration (MBC), the contents of the well that showed higher or equal than 70% of growth inhibition were seeded into MHA. After 24 h of incubation at 37°C, the number of surviving *S. aureus* was determined. The MBC was defined as the lowest extract concentration at which 99.9% of the bacteria have been killed. All experiments were carried out in duplicate on two different days.

2.7. Determination of In Vitro Synergistic Activity. Combinations of *P. guineense* and antimicrobial agents were tested by the checkerboard method. The appropriate dilution of *P. guineense* extract and antimicrobial agents were performed into MHB. From these dilutions, one hundred microliters were added in 96-well microplates to obtain a final concentration equal to MIC or six dilutions lower than MIC to *P. guineense* and nine dilutions lower than MIC to antimicrobial agents. Each well received $5 \,\mu L$ of the bacterial suspensions (10^7 CFU/mL). Plates were incubated for 24 hours. Interpretation of the data was achieved by calculating the fractional inhibitory concentration index (FICI) as follows: (MIC of *P. guineense* in combination with antimicrobial agents/MIC of *P. guineense*) + (MIC of antimicrobial agents in combination with *P. guineense*/MIC of antimicrobial agents). The combination was considered to be synergistic when the FICI was ≤0.5, additive when it was 0.5 to ≤1, and antagonistic when ≥2 [18].

3. Results

3.1. Phytochemical Profile. The phytochemical profile from the *Psidium guineense* extract showed the presence of tannins, flavonoids, condensed proanthocyanidins, leucoanthocyanidins, and sugar. On the other hand, the presence of alkaloids, phenylpropanoid, and saponins was not verified. The aqueous extract of *P. guineense* yielded 3.7% (w/w) over 50 g of plant material.

Concerning the total polyphenol content, the results of this study suggest that the aqueous extract of *P. guineensis* show an important amount of such secondary metabolites. Thus, the total polyphenol content was 21.62 ± 0.40 g% (1.51%) as pyrogallol equivalent.

3.2. Antimicrobial Activity. MIC and MBC values of *P. guineense* extract and of the ampicillin, amoxicillin/clavulanic acid, cefoxitin, ciprofloxacin, gentamicin, and meropenem against twelve *S. aureus* strains are shown in Table 1.

The aqueous extract of *P. guineense* showed a strong activity against all *S. aureus* strains with MIC values between 250 and 500 $\mu g/mL$. The *S. aureus* strains revealed a resistance

profile against most antimicrobial agents tested, in particular to the beta-lactam antibiotics. The MIC values for ampicillin and cefoxitin ranged from 3.12 to $400 \,\mu g/mL$ which showed to be less effective against the *S. aureus* strains tested. For amoxicillin/clavulanic acid, the values ranged from 3.12 to $50 \,\mu g/mL$. Among all *S. aureus* strains tested, seven showed to be resistant to ciprofloxacin. For meropenem, MIC values ranged from 3.12 to $25 \,\mu g/mL$. All strains showed to be sensitive to gentamicin, except LFBM 26, LFBM 28, LFBM 33. Four strains of *S. aureus* (LFBM 01, LFBM 26, LFBM 28, LFBM 33) showed resistance to all antimicrobial agents tested. This resistance profile selected the microorganisms for the study of the interaction between *P. guineense* extract and antimicrobial agents. The values of MBC were higher than MIC in one dilution.

3.3. Determination of In Vitro Synergistic Activity. The MICs obtained by the combination of the aqueous extract of *P. guineense* with ampicillin, amoxicillin/clavulanic acid, cefoxitin, ciprofloxacin, and meropenem, against *S. aureus* strains (LFBM 01, LFBM 26, LFBM 28, LFBM 33) are listed in Table 2. The minimal inhibitory concentrations of *P. guineense* extract ($250 \,\mu g/mL$) enhanced the antistaphylococcal activity of all antimicrobial agents. The synergistic activity was detected by an eight-fold decrease in the MIC of the antimicrobial agents in the combination and determined by the FICI ≤0.5.

The combination of the aqueous extract of *P. guineense* and cefoxitin showed the lowest FICI whose values ranged from 0.125 to 0.5. The MIC of cefoxitin (individual MIC 100–400 $\mu g/mL$) was lowered to 1/512 (combined MIC ranged 0.19–0.78 $\mu g/mL$), when it was used in combination with the aqueous extract of *P. guineense* (MIC 250 $\mu g/mL$). For *S. aureus* LFBM 26 and LFBM 33 strains, this combination was also efficient on reducing the MIC of *P. guineense* extract to 1/8 × MIC (31.25 $\mu g/mL$) or 1/4 × MIC (62.25 $\mu g/mL$), respectively.

The inhibition of the growth of all the *S. aureus* strains by meropenem was enhanced by *P. guineense* extract (FICI 0.5). This combination was more effective against LFBM 28 (FICI 0.25) which individual MIC of *P. guineense* (250 $\mu g/mL$) lowered to 1/4 × MIC.

For *S. aureus* LFBM 33 strain, the MIC value of ciprofloxacin (individual MIC 50 $\mu g/mL$) was lowered to 0.78 $\mu g/mL$ when combined with *P. guineense* extract (1/4 × MIC) enhancing the antistaphylococcal activity of this fluoroquinolone (FICI 0.25). All other combinations of the aqueous extract of *P. guineense* with antimicrobial agents demonstrated synergistic activity, enhancing its activity, with FICI 0.5.

4. Discussion

Phytochemicals have great potential as antimicrobial compounds and have been proven to have great therapeutic potential [19]. Some secondary metabolites have the ability to increase the susceptibility of the microorganism. When used in combination, these metabolites have the potential to either inhibit the modified targets or exhibit a synergy by

TABLE 1: MIC/MBC of the aqueous extract of *Psidium guineense* and antimicrobial agents against MRSA strains.

Staphylococcus aureus	MIC/MBC (μg/mL)							
	AMC	AMP	CEF	CIP	GET	MER	AE	Resistance phenotype[1]
ATCC 25923	3.12/6.24	3.12/6.24	3.12/6.24	3.12/6.24	0.39/0.78	3.12/6.24	250/500	Control strain
LFBM 01	25.0/50.0	400/800	400/800	50.0/100	12.5/25.0	25.0/50.0	250/500	AMC, CFO, CIP, GET
LFBM 05	6.25/12.5	100/200	25/50.0	3.12/6.24	3.12/6.24	12.5/25.0	250/500	AMP, AZI, CFO, PEN
LFBM 08	3.12/6.24	100/200	12.5/25.0	3.12/6.24	3.12/6.24	6.25/12.5	250/500	AMP; AZI, CIP, CFO
LFBM 16	12.5/25.0	25.0/50.0	12.5/25.0	3.12/6.24	3.12/6.24	6.25/12.5	250/500	AMP, AZI, CFO, PEN
LFBM 26	50.0/100	400/800	400/800	50.0/100	25.0/50.0	25.0/50.0	250/500	AMC, AMP, CFO, CIP
LFBM 28	25.0/50.0	400/800	100/200	50.0/100	25.0/50.0	12.5/25.0	250/500	AMC, CIP, CFO, GET
LFBM 29	12.5/25.0	50.0/100	25/50.0	3.12/6.24	3.12/6.24	12.5/25.0	500/1000	AMC, AMP, CFO, PEN
LFBM 30	12.5/25.0	200/400	12.5/25.0	50.0/100	3.12/6.24	12.5/25.0	250/500	AMC, AMP, CFO, CIP
LFBM 31	12.5/25.0	100/100	12.5/25.0	50.0/100	3.12/6.24	6.25/12.5	250/500	AMP, AZI, CFO; CIP;
LFBM 32	12.5/25.0	200/200	12.5/25.0	50.0/100	3.12/6.24	12.5/25.0	250/500	AMP, CFO, CIP, PEN
LFBM 33	12.5/25.0	200/200	200/200	50.0/100	3.12/6.24	12.5/25.0	250/500	AMC, AMP, CFO, CIP

MIC: minimal inhibitory concentration; MBC: minimal bactericidal concentration; ATCC: American Type Culture Collection; LFBM: Laboratório de Fisiologia e Bioquímica de Microorganismos; AMC: amoxicillin/clavulanic acid; AMP: ampicillin; AZI: azithromycin; CFO: cefoxitin; CIP: ciprofloxacin; GET: gentamicin; PEN: penicillin; MER: meropenem; AE: aqueous extract of *P. guineense*.
[1]Resistance phenotype determined by the disk-diffusion method.

TABLE 2: Combination testing of the aqueous extract of *Psidium guineense* with antimicrobial agents against MRSA strains.

Staphylococcus aureus	Combination	Individual MIC (μg/mL)	Combination MIC (μg/mL)	Individual FIC	FIC index (FICI)	Interpretation	% MIC reduced	% reduction of viable cells[1]
LFBM 01	AE + AMC	250/25	125/0.048	0.5/0.002	0.5	Synergistic	50.0/99.8	80.21
	AE + AMP	250/400	125/0.78	0,5/0.002	0.5	Synergistic	50.0/99.8	84.43
	AE + CFO	250/400	125/0.78	0.5/0.002	0.5	Synergistic	50.0/99.8	87.04
	AE + CIP	250/50	125/0.09	0.5/0.002	0.5	Synergistic	50.0/99.8	93.49
	AE + MER	250/25	125/0.048	0.5/0.002	0.5	Synergistic	50.0/99.8	84.24
LFBM 26	AE + AMC	250/50.0	125/0.78	0.5/0.002	0.5	Synergistic	50.0/99.8	88.82
	AE + AMP	250/400	125/0.78	0.5/0.002	0.5	Synergistic	50.0/99.8	83.58
	AE + CFO	250/400	31.25/0.78	0.125/0.002	0.125	Synergistic	87.5/99.8	80.90
	AE + CIP	250/50.0	125/0.09	0.5/0.002	0.5	Synergistic	50.0/99.8	94.59
	AE + MER	250/25	125/0.048	0.5/0.002	0.5	Synergistic	50.0/99.8	80.76
LFBM 28	AE + AMC	250/12.5	125/0.024	0.5/0.002	0.5	Synergistic	50.0/99.8	83.38
	AE + AMP	250/400	125/0.78	0.5/0.002	0.5	Synergistic	50.0/99.8	89.71
	AE + CFO	250/100	125/0.19	0.5/0.002	0.5	Synergistic	50.0/99.8	86.91
	AE + CIP	250/50.0	125/0.09	0.5/0.002	0.5	Synergistic	50.0/99.8	94.45
	AE + MER	250/25.0	62.5/0.048	0.25/0.002	0.25	Synergistic	75.0/99.8	89.08
LFBM 33	AE + AMC	250/12.5	125/0.024	0.5/0.002	0.5	Synergistic	50.0/99.8	78.89
	AE + AMP	250/200	125/0.39	0.5/0.002	0.5	Synergistic	50.0/99.8	76.72
	AE + CFO	250/200	62.5/0.39	0.25/0.002	0.25	Synergistic	75.0/99.8	81.53
	AE + CIP	250/50.0	62.5/0.78	0.25/0.002	0.25	Synergistic	75.0/99.8	94.09
	AE + MER	250/12.5	125/0.024	0.5/0.002	0.5	Synergistic	50.0/99.8	81.84

MIC: minimal inhibitory concentration; FIC: fractional inhibitory concentration; AMC: amoxicillin/clavulanic acid; AMP: ampicillin; CFO: cefoxitin; CIP: ciprofloxacin; MER: meropenem; AE: aqueous extract of *P. guineense*; LFBM: Laboratório de Fisiologia e Bioquímica de Microorganismos.
% of MIC reduced = $(MIC_{alone} - MIC_{combined}) \times 100/MIC_{alone}$.
[1]%reduction of viable cells in wells containing the combination aqueous extract of *P. guineense* (1/2 MIC) and antimicrobials agents (1/512 MIC).

blocking one or more of the targets in the metabolic pathway, acting as a modifier of multidrug resistance mechanisms [6].

Several studies have investigated the interactions between antibiotics and phytochemicals or crude extracts against MRSA strains [4, 7, 8, 20].

In the Myrtaceae family, there is a large variety of phytochemicals with antimicrobial activity which include terpenes present in essential oils, flavonoids [21], and tannins [22].

The genus Psidium is native to America but is distributed worldwide. Their specimens are traditionally used in the prevention or treatment of a large number of diseases. Preparations of twigs or leaves of Psidium species are extensively used for the control of gastrointestinal and respiratory disorders as well as in the treatment of skin damage [23].

Ethnopharmacological evaluation demonstrated the antimicrobial activities of Psidium ssp. that explain its use for the treatment of infectious diseases of the digestive, respiratory, urinary tract as well as the skin and soft tissues. Within this genus, P. guajava is the most studied and used species [24]. Thus, we are interested in P. guineense, an indigenous plant of Brazil, because it is characterized by its ample spectrum of uses including gastrointestinal disorders, infection of the genitourinary tract, treatment of colds, bronchitis, and ulcers [25].

According to Sartoratto et al. [26], strong activity is for MIC values between 50–500 $\mu g/mL$, moderate activity MIC values between 600–1500 $\mu g/mL$, and weak activity above 1500 $\mu g/mL$. Comparing with literature results, the aqueous extract of P. guineense (250–500 $\mu g/mL$) has a strong activity against MRSA strains tested.

The phytochemical analysis from the leaves of P. guineense indicated the presence of flavonoids, hydrolysable, and condensed tannins distributed throughout the leaf tissue [27]. These results are in agreement with those obtained in the present study. Total polyphenol content 21.62 ± 0.40 g% determined in aqueous extract of Psidium guineense plays an important role on biological properties of this extract.

Neira González et al. [11] associated the presence of flavonoids, mainly avicularin, quercetin, and guaijaverin present in the ethanolic extract of P. guineense, with the activity against clinical isolated Streptococcus mutans strains.

Flavonoids have the ability to complex with proteins and bacterial cells forming irreversible complexes mainly with nucleophilic amino acids. This complex often leads to inactivation of the protein and loss of its function [27].

Tannins are not crystallizable substances and when they are in the aqueous system, form colloidal solutions. The antimicrobial activity of tannins can be summarized as follows: (i) binding with proteins and adhesins, inhibiting enzymes, (ii) complexation with the cell wall and metal ions, and (iii) disruption of the plasmatic membrane [28].

Synergism of natural products and antimicrobial agents is a thrust area of phytomedicinal research, developing novel perspective of phytopharmaceuticals. The synergism of plant-derived compounds and antimicrobial agents has been evaluated previously against pathogenic microorganisms. The approach is not exclusive for extract combinations, since effective combinations between single natural products,

essential oils or extracts with chemosynthetics or antibiotics have been described [6, 29].

In addition to achieving these synergistic effects, the combinations of two or more compounds are essential for the following reasons: (1) to prevent or suppress the emergence of resistant strains, (2) to decrease dose-related toxicity, as a result dosage, and (3) to attain a broad spectrum of activity [18].

In this study, a growth inhibitory effect of P. guineense extract on MRSA strains was observed for its combinations with beta-lactams, carbapenems, and fluoroquinolones.

The combination of P. guineense extract and beta-lactam may help to reduce the amount of antimicrobial agents used and deliver a medicine with similar or greater potency as antimicrobial. More importantly, since phytochemicals are structurally different from antimicrobial agents and often have different modes of action, they may provide new means of studying the mechanisms of bacterial control at a molecular level. With the increase prevalence of multidrug resistant S. aureus, synergism testing using various combinations of phytochemicals with antimicrobial agents could be a powerful tool in helping to select appropriate antimicrobial therapy [6, 29].

The indiscriminate use of antimicrobial agents in the treatment of bacterial infections has led to the emergence and spread of resistant strains, and it resulted in a great loss of clinical efficacy of previously effective first-line antimicrobials which results in shifting of antimicrobial treatment regimen to second-line or third-line antimicrobial agents that are often more expensive with many side effects [30]. In fact, studies have showed that crude extracts of plants possess the ability to enhance the activity of antimicrobial agents [4, 7, 8].

There is a wide list of phytochemicals which act as inhibitors, and a few of them are glycosylated flavones suppressing topoisomerase IV activity, myricetin inhibiting DnaB helicase, allicin inhibiting RNA synthesis. Corilagin, a polyphenol from Arctostaphylos uva-ursi is found to markedly reduce the MIC of beta-lactam agents against MRSA. According to Hemaiswarya et al. [6], there are two possibilities regarding the mechanism of action of corilagin, namely, inhibition of penicillin binding protein (PBP2a) activity or inhibiting its production.

The polyphenol epigallocatechin gallate (EGCg) from green tea is believed to be synergic with beta-lactam agents since both attacked the same target site, namely, peptidoglycan. EGCg inhibits the penicillinase produced by S. aureus thereby restoring the activity of penicillin. The combination of EGCg with ampicillin/sulbactam reduced the MIC_{90} to 4 $\mu g/mL$ from its initial value of 16 $\mu g/mL$ [6].

Hatano et al. [31] investigated the synergistic activity of two proanthocyanidins isolated from the fruits of the Zizyphus genus. Although the MICs of these polyphones were 512–1024 $\mu g/moL$, both reduced the MIC for oxacillin to 1/2–1/16 of those in the absence of the polyphenols.

It is interesting to note that most plant secondary metabolites have weak antimicrobial activity, several orders of magnitudes less than that of common antimicrobial agents produced by bacteria and fungi. In spite of the fact that plant-derived antimicrobials are less potent, plants fight infections successfully. Hence, is becomes apparent that plants adopts a synergistic mechanism between their compounds [29].

According to Rosales-Reyes et al. [32], the use of the crude extract in our study was intentional and based on the belief that it would be the closest representation to that of traditional preparations.

Therefore, it is speculated that the efficacy of a crude extract may be due to the interplay between the different active constituents that may be present in the extract leading to better activity and/or decrease in potential toxicity of some individual constituents [33].

5. Conclusion

An antibacterial effect of *P. guineense* extract and a synergistic effect in combination with antimicrobial agents were reproducibly observed and might be an interesting alternative therapy for infectious diseases caused by MRSA strains. In addition, more studies, including toxicity tested *in vivo*, need to be conducted on this plant before therapeutic treatments are implemented.

Acknowledgments

The authors thank Coordenação de Aperfeiçoamento de Pessoal de Nível Superior (CAPES) for granting the fellowship as well as the financial support and infrastructure provided by the Universidade Federal de Pernambuco (UFPE) which made this study possible.

References

[1] M. Guzmán-Blanco, C. Mejía, R. Isturiz et al., "Epidemiology of meticillin-resistant *Staphylococcus aureus* (MRSA) in Latin America," *International Journal of Antimicrobial Agents*, vol. 34, no. 4, pp. 304–308, 2009.

[2] A. Nostro, A. R. Blanco, M. A. Cannatelli et al., "Susceptibility of methicillin-resistant staphylococci to oregano essential oil, carvacrol and thymol," *FEMS Microbiology Letters*, vol. 230, no. 2, pp. 191–195, 2004.

[3] M. E. Mulligan, K. A. Murray-Leisure, B. S. Ribner et al., "Methicillin-resistant *Staphylococcus aureus*: a consensus review of the microbiology, pathogenesis, and epidemiology with implications for prevention and management," *American Journal of Medicine*, vol. 94, no. 3, pp. 313–328, 1993.

[4] G. M. Adwan, B. A. Abu-Shanab, and K. M. Adwan, "In vitro activity of certain drugs in combination with plant extracts against *Staphylococcus aureus* infections," *Pakistan Journal of Medical Sciences*, vol. 24, no. 4, pp. 541–544, 2008.

[5] P. Y. Chung, P. Navaratnam, and L. Y. Chung, "Synergistic antimicrobial activity between pentacyclic triterpenoids and antibiotics against *Staphylococcus aureus* strains," *Annals of Clinical Microbiology and Antimicrobials*, vol. 10, article 25, 2011.

[6] S. Hemaiswarya, A. K. Kruthiventi, and M. Doble, "Synergism between natural products and antibiotics against infectious diseases," *Phytomedicine*, vol. 15, no. 8, pp. 639–652, 2008.

[7] J. E. C. Betoni, R. P. Mantovani, L. N. Barbosa, L. C. Di Stasi, and A. Fernandes, "Synergism between plant extract and antimicrobial drugs used on Staphylococcus aureus diseases," *Memorias do Instituto Oswaldo Cruz*, vol. 101, no. 4, pp. 387–390, 2006.

[8] C. O. Esimone, I. R. Iroha, E. C. Ibezim, C. O. Okeh, and E. M. Okpana, "In vitro evaluation of the interaction between tea extracts and penicillin G against staphylococcus aureus," *African Journal of Biotechnology*, vol. 5, no. 11, pp. 1082–1086, 2006.

[9] M. Pio Corrêa, *Dicionário das plantas úteis do Brasil e das exóticas cultivadas*, Ministério da Agricultura, Instituto Brasileiro de Desenvolvimento Florestal, Rio de Janeiro, Brazil, 2nd edition, 1984.

[10] J. A. Silva, D. B. Silva, N. T. V. Junqueira, and L. R. N. Andrade, *Frutas nativas dos cerrados*, EMBRAPA, Centro de Pesquisa Agropecuária dos Cerrados-CPAC, Brasília, Brazil, 1994.

[11] A. M. Neira González, M. B. Ramírez González, and N. L. Sánchez Pinto, "Phytochemical study and antibacterial activity of Psidium guíneense Sw (choba) against *Streptococcus mutans*, causal agent of dental caries," *Revista Cubana de Plantas Medicinales*, vol. 10, no. 3-4, 2005.

[12] E. A. H. Roberts, R. A. Cartwright, and D. J. Wood, "Flavonols of tea," *Journal of the Science of Food and Agriculture*, vol. 7, pp. 637–646, 1956.

[13] H. Wagner and S. Bladt, *Plant Drug Analysis*, Springer, New York, NY, USA, 2nd edition, 1996.

[14] J. B. Harborne, *Phytochemical Methods*, Chapman & Hall, London, UK, 3rd edition, 1998.

[15] December 2011, http://www.anvisa.gov.br/farmacopeiabrasileira/index.htm.

[16] E. W. Koneman et al., *Diagnóstico microbiológico: texto e atlas colorido*, Guanabara Koogan, Rio de Janeiro, Brazil, 6th edition, 2008.

[17] CLSI—Clinical Laboratory Standards Institute, "Performance Standards for Antimicrobial Susceptibility Testing M100-S20," 2010.

[18] G. M. Eliopoulos and R. C. Moellering Jr., "Antimicrobial combinations," in *Antibiotics in Laboratory Medicine*, pp. 330–338, Williams & Wilkins, Baltimore, Md, USA, 4th edition, 1996.

[19] M. Stavri, L. J. V. Piddock, and S. Gibbons, "Bacterial efflux pump inhibitors from natural sources," *Journal of Antimicrobial Chemotherapy*, vol. 59, no. 6, pp. 1247–1260, 2007.

[20] S. Gibbons, "Anti-staphylococcal plant natural products," *Natural Product Reports*, vol. 21, no. 2, pp. 263–277, 2004.

[21] N. E. Hernández, M. L. Tereschuk, and L. R. Abdala, "Antimicrobial activity of flavonoids in medicinal plants from Tafi del Valle (Tucuman, Argentina)," *Journal of Ethnopharmacology*, vol. 73, no. 1-2, pp. 317–322, 2000.

[22] C. D. Djipa, M. Delmée, and J. Quetin-Leclercq, "Antimicrobial activity of bark extracts of *Syzygium jambos* (L.) Alston (Myrtaceae)," *Journal of Ethnopharmacology*, vol. 71, no. 1-2, pp. 307–313, 2000.

[23] A. Caceres, L. Fletes, L. Aguilar et al., "Plants used in Guatemala for the treatment of gastrointestinal disorders. 3. Confirmation of activity against enterobacteria of 16 plants," *Journal of Ethnopharmacology*, vol. 38, no. 1, pp. 31–38, 1993.

[24] L. C. Di Stasi and C. A. Hiruma-Lima, *Plantas medicinais na Amazônia e Mata Atlântica*, UNESP, São Paulo, Brazil, 2nd edition, 2002.

[25] A. Aguilar, A. Argueta, and L. Cano, *Flora Medicinal Indígena de Mexico*, Instituto indigenista de México, 1994.

In Vitro Synergistic Effect of Psidium guineense (Swartz) in Combination with Antimicrobial Agents against
Methicillin-Resistant Staphylococcus aureus Strains

85

[26] A. Sartoratto, A. L. M. Machado, C. Delarmelina, G. M. Figueira, M. C. T. Duarte, and V. L. G. Rehder, "Composition and antimicrobial activity of essential oils from aromatic plants used in Brazil," *Brazilian Journal of Microbiology*, vol. 35, no. 4, pp. 275–280, 2004.

[27] R. B. Ferreira et al., "Morphoanatomy, histochemistry and phytochemistry of *Psidium guineense* Sw. (Myrtaceae) Leaves," *Journal of Pharmacy Research*, vol. 4, no. 4, pp. 942–944, 2011.

[28] M. M. Cowan, "Plant products as antimicrobial agents," *Clinical Microbiology Reviews*, vol. 12, no. 4, pp. 564–582, 1999.

[29] H. Wagner and G. Ulrich-Merzenich, "Synergy research: approaching a new generation of phytopharmaceuticals," *Phytomedicine*, vol. 16, no. 2-3, pp. 97–110, 2009.

[30] S. Mandal, M. DebMandal, N. K. Pal, and K. Saha, "Synergistic anti-*Staphylococcus aureus* activity of amoxicillin in combination with *Emblica officinalis* and *Nymphae odorata* extracts," *Asian Pacific Journal of Tropical Medicine*, vol. 3, no. 9, pp. 711–714, 2010.

[31] T. Hatano, M. Kusuda, K. Inada et al., "Effects of tannins and related polyphenols on methicillin-resistant *Staphylococcus aureus*," *Phytochemistry*, vol. 66, no. 17, pp. 2047–2055, 2005.

[32] T. Rosales-Reyes, M. de la Garza, C. Arias-Castro et al., "Aqueous crude extract of *Rhoeo discolor*, a Mexican medicinal plant, decreases the formation of liver preneoplastic foci in rats," *Journal of Ethnopharmacology*, vol. 115, no. 3, pp. 381–386, 2007.

[33] T. Birdi, P. Daswani, S. Brijesh, P. Tetali, A. Natu, and N. Antia, "Newer insights into the mechanism of action of *Psidium guajava* L. leaves in infectious diarrhoea," *BMC Complementary and Alternative Medicine*, vol. 10, article no. 33, 2010.

14

A Study on L-Asparaginase of *Nocardia levis* MK-VL_113

Alapati Kavitha and Muvva Vijayalakshmi

Department of Botany and Microbiology, Acharya Nagarjuna University, Guntur 522 510, India

Correspondence should be addressed to Muvva Vijayalakshmi, profmvijayalakshmi@rediffmail.com

Academic Editor: Xavier Thomas

An enzyme-based drug, L-asparaginase, was produced by *Nocardia levis* MK-VL_113 isolated from laterite soils of Guntur region. Cultural parameters affecting the production of L-asparaginase by the strain were optimized. Maximal yields of L-asparaginase were recorded from 3-day-old culture grown in modified asparagine-glycerol salts broth with initial pH 7.0 at temperature 30°C. Glycerol (2%) and yeast extract (1.5%) served as good carbon and nitrogen sources for L-asparaginase production, respectively. Cell-disrupting agents like EDTA slightly enhanced the productivity of L-asparaginase. Ours is the first paper on the production of L-asparaginase by *N. levis*.

1. Introduction

L-asparaginase (L-asparagine amino hydrolase EC 9.5.1.1) is a potent antitumor enzyme that catalyzes the hydrolysis of L-asparagine to L-aspartic acid and ammonium ion. This enzyme has been widely exploited in the treatment of certain kinds of cancer especially acute lymphoblastic leukaemia since the time it was obtained from *Escherichia coli* and its antineoplastic activity demonstrated in guinea pig serum [1–3]. Due to its prompt therapeutic potential, screening of microbial sources for asparaginase activity has been greatly intensified and well documented in *E. coli* [4–6], *Erwinia carotovora* [7, 8], *Enterobacter aerogenes* [9], *Corynebacterium glutamicum* [10], *Candida utilis* [11], *Serratia marcescens* [12], *Staphylococcus aureus* [13], and *Thermus thermophilus* [14]. In addition, actinomycetes act as potential candidates for the production of L-asparaginase. Particularly, *Mycobacterium tuberculosis* [15], *Streptomyces griseus* [16], *S. karnatakensis*, *S. venezuelae* [17], *S. longsporusflavus* F-15 [18], *S. phaeochromogenes* FS-39 [19], and *Nocardia asteroides* [20] were proved to be potential producers of this enzyme. However, very little information is available on the production of L-asparaginase by the genus *Nocardia*. While screening the actinomycetes for asparaginase production, a strain with good asparaginase activity on asparagine-glycerol salts agar medium was identified as *Nocardia levis* by 16S rRNA analysis. In the present study, an attempt has been undertaken to reveal the optimization of L-asparaginase production by *N. levis*.

2. Material and Methods

Nocardia levis MK-VL_113 was isolated from laterite soil samples of Guntur region, and the 16S rRN A gene sequence of the strain has been deposited in NCBI genbank with an accession number FJ209734 [21].

2.1. Production Profile of L-Asparaginase. For determining the production profile of L-asparaginase, culture suspension prepared from one-week-old culture of the strain was inoculated into asparagine-glycerol salts (ISP-5) broth containing 1% asparagine, 1% glycerol, 0.1%K_2HPO_4, 0.1% trace salt solution with initial pH 7.2. The inoculated flasks were incubated at 35°C for 6 days in order to estimate the cell growth of the strain as well as L-asparaginase production for every 24 h interval. Cell growth was expressed in terms of dry weight of biomass (mg/mL).

L-asparaginase assay was performed according to the procedure described by Peterson and Ciegler [22] with slight modifications. Cells were harvested by centrifuging the culture broth at 10,000 rpm for 15 min and were ground in

Tris-HCl buffer by using homogenizer. Later, it was again centrifuged, and the cell free extract (0.2 mL) obtained was mixed with 0.8 mL of 0.05 M Tris-HCl buffer and 1 mL of 0.04 M L-asparagine. After incubating the reaction mixture for 15 min at 37°C in a water bath shaker, the reaction was terminated by the addition of 0.5 mL of 15% (w/v) trichloroacetic acid. Precipitated proteins were removed by centrifugation, and the liberated ammonia was determined spectrometrically at 500 nm by nesslerization. Tubes kept at zero time incubation served as control. Enzyme activity was determined on the basis of liberation of ammonia calculated with reference to a standard curve of ammonium sulphate. One L-asparaginase unit (IU) equals to that amount of enzyme which releases 1 μM of ammonia (ammonium sulphate as standard) in 1 min at 37°C. The cell dry weight was recorded simultaneously by drying the cell debris collected after centrifugation in an oven at 90°C for 24 h.

2.2. Optimization of L-Asparaginase Production.

Influence of different cultural conditions such as initial pH, temperature, carbon, and nitrogen sources on the production of L-asparaginase was determined.

2.2.1. Initial pH and Temperature. Impact of pH on the production of L-asparaginase was examined by culturing the strain in ISP-5 broth adjusted to various pH levels ranging from 4.0 to 10.0. The optimal pH achieved at this step was used for further study. To determine the optimum temperature for L-asparaginase production, the strain was cultured in ISP-5 broth at different temperatures, namely, 20° to 40°C for 72 h of incubation.

2.2.2. Carbon Sources. To investigate the effect of carbon sources on L-asparaginase production by the strain, ISP-5 broth was supplemented with different carbon sources such as arabinose, fructose, galactose, glucose, lactose, maltose, mannitol, sorbitol, starch, sucrose, and xylose each at a concentration of 1% (w/v). Impact of different concentrations of best carbon source (1–4% w/v) on L-asparaginase activity of the strain was studied.

2.2.3. Nitrogen Sources. Different nitrogen sources, namely, L-asparagine, beef extract, L-glutamine, malt extract, peptone, tryptone, and yeast extract were added at a rate of 1% (w/v) to ISP-5 broth consisting an optimal amount of superior carbon source. Besides, the optimal concentration of nitrogen source (1–3% w/v) supporting high yields of L-asparaginase production was determined by maintaining all other conditions at optimum levels [23].

2.2.4. Effect of Cell-Disrupting Agents. The effect of cell disrupting agents like EDTA, lysozyme, penicillin-G, and SDS on the release of L-asparaginase from the 72 h old cultures was tested [24]. Harvested cells suspended in 0.05 M Tris-HCl buffer (pH 7.5) were centrifuged at 10,000 rpm for 15 min. The resulting pellet was washed twice and resuspended in the same buffer. Cell-disrupting agents (50 μg/mL) prepared separately in Tris-HCl were used to treat the cells

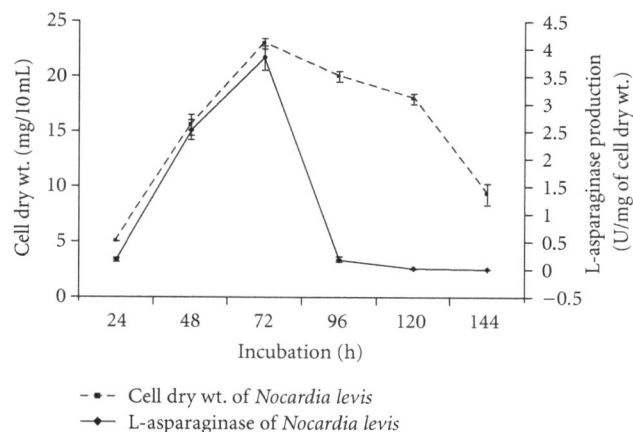

FIGURE 1: Production profile of L-asparaginase by *Nocardia levis* MK-VL_113 grown in modified ISP-5 broth. (Values are the means of three replicates ± SD.)

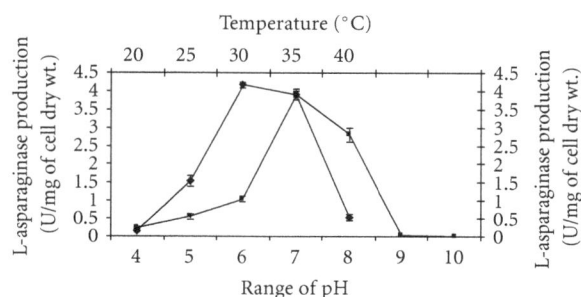

FIGURE 2: Effect of initial pH and temperature on L-asparaginase production by *Nocardia levis* MK-VL_113 cultured in modified ISP-5 broth. (Values are the means of three replicates ± SD.)

twice and centrifuged. The cell debris was discarded, and the supernatant (0.2 mL) thus collected was used as enzyme source.

3. Results and Discussion

3.1. Production Profile of L-Asparaginase. Production of L-asparaginase by *N. levis* started after 24 h of cell growth. Maximum levels of L-asparaginase production as well as cell growth were observed after 72 h of incubation (Figure 1). It revealed a good correlation between enzymatic yields and active cell growth of the strain. A positive correlation between the cell growth and L-asparaginase productivity was reported in *S. karnatakensis* [25] and *S. albidoflavus* [23]. In *Amycolatopsis* CMU-H002, maximum yields of L-asparaginase and biomass were obtained from 72 h old culture [26].

3.2. Optimization of L-Asparaginase Production

3.2.1. Initial pH and Temperature. A study of initial pH levels (4–10) on the production of L-asparaginase by the strain indicated optimal enzyme activity at pH 7.0 (Figure 2).

TABLE 1: L-asparaginase production by *Nocardia levis* grown in modified ISP-5 broth amended with different carbon sources.

Carbon sources (1% w/v)	L-asparaginase production (U/mg of cell dry wt.)
*Control	4.15 ± 0.063
Arabinose	4.03 ± 0.16
Fructose	3.9 ± 0.223
Galactose	3.75 ± 0.221
Glucose	0.32 ± 0.016
Lactose	1.05 ± 0.064
Maltose	0.4 ± 0.057
Mannitol	0.36 ± 0.013
Sorbitol	0.09 ± 0.004
Starch	1.32 ± 0.131
Sucrose	0.19 ± 0.013
Xylose	0.57 ± 0.042

Values are the means of three replicates ± SD. *Modified ISP-5 broth with 1% glycerol (v/v).

TABLE 2: Carbon levels on L-asparaginase production by *Nocardia levis* in modified ISP-5 broth.

Glycerol (v/v) concentrations	L-asparaginase production (U/mg of cell dry wt.)
1%	4.15 ± 0.063
2%	4.28 ± 0.016
3%	1.23 ± 0.009
4%	0.06 ± 0.004

Values are the means of three replicates ± SD.

TABLE 3: L-asparaginase production by *Nocardia levis* grown in modified ISP-5 broth amended with different nitrogen sources.

Nitrogen sources (1% w/v)	L-asparaginase production (U/mg of cell dry wt.)
L-asparagine	4.28 ± 0.016
Beef extract	0.64 ± 0.061
L-glutamine	4.19 ± 0.008
Malt extract	2.47 ± 0.075
Peptone	3.86 ± 0.154
Tryptone	4.34 ± 0.076
Yeast extract	4.56 ± 0.172

Values are the means of three replicates ± SD.

These findings are in conformity with the results of Koshy et al. [27], Abdel-All et al. [28] and Dhevendaran and Anithakumar [29] who recorded pH 7.0 as optimum for L-asparaginase production in *S. longsporusflavus* F-15, *S. plicatus*, *S. phaeochromogenes* FS-39, and *Streptomyces* AQB VC67, respectively. Other actinomycetes species such as *S. aureofasciculus* LA-2, *S. chattanoogenesis* LA-8, *S. hawaiiensis* LA-15, *S. orientalis* LA-20, *S. canus* LA-29 and *S. olivoviridis* LA-35 exhibited optimum growth and L-asparaginase production at pH 7 to 8 [30].

Impact of temperature on the production of L-asparaginase by the strains is presented in Figure 2. Production of L-asparaginase was high with the strain cultured in modified ISP-5 broth at 30°C. In *S. collinus* [17] and *S. longsporusflavus* F-15 [18], the optimum temperature recorded was 30°C for L-asparaginase production. Koshy et al. [27] reported 29 ± 2°C as optimum temperature for L-asparaginase production in *S. plicatus*. The present study revealed that L-asparaginase production by the strain was high when grown in modified ISP-5 broth with initial pH 7.0 for 72 h at 30°C.

3.2.2. Carbon Sources.

Different carbon sources like arabinose, fructose, galactose, glucose, lactose, maltose, mannitol, sorbitol, starch, sucrose, and xylose were amended in ISP-5 broth to determine their impact on L-asparaginase production. As compared to other carbon sources tested, L-asparaginase production was high in the basal medium containing glycerol (Table 1). Krishna Reddy and Reddy [31] also noted glycerol as one of the best carbon sources for L-asparaginase activity of *S. albus*.

The final pH of the fermentation broth consisting glycerol became alkaline whereas in media amended with other carbon sources became acidic that may lead to decline in productivity of L-asparaginase. Acidity of fermentation medium could inhibit the production of L-asparaginase [32], and, because of this nature, glucose is reported as a repressor for L-asparaginase synthesis [9, 23, 33].

Besides, the effect of different concentrations of glycerol (1–4%) on L-asparaginase production is recorded (Table 2). Optimal yields of L-asparaginase by the strain were achieved in the medium amended with 2% glycerol, whereas its biosynthesis greatly declined with further hike in carbon source. The optimal level of carbon source for L-asparaginase production by *S. phaeochromogenes* FS-39 was 2% glycerol [28], while it was 1.5% starch for *S. longsporusflavus* F-15 [34]. In the present work, optimal yields of L-asparaginase were recorded by culturing *N. levis* in the modified ISP-5 broth containing 2% glycerol with initial pH 7.0 at 30°C for 72 h.

3.2.3. Nitrogen Sources.

The effect of nitrogen compounds on the production of L-asparaginase by the strain was studied by incorporating different nitrogen sources to modified ISP-5 broth containing 2% glycerol. L-asparaginase production by the strain varied with different nitrogen compounds tested (Table 3). Among them, culture medium amended with yeast extract favored maximal L-asparaginase production by the strain followed by tryptone. In *Erwinia aroideae*, yeast extract supported high yields of L-asparaginase production [35] while tryptone and yeast extract stimulated the production of this enzyme in *E. carotovora* EC-113 [7]. L-asparagine was reported to be suitable nitrogen source for L-asparaginase production by *S. collinus* [17] and *S. karnatakensis* and *S. venezuelae* [25]. Optimization studies of Abdel-All et al. [28] indicated glycerol L-asparaginase yeast extract (GAY) as a fruitful medium for the synthesis of L-asparaginase by *S. phaeochromogenes* FS-39.

TABLE 4: Impact of yeast extract concentrations on L-asparaginase production by *Nocardia levis* grown in modified ISP-5 broth.

Yeast extract concentrations (w/v)	L-asparaginase production (U/mg of cell dry wt.)
1%	4.56 ± 0.172
1.5%	5.06 ± 0.002
2%	3.42 ± 0.142
2.5%	1.62 ± 0.043
3%	0.97 ± 0.013

Values are the means of three replicates ± SD.

TABLE 5: Effect of cell-disrupting agents on the release of L-asparaginase from 72 h old culture of *Nocardia levis*.

Cell disrupting agents (50 μg/ml)	Release of L-asparaginase from the culture (U/mg of cell dry wt.)
*Control	5.06 ± 0.002
EDTA	5.34 ± 0.04
Lysozyme	5.29 ± 0.019
Penicillin-G	5.11 ± 0.082
SDS	4.01 ± 0.163

Values are the means of three replicates ± SD. *Cell disruption using homogenizer.

The optimal level of yeast extract was found to be 1.5% for obtaining high amounts of L-asparaginase (Table 4). A gradual decline in L-asparaginase production was found with further rise in yeast extract levels. Yeast extract is essential for cell growth and L-asparaginase synthesis, but, in high concentrations, it inhibits the production of L-asparaginase [36]. In *S. albidoflavus*, Narayana et al. [23] recorded 2% yeast extract as optimal concentration for L-asparaginase production. In the present study, enhanced levels of L-asparaginase production by *N. levis* was recorded in the modified ISP-5 broth containing 1.5% yeast extract and 2% glycerol with initial pH 7.0 incubated at 30°C for 72 h.

3.2.4. Effect of Cell-Disrupting Agents on L-Asparaginase Production.

Production of L-asparaginase by the strain was examined for its intracellular or extracellular nature. Enzyme production was observed only in cell extracts. Therefore, the effect of cell-disrupting agents like EDTA, lysozyme, penicillin-G, and SDS on the release of L-asparaginase from the cells was determined. Among the cell disrupting agents tested, EDTA was found to be more effective for the release of L-asparaginase from the cells (Table 5). In *S. griseus* [16] and *S. karnatakensis* [37], the production of L-asparaginase was found to be intracellular. Enhanced levels of L-asparaginase were detected in *S. albidoflavus* when treated with cell disrupting agents like EDTA [23].

In the present study, the optimal cultural and nutritional conditions for the production of L-asparaginase by *N. levis* MK-VL_113 were recorded. Biosynthesis of L-asparaginase by the strain was high when cultured in modified ISP-5 broth composing glycerol (2%) and yeast extract (1.5%) with initial pH 7.0 at temperature 30°C. EDTA was proved to be an efficient cell-disrupting agent in the release of intracellular enzyme from the cells of the strain. This is the new report on the production of L-asparaginase by *N. levis*.

Acknowledgment

Financial support provided by Indian Council of Medical Research (ICMR) to one of the authors (A. Kavitha) in the form of senior research fellowship is greatly acknowledged.

References

[1] J. D. Broome, "Evidence that the L-asparaginase activity of guinea pig serum is responsible for its antilymphoma effects," *Nature*, vol. 191, no. 4793, pp. 1114–1115, 1961.

[2] J. H. Schwartz, J. Y. Reeves, and J. D. Broome, "Two L-asparaginases from *E. coli* and their action against tumors," *Proceedings of the National Academy of Sciences of the United States of America*, vol. 56, no. 5, pp. 1516–1519, 1966.

[3] A. E. Boyse, L. J. Old, and H. A. Campbell, "Suppression of murine leukaemia of various types: comparative inhibition activities of guinea pig serum L-asparaginase and *E. coli*," *Journal of Experimental Medicine*, vol. 125, pp. 17–31, 1967.

[4] L. A. Clavell, R. D. Gelber, and H. J. Cohen, "Four-agent induction and intensive asparaginase therapy for treatment of childhood acute lymphoblastic leukemia," *New England Journal of Medicine*, vol. 315, no. 11, pp. 657–663, 1986.

[5] C. Derst, A. Wehner, V. Specht, and K. H. Rohm, "States and functions of tyrosine residues in *Escherichia coli* asparaginase II," *European Journal of Biochemistry*, vol. 224, no. 2, pp. 533–540, 1994.

[6] L. Mercado and G. Arenas, "*Escherichia coli* L-asparaginase induces phosphorylation of endogenous polypeptides in human immune cells," *Sangre*, vol. 44, no. 6, pp. 438–442, 1999.

[7] N. K. Maladkar, V. K. Singh, and S. R. Naik, "Fermentative production and isolation of L-asparaginase from *Erwinia carotovora*, EC-113," *Hindustan Antibiotics Bulletin*, vol. 35, no. 1-2, pp. 77–86, 1993.

[8] K. Aghaiypour, A. Wlodawer, and J. Lubkowski, "Structural basis for the activity and substrate specificity of *Erwinia chrysanthemi* L-asparaginase," *Biochemistry*, vol. 40, no. 19, pp. 5655–5664, 2001.

[9] J. Mukherjee, S. Majumdar, and T. Scheper, "Studies on nutritional and oxygen requirements for production of L-asparaginase by *Enterobacter aerogenes*," *Applied Microbiology and Biotechnology*, vol. 53, no. 2, pp. 180–184, 2000.

[10] J. M. Mesas, J. A. Gil, and J. F. Martin, "Characterization and partial purification of L-asparaginase from *Corynebacterium glutamicum*," *Journal of General Microbiology*, vol. 136, no. 3, pp. 515–519, 1990.

[11] J. O. Kil, G. N. Kim, and I. Park, "Extraction of extracellular L-asparaginase from *Candida utilis*," *Bioscience, Biotechnology, and Biochemistry*, vol. 59, no. 4, pp. 749–750, 1995.

[12] C. P. Sukumaran, D. V. Singh, and P. F. Mahadeven, "Synthesis of L-asparaginase by *Serratia marcescens* (Nima)," *Journal of Biosciences*, vol. 1, no. 3, pp. 263–269, 1979.

[13] R. G. Muley, S. Sarker, S. Ambedkar et al., "Influence of alkali-treated cornsteep liquor containing medium on protein a production by *Staphylococcus aureus*," *Folia Microbiology*, vol. 43, pp. 31–34, 1998.

[14] A. A. Pritsa and D. A. Kyriakidis, "L-asparaginase of *Thermus thermophilus*: purification, properties and identification of

essential amino acids for its catalytic activity," *Molecular and Cellular Biochemistry*, vol. 216, no. 1-2, pp. 93–101, 2001.

[15] F. Kirchheimer and C. K. Whittaker, "Asparaginase of mycobacteria," *American Review of Tuberculosis*, vol. 70, no. 5, pp. 920–921, 1954.

[16] P. J. DeJong, "L-asparaginase production by *Streptomyces griseus*," *Applied Microbiology*, vol. 23, no. 6, pp. 1163–1164, 1972.

[17] S. A. Mostafa and M. S. Salama, "L-asparaginase producing *Streptomyces* from the soil of Kuwait," *Zentralbl Bakteriol Naturwiss*, vol. 134, pp. 325–334, 1979.

[18] M. K. Abdel-Fatah, "Studies on the asparaginolytic enzymes of *Streptomycetes*. 1—Culture conditions for the production of L-asparaginase enzyme from *Streptomyces longsporusflavus* (F-15) strain," *Egyptian Journal of Microbiology*, vol. 30, pp. 247–260, 1996.

[19] M. K. Abdel-Fatah, A. R. Abdel-Mageed, S.M. Abdel-All et al., "Purification and characterization of L-asparaginase produced by *Streptomyces phaeochromogenes* FS-39," *Journal of Drug Research*, vol. 22, pp. 195–212, 1998.

[20] S. Gunasekaran, L. McDonald, M. Manavathu et al., "Effect of culture media on growth and L-asparaginase production in *Nocardia asteroides*," *Biomedical Letters*, vol. 52, pp. 197–201, 1995.

[21] A. Kavitha, P. Prabhakar, M. Narasimhulu et al., "Isolation, characterization and biological evaluation of bioactive metabolites from *Nocardia levis* MK-VL_113," *Microbiological Research*, vol. 165, no. 3, pp. 199–210, 2010.

[22] R. E. Peterson and A. Ciegler, "L-asparaginase production by various bacteria," *Applied Microbiology*, vol. 17, no. 6, pp. 929–930, 1969.

[23] K. J. P. Narayana, K. G. Kumar, and M. Vijayalakshmi, "L-asparaginase production by *Streptomyces albidoflavus*," *Indian Journal of Microbiology*, vol. 48, no. 3, pp. 331–336, 2008.

[24] S. A. Mostafa and O. K. Ali, "L-asparaginase activity in cell-free extracts of *Thermoactinomyces vulgaris* 13 M.E.S," *Zentralblatt fur Mikrobiologie*, vol. 138, no. 5, pp. 397–404, 1983.

[25] S. A. Mostafa, "Activity of L-asparaginase in cells of *Streptomyces karnatakensis*," *Zentralbl Bakteriol Naturwiss*, vol. 134, pp. 343–351, 1979.

[26] S. Khamna, A. Yokota, and S. Lumyong, "L-asparaginase production by actinomycetes isolated from some Thai medicinal plant rhizosphere soils," *International Journal of Integrative Biology*, vol. 6, no. 1, pp. 22–26, 2009.

[27] A. Koshy, K. Dhevendaran, M. I. Georgekutty, and P. Natarajan, "L-asparaginase in *Streptomyces plicatus* isolated from the alimentary canal of the fish, *Gerres filamentosus* (cuvier)," *Journal of Marine Biotechnology*, vol. 5, no. 2-3, pp. 181–185, 1997.

[28] S. M. Abdel-All, M. K. Abdel-Fatah, A. R. A. Khalil et al., "Studies on the asparaginolytic activity of the brown-pigmented *Streptomycetes*," *Journal of Drug Research*, vol. 22, pp. 171–194, 1998.

[29] K. Dhevendaran and Y. K. Anithakumar, "L-asparaginase activity in growing conditions of *Streptomyces* spp. associated with *Therapon jarbua* and *Villorita cyprinoids* of Veli lake, South India," *Fishery Technology*, vol. 39, p. 155, 2002.

[30] M. K. Sahu, K. Sivakumar, E. Poorani, T. Thangaradjou, and L. Kannan, "Studies on L-asparaginase enzyme of actinomycetes isolated from estuarine fishes," *Journal of Environmental Biology*, vol. 28, no. 2, pp. 465–474, 2007.

[31] V. Krishna Reddy and S. M. Reddy, "Effect of C and N sources on asparaginase production by bacteria," *Indian Journal of Microbiology*, vol. 30, pp. 81–83, 1990.

[32] B. Heinemann and A. J. Howard, "Production of tumor-inhibitory L-asparaginase by submerged growth of *Serratia marcescens*," *Applied Microbiology*, vol. 18, no. 4, pp. 550–554, 1969.

[33] H. Geckil, S. Gencer, B. Ates, U. Ozer, M. Uckun, and I. Yilmaz, "Effect of *Vitreoscilla* hemoglobin on production of a chemotherapeutic enzyme, L-asparaginase, by *Pseudomonas aeruginosa*," *Biotechnology Journal*, vol. 1, no. 2, pp. 203–208, 2006.

[34] M. K. Abdel-Fatah, "Studies on the asparaginolytic enzymes of *Streptomycetes*. II. Purification and characterization of L-asparaginase from *Streptomyces longsporusflavus* (F-15) strain," *Egyptian Journal of Microbiology*, vol. 31, pp. 303–322, 1997.

[35] F. S. Liu and J. E. Zajic, "L-asparaginase synthesis by *Erwinia aroideae*," *Applied Microbiology*, vol. 23, no. 3, pp. 667–668, 1972.

[36] N. Verma, K. Kumar, G. Kaur, and S. Anand, "L-asparaginase: a promising chemotherapeutic agent," *Critical Reviews in Biotechnology*, vol. 27, no. 1, pp. 45–62, 2007.

[37] S. A. Mostafa, "Properties of L-asparaginase in cell-free extracts of *Streptomyces karnatakensis*," *Zentralblatt fur Mikrobiologie*, vol. 137, no. 1, pp. 63–71, 1982.

15

Exploring Marine Cyanobacteria for Lead Compounds of Pharmaceutical Importance

Bushra Uzair, Sobia Tabassum, Madiha Rasheed, and Saima Firdous Rehman

Department of Bioinformatics and Biotechnology, International Islamic University Islamabad, Sector H-10, 44000 Islamabad, Pakistan

Correspondence should be addressed to Bushra Uzair, bushra.uzair@iiu.edu.pk

Academic Editor: Jean-Marc Sabatier

The Ocean, which is called the "mother of origin of life," is also the source of structurally unique natural products that are mainly accumulated in living organisms. Cyanobacteria are photosynthetic prokaryotes used as food by humans. They are excellent source of vitamins and proteins vital for life. Several of these compounds show pharmacological activities and are helpful for the invention and discovery of bioactive compounds, primarily for deadly diseases like cancer, acquired immunodeficiency syndrome (AIDS), arthritis, and so forth, while other compounds have been developed as analgesics or to treat inflammation, and so forth. They produce a large variety of bioactive compounds, including substances with anticancer and antiviral activity, UV protectants, specific inhibitors of enzymes, and potent hepatotoxins and neurotoxins. Many cyanobacteria produce compounds with potent biological activities. This paper aims to showcase the structural diversity of marine cyanobacterial secondary metabolites with a comprehensive coverage of alkaloids and other applications of cyanobacteria.

1. Introduction

Cyanobacteria is a phylum of bacteria that obtain their energy through photosynthesis. The name "cyanobacteria" comes from the color of the bacteria. Cyanobacteria are a major and phylogenetically coherent group of Gram-negative prokaryotes possessing the unifying property of performing oxygenic plantlike photosynthesis with autotrophy as their dominant mode of nutrition [1]. However, in spite of their typically aerobic photosynthetic nature, some of the cyanobacterial species can grow in the dark on organic substrates [2] and others under anaerobic conditions with sulfide as electron donor for photosynthesis [3]. Certain strains have the ability to fix atmospheric dinitrogen into organic nitrogen-containing compounds, so displaying the simplest nutritional requirements of all microorganisms [4]. Cyanobacteria are also characterised by a great morphological diversity, unicellular as well as filamentous species being included with a cell volume ranging over more than five orders of magnitude [5]. Representatives of the group have been found, frequently in abundance, in most of the natural illuminated environments examined so far, both aquatic and terrestrial, including several types of extreme environments [5]. This widespread distribution reflects a large variety of species, covering a broad spectrum of physiological properties and tolerance to environmental stress [6]. Indeed, several cyanobacterial strains such as *chyococcus sp* (Figure 1(a)), *phormidium sp* (Figure 1(b)) possess, outside their outer membrane, additional surface structures, mainly of a polysaccharidic nature, that comprise a wide variety of outermost investments differing in thickness, consistency, and appearance after staining. These structures, in spite of the rather arbitrary terminology sometimes used, can be referred to as three distinct types, namely, sheaths, capsules, and slimes.

Over 300 nitrogen-containing secondary metabolites, represented by diverse structural types, have been reported from the prokaryotic marine cyanobacteria. A majority of these metabolites are biologically active and are products of either the nonribosomal polypeptide (NRP) or the mixed polyketide-NRP biosynthetic pathways. Biomolecules of the NRP and hybrid polyketide-NRP structural types are important subsets of natural products utilized as therapeutic agents.

These include the antibiotic vancomycin, the immunosuppressive agent cyclosporine, and the anticancer agent

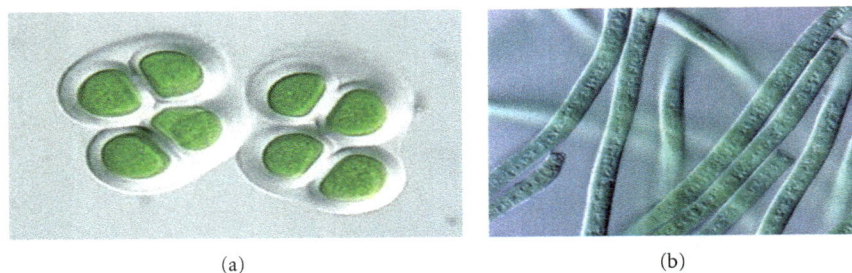

(a)

(b)

FIGURE 1: Nomarski differential interference contrasts photomicrographs of sheathed cyanobacterial strains. (a) *Chroococcus* sp. (1000x); (b) *Phormidium* sp. (1000x).

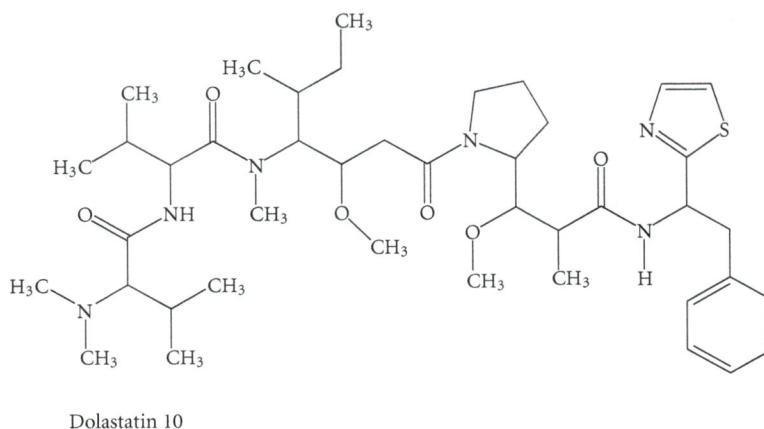

Dolastatin 10

FIGURE 2: Structure of dolastatin 10.

bleomycin [7]. Vancomycin is primarily effective against Gram-positive cocci. *Staphylococcus aureus* and *Staphylococcus epidermidis*, including both methicillin-susceptible (MSSA & MSSE) or resistant species (MRSA & MRSE), are usually sensitive to vancomycin. Vancomycin is also effective against the anaerobes, diphtheroids, and clostridium species, including *C. difficile*, whereas Bleomycin is a glycopeptide antibiotic produced by the bacterium *Streptomyces verticillus*. It works by causing breaks in DNA as anticancer drug. The drug is also used in the treatment of Hodgkin's lymphoma, squamous cell carcinomas, and testicular cancer, as well as in the treatment of plantar warts and as a means of effecting pleurodesis. The discovery of these unique classes of natural products from marine cyanobacteria represents an important source of novel microbial secondary metabolites, in addition to the actinomycetes and fungi, for drug discovery efforts.

1.1. Anticancer Drugs from Marine Cyanobacteria. An increasing number of marine cyanobacterial compounds are found to target tubulin or actin filaments in eukaryotic cells, making them an attractive source of natural products as anticancer agents [8]. Prominent molecules such as the antimicrotubule agents, curacin A (Figure 3) and dolastatin 10 (Figure 2), have been in preclinical and/or clinical trials as potential anticancer drugs [9].

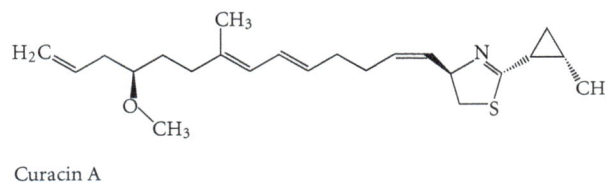

Curacin A

FIGURE 3: Structure of curacin A.

In addition, these molecules served as a drug leading to the development of synthetic analogues, for example, compound 4, TZT-1027 (Figure 4), ILX-651 (Figure 5), and LU-103793 (7), usually with improved pharmacologicaland pharmacokinetic properties for the treatment of different types of cancers. The antitumor activity of TZT-1027 (soblidotin) (Figure 4), a synthetic derivative of dolastatin 10 (Figure 2), was found to be superior to existing anticancer drugs, such as paclitaxel (Figure 6) and vincristine (Figure 7) and is currently undergoing Phase I testing for treating solid tumors [10].

The third generation dolastatin 15 analogue (Figure 5), ILX-651 (or tasidotin) (Figure 5), is another antitumor agent currently undergoing Phase II trials after its successful run in Phase I trials [11]. Pharmacological studies have also showed the mechanistic novelty of certain molecules, such as antillatoxin, in modifying the activity of Nav channels.

(I)

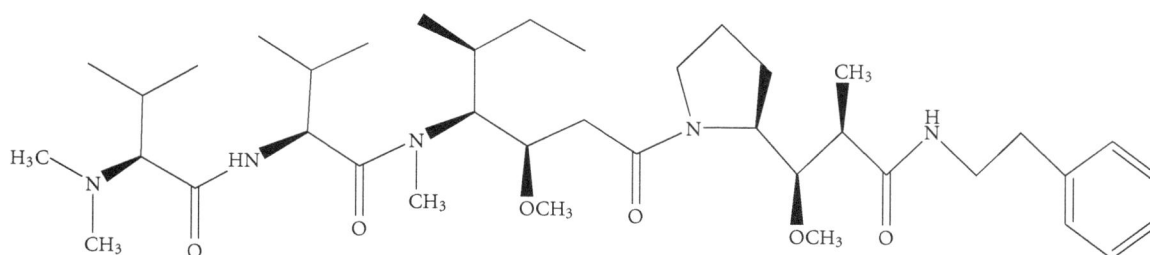

FIGURE 4: Structure of TZT-1027.

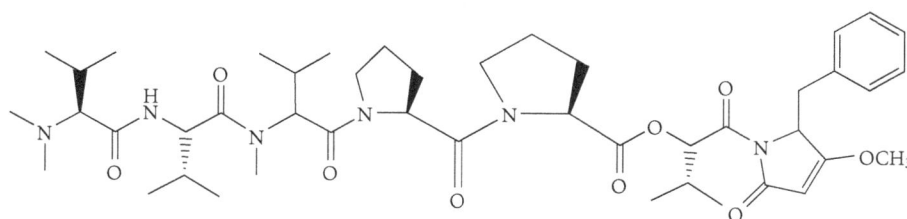

Dolastatin 15

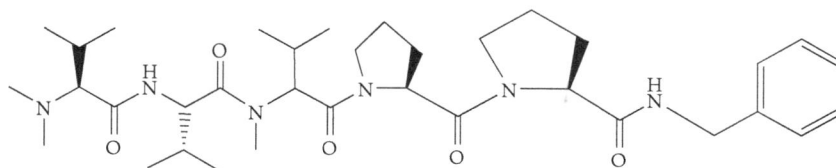

Cemadotin

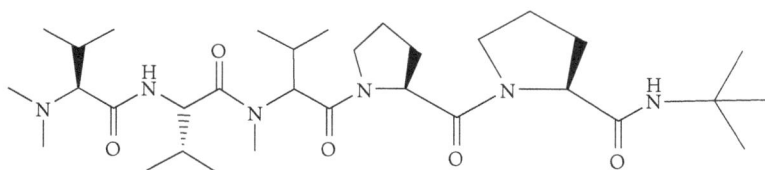

Tasidotin (ILX651)

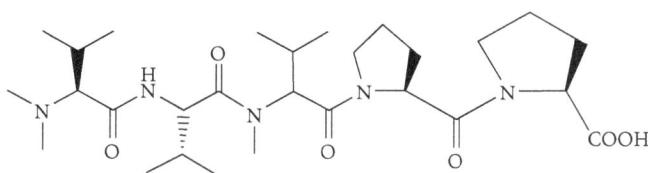

ILX651-C-carboxylate

FIGURE 5: Structures of dolastatin-15, cemadotin, Tasidotin, and ILX651-C-carboxylate.

These cyanobacterial toxins are source of valuable molecular tools in functional characterization of Nav channels as well as potential analgesics and neuroprotectants.

The discovery of tiny, single-celled cyanobacteria as ubiquitous and abundant components of the marine microbiota has radically changed our view of the functioning and composition of marine ecosystems. It is now clear that the two genera *Prochlorococcus* and *Synechococcus* dominate the photoautotrophic picoplankton over vast tracts of the world's oceans where they occupy a key position at the base of the marine food web and contribute significantly to global primary productivity [12]. Cyanobacteria (blue-green algae) are worldwide in distribution, occurring in saline and nonsaline habitats of diverse ionic composition [13]. However, more emphasis is now being placed on the importance of various metabolic features as taxonomic markers in cyanobacteria. Recently it has been suggested that soluble organic compounds, accumulated as internal osmotica in response to salinity stress, may provide a major biochemical character which distinguishes marine

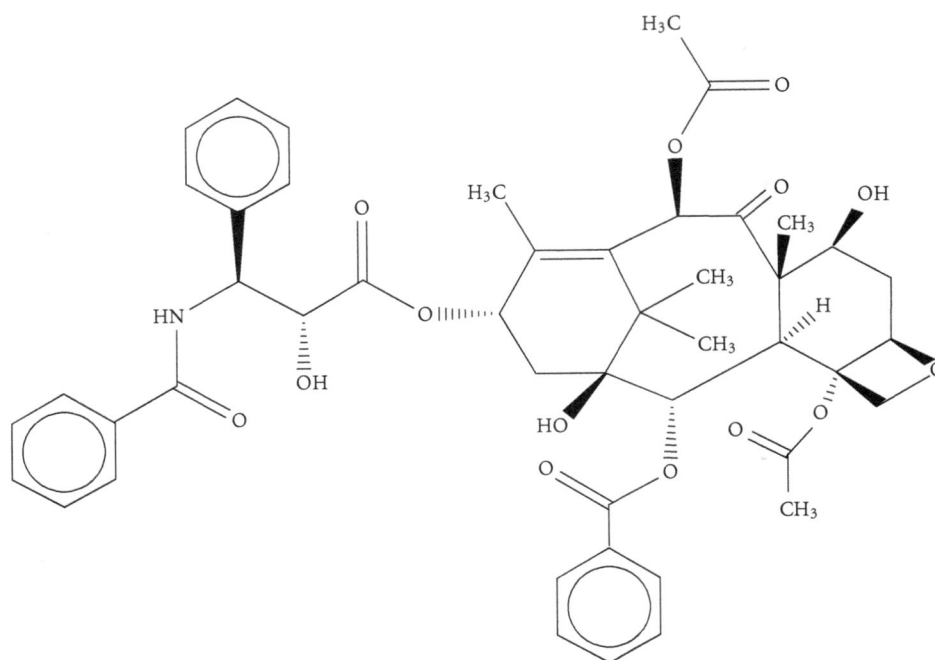

FIGURE 6: Structure of paclitaxel.

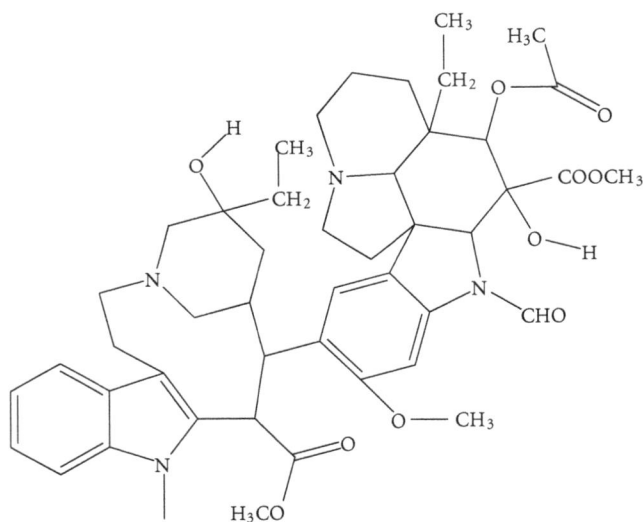

FIGURE 7: Structure of vincristine.

extracts from the large culture collection of marine cyano-bacteria for antiviral, antibacterial, antifungal, and immuno-modulatory activities has resulted in recovery of a compound from marine *Oscillatoria laete*-virians BDU 20801 that shows anti-*Candida* activity. An immunopotentiating compound with male antifertility, without being toxic to other systems in a mice model, was found in the extracts of*Oscillatoria willei* BDU 130511 [16]. Medically important gamma lino-lenic acid (GLA) is relatively rich in cyanobacteria, namely, *Spirulina platensi*s and *Arthrospira sp.* which is easily convert-ed into arachidonic acid in the human body and arachidonic acid into prostaglandin E2 [17].

Prostaglandin E2 has lowering action on blood pressure and the contracting function of smooth muscle and thus plays an important role in lipid metabolism. The bioin-formatic mining of cyanobacterial genomes has led to the discovery of novel cyanobactins. Heterologous expression of these gene clusters provided insights into the role of the genes participating in the biosynthesis of cyanobactins and facilitated the rational design of novel peptides.

and freshwater forms [5]. Thus glucosylglycerol has been considered to be "unique" to marine cyanobacteria [14], while sucrose has been reported to accumulate in response to osmotic stress in freshwater cyanobacteria [14].

1.2. Importance. Cyanobacteria in general and marine forms in particular are one of the richest sources of known and novel bioactive compounds including toxins with wide phar-maceutical applications [15]. Anti-HIV activity of marine cyanobacterial compounds from *Lyngbya lagerheimii* and *Phormidium tenue.* A massive programme of screening of

1.3. Vitamins from Cyanobacteria. Some of the marine cy-anobacteria appear to be potential sources for large-scale production of vitamins of commercial interest such as vita-mins of the B complex group and vitamin E [18]. The carotenoids and phycobiliprotein pigments of cyanobacteria have commercial value as natural food colouring agents, as feed additives, as enhancers of the color of egg yolks, to improve the health and fertility of cattle, as drugs, and in the cosmetic industries. Some anti-HIV activity has been observed with the compounds extracted from *Lyngbya lage-rhaimanii* and *Phormidium tenue* [18].

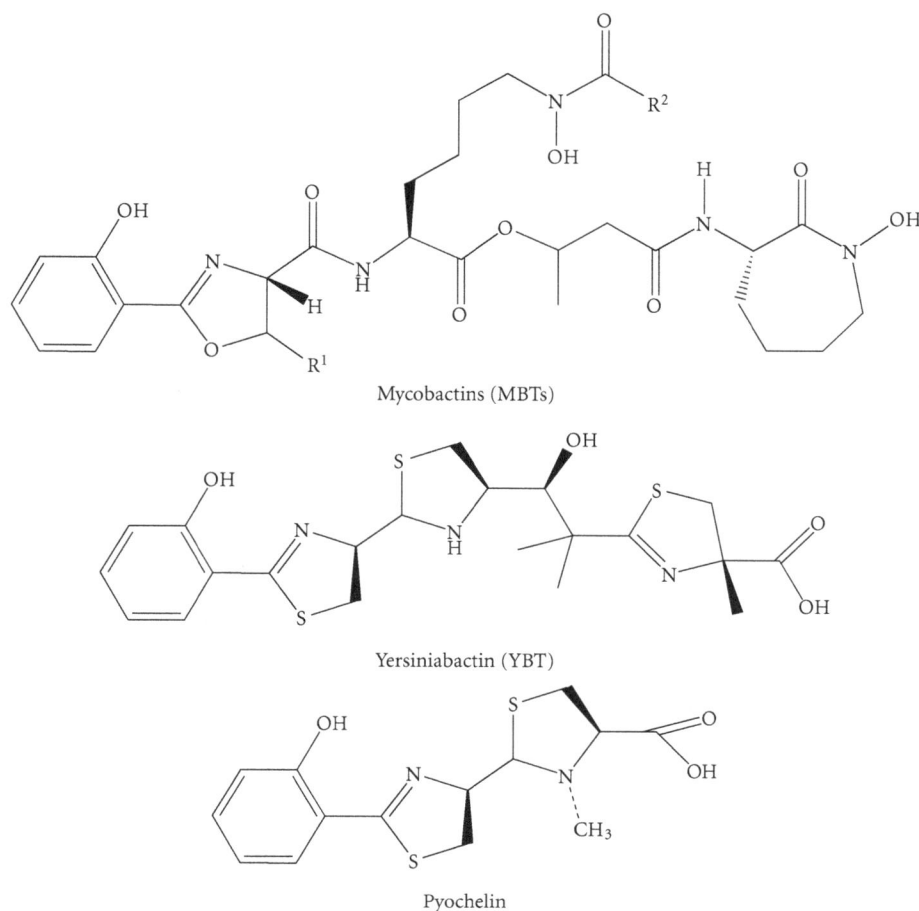

Mycobactins (MBTs)

Yersiniabactin (YBT)

Pyochelin

FIGURE 8: Structures of mycobactins, yersiniabactin, and pyochelin.

1.4. Biotechnology and Applications of Marine Cyanobacteria. The unicellular cyanobacterium *Synechocystis sp.* PCC6803 was the third prokaryote and first photosynthetic organism whose genome was completely sequenced. It continues to be an important model organism. The smallest genomes have been found in *Prochlorococcus spp.* (1.7 Mb) and the largest in *Nostoc punctiforme* (9 Mb) [14]. Some cyanobacteria are sold as food, notably *Aphanizomenon flos-aquae* and *Arthrospira platensis* (Spirulina) [14]. Recent research has suggested the potential application of cyanobacteria to the generation of clean and green energy via converting sunlight directly into electricity. Currently efforts are underway to commercialize algae-based fuels such as diesel, gasoline, and jet fuel [19–24].

2. Secondary Metabolites from Marine Cyanobacteria

2.1. Metabolic Themes and Building Blocks. There are currently some 300 marine cyanobacterial alkaloids. Of these, 128 marine cyanobacterial nitrogen-containing secondary metabolites. The majority of these biomolecules were isolated from the filamentous Order Nostocales, especially members belonging to the genera Lyngbya, Oscillatoria, and Symploca [20]. The locations of the collection sites were mainly from the tropics, including Papua New Guinea and the Pacific islands, in particular Guam and Palau. The predominant metabolic theme of nitrogen-containing marine cyanobacterial compounds is the occurrence of mixed polyketide-nonribosomal polypeptide structural types.

These are molecules containing acetate or propionate units as well as proteinogenic amino acids, forming as either linear or cyclic lipopeptides as found in mycobactins, Yersiniabactin and Pyochelin (Figure 8). The utilization of acetate-derived units in the construction of these hybrid compounds can be seen in several ways. Firstly, acetate-derived fatty acid chain can be coupled through amide bonds with a variety of functionalized amines in linear lipopeptides (e.g. malyngamides, S (41)–W (45)).

Further modifications on the fatty acid chain, such as methylation and halogenation, are common. Lipidation through amide bonds are also common in a number of oligopeptides, such as lyngbyabellin D and somamide. A Single acetate unit or multiple ketides can also be utilized to extend amino acids. For instance, a unit of acetate is used in the extension of a variety of amino acids, such as Ala, Phe, Pro, and Gly. The extension can either be linear or undergo cyclization to form common moieties, such as pyrrolinone ring system in the jamaicamides [21, 22].

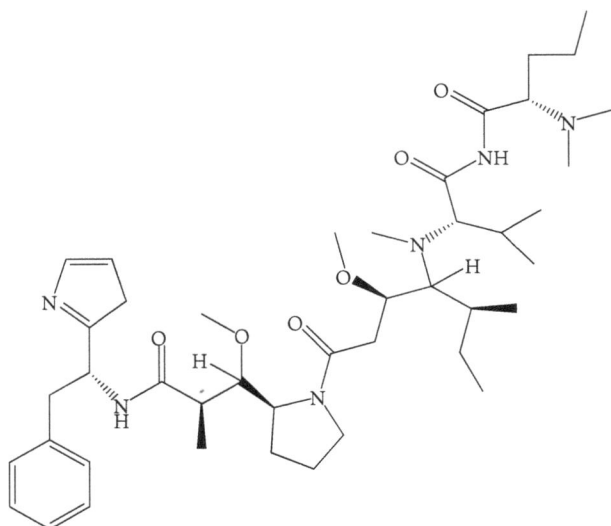

FIGURE 9: Structure of symplostatin 3.

Polyketide-derived moieties occurring as b-hydroxy or amino acid residues are source of nonproteinogenic units in the construction of lipopeptides, especially cyclic depsipeptides [22].

2.2. Nitrogen-Containing Lipids. Two new 2-alkypyridine alkaloids, phormidinines A (10) and B (11), were reported from an Okinawan collection of the marine cyanobacterium, *Phormidium sp.* [21, 22]. The structures and absolute stereochemistry of these compounds were determined based on 2D NMR spectra analysis and Mosher's method, respectively. A series of polychlorinated acetamides (12–16 and 19–29) and its dechlorinated derivatives (17 and 18) have been reported from *Microcoleus lyngbyaceus* and Lyngbyamajuscula/Schizothrix assemblage collected at Chuuk Island and Fiji, respectively [23]. A majority of these unique molecules are characterized by having terminal mono-, di-, or trichlorinated functional groups. Other marine cyanobacterial metabolites, for example, dysidenin-type compounds (e.g., 62) and barbamide (61), having terminal di- and trichloromethyl groups were shown to derive from chlorination of Leu, possibly via free radical mechanism. The biogenesis of the taveuniamides, isolated from Fijian Lyngbya majuscula/Schizothrix assemblage, has been proposed to occur either through the decarboxylation and methylation of an octaketide precursor or the C–C bond formation between the C-1 carboxyl carbon and C-2 of two tetraketide precursors [23].

3. Natural Products from Marine Cyanobacteria

A number of highly potent cyanobacterial natural products have been uncovered as potential lead compounds for further drug development, especially in the area of anticancer agents. An increasing number of lipopeptides, such as symplostatin 3 (Figure 9), lyngbyastatin 3, hectochlorin (Figure 10), and lyngbyabellins (114–116 and 118–123), have been reported

to target eukaryotic cytoskeletal macromolecules, such as actin and microtubule filaments [24].

These are attractive biological features for the development of potential anticancer drugs with specific cellular targets. Apratoxin A (126) is another potent cytotoxic compound worthy of further biological investigation as anticancer agent due to it mechanism of action in attenuating the FGF (fibroblast growth factor) signaling pathway [25]. Synthetic analogues based on the scaffolds of these cyanobacterial natural products can be developed for SAR studies as well as lead optimization for drug development [26].

4. Intramolecular Modulation of Serine Protease Inhibitor Activity in a Marine *Cyanobacterium* with Antifeedant Properties

One prevalent class found in marine and freshwater *cyanobacteria* is comprised of protease inhibitors with a cyclic depsipeptide scaffold that contains a 3-amino-6-hydroxy-2-piperidone (Ahp) moiety as a key feature for inhibition of certain serine proteases [27]. Since many digestive enzymes such as trypsin and chymotrypsin are serine proteases and are inhibited by these compounds, these natural products could function as digestion inhibitors [28]. Serine protease inhibitors also cooccur with microcystins and are linked to an enhanced toxin activity or thought to upregulate biosynthetic genes [29]. The tropical sea urchin *Diadema antillarum*,which is a cyanobacterium , produces a wide array of serine protease inhibitors including lyngbyastatins 4–6 [30, 31], pompanopeptin A [32], and largamides D–G [33]. The antifeedant activity may be a reflection of the secondary metabolite content, known to be comprised of many serine protease inhibitors. Further chemical and NMR spectroscopic investigation led to isolate and structurally characterize a new serine protease inhibitor **1** that is formally derived from an intramolecular condensation of largamide D (**2**) (Figure 11) [33]. The cyclization resulted in diminished activity, but to different extents against two serine proteases tested. This finding suggests that cyanobacteria can endogenously modulate the activity of their protease inhibitors.

5. Cyanobacteria: A Potential Source of New Biologically Active Substances

Cyanobacteria (blue-green algae) provide a potential source of biologically active secondary metabolites [34]. Investigations over the last decades have identified compounds with for instance cytotoxic, antifungal, antibacterial, or antiviral activity. Hydrophilic and lipophilic extracts of cyanobacterial strains, isolated from fresh and brackish water, and water blooms were investigated for their antibiotic activities against microorganisms both Gram negative and Gram positive. Most of the isolated substances belong to groups of polyketides, amides, alkaloids, and peptides [35].

The blue-green algae are among the oldest photoautotrophic organisms. Their cultivation without organic substrates can be an economical advantage over other microorganisms. In view of the growing resistance of bacteria

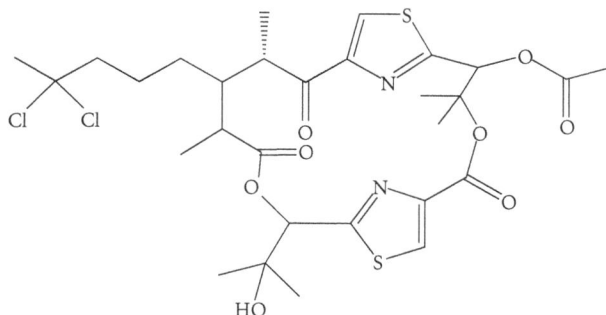

FIGURE 10: Structure of hectochlorin.

to common antibiotics, the search for new antimicrobially active compounds has become increasingly important [36]. Screening programme was made which tested approximately fifty extracts from twelve different cyanobacterial strains and two water blooms against different bacteria and one yeast. The results of the study show the ability of cyanobacteria to produce compounds with antimicrobial effects [36].

5.1. Antimicrobial and Cytotoxic Assessment of Marine Cyanobacteria: Synechocystis and Synechococcus. Aqueous extracts and organic solvent extracts of isolated marine cyanobacteria strains were tested for antimicrobial activity against a fungus, Gram-positive and Gram-negative bacteria and for cytotoxic activity against primary rat hepatocytes and HL-60 cells [37]. Antimicrobial activity was based on the agar diffusion assay. Cytotoxic activity was measured by apoptotic cell death scored by cell surface evaluation and nuclear morphology [38]. A high percentage of apoptotic cells were observed for HL-60 cells when treated with cyanobacterial organic extracts [39]. Slight apoptotic effects were observed in primary rat hepatocytes when exposed to aqueous cyanobacterial extracts [40]. Marine Synechocystis and Synechococcus extracts induce apoptosis in eukaryotic cells and cause inhibition of Gram-positive bacteria. The different activity in different extracts suggests different compounds with different polarities [41].

6. Potential Commercial Development of Insecticides, Algaecides, and Herbicides from Cyanobacteria

Potential commercial development of cyanobacterial compounds for nonbiomedical applications, particularly including herbicides, algaecides, and insecticides poses a potentially important opportunity to utilize the biological activity of these compounds [26].

6.1. Insecticides. Fladmark et al. [42]. screened extracts from 76 isolates of cyanobacteria and found several of these isolates produced compounds that were larvicidal to *Aedes aegypti*. The greatest inhibition, however, was associated with presence of the hepatotoxic microcystins and the neurotoxic anatoxin-a. Humpage and Falconer [43] reported that, while

investigating cyanobacteria as a biofertilizer, several strains were found to inhibit development of mosquito larvae, and subsequently showed that methanolic extracts from an isolate of *Westiellopsis sp.* were larvicidal to several species of mosquito, including representatives of *Aedes aegypti* (a vector for Dengue Fever), *Anopheles stephensi* (a vector for malaria), and *Culex tritaeniorhynchus* and *C. quinquefasciatus* (vectors of encephalitis). The use of genetically engineered cyanobacteria, specifically expressing the insecticidal proteins from *Bacillus thuringiensis* to control mosquito larvae [44, 45]. Likewise, cyanobacteria that produce naturally occurring larvicidal metabolites may eliminate the potential threats associated with release of transgenic organisms [44, 45].

7. Cyanobacterial Cyclopeptides as Lead Compounds to Novel Targeted Cancer Drugs

Cyanobacterial cyclopeptides, including microcystins and nodularins, are considered a health hazard to humans due to the possible toxic effects of high consumption [46]. From a pharmacological standpoint, microcystins are stable hydrophilic cyclic heptapeptides with a potential to cause cellular damage following uptake via organic anion transporting polypeptides (OATPs) [47]. Their intracellular biological effects involve inhibition of catalytic subunits of protein phosphatase 1 (PP1) and PP2, and glutathione depletion andgeneration of reactive oxygen species (ROS) [48]. Interestingly, certain OATPs are prominently expressed in cancers as compared to normal tissues, qualifying MC as potential candidates for cancer drug development. In targeted cancer therapy, cyanotoxins comprise a rich source of natural cytotoxic compounds with a potential to target cancers expressing specific uptake transporters [49]. Their structure offers opportunities for combinatorial engineering to enhance the therapeutic index and resolve organ-specific toxicity issues [50].

8. Conclusion

The fact that cyanobacteria are one of the richest sources of known and novel bioactive compounds including toxins with wide pharmaceutical applications is unquestionable. Many compounds from cyanobacteria are useful for welfare of mankind. Because of high discovery rate, research

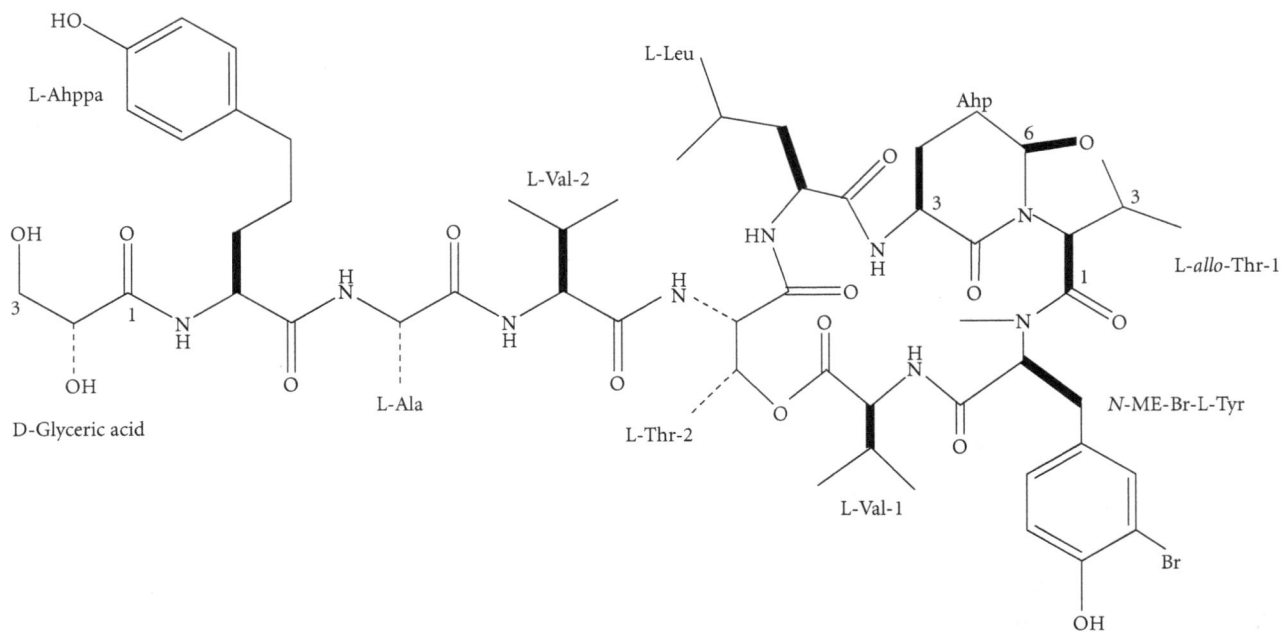

Largamide D oxazolidine (1)

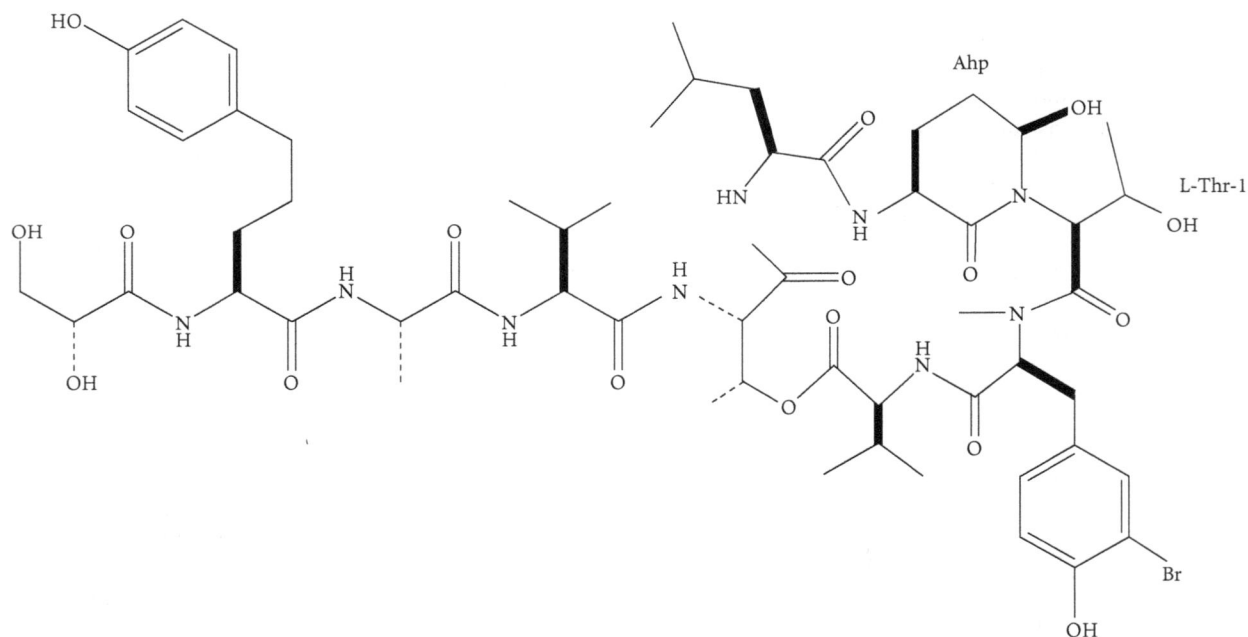

Largamide D (2)

FIGURE 11: Structures of largamide D oxazolidihe (1) , largamide D (2).

should be done to unfold other hidden aspects of marine cyanobacteria. An advantage of natural products research on marine cyanobacteria is the high discovery rate (>95%) of novel compounds as compared to other traditional microbial sources. This is due largely to the unexplored nature of this group of microalgae. One of the key areas to further tap these microalgae for new chemical entities is the collection of cyanobacterial strains from unexplored localities, especially from Africa and Asia. In addition to the procurement of marine cyanobacteria from unexplored locales, the amenability of field collected strains to laboratory culture is an important factor in the drug discovery process.

Acknowledgments

The authors would like to acknowledge Higher Education Commission of Pakistan for providing financial support for the project (PM IPFP/HRD/HEC/2010/1815) and International Islamic University Islamabad for supporting the project.

References

[1] R. W. Castenholz and J. B. Waterbury, "Cyanobacteria," in *Bergey's Manual of Systematic Bacteriology*, vol. 3, pp. 171–179, 1989.

[2] A. J. Smith, "Modes of cyanobacterial carbon metabolism," *Annales de Microbiologie*, vol. 134B, no. 1, pp. 93–113, 1983.

[3] Y Cohen, B. B. Jrgensen, N. P. Revsbech, and R. Paplawski, "Adaptation to hydrogen sulfide of oxygenic and anoxygenic photosynthesis among cyanobacteria," *Applied and Environmental Microbiology*, vol. 51, no. 2, pp. 398–407, 1986.

[4] P. Fay, "Oxygen relations of nitrogen fixation in cyanobacteria," *Microbiological Reviews*, vol. 56, no. 2, pp. 340–373, 1992.

[5] B. A. Whitton, "Diversity, ecology and taxonomy of the cyanobacteria," in *Photosynthetic Prokaryotes*, pp. 1–51, Plenum Press, New York, NY, USA, 1992.

[6] N. Tandeau de Marsac and J. Houmard, "Adaptation of cyanobacteria to environmental stimuli: new steps towards molecular mechanisms," *FEMS Microbiology Reviews*, vol. 104, no. 1-2, pp. 119–189, 1993.

[7] D. Schwarzer, R. Finking, and M. A. Marahiel, "Nonribosomal peptides: from genes to products," *Natural Product Reports*, vol. 20, no. 3, pp. 275–287, 2003.

[8] M. A. Jordan and L. Wilson, "Microtubules and actin filaments: dynamic targets for cancer chemotherapy," *Current Opinion in Cell Biology*, vol. 10, no. 1, pp. 123–130, 1998.

[9] W. H. Gerwick, L. T. Tan, and N. Sitachitta, "Nitrogen-containing metabolites from marine cyanobacteria," in *The Alkaloids: Chemistry and Biology*, vol. 57, pp. 75–184, Academic Press, San Diego, Calif, USA, 2001.

[10] J. Watanabe, M. Minami, and M. Kobayashi, "Antitumor activity of TZT-1027 (soblidotin)," *Anticancer Research*, vol. 26, no. 3, pp. 1973–1981, 2006.

[11] A. C. Mita, L. A. Hammond, P. L. Bonate et al., "Phase I and pharmacokinetic study of tasidotin hydrochloride (ILX651), a third-generation dolastatin-15 analogue, administered weekly for 3 weeks every 28 days in patients with advanced solid tumors," *Clinical Cancer Research*, vol. 12, no. 17, pp. 5207–5215, 2006.

[12] K. M. Blumenthal and A. L. Seibert, "Voltage-gated sodium channel toxins: poisons, probes, and future promise," *Cell Biochemistry and Biophysics*, vol. 38, no. 2, pp. 215–237, 2003.

[13] R. H. Reed, L. J. Borowitzka, and M. A. Mackay, "Organic solute accumulation in osmotically stressed cyanobacteria," *FEMS Microbiology Reviews*, vol. 39, no. 1-2, pp. 51–56, 1986.

[14] A. Dufresne, O. Martin, J. S. David et al., "Unraveling the genomic mosaic of a ubiquitous genus of marine cyanobacteria," *Genome Biology*, vol. 9, no. 5, article R90, 2008.

[15] C. Raghavan, B. Kadalmani, T. Thirunalasundari, G. Subramanian, and M. A. Akbarsha, *Biological and Comparative Endocrinology*, Bharathidasan University, Tiruchirapalli, India, 2002.

[16] S. K. Deth, *Antimicrobial compounds from marine cyanobacteria with special reference to the bioactivity of a purified compound from* Oscillatoria laete-virens *BDU 20801*, Ph.D. thesis, Bharathidasan University, Thiruchirappalli, India, 1999.

[17] U. S. Euler and R. Eliassen, *Prostaglandins*, Academic Press, New York, NY, USA, 1967.

[18] D. J. Schaeffer and V. S. Krylov, *Anti-HIV Activity of Extracts and Compounds from Algae and Cyanobacteria Department of Veterinary Biosciences*, University of Illinois, 2001.

[19] T. Kaneko, A. Tanaka, S. Sato et al., "Sequence analysis of the genome of the unicellular cyanobacterium *Synechocystis* sp. strain PCC6803. I. Sequence features in the 1 Mb region from map positions 64% to 92% of the genome," *DNA Research*, vol. 2, no. 4, pp. 153–166, 1995.

[20] W. H. Gerwick, L. T. Tan, and N. Sitachitta, "Nitrogen-containing metabolites from marine cyanobacteria," *Alkaloids: Chemistry and Biology*, vol. 57, pp. 75–184, 2001.

[21] T. Teruya, K. Kobayashi, K. Suenaga, and H. Kigoshi, "Phormidines A and B, novel 2-alkylpyridine alkaloids from the cyanobacterium *Phormidium* sp," *Tetrahedron Letters*, vol. 46, no. 23, pp. 4001–4003, 2005.

[22] M. A. Orsini, L. K. Pannell, and K. L. Erickson, "Polychlorinated acetamides from the cyanobacterium *Microcoleus lyngbyaceus*," *Journal of Natural Products*, vol. 64, no. 5, pp. 572–577, 2001.

[23] J. C. Meeks, "An overview of the genome of *Nostoc punctiforme*, a multicellular, symbiotic cyanobacterium," *Current Science*, vol. 89, no. 1, 2005.

[24] P. Spolaore, C. Joannis-Cassan, E. Duran, and A. Isambert, "Commercial applications of microalgae," *Journal of Bioscience and Bioengineering*, vol. 101, no. 2, pp. 87–96, 2006.

[25] L. Lehane and R. J. Lewis, "Ciguatera: recent advances but the risk remains," *International Journal of Food Microbiology*, vol. 61, no. 2-3, pp. 91–125, 2000.

[26] Y. Shimizu, "Microalgal metabolites," *Current Opinion in Microbiology*, vol. 6, no. 3, pp. 236–243, 2003.

[27] S. P. Gunasekera, M. W. Miller, J. C. Kwan, H. Luesch, and V. J. Paul, "Molassamide, a depsipeptide serine protease inhibitor from the marine cyanobacterium *Dichothrix utahensis*," *Journal of Natural Products*, vol. 73, no. 3, pp. 459–462, 2010.

[28] J. C. Kwan, K. Taori, V. J. Paul, and H. Luesch, "Lyngbyastatins 8-10, elastase inhibitors with cyclic depsipeptide scaffolds isolated from the marine cyanobacterium *Lyngbya semiplena*," *Marine Drugs*, vol. 7, no. 4, pp. 528–538, 2009.

[29] G. Radau, "Cyanopeptides: a new and nearly inexhaustible natural resource for the design and structure-activity relationship studies of the new inhibitors of trypsin-like serine proteases," *Current Enzyme Inhibition*, vol. 1, pp. 295–307, 2005.

[30] K. Taori, S. Matthew, C. Ross, R. R. James, V. J. Paul, and H. Luesch, "Lyngbyastatins 5-7, potent elastase inhibitors from Floridian marine cyanobacteria, *Lyngbya* spp," *Journal of Natural Products*, vol. 70, no. 10, pp. 1593–1600, 2007.

[31] S. Matthew, C. Ross, V. J. Paul, and H. Luesch, "Pompanopeptins A and B, new cyclic peptides from the marine cyanobacterium *Lyngbya confervoides*," *Tetrahedron*, vol. 64, no. 18, pp. 4081–4089, 2008.

[32] A. Plaza and C. A. Bewley, "Largamides A-H, unusual cyclic peptides from the marine cyanobacterium *Oscillatoria* sp," *Journal of Organic Chemistry*, vol. 71, no. 18, pp. 6898–6907, 2006.

[33] S. P. Gunasekera, R. Ritson-Williams, and V. J. Paul, "Carriebowmide, a new cyclodepsipeptide from the marine cyanobacterium *Lyngbya polychroa*," *Journal of Natural Products*, vol. 71, no. 12, pp. 2060–2063, 2008.

[34] E. Cruz-Rivera and V. J. Paul, "Chemical deterrence of a cyanobacterial metabolite against generalized and specialized grazers," *Journal of Chemical Ecology*, vol. 33, no. 1, pp. 213–217, 2007.

[35] R. W. Thacker, D. G. Nagle, and V. J. Paul, "Effects of repeated exposures to marine cyanobacterial secondary metabolites on feeding by juvenile rabbitfish and parrotfish," *Marine Ecology Progress Series*, vol. 147, no. 1–3, pp. 21–29, 1997.

[36] V. J. Paul, R. W. Thacker, K. Banks, and S. Golubic, "Benthic cyanobacterial bloom impacts the reefs of South Florida (Broward County, USA)," *Coral Reefs*, vol. 24, no. 4, pp. 693–697, 2005.

[37] L. T. Tan, N. Sitachitta, and W. H. Gerwick, "The guineamides, novel cyclic depsipeptides from a Papua New Guinea collection of the marine cyanobacterium Lyngbya majuscula," *Journal of Natural Products*, vol. 66, no. 6, pp. 764–771, 2003.

[38] S. Bunyajetpong, W. Y. Yoshida, N. Sitachitta, and K. Kaya, "Trungapeptins A-C, cyclodepsipeptides from the marine cyanobacterium *Lyngbya majuscula*," *Journal of Natural Products*, vol. 69, no. 11, pp. 1539–1542, 2006.

[39] L. T. Tan, B. L. Márquez, and W. H. Gerwick, "Lyngbouilloside, a novel glycosidic macrolide from the marine cyanobacterium *Lyngbya bouillonii*," *Journal of Natural Products*, vol. 65, no. 6, pp. 925–928, 2002.

[40] F. Sponga, L. Cavaletti, A. Lazzarini et al., "Biodiversity and potentials of marine-derived microorganisms," *Journal of Biotechnology*, vol. 70, no. 1–3, pp. 65–69, 1999.

[41] A. M. S. Mayer and K. R. Gustafson, "Marine pharmacology in 2000: antitumor and cytotoxic compounds," *International Journal of Cancer*, vol. 105, no. 3, pp. 291–299, 2003.

[42] K. E. Fladmark, M. H. Serres, N. L. Larsen, T. Yasumoto, and T. Aune, "Sensitive detection of apoptogenic toxins in suspension cultures of rat and salmon hepatocytes," *Toxicon*, vol. 36, no. 8, pp. 1101–1114, 1998.

[43] A. R. Humpage and I. R. Falconer, "Microcystin-LR and liver tumor promotion: effects on cytokinesis, ploidy, and apoptosis in cultured hepatocytes," *Environmental Toxicology*, vol. 14, no. 1, pp. 61–75, 1999.

[44] C. Angsuthanasombat and S. Panyim, "Biosynthesis of 130-kilodalton mosquito larvicide in the cyanobacterium *Agmenellum quadruplicatum* PR-6," *Applied and Environmental Microbiology*, vol. 55, no. 9, pp. 2428–2430, 1989.

[45] R. C. Murphy and S. E. Stevens, "Cloning and expression of the cryIVD gene of Bacillus thuringiensis subsp. israelensis in the cyanobacterium *Agmenellum quadruplicatum* PR-6 and its resulting larvicidal activity," *Applied and Environmental Microbiology*, vol. 58, no. 5, pp. 1650–1655, 1992.

[46] B. Soni, U. Trivedi, and D. Madamwar, "A novel method of single step hydrophobic interaction chromatography for the purification of phycocyanin from *Phormidium fragile* and its characterization for antioxidant property," *Bioresource Technology*, vol. 99, no. 1, pp. 188–194, 2008.

[47] G. Francis, "Poisonous Australian lake," *Nature*, vol. 18, no. 444, pp. 11–12, 1878.

[48] N. Gupta, S. C. Pant, R. Vijayaraghavan, and P. V. L. Rao, "Comparative toxicity evaluation of cyanobacterial cyclic peptide toxin microcystin variants (LR, RR, YR) in mice," *Toxicology*, vol. 188, no. 2-3, pp. 285–296, 2003.

[49] I. R. Falconer and A. R. Humpage, "Health risk assessment of cyanobacterial (blue-green algal) toxins in drinking water," *International Journal of Environmental Research and Public Health*, vol. 2, no. 1, pp. 43–50, 2005.

[50] K. Vareli, E. Briasoulis, G. Pilidis, and I. Sainis, "Molecular confirmation of Planktothrix rubescens as the cause of intense, microcystin-Synthesizing cyanobacterial bloom in Lake Ziros, Greece," *Harmful Algae*, vol. 8, no. 3, pp. 447–453, 2009.

16

Current View on Phytoplasma Genomes and Encoded Metabolism

Michael Kube,[1,2] Jelena Mitrovic,[3] Bojan Duduk,[3] Ralf Rabus,[4,5] and Erich Seemüller[6]

[1] Department of Crop and Animal Sciences, Humboldt-University of Berlin, Lentzeallee 55/57,
 14195 Berlin, Germany
[2] Max Planck Institute for Molecular Genetics, Ihnestr. 63, 14195 Berlin, Germany
[3] Department of Plant Pathology, Institute of Pesticides and Environmental Protection, Banatska 31b, P.O. Box 163,
 11080 Belgrade, Serbia
[4] Institute for Chemistry and Biology of the Marine Environment, Carl von Ossietzky University of Oldenburg,
 Carl-von-Ossietzky Straße 9-11, 26111 Oldenburg, Germany
[5] Department for Microbiology, MaxPlanck Institute for Marine Microbiology, Celsiusstraße 1, 28359 Bremen, Germany
[6] Institute for Plant Protection in Fruit Crops and Viticulture, Federal Research Centre for Cultivated Plants,
 Schwabenheimer Straße 101, 69221 Dossenheim, Germany

Correspondence should be addressed to Michael Kube, michael.kube@agrar.hu-berlin.de

Academic Editors: M. J. Paul and T. Tanisaka

Phytoplasmas are specialised bacteria that are obligate parasites of plant phloem tissue and insects. These bacteria have resisted all attempts of cell-free cultivation. Genome research is of particular importance to analyse the genetic endowment of such bacteria. Here we review the gene content of the four completely sequenced 'Candidatus Phytoplasma' genomes that include those of 'Ca. P. asteris' strains OY-M and AY-WB, 'Ca. P. australiense,' and 'Ca. P. mali'. These genomes are characterized by chromosome condensation resulting in sizes below 900 kb and a G + C content of less than 28%. Evolutionary adaption of the phytoplasmas to nutrient-rich environments resulted in losses of genetic modules and increased host dependency highlighted by the transport systems and limited metabolic repertoire. On the other hand, duplication and integration events enlarged the chromosomes and contribute to genome instability. Present differences in the content of membrane and secreted proteins reflect the host adaptation in the phytoplasma strains. General differences are obvious between different phylogenetic subgroups. 'Ca. P. mali' is separated from the other strains by its deviating chromosome organization, the genetic repertoire for recombination and excision repair of nucleotides or the loss of the complete energy-yielding part of the glycolysis. Apart from these differences, comparative analysis exemplified that all four phytoplasmas are likely to encode an alternative pathway to generate pyruvate and ATP.

1. Introduction

Phytoplasmas are bacteria which colonise plants and some groups of insects. They are associated with diseases of several hundred plant species including important crops [1]. Just for apple, economical losses caused by 'Candidatus Phytoplasma mali' and related phytoplasmas of 100 million €/year in Italy and Germany were estimated [2]. Phytoplasmas are pleomorphic but nonhelical, 0.2–0.8 μm in diameter, and lack, as members of the class *Mollicutes*, a firm cell wall. Phytoplasmas were first called mycoplasma-like organisms on discovery in 1967 [3]. Today, they are grouped in the provisional genus 'Candidatus Phytoplasma' that is assigned to a family of *incertae sedis* [4]. Thirty-two 'Candidatus Phyto-plasma' species have been described [5–7]. The phytoplasma clade is divided into two major subclades and numerous subgroups based on the 16S rRNA marker gene (Figure 1). Based on their 16S rDNA sequence they are phylogenetically assigned to a distinct monophyletic taxon in the *Mollicutes* as member of the order *Acholeplasmatales* [8]. *Mollicutes* are deeply branching from the clostridial lineage leading to the genera *Bacillus* and *Lactobacillus* [9–11]. They are assigned to the phylum *Tenericutes*. Within the mollicutes, phytoplasmas are most closely related to the genus *Acholeplasma* [12, 13] with which they share, in contrast to other mollicutes, the usage of UGA as a stop codon [14]. Based on 16S rDNA analysis, Weisburg et al. [15] proposed that acholeplasmas were formed early at the initial divergence of the *Mollicutes*

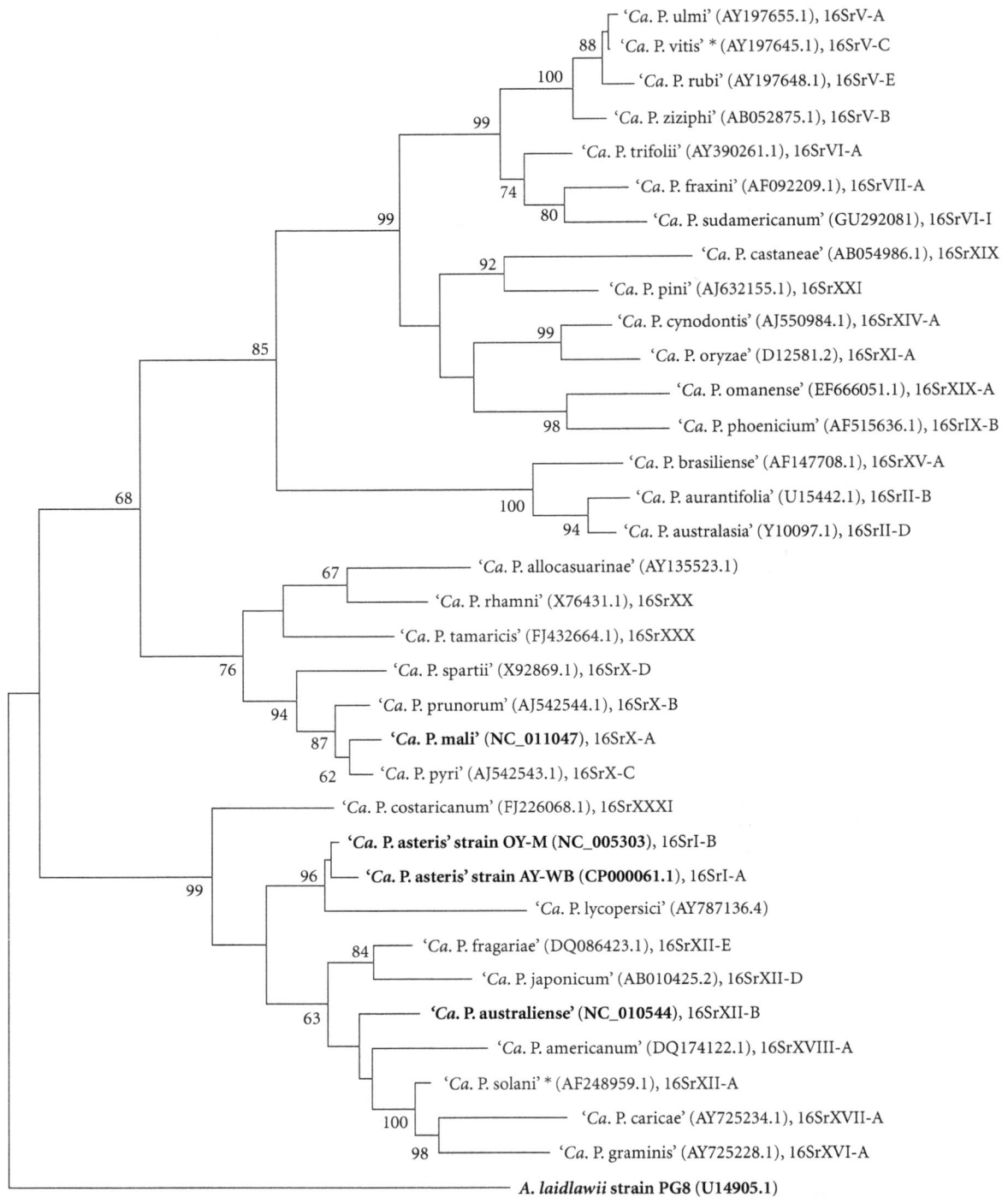

FIGURE 1: Phylogenetic tree constructed by parsimony analyses of 16S rDNA sequences of 34 phytoplasma strains, belonging to all 'Candidatus Phytoplasma' species described employing A. laidlawii as outgroup. The completely sequenced strains are in bold. CLUSTAL W from the Molecular Evolutionary Genetics Analysis program-MEGA5 [26] was used for multiple alignment. The maximum parsimony tree was obtained using the Close-Neighbour-Interchange algorithm, implemented in the MEGA5, with search level 3 in which the initial trees were obtained with the random addition of sequences (10 replicates). One of nine equally parsimonious trees is shown in Figure 1. Numbers on the branches are bootstrap values obtained for 2000 replicates (only values above 60% are shown). GenBank accession numbers are given in parentheses, and 16Sr group classification (when available) is shown on the right side of the tree. Asterisk indicates 'Candidatus Phytoplasma' proposed by the IRPCM Phytoplasma/Spiroplasma Working Team—phytoplasma taxonomy group [5], not yet formally described. The tree is drawn to scale with branch lengths calculated using the average pathway method and represents the number of changes over the whole sequence. The scale bar represents 20 nucleotide substitutions.

from clostridial ancestors, followed by genome condensation and loss of the cell wall. Another divergence may have occurred that led to the separation of the phytoplasmas from the acholeplasmas. Genome condensation took place in both groups resulting in genomes sizes of about 1.215–2.095 kb and 27–38 mol% GC for the acholeplasmas [16, 17] and 530–1.350 kb and 21–33 mol% GC for the phytoplasmas [4, 17–20].

Information on the phytoplasma host environment is essential for understanding their adaptation and genome condensation. In plants, phytoplasmas mainly colonise the phloem sieve elements, but have occasionally also been reported to occur in companion and parenchyma cells as well [21–23]. The sieve tube environment is highly appropriate for the pathogens because it offers excellent conditions for the spread in the plant host. The phytoplasmas can readily pass the sieve pores of the sieve tubes and may be passively translocated according to the source/sink principle of the phloem sap flow in the various growing seasons [1, 24, 25].

The phloem fulfills functions in long-distance transport and allocation of nutrients and emits signals related to nutrition, development, and stress response. Probably the phloem transport functions by mass flow through the sieve elements and by cell-to-cell transport by surrounding companion cells and phloem parenchyma cells that provide metabolites to the sap [27].

Phloem sap is unique for several reasons. It is under high hydrostatic pressure and rich in nutrients. The content varies depending on the plant species but always contains large amounts of carbohydrates. The most abundant one is often sucrose (12–30%) beside other sugars such as glucose and fructose, minerals, proteins, amino acids, and ATP [28, 29]. Other sugars such as polyols and oligosaccharides of the raffinose family may be also abundant, depending on the species [30]. Sugar alcohols such as mannitol and sorbitol may also be present in considerable amount [31, 32].

The osmotic pressure of the sieve tube sap ranges from 0.8–2.5 MP and the pH is usually between 7.3 and 8.5 [30]. Bicarbonate and malate may play a role in the control of phloem sap pH [33, 34]. It is notable that malate and citrate represent the predominant organic acids in phloem and xylem sap [35], and one may speculate whether the content of these organic acids is regulated as a consequence of phytoplasma infection. It is known that the pH value of media used for cultivation of plant-derived isolates of *Acholeplasma brassicae* and *A. palmae* decreases significantly during growth.

Other nutrients such as nitrate, inorganic phosphate, glutathione, glutamate, and amino acids were also detected in phloem sap [36, 37]. For lettuce (*Lactuca sativa* L.) it was shown [28] that fourteen amino acids are present (Asp, Thr, Ser, Asn, Glu, Gln, Gly, Ala, Val, Ile, Leu, Tyr, Phe, and gamma-isobutyric acid/GABA). Glutamine, glutamic acid, serine, and GABA were often predominant. The concentration in lettuce was 125 mM for all amino acids. The presence of high ammonia concentrations (6 mM) is also remarkable. However, depending on the method also arginine and trace quantities of alpha-aminobutyric-acids, methionine, tryptophan, and ornithine were detected. Secondary sub-

stances such as phenolics, flavonoids, sesquiterpene lactones, alkaloids, and sterols have also been reported [28].

Some of the listed compounds and others such as cytokinins, auxins, abscisic acid, gibberellins, jasmonates, and methyl salicylates [27, 36, 37] and small RNAs [38, 39] highlight the importance of the phloem for signalling. It is still under discussion if the phloem is also a location for protein synthesis [27]. The determination of pumpkin sap proteome supports this idea [40].

Phytoplasmas are mainly transmitted by phloem-feeding insects such as leafhoppers, planthoppers, and psyllids [41, 42]. Under certain conditions, vectors may be attracted by plants releasing plant allomones manipulated by the phytoplasma infection [42]. They have to acquire the pathogen by phloem feeding [43, 44]. Subsequently, the phytoplasmas multiply in the hemolymph in which the concentration of organic compounds is similarly high than in phloem sap [45]. It is remarkable that trehalose is the predominant sugar in the hemolymph of aphids while glucose appeared as an artefact, and sucrose was not detected [45]. The trehalose content of the hemolymph of the phloem-sucking insects is correlated to the sucrose content of the aphid's host plants [45]. Interaction of phytoplasmas with their insect vectors has been reported for 'Ca. P. asteris' strains. Dominant membrane proteins such as the major antigenic protein Amp interact with the vector proteins such as actin and the alpha and beta subunits of the ATP synthase [46]. The insect environment in combination with the ability to migrate from the insect gut to the salivary glands allows the spread of the pathogen [47].

Four complete phytoplasma genomes have been published. However, there is no recent publication summarizing the encoded repertoire of these genomes (see also ST, supplementary table in Supplementary Material available online at doi: 10.1100/2012/185942) and discussing the deduced metabolism. For this reason this paper attempts to fill this gap by reviewing published results and re-evaluating the genomic data.

2. Genomic Benchmarks

The first fully sequenced genome was that of the onion yellows phytoplasma strain OY-M of subgroup 16SrI-B [48]. It was followed by the closely related aster yellows witches'-broom phytoplasma strain AY-WB of subgroup 16SrI-A that showed large regions of conserved synteny [49]. In addition to the information on this two 'Ca. P. asteris' genomes the sequences of two members from other phylogenetic groups were published in 2008, namely, that of 'Ca. P. australiense' [50] and that of 'Ca. P. mali' [20] (Table 1). With a size of 880 kb 'Ca. P. australiense' shows the largest phytoplasma chromosome deciphered so far while 'Ca. P. mali' with a size of 602 kb stands at the opposite end belonging to the second major subclade. 'Ca. P. mali' is characterized by the extremely low G + C content of less than 22%, while the other three genomes show a similar G + C content of 27-28%. The number of protein-encoding genes ranges corresponding to their genome sizes from 481 to 776 but may be also influenced by different annotation styles. The

TABLE 1: Overview of the four complete phytoplasma genomes and *A. laidlawii*.

| Strain | 'Candidatus Phytoplasma' species | | | | Acholeplasma laidlawii |
	asteris OY-M	asteris AY-WB	australiense Rp-A	mali AT	PG-8A
Chromosome organisation	Circular	Circular	Circular	Linear	Circular
Chromosome size	853,092	706,569	879,959	601,943	1,496,992
G + C content (%)	27.76	26.89	27.42	21.39	31.93
G + C % of protein-coding genes[1]	29.09	28.54	28.72	22.58	32.23
Protein-coding genes[1,2]	752	776	684 (155)	481 (16)	1,380 (11)
Protein coding (%)[1]	73.1	73.7	64.1	76.3	90.7
Average ORF size[1]	829	776	825	955	984
Protein-coding genes/kb[1]	0.881	0.949	0.777	0.799	0.921
rRNA operons	2	2	2	2	2
tRNAs	32	31	35	32	34
Accession no.	AP006628.2	CP000061.1	AM422018.1	CU469464.1	CP000896.1
Plasmid-like elements	2	4	1	0	—

[1]Without pseudo genes; [2]number of annotated pseudo genes is given in brackets.

highest number of strain-specific genes is present in 'Ca. P. australiense' [50]. The only available genome sequence from the related genus *Acholeplasma* is that of *A. laidlawii* strain PG-8A (Acc. no. CP000896), which differs from phytoplasma genomes by a larger size (1.5 Mb), higher number of protein-encoding genes (1,380), and a higher G + C content (32%). However, all sequenced phytoplasmas belong to three subgroups leaving most subgroups out of this analyses (Figure 1).

The reason for the low number of completely sequenced phytoplasma genomes is not surprising if one considers the difficulties encountered. Mainly due to the problem in obtaining suitable DNA for sequencing, it took years to complete 'Ca. P. asteris' strain OY-M and 'Ca. P. mali' strain AT sequencing. Because phytoplasma DNA usually has to be extracted from infected plants in which it occurs at very low concentration, elaborate enrichment and purification procedures were required to obtain it in amounts of DNA necessary for a high coverage of each position of the genome in shotgun sequencing. Flowers of *Catharanthus roseus* (periwinkle) and phloem of greenhouse-grown *Nicotiana tabacum* (tobacco) plants were preferentially used to obtain DNA templates for 'Ca. P. australiense' and 'Ca. P. mali' sequencing, respectively. The amount of phytoplasma DNA obtained depends on the enrichment procedure used. For example, when 'Ca. P. mali' DNA was, after extraction from *N. occidentalis*, treated by repeated bisbenzimide-CsCl buoyant density gradient centrifugation, a 30% enrichment of phytoplasma DNA was obtained. In contrast, purifying of phloem extract from *N. tabacum* using pulsed-field gel electrophoresis resulted in 80% enrichment [20]. Starting from an enriched phytoplasma DNA template, the genome sequences were determined by a shotgun approach [51]. In theory, shotgun fragments and the derived sequences represent the whole genome equally. However, AT-rich regions show a decreased physical coverage by recombinant clones/shotgun sequences and a lower sequence quality. Low coverage regions in the genomes are the result. This problem

also arises if new strategies from next generation technologies such as pyrosequencing [52] are used. The effect is reduced by these clone-independent approaches and high sequence coverage.

The assembly and finishing of the repeat-rich genomes represents an additional problem. It remains time consuming because clone-based large-insert libraries of phytoplasmas (e.g. fosmid libraries) do show a high cloning bias and are thus limited in their suitability for finishing experiments. The generation of additional complete genome sequence will remain a challenge in near future, but, hopefully, decreasing sequencing costs, high sequencing coverage, mate-pair libraries, and enlarged read lengths averaging 800 bases derived from pyrosequencing will help to overcome these problems.

The four phytoplasma genomes are organized in chromosomes and often contain extrachromosomal elements, which are reported for several phytoplasma strains [53]. The extrachromosomal elements, often called plasmids, allow the integration of their genetic material into the chromosome [49] and remarkably influence the vector transmissibility as shown for some strains [54]. Transmission experiments also indicate that plasmid-like extrachromosomal elements may not belong to the stable content of the genome.

The chromosomes of nearly all *Mollicutes* including those of the two 'Ca. P. asteris' strains [48, 49], 'Ca. P. australiense' [50], and *A. laidlawii* are circular. In contrast, 'Ca. P. mali' and the closely related species 'Ca. P. pyri' (pear decline phytoplasma) and 'Ca. P. prunorum' (European stone fruit yellows phytoplasmas) have a linear chromosome [20]. It may indicate a characteristic of this phylogenetic cluster. The chromosome of 'Ca. P. mali' has terminal inverted repeats (TIRs) with a size of 43 kb that have covalently closed hairpin ends protecting the chromosome [20]. TIRs are also known from *Streptomyces* genomes and covalently closed hairpin ends occur in genera *Borellia* and *Coxiella*, and several linear plasmids and phages [55]. A candidate gene encoding a required telomere resolvase (ResT) has been suggested to occur in 'Ca. P. mali' chromosome (ATP_00103) [20]. However, further experiments are needed to clarify the proposed function. Within the TIR region neither tRNAs nor rRNAs genes are encoded. The origin of this unusual chromosome organisation within the *Mollicutes* and its influence on the replication process in phytoplasmas remains unclear. One may speculate whether these structures have a virus-related origin.

Viruses influence the phytoplasma genomes. Wei and colleagues proposed that phages of the order *Caudovirales* have a major impact on the two fully sequenced 'Ca. P. asteris' genomes [56]. They analysed the deviations in the conserved synteny of these two closely related genomes and the encoded cryptic phage-derived genes and calculated that 264.2 kb (~31% of the genome) of strain OY-M and 160.2 kb (~22.7% of the genome) of strain AY-WB are putatively of viral origin. The Prophage Finder was one of the central bioinformatical tools used [56, 57]. Employing the Prophage Finder approach to examine the four fully sequenced phytoplasma genomes results in similar percentages of phage-derived sequences and shows 32% for strain OY-M, 20% for strain AY-WB, 26% for 'Ca. P. australiense' and 26% for 'Ca. P. mali'. Even if such an approach can just give a raw estimation, it is obvious that these genomes are heavily influenced by integration events. Genome diversity of phytoplasmas was also shown in other studies, for example, by PFGE analysis and mapping of marker genes of the sweet potato little leaf (SPLL) disease strain V4 (SPLL-V4) and of the closely related tomato big bud (TBB) phytoplasma [58].

Genome instability of phytoplasmas is not restricted to phage integration and recombination events or exchangeable plasmid material. Complex transposons, called potential mobile units (PMUs) [49], have significant impact in duplication of parts of phytoplasma genomes. PMUs are suggested to act as a tool to generate genetic variability [49]. Recently, it was shown that they might form extrachromosomal elements that may replicate and integrate again into phytoplasma genomes [59]. A high portion of the encoded proteins in the four genomes was assigned to these elements. Hogenhout and Musić [60] assigned 486 (OY-M) to 408 ('Ca. P. mali') proteins of the four phytoplasma genomes as single copy genes. This means that 35% of all protein-encoding genes in OY-M are multicopy genes. The corresponding proteins encoded by multicopy genes were assigned to PMUs. This indicates that 65–70% of them of 'Ca. P. asteris' strains AY-WB and OY-M and of 'Ca. P. australiense' but only 4% of 'Ca. P. mali' genes are encoded in such structures. Even if this provisional assignment of genes to phages or PMUs will show overlaps, there is no doubt that both events took place in phytoplasma genomes and that it seems unlikely that PMUs are the major driving force for duplicated genes in 'Ca. P. mali' that encodes only one *tra5* transposase as central element of PMUs. Interestingly, no PMUs have been identified in *A. laidlawii*.

The impact of other genetic elements such as reverse transcriptases remains unclear. At least 'Ca. P. asteris' strain OY-M and 'Ca. P. australiense' encode a reverse transcriptase (ST1). The presence of such a gene usually indicates a mobile element such as a retrotransposon or retrovirus [61]. The

result of these integration and recombination events can be visualized by cumulative GC analysis that shows irregular skews for the 'Ca. P. asteris' strains and 'Ca. P. australiense' but only in parts for 'Ca. P. mali' and not for *A. laidlawii* (Figure 2).

All these events enlarge the condensed phytoplasma genomes [49], but the enlarged genome size does not necessarily expand the number of encoded functions.

Different levels of genome condensation [20] and the putative phage-derived horizontal transfers of genetic material [56] or transposon-driven duplication events [49] result in different genome sizes in phytoplasmas. Insights for the COG (cluster of orthologous groups) assignment [62] of the protein-coding content of the four phytoplasmas show (Figure 3) that the majority of the functional categories is affected by these driving mechanisms and gives the impression that they share only for few categories a similar content. The largest category without any COG assignment may indicate the individual evolution of each phytoplasma but also an amount of genes, which are resulting from over-prediction or by representing pseudogenes. The analysis of the shared genetic content provides insights into the general demands on the host while differences may reflect strain-specific capabilities.

3. Replication, Recombination, and Repair

3.1. Replication Proteins and oriC. All essential genes necessary to generate the bacterial replication fork are encoded in the four genomes. These compromise the proteins forming the prereplisome such as DnaA, which binds at the *oriC*, the interacting DnaB/DnaC complex, and DnaD [63]. The phytoplasmas also share the DNA gyrase (GyrAB) and the DnaB helicase performing the bidirectional DNA unwinding [64]. The putative modulator PmbA of the DNA gyrase can influence this process.

Other encoded proteins, such as DnaG, probably synthesise initial primers at the site of the prereplisome [65], which is needed for the polymerase III heteromultimer. The genes encoding the proteins forming the DNA polymerase III holoenzyme are also present. This protein set contains PolC (alpha subunit, leading strand), DnaE (alpha subunit, lagging strand), DnaN (beta subunit), DnaX (tau subunit), and HolAB (delta and delta prime subunit). The putative assignment of PolC to the leading strand and the DnaE subunit to the lagging strand appears to be likely with respect to the Gram-positive ancestors [15, 66].

All phytoplasma genomes encode a NAD-dependent DNA ligase (Lig) for sealing breaks in DNA that occur, for example, during the discontinuous synthesis of the lagging strand. An ATP-dependent DNA ligase has not been annotated, a feature shared with *A. laidlawii* (Acc. no. CP000896). Other proteins involved in replication such as PriA responsible for the restart of stalled replication forks or acting at the occurring RNA/DNA hybrids (RnhC, NrdFA) are also present.

Some of the genes including *dnaB*, *dnaG*, and others involved in replication such as the single-stranded binding protein (*ssb*), the dna primase (*dnaG*), and the delta subunit

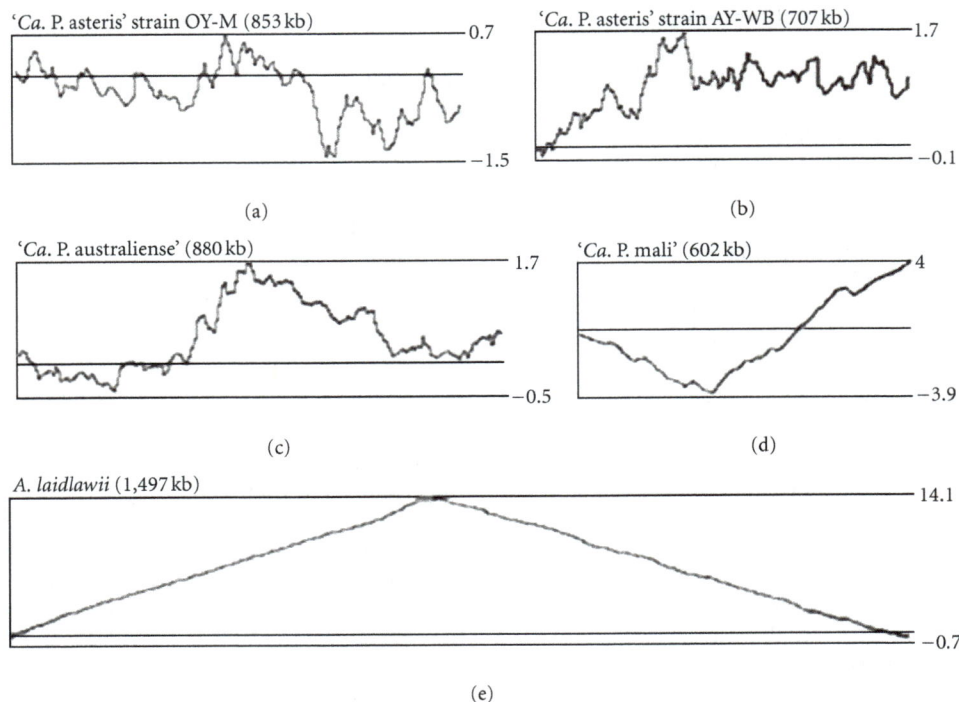

FIGURE 2: Cumulative GC skew analysis $[(C - G)/(G + C)]$ of the chromosomes of the two 'Ca. P. asteris' strains OY-M and AY-WB, 'Ca. P. australiense', 'Ca. P. mali,' and A. laidlawii. A window size of 5000 bases was used for calculation. Maxima and minima values obtained are indicated.

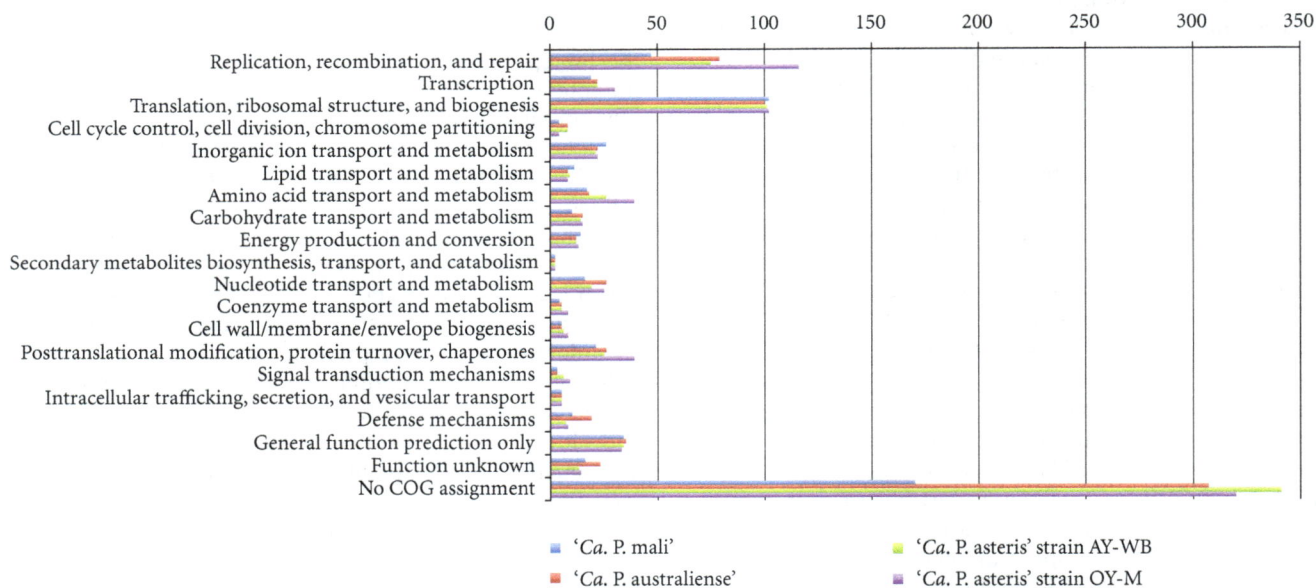

FIGURE 3: COG assignment of the deduced proteins of the four phytoplasma genomes. Single deduced proteins can show more than one assignment to the COGs database.

of the DNA polymerase III (*polC*) are present as multicopy genes. The origin of these multiple copies is still in discussion but they can be assigned to PMU or phage origin (see Section 2). In spite of this discussion it is notable that the *dnaB* gene is annotated at least twice in phytoplasmas. A situation that is also present in A. laidlawii (Acc. no. CP000896). The

annotated DnaB2 shows significant differences to DnaB1 in phytoplasmas. DnaB1 contains the DnaB-like helicase N-terminal domain (Pfam entry: PF00772.15) and the corresponding C-terminal domain (Pfam entry: PF03796.9) while *dnaB2* just partially matches the DnaB 2 family entry for replication initiation and membrane attachment

(Pfam entry: PF07261.5). A tandem arrangement of *dnaB2* and *dnaC* is encoded in all four genomes. Other tandem arrangements of single copy genes such as *dnaAN* or *gyrAB* and *nrdA/nrdF* are also present in phytoplasmas.

The coding for the suggested telomere resolvase is restricted to 'Ca. P. mali' [20]. Apart from that *orf*, the genetic environment of the genes involved in replication varies only in the copy number (ST2).

The *oriC* of the majority of bacterial chromosomes lies close to the overall minimum of the cumulative GC-skew [67, 68]. As phytoplasma chromosomes are heavily influenced by rearrangements and integration events that result in irregular cumulative skews and thus do not show a distinct minimum, an exception is the linear chromosome of 'Ca. P. mali' (Figure 2) [20, 49] that seems to be less influenced by such events and shows a shifted cumulative GC-skew with a minimum close to the sequence upstream the *dnaA* gene (Figure 2) [20]. The other three chromosomes show little pronounced minima close to the location of *dnaA*, indicating the presence of a putative *oriC* [20, 49]. This assignment is supported by several other approaches such as Oriloc [20, 49, 69] or OligoSkew analyses [20] and the typically decreased G + C content at the proposed *oriC*, which is present at the intergenic regions of the phytoplasmas flanking *dnaA*. They contain long AT-stretches of at least 30 bases (31, 33, and 48 bases in strain OY-M; 32 and 48 bases in strain AY-WB; 161 and 39 bases present upstream PAa_0839 in 'Ca. P. australiense'; 30, 33, 43, 87, and 53 bases in 'Ca. P. mali'). AT-stretches were limited in 'Ca. P. australiense' and 'Ca. P. mali' to positions upstream *dnaA* due to the coding density.

The *oriC* regions show only a weakly conserved synteny (Figure 4), a situation similar to that of other *Mollicutes* [70]. A typical switch in the coding strand preference at the *oriC* is present within this region and corresponds to a bidirectional replication. However, 'Ca. P. mali' shows a coding strand preference all over the chromosome [20].

The DnaA protein of *Escherichia coli* initially binds a variant of the DnaA-box consensus 5'-TTAT(C/A)CA(C/A)A-3' at the *oriC* region [71]. DnaA-box motifs flank the *dnaA* gene in several *Mollicutes* [70]. The consensus of the DnaA-box motif of *Mollicutes* such as *M. pulmonis*, *Spiroplasma citri*, *M. capricolum*, and *M. genitalium* (5'-TT(A/T)TC(C/A)ACA-3') differs from the consensus of *E. coli* in the positions of the variable nucleotides (Table 2) [70]. The seven annotated DnaA-boxes of *A. laidlawii* (Acc. no. CP000896) result in a consensus motif (5'-TT(A/T)T(C/T)(C/A)ACA-3'), which largely matches the suggested motif for other mollicutes [70], but also shows that the fifth position is more variable (A/C/T). Such a consensus sequence of the DnaA-box motif was not identified at the *oriC*-region of the phytoplasma chromosomes. The suggested DnaA-box motif for *Mollicutes* occurs in the neighbourhood of *dnaA* of 'Ca. P. asteris' strain OY-M (position 4878–4886, within the disrupted *dotG* gene; complement 848'134–848'142, encoding also *ugpA* at this position) but not within the intergenic regions flanking the *dnaA* gene. It is absent within the flanking intergenic regions of *dnaA* in strain AY-WB and also in 'Ca. P. australiense', where the occurrence is limited to a region (position 1503–1511) encoding *dnaN*. A similar situation can be found in

TABLE 2: Consensus of the DnaA-box motives.

Species/group	DnaA-box	Reference
Escherichia coli	5'-TTAT(C/A)CA(C/A)A-3'	[71]
Mollicutes	5'-TT(T/A)TC(C/A)ACA-3'	[70]
A. laidlawii	5'-TT(T/A)T(C/T)(C/A)ACA-3'	(CP000896)
Candidate motif proposed for phytoplasmas	5'-TTA**GG**AACA-3'	(this study)

'Ca. P. mali'. The mollicute motif is present within the *dnaN* gene (position 252'493–252'501) and within the sequence encoding a putative ABC transporter subunit for methionine (position complement 248'550–248'558). Strain OY-M and 'Ca. P. australiense' share the motif 5'-TTTTCAACA-3' while 'Ca. P. mali' contains 5'-TTATCAACA-3'. However, it was not possible to identify the proposed mollicute motif in strain AY-WB, its occurrence within sequences encoding genes, and the rare occurrence within this region which raises the question if the consensus (5'-TT(T/A)TCAACA-3') covers the sequence of the DnaA-box motif. It seems to be likely that the DnaA-box motif shows a higher variability. Other motifs occur with higher frequency such as 5'-ATAGGAACA-3' (eight times), more interesting 5'-TTAGGAACA-3' (three times) for 'Ca. P. mali,' and are located within the intergenic regions flanking *dnaA*. The latter is a promising candidate present in all four genomes within the *oriC* region with at least one copy (strain AY-WB) but contains an unusual GG-coding at positions four and five. It is present in the intergenic region of OY-M for four times, of AY-WB once, 'Ca. P. australiense' five times, and 'Ca. P. mali' three times. However, experimental evidence is needed to verify this.

3.2. Proteins Involved in Chromosome Partitioning. Segregation and partitioning of the chromosomal DNA start immediately after the initiation of the replication. However, the identified repertoire for this process is sparse, and it is likely that some proteins assigned as conserved hypothetical proteins will be involved. Topoisomerases will decatenate the two chromosome copies.

All four genomes also encode a regulator belonging to stage V sporulation protein family G (ST3). It is essential for sporulation and specific to stage V sporulation in *Bacillus megaterium* and *B. subtilis* [72]. However, the function in phytoplasma is unknown, and the protein may be involved in other processes.

In OY-M *smc*-genes were annotated that encode chromosome segregation of ATPase-like proteins. This assignment remains unclear, and the assignment to a glycoprotein appears to be more likely (see below).

3.3. Recombination. The complement of genes necessary for resolving Holliday junction and recombination differs in phytoplasmas. 'Ca. P mali' shows the most complete set for *rec*-dependent repair system. It encodes *ruvAB* but the endonuclease *ruvC* was not identified. Instead, RecU may have this function. The RuvABC enzymatic complex binds

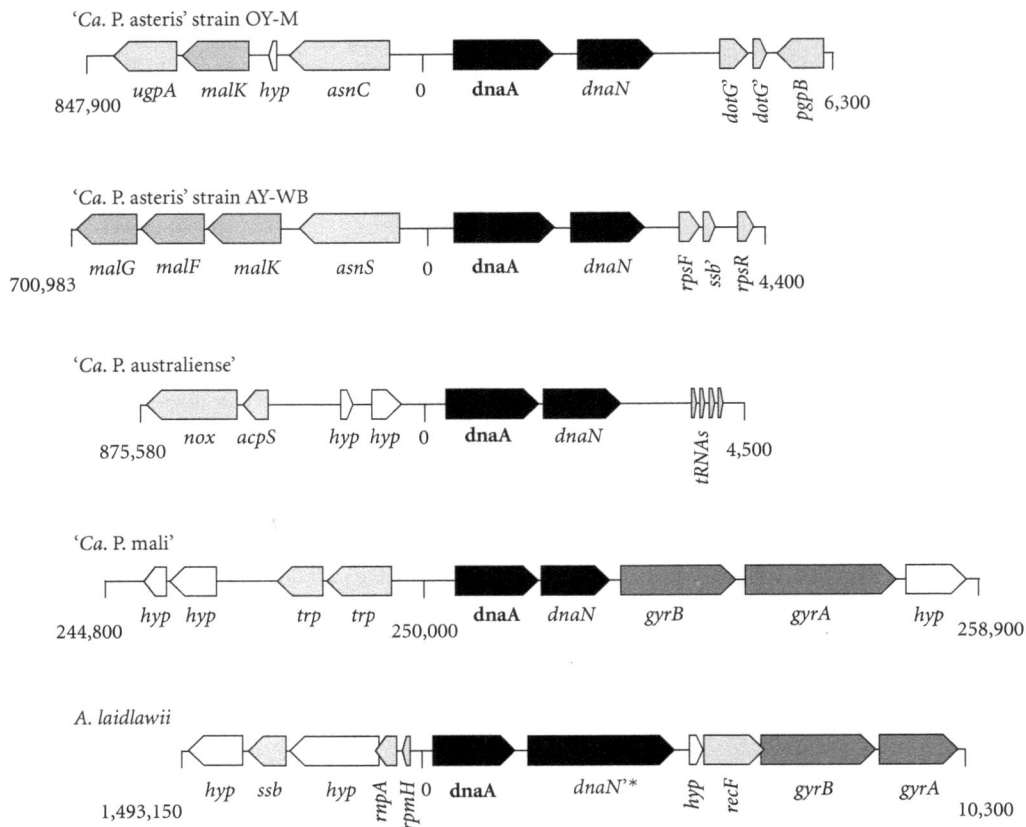

FIGURE 4: Genomic context at the suggested *oriC*. The *oriC* region is shown for the four phytoplasma chromosomes and *A. laidlawii* illustrating the weak conserved synteny within this region. Several genes are truncated (" ' ") and one gene probably destroyed by transposase integration ("*"). Abbreviation of genes: hyp: hypothetical protein; trp: ABC transporter subunit. Genome positions were given for each region taking in account the circular organisation of four chromosomes.

to recombinational junctions catalysing strand cleavage and branch migration. This is needed for the repair of double-strand breaks, which result from the RecA-dependent pathway. A basic system encoded in 'Ca. P. mali' but not in the other three phytoplasmas consists of recombinase A, the ATP-dependent zinc-containing RecG-like helicase, and the recombinational DNA repair proteins O and R in contrast to the other three phytoplasmas. However, genes *recA*, *recU*, *ruvA*, and *ruvB*, which are absent in three phytoplasmas, do not belong to the essential gene set of *M. genitalium* but may have long-term influence on the chromosome maintenance [73].

The function of the RmuC-family protein containing an endonuclease domain remains unclear in this context [74]. It is annotated in phytoplasmas except for 'Ca. P. mali' (ST4).

Due to dispersed locations within the genome of 'Ca. P. mali' it seems to be unlikely that the *ruv*- and *rec*-genes derived from a horizontal transfer. One may speculate that if this gene cluster is still functional as a unit. In comparison to *A. laidlawii*, similarities such as the lack of *ruvC* are obvious but *refN*, *recF*, and *refD* are encoded.

Functional recombination machinery provides an advantage, if selection pressure induced by attacks from the environment or competition in the same ecological niche exists.

This aspect is addressed in the statement of "living with instability" of phytoplasmas [49], so differences in coding capacity would enable response to such selection pressure.

However, plasmid-derived proteins involved in replication such as Rep are encoded in asteris yellows phytoplasma chromosomes (PAM_667, AYWB_403).

3.4. DNA Repair and Degradation. UvrABCD proteins represent the genetic components of the nucleotide excision repair DNA repair system (ST5), which also is involved in SOS response. UvrA protein forms homodimers in the presence of ATP, binds to the DNA, and associates with UvrB. The two other units are formed by the nuclease UvrC and the helicase UvrD. They act jointly with the DNA ligase and complete the excision-repair process. The complete genetic repertoire is encoded in three phytoplasma genomes and *A. laidlawii* but is absent in 'Ca. P. mali' with exception of UvrD, which is probably also involved in other reactions due to its helicase activity.

All four phytoplasmas encode parts of the base excision repair pathway to repair certain types of premutagenic lesion (ST5). The genes *mutM* and *mutT* are present in all four phytoplasma genomes but additional genes of the *mut*-system such as the ATPase *mutLS* are missing. In contrast,

TABLE 3: Localization of the rRNA operons.

Product	OY-M	AY-WB	'Ca. P. australiense'	'Ca. P. mali'
16S rRNA	279401–280921	271740–273260	Complement (682149–683667)	264151–265657
23S rRNA	281174–284036	273513–276376	Complement (679085–681932)	265879–268741
5S rRNA	284074–284181	276414–276521[2]	Complement (678945–679050)	268777–268888
16S rRNA	Complement (555991–557511)	Complement (496364–497884)[1]	Complement (863584–865102)	450392–451898
23S rRNA	Complement (552876–555738)	Complement (493248–496111)[1]	Complement (860519–863366)	452120–454982
5S rRNA	Complement (552731–552838)	Complement (493103–493210)[2]	Complement (860380–860484)	455018–455129

rRNA: ribosomal RNA.
[1]Gene was assigned to the opposite strand in contrast to CP000061.
[2]5S-rRNA gene was not annotated in CP000061.

they are present in *A. laidlawii*, which also carries *mutT* but lacks *mutM*. It is possible that other genes compensate the function of MutLS and enable the hydrolase and the glycolyse/lyase to function in combination with the other encoded proteins required. The annotated uracil-DNA glycolase would support the presence of such a basic system.

The genetic repertoire for the nucleotide excision repair DNA for damaged oligonucleotides is encoded in 'Ca. P. mali', while misincorporated bases cannot be corrected by the nucleotide excision repair system. This will result in uncorrected bases, which include the deamination leading to a C to A transversion and may affect the G + C content of 'Ca. P. mali'.

An incomplete restriction-modification system is encoded in strain OY-M but is absent in the closely related strain AY-WB and 'Ca. P. australiense' and 'Ca. P. mali'. This indicates that this system does not belong to the shared core set. It may thus be derived from a horizontal transfer.

3.5. DNA Modification and Structure. All phytoplasmas contain the bacterial core proteins encoding the type I topoisomerase TopA and the two gyrase subunits forming the type II topoisomerase (ST6). This protein content enables the phytoplasmas to relax supercoiled DNA by introducing single-stranded or double-stranded DNA breaks.

The high copy number of methyltransferases and of the nucleotide-binding proteins is remarkable. The methylation protects the DNA, for example, from restriction by endonucleases, a strategy, which is used to protect the integrity of the chromosomal DNA and to decrease the number of integration events. The high number of nucleotide-binding proteins corresponds to the latter scenario. This is in agreement with the high number of copies of the integration host factor HimA, which is required in site-specific phage recombination [75] and is part of the complex transposons such as PMU1 [59]. The copy numbers of these genes differ even between closely related strains. Only a single copy of *himA* is present in 'Ca. P. mali', corresponding to the putative single PMU [60].

4. Transcription

4.1. RNA Polymerase and Sigma Factors. RNA polymerase subunits for alpha, beta, beta', and omega chain are encoded in all 4 genomes (ST7). The sigma subunit is present with several paralogs, which is not unusual for bacterial genomes. The regulators RseABC and the sigma factors related to heat shock (RpoH), metabolic functions (RpoN), or stress (RpoS) were not identified. It is remarkable that the genomes of strain OY-M and of 'Ca. P. australiense' encode a high number of *fliA* genes whereas 'Ca. P. mali' encodes only a single gene. The small sigma factor FliA is known to be involved in transcriptions of operons involved in chemotaxis, motility and biofilm formation [76, 77]. However, there are no hints available on these abilities except for transcription in phytoplasmas.

4.2. Factors Affecting the RNA Polymerase. A common set of genes encoding proteins for transcription regulation and elongation is shared (ST8). It consists of the transcription elongation factor GreA and the N utilization substance protein involved in prevention and enhancement of transcription termination [78]. The transcription coupling factor Mfd [79] and the transcriptional termination factor Rho [80], like in *Mycoplasma* species, have not been identified in phytoplasmas.

4.3. Transcriptional Regulators. Only a sparse repertoire of transcriptional regulators is present in phytoplasmas. The negative regulator of class I heat shock proteins HrcA and putative cold shock proteins are encoded in all four phytoplasma genomes (ST9). Regulators of two-component systems were not annotated in these genomes.

5. Translation

5.1. rRNA-Operons. Two rRNA operons are encoded in the complete determined phytoplasma genomes. This applies for all phytoplasmas [81] and *A. laidlawii* (Acc. no. CP000896). The positions of the rRNA genes (Table 3) were recalculated using the actual version of the RNAmmer (v.1.2) software [82]. However, it should be taken in account that the prediction by RNAammer, which is based on HMM and BLAST, of start and stop position can differ by about 10 nt. A switch in strand preference of the rRNA operons with respect to the proposed *oriC* and *terC* is only present in the two aster yellows strains.

5.2. tRNA Synthetases. Aminoacyl tRNA synthetases catalyse the esterification of a specific amino acid or its precursor to one of all its compatible cognate tRNAs to form an aminoacyl-tRNA. They are essential for the translation process.

In contrast to the other tRNA synthetases (ST10), the aspartyl/glutamyl-tRNA(Asn/Gln) amidotransferase subunits ABC (*gatABC*) are absent in all phytoplasma genomes and *A. laidlawii*. This absence separates the four phytoplasma genomes from other *Mycoplasma*, *Ureaplasma*, and *Mesoplasma* species.

5.3. Transfer RNAs. Transfer RNAs are essential for the transfer of a specific active amino acid to a growing polypeptide chain during translation. The complete set of tRNAs was recalculated for the four genomes by tRNAscan-SE v.1.23 using the bacterial model [83]. The qualitative endowment of tRNAs is shared by the phytoplasmas but there are quantitative differences. Most different is '*Ca*. P. australiense' by the presence of an additional gene for tRNA-Gln/His/Leu and only a single copy for tRNA-Val (ST11). Transfer RNA coding for the usage of selenocysteine was not identified in the four genomes and *A. laidlawii*.

5.4. tRNA/rRNA Modification Factors. All phytoplasmas share a similar set of modification factors (ST12), except for RimM, which was identified only in '*Ca*. P. mali'. It is also present in *A. laidlawii*. RimM is essential for the efficient processing of 16S rRNA in *E. coli* [84]. It has affinity for free ribosomal 30S subunits but not for 30S subunits in the 70S ribosomes. The absence in the other three phytoplasmas and *Mycoplasma* species indicates the distinct phylogenetic position of '*Ca*. P. mali' but also the higher relatedness to the acholeplasmas than the other three phytoplasmas.

5.5. Translation Factors. All four genomes share a common set of translation factors including LepA (elongation factor 4) (ST13). LepA is a ribosomal back translocase related to EF-G and EF-Tu and is supposed to recognize ribosomes after a defective translocation reaction. This reaction allows EF-G to translocate the tRNAs correctly [85].

PrfA and PrfB mediate translation termination. The presence of these proteins is in accordance with the observation that not only the four fully sequenced phytoplasma genomes but all phytoplasmas use the bacterial code. PrfB is absent in *Mycoplasma* species, which translate UGA as tryptophan rather than using it as a stop codon [86].

5.6. Ribosomal Proteins and Modifying Factors. The environment for genes encoding ribosomal proteins consisting of 32 large subunits and 20 small subunits was identified in all phytoplasma genomes (ST14), in addition to some modifying enzymes (ST15). The genes encoding the large subunit protein RplY (L25) and the small subunit protein RpsA (S1) were not identified. However, RpsA was identified in *A. laidlawii*.

Some genes seem to be absent in phytoplasmas such as the ribosomal protein L11 methyltransferase (PrmA) or FtsJ. They also are absent in other mollicutes. The glutathione synthetase RimK present in *M. pneumoniae* and *M. genitalium* is also absent in all phytoplasmas. 50S ribosomal

stability factor encoded by the subunits *engABCD* is present in all phytoplasmas.

5.7. mRNA Degradation. All four genomes encode proteins similar to YkqC from *Bacillus licheniformis* (YP_078844.2) encoding the essential RNase J1/J2 involved in mRNA degradation (ST16).

5.8. Heat Shock Proteins and Chaperons. The four phytoplasmas share a basic gene set containing the Hsp70-type (*dnaK*, *grpE*), the Hsp60-type (*groEL*, *groES*) chaperone system [87], the chaperone *dnaJ*, and the trigger factor (ST17). The small heat shock protein IbpA (Hsp20) that binds to aggregated proteins and is present in other phytoplasmas [88] was not identified in '*Ca*. P. mali'.

Other encoded cytosolic proteases such as Lon-protease that hydrolyses ATP and unfolds bound substrates are encoded in all four phytoplasmas. These proteins are involved in various cellular activities [89].

6. Cell Envelope and Cell Division

6.1. Proteins Involved in N-Acetylglucosamine and Murein Biosynthesis. Proteins involved in N-acetylglucosamine and murein biosynthesis have not been identified in phytoplasma genomes. The absence of these genes is in accordance with the lack of a cell wall. However, all four genomes encode MraW (ST19), which is a S-adenosyl-dependent methyltransferase also described to show activity toward substrates associated with membrane components [90].

6.2. Proteins Involved in Lipopolysaccharide Synthesis. Only *rfaG* encoding for core functionality in lipopolysaccharide (LPS) synthesis was identified in '*Ca*. P. asteris' strain OY-M and '*Ca*. P. mali'. RgaG acts as a UDP-glucose: (heptosyl) LPS α-1,3-glucosyltransferase in *E. coli* [91] releasing UDP and H$^+$. This enzyme is involved in lipid A-core biosynthesis (ST20). Other proteins of this pathway were not identified.

6.3. Cell Division Proteins and Regulators. In enterobacter, cell division is mainly mediated by *fts* and *min* genes that determine cell shape. In phytoplasmas cell division proteins FtsY and FtsH were identified. Two additional genes associated with cell division were detected (ST21). The impact of the glucose-inhibited division (Gid) proteins encoded by phytoplasmas on the regulation of the cell division processes remains unclear. GidA is reported to be involved in 5-carboxymethylaminomethyl modification of the wobble uridine base in some tRNAs [92, 93] whereas GidB, which is missing in '*Ca*. P. mali', is suggested to act as S-adenosyl-L-methionine- (SAM-) dependent methyltransferase [94]. Both genes are encoded in the majority of bacterial genomes and also encoded in the genome of *A. laidlawii*. However, it was shown by mutagenesis of *M. genitalium* and *M. pneumoniae* that GidB is not essential [95]. It was suggested that the GidB of *E. coli* is specific for sterol and/or lipid substrates, but it also seems to be possible that GidB is specific for nucleic acids [96].

7. Membrane Proteins, Secretion, and Transport

7.1. Porins and Outer Membrane Proteins.

Phytoplasma membrane proteins are in direct contact with their environment due to the lack of a cell wall. Membrane proteins such as Vmp1 show a wide diversity in accordance with their importance for phytoplasmas [97]. A subset of the membrane proteins are the abundant immunodominant proteins, which also show a high diversity [98]. Proteins such as the antigenic membrane protein AmP interact with the insect cell microfilaments and contribute to the insect vector specifity [99]. They are under positive selection pressure [100]. AmP is encoded in the genomes of the two 'Ca. P. asteris' strains. A putative homolog is also present in 'Ca. P. australiense', while 'Ca. P. mali' encodes the immunodominant protein ImP (ST22) [101]. However, imp and amp were also identified in 'Ca. P. asteris' strain OY-W [102].

These proteins represent only a small part of the predicted membrane proteins that are characterized by at least one transmembrane (TM) helix (ST23). Integral membrane proteins and secreted proteins have been efficiently predicted by Phobius [103] from the annotated proteins in all four genomes. The 'Ca. P. asteris' strains OY-M and AY-WB carry 184 and 169 proteins containing at least one TM domain but no signal peptide (SP). The large chromosome of 'Ca. P. australiense' encodes 181 of such proteins and 'Ca. P. mali' with the shortest chromosome 145. Some membrane proteins carry a SP in addition that indicates a cleavage during protein translocation [98]. A low number of these proteins have been predicted in phytoplasmas: 12 in 'Ca. P. asteris' strain OY-M, 9 in strain AY-WB, 10 in 'Ca. P. australiense', and 5 in 'Ca. P. mali' (ST24).

The majority of membrane-associated proteins involved in transport and metabolism will be treated within the following sections.

7.2. Sec-Dependent Pathway (General Signal-Dependent Export Pathway).

The sec-dependent pathway represents the best-characterized secretion system of phytoplasmas. The four phytoplasma genomes share a common gene set of sec-genes. (ST25). Genes encoding proteins of the general secretion pathways (T2SS) were not identified and also the twin arginine pathway for the secretion of folded proteins and their cofactors is absent. This can be explained by the absence of a second outer membrane. Other secretion pathways of phytopathogenic bacteria such as the type III secretion system (T3SS), T4SS, or pili have not been identified [104]. There is weak indication for a deduced IcmE-like membrane protein involved in a T4SS (orf552 in 'Ca. P. australiense') [50] but further analysis is needed.

Secreted phytoplasma proteins may directly interact with their hosts and may in this way manipulate or weaken the plant host or insect vector without a needled system as provided by T3SS. Examples for the phytoplasmas are the secreted proteins tengu and SAP11 of 'Ca. P. asteris' strains OY-M and AY-WB, respectively [105]. For tengu it has been shown that the protein expression in *Arabidopsis thaliana* results in dwarfism [106]. The sec-dependent pathway is

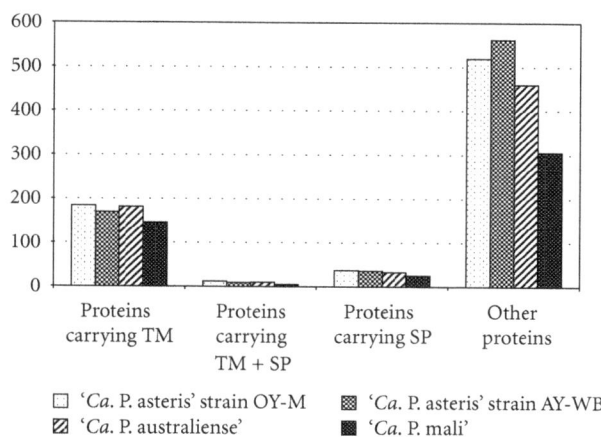

FIGURE 5: Phobius predictions of the four phytoplasma genomes. Abbreviations: TM: transmembrane helices; SP: signal peptide.

encoded in all four genomes by secYEG forming the integral membrane pore complex (Figure 5) and completed by secA and yidC [107]. YidC mediates the membrane insertion/assembly of the inner membrane proteins [108].

Genes encoding SecB, SecD/F, SecG, and YajC were not identified in the four genomes (ST25). GroEL may fulfil the function of the chaperone SecB. The effect of the absence of bifunctional SecD/F protein remains unclear due to the lack of information on their exact function. They might act to facilitate the protein export [109, 110]. The gene encoding the integral membrane protein secG is present in *A. laidlawii* but absent in phytoplasmas. A similar situation is present for YajC, which is supposed to stabilize the insertion of the SecA-preprotein complex in other bacteria.

The signal peptide mechanism seems to be similar to *E. coli*, and prediction software such as SignalP [111] or PSORT developed for Gram-negative bacteria [112] can be applied for the prediction of proteins secreted by the sec-dependent pathway [107] and other software such as Phobius [103] limiting the analysis on proteins carrying a signal peptide but no transmembrane regions (Figure 5). The 'Ca. P. asteris' strains OY-M and AY-WB carry 37 and 36 predicted secreted proteins containing a SP, while 'Ca. P. australiense' encodes 33 and 'Ca. P. mali' 25 (ST26).

7.3. Signal Recognition Particle (SRP) Pathway.

The SRP is a ribonucleoprotein, which was identified in all organisms. It acts in targeting translating ribosomes to the SecYEG translocation complex for cotranslational protein translocation over the membrane [113]. The SRP complex consists of the RNA component (4.5S RNA, ffs gene product) and the protein component (the ffh gene product). This complex is required for targeting some integral membrane proteins to the membrane for cotranslational integration [114]. Additional proteins are required for the functional translocation system. They comprise YidC (see above), which is bound to the SecYEG translocation complex and FtsY. The latter serves as docking protein for the SRP-ribosome complex synthesising the nascent peptide [115].

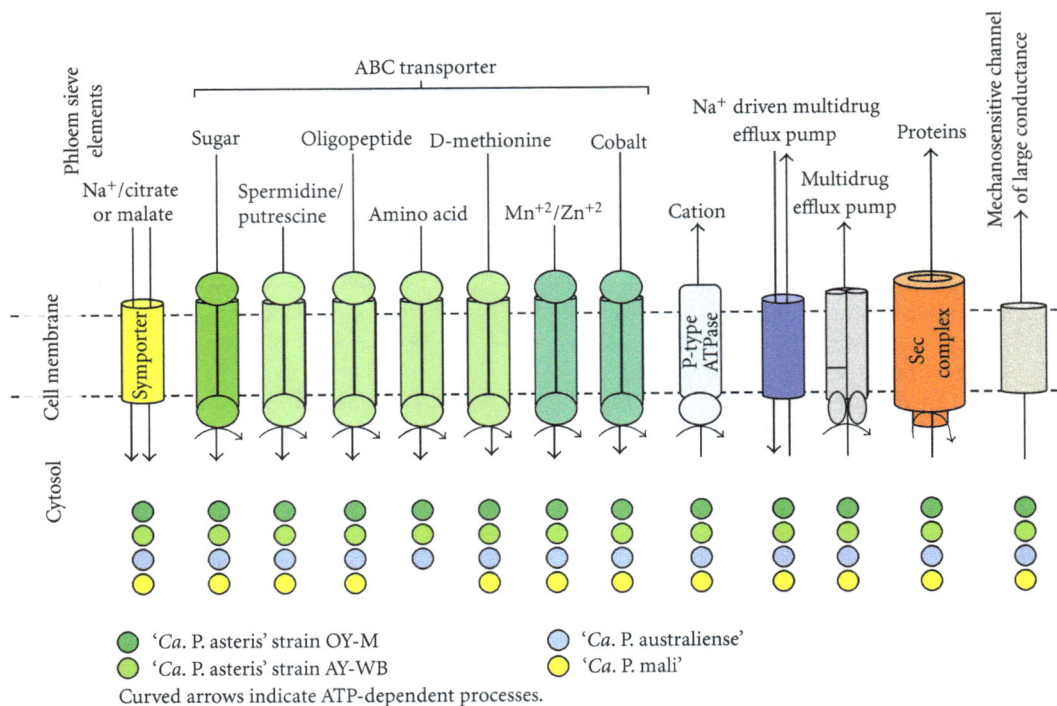

FIGURE 6: Symporter, transporter, ion translocation, and secretion.

All units forming this protein complex were encoded in the four phytoplasmas (ST27) except the *secG* and the *ffs* gene for 4.5S RNA, which were also not identified in acholeplasmas. However, it is difficult to identify 4.5S RNA that is encoded by eubacteria, archebacteria, and *Mycoplasma* species [116].

7.4. ABC Transporters and Symporters. All four phytoplasmas depend on the uptake of essential compounds from their hosts (Figure 6) [48]. They encode a common core set of ABC transporters (ST28). This includes the clustered spermidine/putrescine transport system (*potABCD*), a Mn/Zn transport system (*znuACB*), sugar transport system (assigned as *malEGFK*), a dipeptide/oligopeptide transport system (assigned as *dppCBADF*), and a methionine transport system (*metNQI*). An additional amino acid transport system (*artQPIM*) is encoded in OY-M (probably duplicated), AY-WB, and 'Ca. P. australiense' but is absent in 'Ca. P. mali'. The genes for *znuACB* are present in two copies in 'Ca. P. mali' due to the location in the terminal inverted repeats.

The template of the sugar transport system remains unclear. The uptake of maltose, trehalose, sucrose, and palatinose was suggested, and sucrose appears to be most likely [50, 117] with respect to the presence in the phloem and the corresponding protein machinery. *Spiroplasma citri*, another phloem colonizing parasite, is able to utilize glucose, fructose, and trehalose [118].

The subunits in phytoplasmas show similarities to maltose/maltodextrin transporters but also to glycerol-3-phosphate transporters (*ugpBEAC*). Both protein sets are highly homologous and might be functionally exchangeable, but show differences with respect to the growth substrates [119]. The import of glycerol-3-phosphate would also offer a phosphate source for the phytoplasmas.

The phytoplasma dipeptide/oligopeptide transport system shows sequence differences in the substrate-binding unit that may indicate differences in substrate specifity.

In addition, several incomplete ABC transporters (ST29) are present in phytoplasmas. Subunits of the above mentioned transporters of dipeptide/oligopeptide transporters, amino acids, and cobalt are present in the genome. Incomplete transporter units might be completed by subunits of other peptide and amino acid transporters. The functionality of cobalt import system remains unclear because the periplasmatic subunit was not identified. However, the permease (*cbiQ*) and ATPase subunits (*cbiO*) are present in all phytoplasma genomes.

A putative thiamine transporter subunit is encoded in all four phytoplasma genomes (ST30). The corresponding peptide sequences were assigned by InterPro [120] to the YuaJ-family (IPR012651). Many YuaJ-family members have been assigned as ATP-independent thiamine transporter, which are involved in regulation of the thiamine pyrophosphate (TPP) concentration [121]. The import is probably proton coupled. Thiamine is phosphorylated at the membrane by thiamine kinase (ThiK) in *Bacillus cereus* [122]. ThiK has not yet been identified in mollicute genomes. This raises the question if thiamine is imported or if it is an already phosphorylated substrate. The thiamine monophosphate kinase ThiL (syn. ThiJ) is encoded in the four phytoplasma genomes (PAM_137, AYWB_584, PAa_0495, ATP_00237). This finding supports the idea that a phosphorylated thiamine substrate or an unassigned protein has to fulfil the function of ThiK in phytoplasmas.

Beside the ABC transporters, malate or citrate/Na⁺ symporters are encoded in all four genomes that may provide an important carbon source [49] (ST31).

Several other membrane-located proteins are involved in transport in all four genomes and belong to the core gene repertoire of phytoplasmas (ST32). They contain a multidrug efflux pump such as MdlAB and the Na⁺-driven multidrug efflux pump NorM. The genes assigned as *norM* differ in sequence similarity. The NorM protein from 'Ca. P. mali' has only one ortholog in 'Ca. P. australiense' (PAa_0171).

Another common feature is the P-type ATPases exporting cations such as magnesium, calcium, and cadmium (ST33). Notably, mutants deficient in the single P-type ATPase encoded in *Spiroplasma citri* are affected in their growth capacity [123].

Finally, the regulation of the osmotic pressure by opening the membrane lipid bilayer to prevent cell disruption and death [124] is given by the mechanosensitive channel formed by the MscL protein in phytoplasmas (ST34).

8. Metabolism

8.1. Carbohydrate Metabolism Glycolysis. Embden-Meyerhof-Parnas pathway was suggested to be the major energy-yielding pathway in phytoplasmas [48] despite the apparent lack of hexokinase (glucose phosphorylating) and a sugar-specific phosphotransferase system (PTS) mediating a phosphorylated hexose to enter glycolysis. One promising candidate is the glucosyltransferase GtfA, which was predicted in 'Ca. P. australiense' first. The assignment of the deduced protein sequence and the ortholog of OY-M were confirmed by InterPro (IPR022527 sucrose phosphorylase, GftA). The GtfA, which is probably better described as disaccharide glucosyltransferase or sucrose phosphorylase, allows the formation of α-D-glucose-1-phosphate from phosphate and sucrose [125], which is often a predominant sugar in the phloem. α-D-glucose-1-phosphate is the entry compound of glycolysis. GtfA may compensate the absence of a hexokinase and PTS system. However, GtfA or a similar phosphorylase are not a general trait of phytoplasmas, since they are absent in the genomes of closely related AY-WB and in 'Ca. P. mali'. Thus, the observed differences in glycolysis among phytoplasmas may arise from genome plasticity as suggested by the close proximity of prophage-related elements and *gtfA* in 'Ca. P. australiense'.

Theoretically, the uptake of phosphorylated hexoses would overcome this problem. Candidates for such sugar phosphates are trehalose-6-phosphate, sucrose-6-phosphate, and β-D-fructose-6-phosphate. The transporter complements of the phytoplasmas may contain uptake systems for importing these phosphorylated di- and monosaccharides from the environment. For example trehalose-6-phosphate has to be monomerized prior to entry into the glycolysis. The breakdown would occur within the phytoplasmas mediated by hydrolases/phosphatases. At least one phosphatase each of subfamily IIIa and of subfamily IIb is encoded within the four phytoplasma genomes (ST35). These phosphatases are poorly characterized so far. A single copy of the IIIA subfamily of the haloacid dehalogenase (HAD) superfamily of hydrolases representing hypothetical proteins is encoded in each phytoplasma genome. Most characterised members of this subfamily and of the HAD superfamily are phosphatases. This protein family consists of sequences from fungi, plants, cyanobacteria, Gram-positive bacteria, and *Deinococcus* (according to IPR010021 entry).

Functional interpretation of the second hydrolase of the Had superfamily hydrolase subfamily IIb may be more straight forward. Members of the Had superfamily hydrolase subfamily IIb are encoded by at least one gene ('Ca. P. mali') in each genome. They encompass trehalose-6-phosphatase, plant and cyanobacterial sucrose phosphatise, and a closely related group of bacterial and archaeal orthologs, eukaryotic phosphomannomutase (according to IPR006379 entry). If these proteins function as trehalose-6-phosphatase, phytoplasmas could use α, α-trehalose-6-phosphate, and phosphate to produce glucose-6-phosphate and β-D-glucose-1-phosphate, which are the entry molecules of glycolysis.

If sucrose-6-phosphate is used as a substrate, sucrose-6-phosphatase may generate glucose-6-phosphate and fructose [126]. The utilization of trehalose-6-phosphate and/or sucrose-6-phosphate appears to be likely due to the phosphoglucose isomerase encoded in all four genomes. This step would be unnecessary, if fructose-6-phosphate is available. However, it should be considered that only trace amounts of trehalose-6-phosphate are present in higher plants on average. The impact of a phytoplasma infection on plant metabolism cannot be estimated so far, but trehalose-6-phosphate is a signaling molecule in plants with strong regulatory effects on metabolism, growth, and development [127–129]. The sucrose-6-phosphate concentration in the phloem is unclear. It is an interesting scenario that sucrose and trehalose compounds could be utilized depending on their availability in phloem and hemolymph.

The general upper part of the glycolysis (energy demanding) is encoded within all four phytoplasma genomes [20, 48–50] starting with α-D-glucose-6-phosphate converted to β-D-fructose-6-phosphate by phosphoglucose-isomerase (Pgi) and ATP-dependent formation of two molecules β-D-fructose-1,6-bisphosphate by phosphofructose kinase (PfkA). Subsequently, fructose-bisphosphate-aldolase (Fba) catalyses the formation of D-glyceraldehyde-3-phosphate and dihydroxyacetone-phosphate. The latter is suggested to enter glycerophospholipid metabolism, while D-glyceraldehyde-3-phosphate is channelled into the energy yielding part of the glycolysis (Figure 7). The interconversion of these C₃-intermediates is performed by triosephosphate isomerase (TpiA).

Dihydroxyacetone phosphate can also be generated by a conserved kinase related to dihydroxyacetone kinase (DhaK). However, it remains unclear if this kinase can also act in the opposite direction as a transferase.

Except for 'Ca. P. mali' [20] the protein components of the lower part of glycolysis (energy yielding) from D-glyceraldehyde-3-phosphate to pyruvate are encoded in the analyzed phytoplasma genomes (ST35): glycerinaldehyd-phosphate dehydrogenase, phosphoglycerate kinase, and

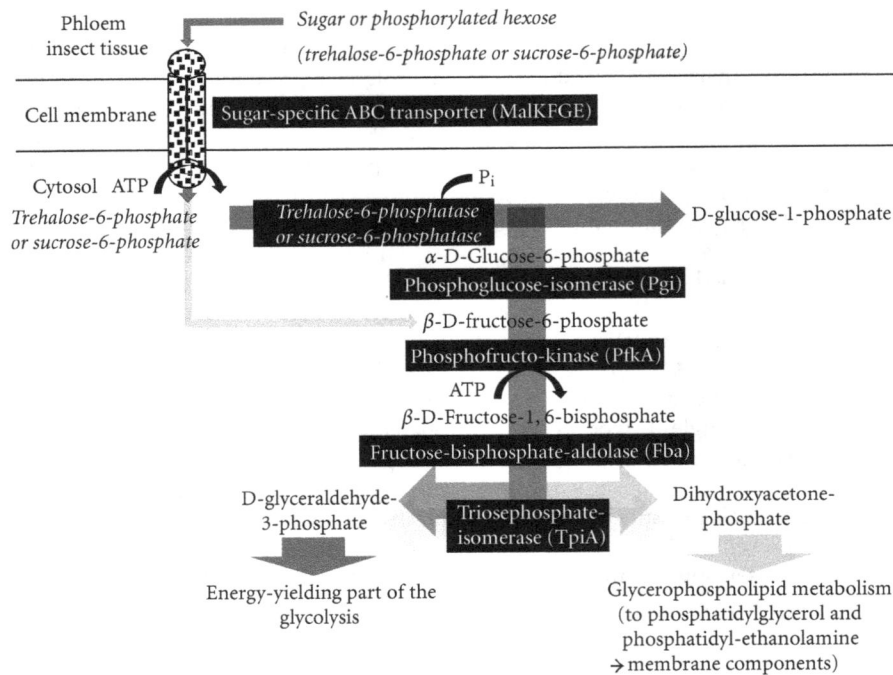

FIGURE 7: The sugar uptake and upper part of the glycolysis (energy investment) including trehalose-6-phosphate and β-D-fructose-6-phosphate as suggested entry substrates (steps indicated in italics).

mutase, enolase, and pyruvate kinase (Figure 8). Notably, gluconeogenic phosphoenolpyruvate synthase (PpsA) is lacking in all four phytoplasmas (Figure 8). Other strategies to obtain host-derived ATP such as ATP/ADP translocase known from *Chlamydia* species [20, 48, 130] and also the arginine dihydrolase pathway [131] have not been identified in phytoplasmas.

8.2. An Alternative Energy-Yielding Pathway Deduced from the Genome Sequences. Carbohydrate metabolism is one of the most important physiological traits of the phytoplasmas. Alternative pathways have to be considered that could compensate for the lack of a complete glycolysis in '*Ca.* P. mali' (Figure 9, ST36). A pathway for malate conversion to acetate is potentially encoded in all four genomes. Uptake of malate is enabled by the symporter MleP [20, 48]. Malate can be oxidatively decarboxylated to pyruvate by the malic enzyme ScfA. Pyruvate would then also be oxidatively decarboxylated to acetyl-CoA by the pyruvate dehydrogenase multienzyme complex.

In the case of '*Ca.* P. mali' pyruvate might be generated by an additional way. Here, an aldolase (Eda) is predicted, serving two possible functions. A 4-hydroxy-2-oxoglutarate aldolase (EC: 4.1.3.16) (KHG-aldolase) would catalyze the interconversion of 4-hydroxy-2-oxoglutarate into pyruvate and glyoxylate. Phospho-2-dehydro-3-deoxygluconate aldolase (EC: 4.1.2.14) (KDPG-aldolase) would catalyse the interconversion of 6-phospho-2-dehydro-3-deoxy-D-gluconate into pyruvate and glyceraldehyde-3-phosphate. In both cases, pyruvate would be formed independently from glycolysis.

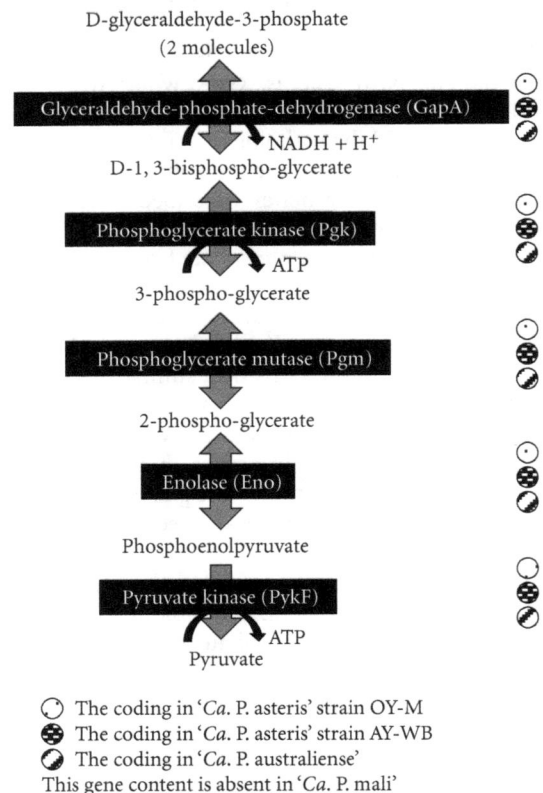

○ The coding in '*Ca.* P. asteris' strain OY-M
⊕ The coding in '*Ca.* P. asteris' strain AY-WB
◑ The coding in '*Ca.* P. australiense'
This gene content is absent in '*Ca.* P. mali'

FIGURE 8: The energy-yielding part of the glycolysis encoded in the asteris strains and '*Ca.* P. australiense' but absent in '*Ca.* P. mali'.

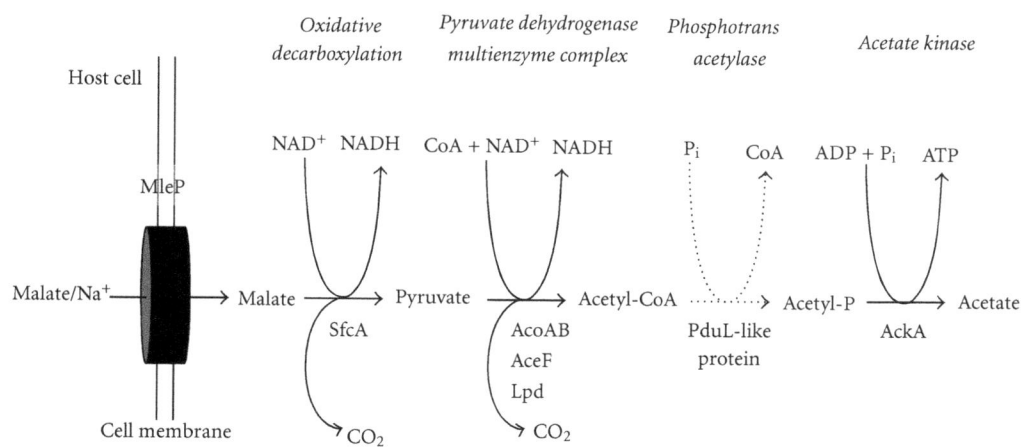

| Oxidative decarboxylation | Pyruvate dehydrogenase multienzyme complex | Phosphotrans acetylase | Acetate kinase |

FIGURE 9: Proposed energy-yielding pathway from malate to acetate.

Protein names and products:

MleP	Malate/Na$^+$ symporter
SfcA	Phosphate acetyltransferase
AcoAB(syn. PdhAB)	Pyruvate dehydrogenase E1 component, α and β subunit
AceF(syn. PdhC)	Pyruvate dehydrogenase E2 comp. (dihydrolipoamide acetyltransferase)
Lpd(syn. PdhD)	Dihydrolipoamide dehydrogenase
PduL-like protein	Phytoplasmas encode a phosphotransacetylase similar to PduL but lack the common phosphotransacetylase Pta
AckA	Acetate kinase

It is also notable that the gene set necessary for the generation of coenzyme A (CoA) is only partially annotated in phytoplasmas. In *E. coli*, CoA is synthesized from (R)-pantothenate. This requires involvement of 4′-phosphopantothenoylcysteine decarboxylase and phospho-pantothenoylcysteine synthetase (fused in *E. coli*, CoaBC), the phosphopantetheine adenylyltransferase (CoaD), and the dephosphocoenzyme A kinase (CoaE). CoaBC was identified in some *Mollicutes*, while CoaD is annotated in the majority of phytoplasmas and *A. laidlawii*. While genes encoding CoaBC and CoaD were not identified in phytoplasmas, CoaE (annotated as formamidopyrimidine-DNA glycosylase) is encoded in all four genomes. The intermediate dephospho-CoA is the substrate of CoaE. This finding supports the possibility that at least the last steps in CoA-biosynthesis are performed by the phytoplasmas, but it remains unclear whether alternative reactions for the other steps in the pathway are encoded in the genomes.

Hydrogen peroxide (H_2O_2) could be formed by HcaD, oxidizing NADH. H_2O_2 may also be generated by the encoded superoxide dismutase and represent a potential virulence factor, as known from *M. pneumoniae* and also encoded in *A. laidlawii* [132–134].

Acetyl CoA can be converted to acetyl phosphate phosphotransacetylase (Pta) in many mycoplasmas. Subsequently, acetyl phosphate can be transformed by acetate kinase (AckA) to acetate and ATP (Figure 9). However, Pta, which is encoded on many mycoplasma chromosomes close to *ackA*, is absent from all four phytoplasma genomes [14, 20]. Lacking Pta could be substituted by another phosphotransacetylase (PduL) described for *Salmonella*

enterica subsp. *enterica* serovar typhimurium (Acc. no. AAD39011), since it is encoded in all four phytoplasma chromosomes. This assignment of PduI is in accordance with the kinetic parameter with a KM value of 0.97 mM for acetyl phosphate and a V_{max} of 13.4 μM/min/mg enzyme with acetyl phosphate as substrate (according to Uniprot entry Q9XDN5).

The assignment of the phosphotransacetylase as propanediol utilisation-like protein PduL in phytoplasmas remains ambiguous. Propanediol is produced by fermentation of the common plant sugars rhamnose and fucose. In *S. enterica*, PduL is part of the *pdu*-operon and involved in the coenzyme-B$_{12}$-dependent degradation of 1,2-propanediol [135–137]. It seems to be likely that in phytoplasmas, phosphotransacylase PduL does not function in propanediol degradation, since other genes of this pathway (e.g., *pduCDE*, *pduQ*, *pduP*, and *pduW*) are absent in phytoplasmas and *A. laidlawii* (except for *pduL*). Thus, PduL could use acetyl-CoA instead of propionyl-CoA as substrate.

8.3. Associated Processes

8.3.1. Proteins Involved in NAD Synthesis. All four phytoplasma chromosomes encode the glutamine-dependent NAD$^+$ synthetase NadE (ST49). In this ATP-dependent process nicotinate adenine dinucleotide (deamido-NAD$^+$), L-glutamine, and H_2O are used to form L-glutamate, AMP, diphosphate, H$^+$, and NAD$^+$. Other proteins involved in the generation of NAD$^+$ were not identified in the phytoplasma genomes. In contrast, *A. laidlawii* contains NadD and PncB allowing the formation of NAD$^+$ from nicotinate. It appears

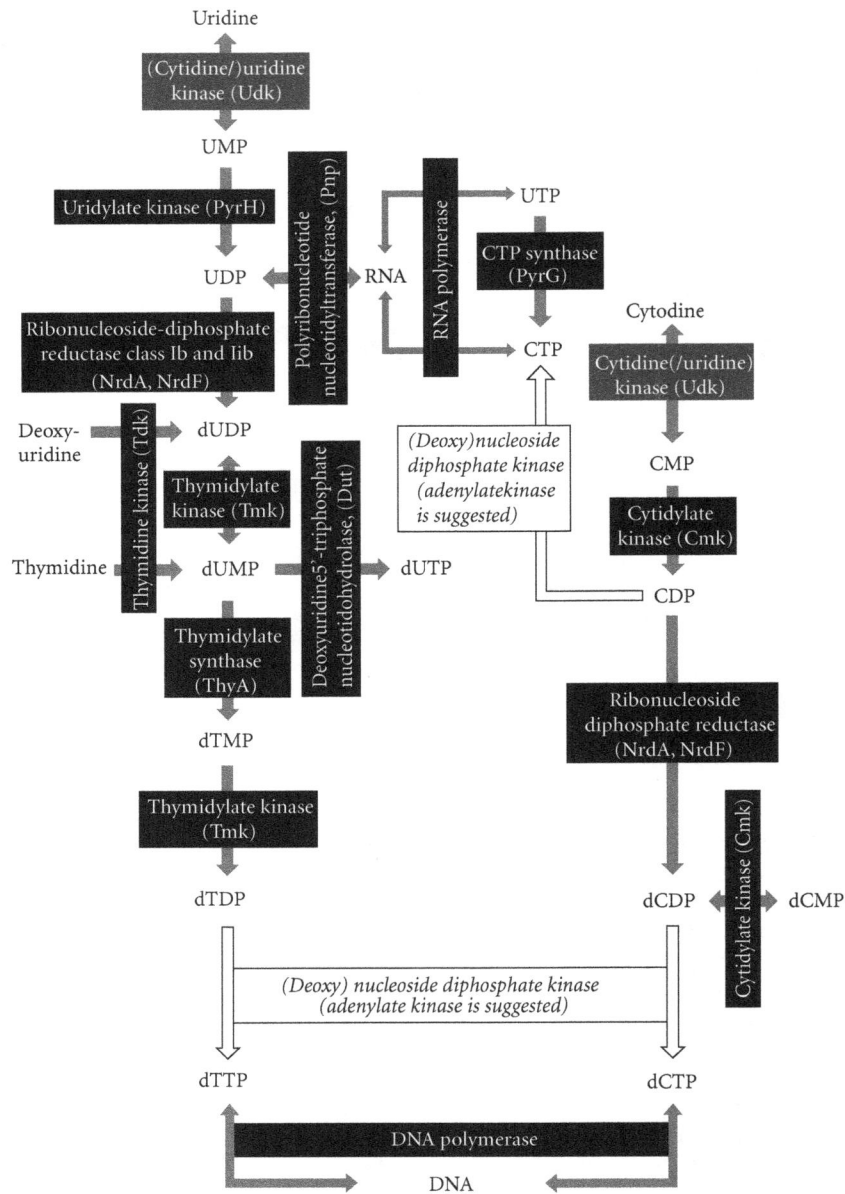

FIGURE 10: Principle of pyrimidine metabolism in phytoplasmas.

likely that phytoplasmas import deamido-NAD$^+$ or a precursor from their environment.

8.3.2. *Proteins Involved in Oxygen Detoxification.* All four genomes encode a superoxide dismutase (SOD), which converts O_2^- to the less toxic H_2O_2 (ST40). SOD activity is documented for several *Mollicutes* including the genera *Mycoplasma, Ureaplasma,* and *Acholeplasma* [138]. SOD requires metal cations such as copper, manganese, iron, or nickel. Prokaryotes were assigned to form cytoplasmatic MnSOD and/or FeSOD [139], but for *M. hyopneumoniae* the production of Cu/ZnSOD was shown [140]. The formation of MnSOD is reported for the acholeplasmas [141]. The catalase dismutase responsible for the conversion of H_2O_2

to nontoxic products was not identified in phytoplasmas so far. This is also reported for several other mycoplasmas and may contribute to virulence of these organisms by release of reactive H_2O_2 [142]. It remains unclear, whether the release of H_2O_2 will weaken or damage the plant host upon presence of phytoplasmas inside the sieve elements.

8.3.3. *Lipoyl-Protein Ligase.* Lipoic acid derivates act as cofactors in enzymatic systems, such as the pyruvate dehydrogenase [143] in the phytoplasmas. All four chromosomes encode the ATP-dependent lipoyl-protein ligase LplA (ST50), which preferentially utilizes imported lipoate to form lipoyl adenylate or an octanylated protein from octanylate [144].

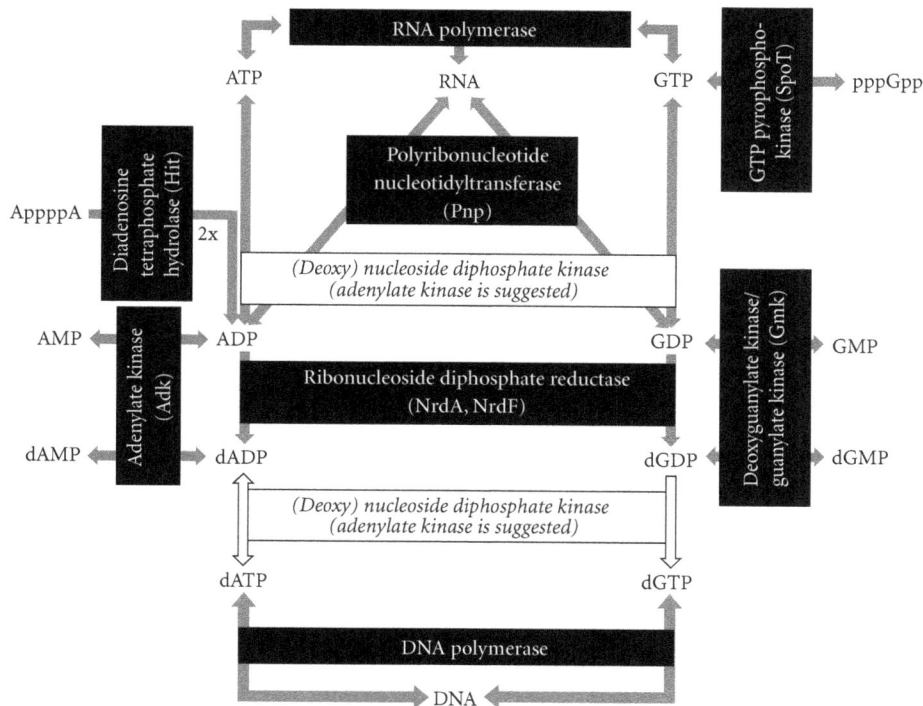

FIGURE 11: Principle of purine metabolism in phytoplasmas.

9. Purine and Pyrimidine Synthesis

Purine and pyrimidine metabolism in *Mollicutes* was reviewed recently [145]. Therefore, the following focuses on the genetic key elements or aspects of general interest for the phytoplasmas (ST42). Since phytoplasmas possess no genetic repertoire for *de novo* synthesis of purine or pyrimidine bases [48], they depend like many other mollicutes on environmentally derived nucleotide precursors [145]. However, nucleobase or nucleoside transporters were not identified in phytoplasmas so far [48], but it was suggested for mycoplasmas that the limited repertoire of transporters could be tuned to a wider variety of substrates [146]. Membrane-associated nucleases are suggested for the nucleotide precursor uptake and were identified in 20 mycoplasma species [147]. Membrane nucleases such as MnuA of *M. pulmonis* [147] were characterized, and putative orthologs were identified in *M. hyopneumoniae*, *M. gallisepticum*, *M. pneumonia*, *M. penetrans*, and *U. urealyticum* [145]. However, it was not possible to identify candidates in phytoplasmas so far. The uptake and incorporation of dNMP without prior dephosphorylation was shown for *M. mycoides* subsp. *mycoides* [148]. The presence of such a strategy in phytoplasmas is important for nucleotide and energy metabolism. However, all four genomes encode common pathways for purine and pyrimidine synthesis with one exception (Figure 10). The usage of uridine and cytodine in the '*Ca*. P. australiense' and '*Ca*. P. mali' remains questionable, because the cytidine/uridine kinase (Udk) mediating the formation of CMP/UMP is absent in both chromosomes. However, it is present in *A. laidlawii* (YP_001620379).

AMP and dAMP as well as GMP and dGMP represent the entry points into purine metabolism in all four genomes (Figure 11). The (deoxy)nucleoside diphosphate kinase (Ndk) is absent in the four phytoplasma genomes. This was also observed with many other mycoplasmas, the 6-phosphofructo-, phosphoglycerate-, pyruvate-, and acetate-kinases which could use other ribo- and deoxyribopurine and pyrimidine NDPs and NTPs besides ADP/ATP [146]. Similarly it was shown for *Mycobacterium tuberculosis* that the adenylate kinase (AdK) acts as general (deoxy)nucleoside diphosphate kinase (NdK) [149], a scenario that was also proposed for mycoplasmas [145].

CMP and CTP are also produced as byproducts within the phospholipid metabolism. However, it should be also noted that such function of the Adk might also result in an imbalance of the nucleotide pool and in the low G + C content of phytoplasma chromosomes in evolution. Notably, '*Ca*. P. mali' with the lowest G + C content encodes one *adk* on each terminal repeat of the chromosome.

10. Miscellaneous

10.1. Lipid Synthesis. All four phytoplasma genomes encode a common biosynthetic pathway for essential phospholipids (Figure 12, ST44), that is, for the formation of CDP-diacylglycerol from an acyl phosphate and dihydroxyacetone phosphate. Moreover, the genetic modules for the generation of L-1-phosphatidyl-glycerol and L-1-phosphatidyl-ethanolamine are present. The potential to form cardiolipin from L-1-phosphatidyl-glycerol remains unclear, since a cardiolipin synthase is apparently not encoded.

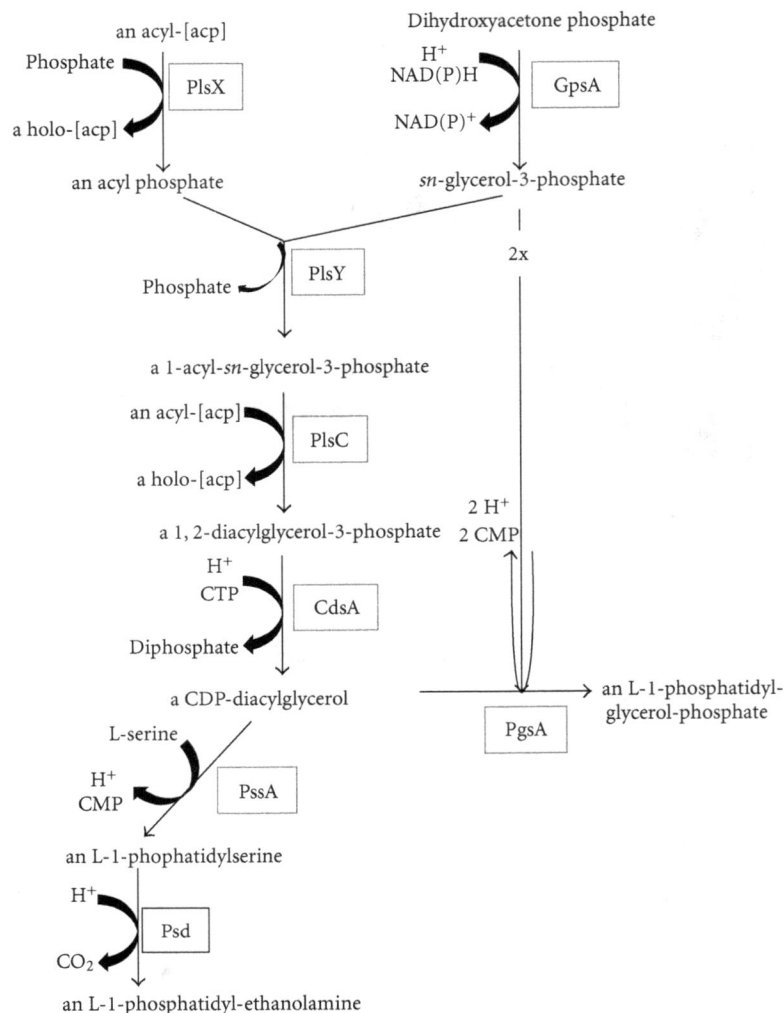

FIGURE 12: Key elements of the lipid metabolism.

10.2. Proteolysis. Diverse proteases involved in the breakdown of proteins are encoded in the phytoplasma genomes. Several encoded peptidases enable the degradation of essential imported peptides and regulation of intracellular products.

A remarkable number of zinc-dependent proteases are predicted from the phytoplasma genomes. For example the four genomes share the protease PmbA and Zn-dependent protease TldD (ST38), and the exact role in phytoplasmas remains elusive.

The annotated zinc-dependent HflB metalloproteases may play a prominent role in the 'Ca. P. asteris', due to the high number of paralogs: 19 in strain OY-M and 10 in strain AY-WB. High numbers of paralogs are also present in 'Ca. P. australiense' [6] and 'Ca. P. mali' [11]. These *hflB* gene products show differences in the encoded peptide length and domain composition. In *E. coli*, HflB is affiliated with multiple cellular functions, including a putative involvement in lysogeny of lambda phage [150]. Thus, one may speculate that the high number of *hflB* genes resulted from a high pressure of phage attacks. Studies on the various *hflB* genes

and their presence within the genomes of differently virulent (mild to severe) strains are on the way.

10.3. Amino Acid Synthesis and Modification. It was not possible to detect proteins involved in the synthesis of amino acids, except for *S*-adenosyl-L-methionine (MetK), L-asparagine (AsnB), and L-ornithine (ArgE) (ST41, supplementary Figure 1 (SF1)). However, these enzymes are neither encoded in all four genomes nor do they enable the phytoplasmas to synthesise any amino acid [48]. Thus, all necessary amino acids or peptides have to be imported.

10.4. Riboflavin Synthesis. The only detected putative protein involved in riboflavin syntheses was ATP-dependent riboflavin kinase (ST45) in strain OY-M.

10.5. Folate Synthesis. Folic acid (or vitamin B9) is a precursor of the coenzyme tetrahydrofolate (THF). All four phytoplasma genomes (ST46) encode dihydrofolate reductase (FolA), which uses 7,8-dihydrofolate monoglutamate

6-hydroxymethyl-7, 8-dihydropterin

ATP

FolK

AMP
H+

6-hydroxymethyl-dihydropterin diphosphate   4-aminobenzoate

Diphosphate   FolP

7, 8-dihydropteroate

ATP
L-glutamate

FolC

ADP
Phosphate
H+

7, 8-dihydrofolate monoglutamate

NADPH
H+

FolA

NADPH+

Tetrahydrofolate

L-serine   GlyA

Glycine
H2O

5, 10-methylenetetrahydrofolate

⊙ The coding in 'Ca. P. asteris' strain OY-M
⊛ The coding in 'Ca. P. asteris' strain AY-WB
⊘ The coding in 'Ca. P. australiense'
● The coding in 'Ca. P. mali'

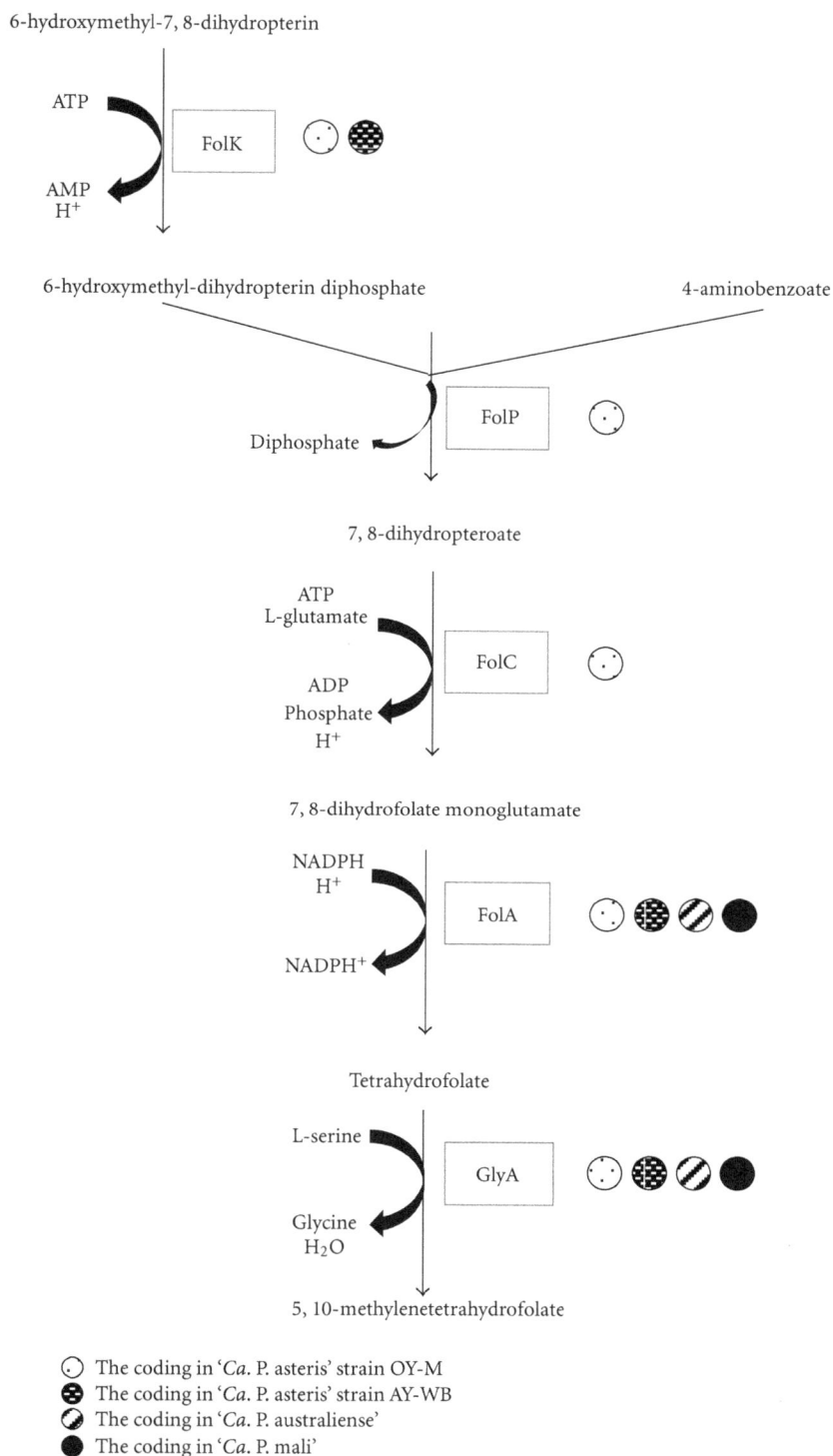

FIGURE 13: Folate synthesis.

and NADPH + H+ for the generation of THF (Figure 13). Only strain OY-M encodes a gene set, which may allow generating THF from other precursors [49]. In the clover phyllody phytoplasma, genes assigned as *folP* and *folK* encode frameshifts and represent pseudogenes [151]. It remains unclear how phytoplasmas obtain 7,8-dihydrofolate monoglutamate and how they process 5,10-methylenetetrahydrofolate. The incomplete folate synthesis may indicate the loss of this genetic modules and folate-dependance on the host. This may influence the host because folate is also involved in photorespiration, amino acid metabolism, and chloroplastic protein biosynthesis in plants [152].

10.6. Thiamine Synthesis. All four phytoplasma genomes encode the ATP-dependent thiamine-phosphate phosphotransferase ThiJ (syn. ThiL) (ST47), which forms the coenzyme thiamine pyrophosphate (syn. thiamine diphosphate) from thiamine phosphate. ThiI, which is encoded in the chromosomes except for 'Ca. P. mali', is required for thiazole synthesis in the thiamine biosynthesis pathway [153].

10.7. Pyridoxal phosphate Synthesis. The kinase PdxK of the vitamin B6 metabolism is only encoded in the genome of 'Ca. P. asteris' OY-M (ST48). This ATP-dependent kinase catalyses the formation of pyridoxine-5′-phosphate from pyridoxal, pyroxidine, and pyridoxamine.

10.8. Iron-Sulfur Cluster Biosynthesis. NifU-like proteins are predicted to occur in all four phytoplasmas (ST39). Proteins such as NifS and NifU are required for the formation of metalloclusters of nitrogenase in *Azotobacter vinelandii* and the maturation of other FeS proteins. No further genes associated with nitrogen fixation were found in the phytoplasma genomes. Notably, *nifU*-like genes are also encoded in genomes of other organisms lacking the ability to fix nitrogen [154].

10.9. Acyl Carrier Protein Metabolism. Two proteins assigned to the acyl carrier metabolism are encoded in all four genomes (ST43): the acyl carrier protein AcpP, which is involved in the biosynthesis of fatty acids and membrane-derived oligosaccharides in *E. coli* [155] and the holo-[ACP] synthase (AcpS) involved in carrier formation in lipid synthesis [156].

10.10. Phosphate Metabolism. A set of six proteins involved in the metabolism of phosphorous compounds is encoded in all four genomes (ST51). The inorganic pyrophosphatases (Ppa) catalyse the hydrolysis of inorganic pyrophosphate (PPi) to two orthophosphates (Pi), regulating the PPi pool which is replenished by various metabolic processes [157]. Additional predicted proteins related to phosphate metabolism are phosphohydrolases, metallophosphoesterases, and lysophospholipases.

10.11. beta-Glucanase. Several genes with a weak catalytic or unclear pathway assignment are shared between some of the four phytoplasmas. An example is the predicted endoglucanase FrvX (M42 peptidase family). Considering that endo-1,4-beta-glucanase proteins in *Arabidopsis thaliana* are associated with plant growth (in particular elongation), xylem development, and cell wall thickening [158] the phytoplasma FrvX protein could potentially contribute to virulence (ST52).

11. Outlook

Future research has to experimentally validate the functions of the key proteins in phytoplasmas. The labelling of substrates and heterologous expression in cultivable hosts are possible strategies as long as axenic cultures of phytoplasmas are unavailable. Beside this basic molecular research at least two further fields of research have to be developed for phytoplasmas.

First, a 'Ca. Phytoplasma species' with a less reduced genome and/or a 'Ca. Phytoplasma species' closer affiliating with the phylogenetic separation of *Acholeplasma* and 'Ca. Phytoplasma' has to be identified. Studies on such an ancestor will provide insights into the evolution of the genomes and the host relationship.

Second, gene expression data from a strain with complete and a host with at least a draft genome sequence are needed. To date, most insights were obtained from pathogens, but these data need to be integrated with the situation in the healthy plant, to learn more about the overall plant response.

Acknowledgment

This work was supported by COST Action FA0807 (Integrated Management of Phytoplasma Epidemics in Different Crop Systems) and the Deutsche Forschungsgemeinschaft through Projects KU 2679/2-1 and BU 890/21-1.

References

[1] R. E. McCoy, A. Caudwell, C. J. Chang et al., "Plant diseases associated with mycoplasma-like organisms," in *The Mycoplasmas*, R. F. Whitcomb and J. G. Tully, Eds., vol. 5, pp. 546–640, Academic Press, New York, NY, USA, 1989.

[2] E. Strauss, "Phytoplasma research begins to bloom," *Science*, vol. 325, no. 5939, pp. 388–390, 2009.

[3] Y. T. M. Doi, K. Yora, and H. Asuyama, "Mycoplasma or PLT-group-like organisms found in the phloem elements of plants infected with mulberry dwarf, potato witches' broom, aster yellows or paulownia witches broom," *Annals of the Phytopathological Society of Japan*, vol. 33, pp. 259–266, 1967.

[4] G. Firrao, M. Andersen, A. Bertaccini et al., "'Candidatus Phytoplasma', a taxon for the wall-less, non-helical prokaryotes that colonize plant phloem and insects," *International Journal of Systematic and Evolutionary Microbiology*, vol. 54, no. 4, pp. 1243–1255, 2004.

[5] D. Harrison and R. E. Davis, "Genus I. "Candidatus Phytoplasma" gen. nov. IRPCM phytoplasma/spiroplasma working team 2004, 1244," in *The Bacteroidetes, Spirochaetes, Tenericutes (Mollicutes), Acidobacteria, Fibrobacteres, Fusobacteria, Dictyoglomi, Gemmatimonadetes, Lentisphaerae, Verrucomicrobia, Chlamydiae, and Planctomycetes*, N. R. Krieg, W. Ludwig, W. B. Whitman et al., Eds., vol. 4, pp. 696–718, Springer, New York, NY, USA, 2011.

[6] R. E. Davis et al., "'Candidatus Phytoplasma sudamericanum', a novel taxon, and strain PassWB-Br4, a new subgroup 16SrIII-V phytoplasma, from diseased passion fruit (Passiflora edulis f. flavicarpa Deg.)," *International Journal of Systematic and Evolutionary Microbiology*. In press.

[7] I. M. Lee, K. D. Bottner-Parker, Y. Zhao, W. Villalobos, and L. Moreira, "'Candidatus Phytoplasma costaricanum' a new phytoplasma associated with a newly emerging disease in soybean in Costa Rica," International Journal of Systematic and Evolutionary Microbiology., vol. 61, no. 12, pp. 2822–2826, 2011.

[8] I. M. Lee, R. E. Davis, and D. E. Gundersen-Rindal, "Phytoplasma: phytopathogenic mollicutes," *Annual Review of Microbiology*, vol. 54, pp. 221–255, 2000.

[9] H. Neimark, *Phylogenetic Relationships between Mycoplasmas and Other Prokaryotes*, Academic Press, New York, NY, USA, 1979.

[10] C. R. Woese, J. Maniloff, and L. B. Zablen, "Phylogenetic analysis of the mycoplasmas," *Proceedings of the National Academy of Sciences of the United States of America*, vol. 77, no. 1, pp. 494–498, 1980.

[11] M. J. Rogers, J. Simmons, and R. T. Walker, "Construction of the mycoplasma evolutionary tree from 5S rRNA sequence data," *Proceedings of the National Academy of Sciences of the United States of America*, vol. 82, no. 4, pp. 1160–1164, 1985.

[12] P. O. Lim and B. B. Sears, "Evolutionary relationships of a plant-pathogenic mycoplasmalike organism and Acholeplasma laidlawii deduced from two ribosomal protein gene sequences," *Journal of Bacteriology*, vol. 174, no. 8, pp. 2606–2611, 1992.

[13] K. F. Toth, N. Harrison, and B. B. Sears, "Phylogenetic relationships among members of the class Mollicutes deduced from rps3 gene sequences," *International Journal of Systematic Bacteriology*, vol. 44, no. 1, pp. 119–124, 1994.

[14] S. Razin, D. Yogev, and Y. Naot, "Molecular biology and pathogenicity of mycoplasmas," *Microbiology and Molecular Biology Reviews*, vol. 62, no. 4, pp. 1094–1156, 1998.

[15] W. G. Weisburg, J. G. Tully, D. L. Rose et al., "A phylogenetic analysis of the mycoplasmas: basis for their classification," *Journal of Bacteriology*, vol. 171, no. 12, pp. 6455–6467, 1989.

[16] P. Carle, F. Laigret, J. G. Tully, and J. M. Bove, "Heterogeneity of genome sizes within the genus Spiroplasma," *International Journal of Systematic Bacteriology*, vol. 45, no. 1, pp. 178–181, 1995.

[17] H. Neimark and B. C. Kirkpatrick, "Isolation and characterization of full-length chromosomes from non-culturable plant-pathogenic mycoplasma-like organisms," *Molecular Microbiology*, vol. 7, no. 1, pp. 21–28, 1993.

[18] C. Marcone and E. Seemüller, "A chromosome map of the European stone fruit yellows phytoplasma," *Microbiology*, vol. 147, no. 5, pp. 1213–1221, 2001.

[19] C. Marcone, H. Neimark, A. Ragozzino, U. Lauer, and E. Seemüller, "Chromosome sizes of phytoplasmas composing major phylogenetic groups and subgroups," *Phytopathology*, vol. 89, no. 9, pp. 805–810, 1999.

[20] M. Kube, B. Schneider, H. Kuhl et al., "The linear chromosome of the plant-pathogenic mycoplasma "*Candidatus* Phytoplasma mali"," *BMC Genomics*, vol. 9, article 306, 2008.

[21] N. M. Christensen, M. Nicolaisen, M. Hansen, and A. Schulz, "Distribution of phytoplasmas in infected plants as revealed by real-time PCR and bioimaging," *Molecular Plant-Microbe Interactions*, vol. 17, no. 11, pp. 1175–1184, 2004.

[22] K. Oshima, T. Shiomi, T. Kuboyama et al., "Isolation and characterization of derivative lines of the onion yellows phytoplasma that do not cause stunting or phloem hyperplasia," *Phytopathology*, vol. 91, no. 11, pp. 1024–1029, 2001.

[23] D. R. Webb, R. G. Bonfiglioli, L. Carraro, R. Osler, and R. H. Symons, "Oligonucleotides as hybridization probes to localize phytoplasmas in host plants and insect vectors," *Phytopathology*, vol. 89, no. 10, pp. 894–901, 1999.

[24] K. J. Oparka and R. Turgeon, "Sieve elements and companion cells—traffic control centers of the phloem," *Plant Cell*, vol. 11, no. 4, pp. 739–750, 1999.

[25] R. D. Sjölund, "The phloem sieve element: a river runs through it," *Plant Cell*, vol. 9, no. 7, pp. 1137–1146, 1997.

[26] K. Tamura, D. Peterson, N. Peterson, G. Stecher, M. Nei, and S. Kumar, "MEGA5: molecular evolutionary genetics analysis using maximum likelihood, evolutionary distance, and maximum parsimony methods," *Molecular Biology and Evolution*, vol. 28, no. 10, pp. 2731–2739, 2011.

[27] S. Dinant and R. Lemoine, "The phloem pathway: new issues and old debates," *Comptes Rendus—Biologies*, vol. 333, no. 4, pp. 307–319, 2010.

[28] M. van Helden, W. F. Tjallingii, and T. A. van Beek, "Phloem sap collection from lettuce (*Lactuca sativa* L.): chemical comparison among collection methods," *Journal of Chemical Ecology*, vol. 20, no. 12, pp. 3191–3206, 1994.

[29] A. J. E. van Bel and P. H. Hess, "Hexoses as phloem transport sugars: the end of a dogma?" *Journal of Experimental Botany*, vol. 59, no. 2, pp. 261–272, 2008.

[30] S. Dinant, J. L. Bonnemain, C. Girousse, and J. Kehr, "Phloem sap intricacy and interplay with aphid feeding," *Comptes Rendus—Biologies*, vol. 333, no. 6-7, pp. 504–515, 2010.

[31] N. Noiraud, L. Maurousset, and R. Lemoine, "Identification of a mannitol transporter, AgMaT1, in celery phloem," *Plant Cell*, vol. 13, no. 3, pp. 695–705, 2001.

[32] M. Ramsperger-Gleixner, D. Geiger, R. Hedrich, and N. Sauer, "Differential expression of sucrose transporter and polyol transporter genes during maturation of common plantain companion cells," *Plant Physiology*, vol. 134, no. 1, pp. 147–160, 2004.

[33] M. Rokitta, A. D. Peuke, U. Zimmermann, and A. Haase, "Dynamic studies of phloem and xylem flow in fully differentiated plants by fast nuclear-magnetic-resonance microimaging," *Protoplasma*, vol. 209, no. 1-2, pp. 126–131, 1999.

[34] L. A. Cernusak, D. J. Arthur, J. S. Pate, and G. D. Farquhar, "Water relations link carbon and oxygen isotope discrimination to phloem sap sugar concentration in *Eucalyptus globulus*," *Plant Physiology*, vol. 131, no. 4, pp. 1544–1554, 2003.

[35] H. Ziegler, "Nature of substances in phloem," in *Encyclopedia of Plant Physiology, Transport in Plants*, M. H. Zimmermann and J. A. Milburn, Eds., vol. 1, pp. 57–100, Springer, Berlin, Germany, 1975.

[36] A. Gojon, P. Nacry, and J. C. Davidian, "Root uptake regulation: a central process for NPS homeostasis in plants," *Current Opinion in Plant Biology*, vol. 12, no. 3, pp. 328–338, 2009.

[37] T. Y. Liu, C. Y. Chang, and T. J. Chiou, "The long-distance signaling of mineral macronutrients," *Current Opinion in Plant Biology*, vol. 12, no. 3, pp. 312–319, 2009.

[38] J. Kehr and A. Buhtz, "Long distance transport and movement of RNA through the phloem," *Journal of Experimental Botany*, vol. 59, no. 1, pp. 85–92, 2008.

[39] B. D. Pant, M. Musialak-Lange, P. Nuc et al., "Identification of nutrient-responsive Arabidopsis and rapeseed microRNAs by comprehensive real-time polymerase chain reaction profiling and small RNA sequencing," *Plant Physiology*, vol. 150, no. 3, pp. 1541–1555, 2009.

[40] M. K. Lin, Y. J. Lee, T. J. Lough, B. S. Phimney, and W. J. Lucas, "Analysis of the pumpkin phloem proteome provides insights into angiosperm sieve tube function," *Molecular and Cellular Proteomics*, vol. 8, no. 2, pp. 343–356, 2009.

[41] P. Jones, "Phytoplasma plant pathogens," in *Plant Pathologist's Pocketbook*, J. M. Waller, J. M. Lenné, and S. J. Waller, Eds., CABI, New York, NY, USA, 3rd edition, 2002.

[42] P. G. Weintraub and L. Beanland, "Insect vectors of phytoplasmas," *Annual Review of Entomology*, vol. 51, pp. 91–111, 2006.

[43] C. J. Mayer, A. Vilcinskas, and J. Gross, "Pathogen-induced release of plant allomone manipulates vector insect behavior," *Journal of Chemical Ecology*, vol. 34, no. 12, pp. 1518–1522, 2008.

[44] B. C. Kirkpatrick, D. C. Stenger, T. Jack Morris, and A. H. Purcell, "Cloning and detection of DNA from a nonculturable plant pathogenic mycoplasma-like organism," *Science*, vol. 238, no. 4824, pp. 197–200, 1987.

[45] N. Moriwaki, K. Matsushita, M. Nishina, and Y. Kono, "High concentrations of trehalose in aphid hemolymph," *Applied Entomology and Zoology*, vol. 38, no. 2, pp. 241–248, 2003.

[46] L. Galetto, D. Bosco, R. Balestrini, A. Genre, J. Fletcher, and C. Marzachì, "The major antigenic membrane protein of "*Candidatus* Phytoplasma asteris" selectively interacts with ATP synthase and actin of leafhopper vectors," *PLoS One*, vol. 6, no. 7, Article ID e22571, 2011.

[47] J. P. Gourret, P. L. Maillet, and J. Gouranton, "Virus-like particles associated with the mycoplasmas of clover phyllody in the plant and in the insect vector," *Journal of General Microbiology*, vol. 74, no. 2, pp. 241–249, 1973.

[48] K. Oshima, S. Kakizawa, H. Nishigawa et al., "Reductive evolution suggested from the complete genome sequence of a plant-pathogenic phytoplasma," *Nature Genetics*, vol. 36, no. 1, pp. 27–29, 2004.

[49] X. Bai, J. Zhang, A. Ewing et al., "Living with genome instability: the adaptation of phytoplasmas to diverse environments of their insect and plant hosts," *Journal of Bacteriology*, vol. 188, no. 10, pp. 3682–3696, 2006.

[50] L. T. T. Tran-Nguyen, M. Kube, B. Schneider, R. Reinhardt, and K. S. Gibb, "Comparative genome analysis of "*Candidatus* Phytoplasma australiense" (subgroup tuf-Australia I; rp-A) and "Ca. phytoplasma asteris" strains OY-M and AY-WB," *Journal of Bacteriology*, vol. 190, no. 11, pp. 3979–3991, 2008.

[51] R. D. Fleischmann, M. D. Adams, O. White et al., "Whole-genome random sequencing and assembly of Haemophilus influenzae Rd," *Science*, vol. 269, no. 5223, pp. 496–521, 1995.

[52] M. Ronaghi, S. Karamohamed, B. Pettersson, M. Uhlén, and P. Nyrén, "Real-time DNA sequencing using detection of pyrophosphate release," *Analytical Biochemistry*, vol. 242, no. 1, pp. 84–89, 1996.

[53] L. T. T. Tran-Nguyen and K. S. Gibb, "Extrachromosomal DNA isolated from tomato big bud and *Candidatus* Phytoplasma australiense phytoplasma strains," *Plasmid*, vol. 56, no. 3, pp. 153–166, 2006.

[54] H. Nishigawa, K. Oshima, S. Kakizawa et al., "A plasmid from a non-insect-transmissible line of a phytoplasma lacks two open reading frames that exist in the plasmid from the wild-type line," *Gene*, vol. 298, no. 2, pp. 195–201, 2002.

[55] G. C. Chaconas and C. W. Chen, "Replication of linear bacterial chromosomes: no longer going around the circle," in *The Bacterial Chromosome*, N. Higgins, Ed., pp. 525–539, ASM Press, Washington, DC, USA, 2005.

[56] W. Wei, R. E. Davis, R. Jomantiene, and Y. Zhao, "Ancient, recurrent phage attacks and recombination shaped dynamic sequence-variable mosaics at the root of phytoplasma genome evolution," *Proceedings of the National Academy of Sciences of the United States of America*, vol. 105, no. 33, pp. 11827–11832, 2008.

[57] M. Bose and R. D. Barber, "Prophage Finder: a prophage loci prediction tool for prokaryotic genome sequences," *In Silico Biology*, vol. 6, no. 3, pp. 223–227, 2006.

[58] A. C. Padovan, G. Firrao, B. Schneider, and K. S. Gibb, "Chromosome mapping of the sweet potato little leaf phytoplasma reveals genome heterogeneity within the phytoplasmas," *Microbiology*, vol. 146, no. 4, part 4, pp. 893–902, 2000.

[59] T. Y. Toruño, M. Seruga Musić, S. Simi, M. Nicolaisen, and S. A. Hogenhout, "Phytoplasma PMU1 exists as linear chromosomal and circular extrachromosomal elements and has enhanced expression in insect vectors compared with plant hosts," *Molecular Microbiology*, vol. 77, no. 6, pp. 1406–1415, 2010.

[60] S. A. Hogenhout and S. A. Musić, "Phytoplasma genomics, from sequencing to comparative and functional genomics-what have we learnt?" in *Phytoplasmas: Genomes, Plant Hosts and Vectors*, P. G. Weintraub and P. Jones, Eds., pp. 19–36, CABI, Wallingford, UK, 2010.

[61] Y. Xiong and T. H. Eickbush, "Origin and evolution of retroelements based upon their reverse transcriptase sequences," *The EMBO Journal*, vol. 9, no. 10, pp. 3353–3362, 1990.

[62] R. L. Tatusov, M. Y. Galperin, D. A. Natale, and E. V. Koonin, "The COG database: a tool for genome-scale analysis of protein functions and evolution," *Nucleic Acids Research*, vol. 28, no. 1, pp. 33–36, 2000.

[63] J. M. Kaguni, L. L. Bertsch, D. Bramhill et al., "Initiation of replication of the Escherichia coli chromosomal origin reconstituted with purified enzymes," *Basic Life Sciences*, vol. 30, pp. 141–150, 1985.

[64] T. A. Baker, B. E. Funnell, and A. Kornberg, "Helicase action of dnaB protein during replication from the Escherichia coli chromosomal origin in vitro," *The Journal of Biological Chemistry*, vol. 262, no. 14, pp. 6877–6885, 1987.

[65] W. Seufert, B. Dobrinski, R. Lurz, and W. Messer, "Functionality of the dnaA protein binding site in DNA replication is orientation-dependent," *The Journal of Biological Chemistry*, vol. 263, no. 6, pp. 2719–2723, 1988.

[66] C. S. McHenry, "Chromosomal replicases as asymmetric dimers: studies of subunit arrangement and functional consequences," *Molecular Microbiology*, vol. 49, no. 5, pp. 1157–1165, 2003.

[67] J. R. Lobry, "Origin of replication of Mycoplasma genitalium," *Science*, vol. 272, no. 5262, pp. 745–746, 1996.

[68] J. R. Lobry, "Asymmetric substitution patterns in the two DNA strands of bacteria," *Molecular Biology and Evolution*, vol. 13, no. 5, pp. 660–665, 1996.

[69] A. C. Frank and J. R. Lobry, "Oriloc: prediction of replication boundaries in unannotated bacterial chromosomes," *Bioinformatics*, vol. 16, no. 6, pp. 560–561, 2000.

[70] C. M. M. Cordova, C. Lartigue, P. Sirand-Pugnet, J. Renaudin, R. A. F. Cunha, and A. Blanchard, "Identification of the origin of replication of the *Mycoplasma pulmonis* chromosome and its use in oriC replicative plasmids," *Journal of Bacteriology*, vol. 184, no. 19, pp. 5426–5435, 2002.

[71] R. S. Fuller, B. E. Funnell, and A. Kornberg, "The dnaA protein complex with the E. coli chromosomal replication origin (*oriC*) and other DNA sites," *Cell*, vol. 38, no. 3, pp. 889–900, 1984.

[72] D. S. S. Hudspeth and P. S. Vary, "*spoVG* sequence of *Bacillus megaterium* and *Bacillus subtilis*," *Biochimica et Biophysica Acta*, vol. 1130, no. 2, pp. 229–231, 1992.

[73] J. I. Glass, N. Assad-Garcia, N. Alperovich et al., "Essential genes of a minimal bacterium," *Proceedings of the National Academy of Sciences of the United States of America*, vol. 103, no. 2, pp. 425–430, 2006.

[74] L. N. Kinch, K. Ginalski, L. Rychlewski, and N. V. Grishin, "Identification of novel restriction endonuclease-like fold

families among hypothetical proteins," *Nucleic Acids Research*, vol. 33, no. 11, pp. 3598–3605, 2005.

[75] H. I. Miller and D. I. Friedman, "An E. coli gene product required for λ site-specific recombination," *Cell*, vol. 20, no. 3, pp. 711–719, 1980.

[76] D. N. Arnosti and M. J. Chamberlin, "Secondary σ factor controls transcription of flagellar and chemotaxis genes in Escherichia coli," *Proceedings of the National Academy of Sciences of the United States of America*, vol. 86, no. 3, pp. 830–834, 1989.

[77] A. F. González Barrios, R. Zuo, D. Ren, and T. K. Wood, "Hha, YbaJ, and OmpA regulate Escherichia coli K12 biofilm formation and conjugation plasmids abolish motility," *Biotechnology and Bioengineering*, vol. 93, no. 1, pp. 188–200, 2006.

[78] D. F. Ward and M. E. Gottesman, "The nus mutations affect transcription termination in Escherichia coli," *Nature*, vol. 292, no. 5820, pp. 212–215, 1981.

[79] J. Roberts and J. S. Park, "Mfd, the bacterial transcription repair coupling factor: translocation, repair and termination," *Current Opinion in Microbiology*, vol. 7, no. 2, pp. 120–125, 2004.

[80] J. W. Roberts, "Termination factor for RNA synthesis," *Nature*, vol. 224, no. 5225, pp. 1168–1174, 1969.

[81] B. Schneider and E. Seemuller, "Presence of two sets of ribosomal genes in phytopathogenic mollicutes," *Applied and Environmental Microbiology*, vol. 60, no. 9, pp. 3409–3412, 1994.

[82] K. Lagesen, P. Hallin, E. A. Rødland, H. H. Stærfeldt, T. Rognes, and D. W. Ussery, "RNAmmer: consistent and rapid annotation of ribosomal RNA genes," *Nucleic Acids Research*, vol. 35, no. 9, pp. 3100–3108, 2007.

[83] T. M. Lowe and S. R. Eddy, "tRNAscan-SE: a program for improved detection of transfer RNA genes in genomic sequence," *Nucleic Acids Research*, vol. 25, no. 5, pp. 955–964, 1997.

[84] G. O. Bylund, L. C. Wipemo, L. A. C. Lundberg, and P. M. Wikström, "RimM and RbfA are essential for efficient processing of 16S rRNA in Escherichia coli," *Journal of Bacteriology*, vol. 180, no. 1, pp. 73–82, 1998.

[85] Y. Qin, N. Polacek, O. Vesper et al., "The highly conserved LepA is a ribosomal elongation factor that back-translocates the ribosome," *Cell*, vol. 127, no. 4, pp. 721–733, 2006.

[86] G. Grentzmann, D. Brechemier-Baey, V. Heurgue-Hamard, and R. H. Buckingham, "Function of polypeptide chain release factor RF-3 in Escherichia coli. RF-3 action in termination is predominantly at UGA-containing stop signals," *The Journal of Biological Chemistry*, vol. 270, no. 18, pp. 10595–10600, 1995.

[87] B. Bukau and A. L. Horwich, "The Hsp70 and Hsp60 chaperone machines," *Cell*, vol. 92, no. 3, pp. 351–366, 1998.

[88] E. Laskowska, A. Wawrzynów, and A. Taylor, "IbpA and IbpB, the new heat-shock proteins, bind to endogenous Escherichia coli proteins aggregated intracellularly by heat shock," *Biochimie*, vol. 78, no. 2, pp. 117–122, 1996.

[89] V. Tsilibaris, G. Maenhaut-Michel, and L. van Melderen, "Biological roles of the Lon ATP-dependent protease," *Research in Microbiology*, vol. 157, no. 8, pp. 701–713, 2006.

[90] M. Carrión, M. J. Gómez, R. Merchante-Schubert, S. Dongarrá, and J. A. Ayala, "mraW, an essential gene at the dcw cluster of Escherichia coli codes for a cytoplasmic protein with methyltransferase activity," *Biochimie*, vol. 81, no. 8-9, pp. 879–888, 1999.

[91] C. T. Parker, A. W. Kloser, C. A. Schnaitman, M. A. Stein, S. Gottesman, and B. W. Gibson, "Role of the rfaG and rfaP genes in determining the lipopolysaccharide core structure and cell surface properties of Escherichia coli K-12," *Journal of Bacteriology*, vol. 174, no. 8, pp. 2525–2538, 1992.

[92] N. Umeda, T. Suzuki, M. Yukawa et al., "Mitochondria-specific RNA-modifying enzymes responsible for the biosynthesis of the wobble base in mitochondrial tRNAs: implications for the molecular pathogenesis of human mitochondrial diseases," *The Journal of Biological Chemistry*, vol. 280, no. 2, pp. 1613–1624, 2005.

[93] D. Brégeon, V. Colot, M. Radman, and F. Taddei, "Translational misreading: a tRNA modification counteracts a +2 ribosomal frameshift," *Genes and Development*, vol. 15, no. 17, pp. 2295–2306, 2001.

[94] K. von Meyenburg, B. B. Jorgensen, J. Nielsen, and F. G. Hansen, "Promoters of the atp operon coding for the membrane-bound ATP synthase of Escherichia coli mapped by Tn10 insertion mutations," *Molecular and General Genetics*, vol. 188, no. 2, pp. 240–248, 1982.

[95] C. A. Hutchison, S. N. Peterson, S. R. Gill et al., "Global transposon mutagenesis and a minimal mycoplasma genome," *Science*, vol. 286, no. 5447, pp. 2165–2169, 1999.

[96] M. J. Romanowski, J. B. Bonanno, and S. K. Burley, "Crystal structure of the *Escherichia coli* glucose-inhibited division protein B (GidB) reveals a methyltransferase fold," *Proteins*, vol. 47, no. 4, pp. 563–567, 2002.

[97] A. Cimerman, D. Pacifico, P. Salar, C. Marzachi, and X. Foissac, "Striking diversity of vmp1, a variable gene encoding a putative membrane protein of the stolbur phytoplasma," *Applied and Environmental Microbiology*, vol. 75, no. 9, pp. 2951–2957, 2009.

[98] S. Kakizawa, K. Oshima, and S. Namba, "Diversity and functional importance of phytoplasma membrane proteins," *Trends in Microbiology*, vol. 14, no. 6, pp. 254–256, 2006.

[99] S. Suzuki, K. Oshima, S. Kakizawa et al., "Interaction between the membrane protein of a pathogen and insect microfilament complex determines insect-vector specificity," *Proceedings of the National Academy of Sciences of the United States of America*, vol. 103, no. 11, pp. 4252–4257, 2006.

[100] S. Kakizawa, K. Oshima, H. Y. Jung et al., "Positive selection acting on a surface membrane protein of the plant-pathogenic phytoplasmas," *Journal of Bacteriology*, vol. 188, no. 9, pp. 3424–3428, 2006.

[101] S. A. Hogenhout, K. Oshima, E. D. Ammar, S. Kakizawa, H. N. Kingdom, and S. Namba, "Phytoplasmas: bacteria that manipulate plants and insects," *Molecular Plant Pathology*, vol. 9, no. 4, pp. 403–423, 2008.

[102] S. Kakizawa, K. Oshima, Y. Ishii et al., "Cloning of immunodominant membrane protein genes of phytoplasmas and their in planta expression," *FEMS Microbiology Letters*, vol. 293, no. 1, pp. 92–101, 2009.

[103] L. Käll, A. Krogh, and E. L. L. Sonnhammer, "A combined transmembrane topology and signal peptide prediction method," *Journal of Molecular Biology*, vol. 338, no. 5, pp. 1027–1036, 2004.

[104] S. O. Kakizawa, K. Oshima, and S. Namba, "Functional genomics of phytoplasmas," in *Phytoplasmas: Genomes, Plant Hosts and Vectors*, P. J. Weintraub and P. Jones, Eds., pp. 37–50, CABI, Wallingford, UK, 2010.

[105] X. Bai, V. R. Correa, T. Y. Toruño, E. D. Ammar, S. Kamoun, and S. A. Hogenhout, "AY-WB phytoplasma secretes a protein that targets plant cell nuclei," *Molecular Plant-Microbe Interactions*, vol. 22, no. 1, pp. 18–30, 2009.

[106] A. Hoshi, K. Oshima, S. Kakizawa et al., "A unique virulence factor for proliferation and dwarfism in plants identified

from a phytopathogenic bacterium," *Proceedings of the National Academy of Sciences of the United States of America*, vol. 106, no. 15, pp. 6416–6421, 2009.

[107] S. Kakizawa, K. Oshima, H. Nishigawa et al., "Secretion of immunodominant membrane protein from onion yellows phytoplasma through the Sec protein-translocation system in Escherichia coli," *Microbiology*, vol. 150, no. 1, pp. 135–142, 2004.

[108] J. C. Samuelson, M. Chen, F. Jiang et al., "YidC mediates membrane protein insertion in bacteria," *Nature*, vol. 406, no. 6796, pp. 637–641, 2000.

[109] K. J. Pogliano and J. Beckwith, "Genetic and molecular characterization of the Escherichia coli secD operon and its products," *Journal of Bacteriology*, vol. 176, no. 3, pp. 804–814, 1994.

[110] J. Fang and Y. Wei, "Expression, purification and Characterization of the Escherichia coli integral membrane protein YajC," *Protein and Peptide Letters*, vol. 18, no. 6, pp. 601–608, 2011.

[111] H. Nielsen, J. Engelbrecht, S. Brunak, and G. von Heijne, "A neural network method for identification of prokaryotic and eukaryotic signal peptides and prediction of their cleavage sites," *International Journal of Neural Systems*, vol. 8, no. 5-6, pp. 581–599, 1997.

[112] K. Nakai and M. Kanehisa, "Expert system for predicting protein localization sites in gram-negative bacteria," *Proteins*, vol. 11, no. 2, pp. 95–110, 1991.

[113] R. Rabus, M. Kube, J. Heider et al., "The genome sequence of an anaerobic aromatic-degrading denitrifying bacterium, strain EbN1," *Archives of Microbiology*, vol. 183, no. 1, pp. 27–36, 2005.

[114] H. G. Koch, T. Hengelage, C. Neumann-Haefelin et al., "In vitro studies with purified components reveal signal recognition particle (SRP) and SecA/SecB as constituents of two independent protein- targeting pathways of *Escherichia coli*," *Molecular Biology of the Cell*, vol. 10, no. 7, pp. 2163–2173, 1999.

[115] H. G. Koch, M. Moser, and M. Müller, "Signal recognition particle-dependent protein targeting, universal to all kingdoms of life," *Reviews of Physiology, Biochemistry and Pharmacology*, vol. 146, pp. 55–94, 2003.

[116] S. Brown, "4.5S RNA: does form predict function?" *New Biologist*, vol. 3, no. 5, pp. 430–438, 1991.

[117] Z. Silva, M. M. Sampaio, A. Henne et al., "The high-affinity maltose/trehalose ABC transporter in the extremely thermophilic bacterium Thermus thermophilus HB27 also recognizes sucrose and palatinose," *Journal of Bacteriology*, vol. 187, no. 4, pp. 1210–1218, 2005.

[118] A. André, W. Maccheroni, F. Doignon, M. Garnier, and J. Renaudin, "Glucose and trehalose PTS permeases of Spiroplasma citri probably share a single IIA domain, enabling the spiroplasma to adapt quickly to carbohydrate changes in its environment," *Microbiology*, vol. 149, no. 9, pp. 2687–2696, 2003.

[119] D. Hekstra and J. Tommassen, "Functional exchangeability of the ABC proteins of the periplasmic binding protein-dependent transport systems Ugp and Mal of Escherichia coli," *Journal of Bacteriology*, vol. 175, no. 20, pp. 6546–6552, 1993.

[120] S. Hunter, R. Apweiler, T. K. Attwood et al., "InterPro: the integrative protein signature database," *Nucleic Acids Research*, vol. 37, no. 1, pp. D211–D215, 2009.

[121] D. A. Rodionov, A. G. Vitreschak, A. A. Mironov, and M. S. Gelfand, "Comparative genomics of thiamin biosynthesis in procaryotes. New genes and regulatory mechanisms," *The Journal of Biological Chemistry*, vol. 277, no. 50, pp. 48949–48959, 2002.

[122] I. Tobueren Bots and H. Hagedorn, "Studies on the thiamine transport system in Bacillus cereus," *Archives of Microbiology*, vol. 113, no. 1-2, pp. 23–31, 1977 (German).

[123] X. Foissac, J. L. Danet, C. Saillard et al., "Mutagenesis by insertion of Tn4001 into the genome of Spiroplasma citri: characterization of mutants affected in plant pathogenicity and transmission to the plant by the leafhopper vector Circulifer haematoceps," *Molecular Plant-Microbe Interactions*, vol. 10, no. 4, pp. 454–461, 1997.

[124] S. I. Sukharev, P. Blount, B. Martinac, F. R. Blattner, and C. Kung, "A large-conductance mechanosensitive channel in E. coli encoded by mscL alone," *Nature*, vol. 368, no. 6468, pp. 265–268, 1994.

[125] J. G. Voet and R. H. Abeles, "The mechanism of action of sucrose phosphorylase. Isolation and properties of a beta-linked covalent glucose-enzyme complex," *The Journal of Biological Chemistry*, vol. 245, no. 5, pp. 1020–1031, 1970.

[126] R. Bruckner, E. Wagner, and F. Gotz, "Characterization of a sucrase gene from Staphylococcus xylosus," *Journal of Bacteriology*, vol. 175, no. 3, pp. 851–857, 1993.

[127] P. J. Eastmond, Y. Li, and I. A. Graham, "Is trehalose-6-phosphate a regulator of sugar metabolism in plants?" *Journal of Experimental Botany*, vol. 54, no. 382, pp. 533–537, 2003.

[128] Y. Zhang, L. F. Primavesi, D. Jhurreea et al., "Inhibition of SNF1-related protein kinase activity and regulation of metabolic pathways by trehalose-6-phosphate," *Plant Physiology*, vol. 149, no. 4, pp. 1860–1871, 2009.

[129] M. J. Paul, D. Jhurreea, Y. Zhang et al., "Upregulation of biosynthetic processes associated with growth by trehalose 6-phosphate," *Plant Signaling and Behavior*, vol. 5, no. 4, pp. 386–392, 2010.

[130] R. S. Stephens, S. Kalman, C. Lammel et al., "Genome sequence of an obligate intracellular pathogen of humans: chlamydia trachomatis," *Science*, vol. 282, no. 5389, pp. 754–759, 1998.

[131] J. D. Pollack, M. V. Williams, and R. N. McElhaney, "The comparative metabolism of the mollicutes (Mycoplasmas): the utility for taxonomic classification and the relationship of putative gene annotation and phylogeny to enzymatic function in the smallest free-living cells," *Critical Reviews in Microbiology*, vol. 23, no. 4, pp. 269–354, 1997.

[132] G. Cohen and N. L. Somerson, "Glucose-dependent secretion and destruction of hydrogen peroxide by Mycoplasma pneumoniae," *Journal of Bacteriology*, vol. 98, no. 2, pp. 547–551, 1969.

[133] G. Cohen and N. L. Somerson, "Mycoplasma pneumoniae: hydrogen peroxide secretion and its possible role in virulence," *Annals of the New York Academy of Sciences*, vol. 143, no. 1, pp. 85–87, 1967.

[134] N. L. Somerson, B. E. Walls, and R. M. Chanock, "Hemolysin of mycoplasma pneumoniae: tentative identification as a peroxide," *Science*, vol. 150, no. 3693, pp. 226–227, 1965.

[135] T. A. Bobik, G. D. Havemann, R. J. Busch, D. S. Williams, and H. C. Aldrich, "The propanediol utilization (pdu) operon of Salmonella enterica serovar Typhimurium LT2 includes genes necessary for formation of polyhedral organelles involved in coenzyme B12-dependent 1,2-propanediol degradation," *Journal of Bacteriology*, vol. 181, no. 19, pp. 5967–5975, 1999.

[136] D. Walter, M. Ailion, and J. Roth, "Genetic characterization of the pdu operan: use of 1,2-propanediol in Salmonella typhimurium," *Journal of Bacteriology*, vol. 179, no. 4, pp. 1013–1022, 1997.

[137] Y. Liu, N. A. Leal, E. M. Sampson, C. L. V. Johnson, G. D. Havemann, and T. A. Bobik, "PduL is an evolutionarily distinct phosphotransacylase involved in B 12-dependent 1,2-propanediol degradation by Salmonella enterica serovar typhimurium LT2," *Journal of Bacteriology*, vol. 189, no. 5, pp. 1589–1596, 2007.

[138] B. Meier and G. G. Habermehl, "Evidence for superoxide dismutase and catalase in mollicutes and release of reactive oxygen species," *Free Radical Research Communications*, vol. 12-13, pp. 451–454, 1991.

[139] I. Fridovich, "Superoxide radical and superoxide dismutases," *Annual Review of Biochemistry*, vol. 64, pp. 97–112, 1995.

[140] J. R. Chen, C. N. Weng, T. Y. Ho, I. C. Cheng, and S. S. Lai, "Identification of the copper-zinc superoxide dismutase activity in *Mycoplasma hyopneumoniae*," *Veterinary Microbiology*, vol. 73, no. 4, pp. 301–310, 2000.

[141] S. J. O'Brien, J. M. Simonson, and M. W. Grabowski, "Analysis of multiple isoenzyme expression among twenty-two species of *Mycoplasma* and Acholeplasma," *Journal of Bacteriology*, vol. 146, no. 1, pp. 222–232, 1981.

[142] B. Meier and G. G. Habermehl, "Evidence for superoxide dismutase and catalase in mollicutes and release of reactive oxygen species," *Archives of Biochemistry and Biophysics*, vol. 277, no. 1, pp. 74–79, 1990.

[143] K. E. Reed and J. E. Cronan Jr., "Lipoic acid metabolism in Escherichia coli: sequencing and functional characterization of the lipA and lipB genes," *Journal of Bacteriology*, vol. 175, no. 5, pp. 1325–1336, 1993.

[144] T. W. Morris, K. E. Reed, and J. E. Cronan Jr., "Identification of the gene encoding lipoate-protein ligase A of Escherichia coli. Molecular cloning and characterization of the lplA gene and gene product," *The Journal of Biological Chemistry*, vol. 269, no. 23, pp. 16091–16100, 1994.

[145] C. V. Bizarro and D. C. Schuck, "Purine and pyrimidine nucleotide metabolism in Mollicutes," *Genetics and Molecular Biology*, vol. 30, no. 1, pp. 190–201, 2007.

[146] J. D. Pollack, M. A. Myers, T. Dandekar, and R. Herrmann, "Suspected utility of enzymes with multiple activities in the small genome Mycoplasma species: the replacement of the missing "household" nucleoside diphosphate kinase gene and activity by glycolytic kinases," *OMICS A Journal of Integrative Biology*, vol. 6, no. 3, pp. 247–258, 2002.

[147] F. C. Minion, K. J. Jarvill-Taylor, D. E. Billings, and E. Tigges, "Membrane-associated nuclease activities in Mycoplasmas," *Journal of Bacteriology*, vol. 175, no. 24, pp. 7842–7847, 1993.

[148] G. A. M. Neale, A. Mitchell, and L. R. Finch, "Pathways of pyrimidine deoxyribonucleotide biosynthesis in Mycoplasma mycoides subsp. mycoides," *Journal of Bacteriology*, vol. 154, no. 1, pp. 17–22, 1983.

[149] L. S. Meena, P. Chopra, R. S. Bedwal, and Y. Singh, "Nucleoside diphosphate kinase-like activity in adenylate kinase of Mycobacterium tuberculosis," *Biotechnology and Applied Biochemistry*, vol. 38, no. 2, pp. 169–174, 2003.

[150] Y. Shotland, A. Shifrin, T. Ziv et al., "Proteolysis of bacteriophage lambda CII by *Escherichia coli* FtsH (HflB)," *Journal of Bacteriology*, vol. 182, no. 11, pp. 3111–3116, 2000.

[151] R. E. Davis, R. Jomantiene, Y. Zhao, and E. L. Dally, "Folate biosynthesis pseudogenes, PsifolP and PsifolK, and an O-sialoglycoprotein endopeptidase gene homolog in the phytoplasma genome," *DNA and Cell Biology*, vol. 22, no. 11, pp. 697–706, 2003.

[152] A. D. Hanson and J. F. Gregory, "Synthesis and turnover of folates in plants," *Current Opinion in Plant Biology*, vol. 5, no. 3, pp. 244–249, 2002.

[153] E. Webb, K. Claas, and D. M. Downs, "Characterization of thiI, a new gene involved in thiazole biosynthesis in Salmonella typhimurium," *Journal of Bacteriology*, vol. 179, no. 13, pp. 4399–4402, 1997.

[154] D. M. Hwang, A. Dempsey, K. T. Tan, and C. C. Liew, "A modular domain of NiFu, a nitrogen fixation cluster protein, is highly conserved in evolution," *Journal of Molecular Evolution*, vol. 43, no. 5, pp. 536–540, 1996.

[155] M. Rawlings and J. E. Cronan Jr., "The gene encoding Escherichia coli acyl carrier protein lies within a cluster of fatty acid biosynthetic genes," *The Journal of Biological Chemistry*, vol. 267, no. 9, pp. 5751–5754, 1992.

[156] R. H. Lambalot and C. T. Walsh, "Cloning, overproduction, and characterization of the Escherichia coli holo-acyl carrier protein synthase," *The Journal of Biological Chemistry*, vol. 270, no. 42, pp. 24658–24661, 1995.

[157] B. S. Cooperman, A. A. Baykov, and R. Lahti, "Evolutionary conservation of the active site of soluble inorganic pyrophosphatase," *Trends in Biochemical Sciences*, vol. 17, no. 7, pp. 262–266, 1992.

[158] Z. Shani, M. Dekel, L. Roiz et al., "Expression of endo-1,4-β-glucanase (cel1) in Arabidopsis thaliana is associated with plant growth, xylem development and cell wall thickening," *Plant Cell Reports*, vol. 25, no. 10, pp. 1067–1074, 2006.

17

Determination of Mould and Aflatoxin Contamination in Tarhana, a Turkish Fermented Food

Hilal Colak,[1] Hamparsun Hampikyan,[2] Enver Baris Bingol,[1] Omer Cetin,[1]
Meryem Akhan,[3] and Sumeyre Ipek Turgay[3]

[1] Department of Food Hygiene and Technology, Faculty of Veterinary Medicine, Istanbul University, Avcilar, 34320 Istanbul, Turkey
[2] The School of Vocational Studies, Beykent University, Buyukcekmece, 34500 Istanbul, Turkey
[3] Academic Hygiene KGaA, Training, Audit and Consulting Services, Kuştepe Mahallesi, Tomurcuk Sokak, İzmen Sitesi, Sisli, 34387 Istanbul, Turkey

Correspondence should be addressed to Hilal Colak, hcolak@istanbul.edu.tr

Academic Editors: K. Ohmiya, G. K. Paterson, and J. Qiu

Tarhana is a popular traditional Turkish cereal-based fermented food product mainly produced at home or at home-scale level. Some certain mould species can grow even at low moisture and pH values and produce aflatoxins in food. This study was conducted to determine aflatoksin levels in tarhana. For this purpose, a total of 138 tarhana powder samples were collected from bazaars in Istanbul and analyzed for aflatoxins, mould contamination, and some physco-chemical parameters. As a result, 32 out of 138 tarhana samples (23.2%) were found to be contaminated with aflatoxins in the range of 0.7–16.8 μg/kg, whereas 29 samples contained Aflatoxin B1 (AFB1) ranging from 0.2–13.2 μg/kg. All samples (100%) contaminated with moulds in the range of 1.4×10^1–5.8×10^7 cfu/g. The average pH, moisture and a_w results were detected as 3.82, 12.71%, and 0.695, respectively.

1. Introduction

Fermented cereal-yoghurt mixtures play an important role in the diets of many people in the Middle East, Asia, Africa, and some parts of Europe [1–3]. Safety and some nutritional benefits such as improvement of protein digestibility, degradation of antinutritional factors have been attributed to fermented foods, and therefore they have promoted for safety and nutritional purpose [4]. Tarhana is a popular traditional Turkish cereal-based fermented food product mainly produced at home or at home-scale level [5, 6]. Tarhana is prepared by mixing wheat flour, yoghurt, yeast, salt, some raw or cooked vegetables (tomato, pepper, and onion), and spices (mint, basil, dill, paprika, tarhana herb, etc.) followed by lactic and alcoholic fermentation for one to seven days. The dough at fermentation is called as wet tarhana. After fermentation, the mixture is dried in the sun as a lump, nugget, or thin layers to obtain dry tarhana. Finally, it is ground to powders smaller than 1 mm [1, 4, 5, 7]. Since there is no standard procedure in the production method of tarhana, its nutritional properties depend heavily on the ingredients and the amount used in the recipe [4, 8].

Methods for production of tarhana may vary from one place to another, but cereals and yoghurt are always the major component [2, 3, 9, 10]. Production process of traditional tarhana is shown in Figure 1.

Tarhana is mainly used in the form of a thick and creamy soup reconstituting with water followed by simmering and is consumed at lunch or dinner especially on cold days in Turkey [9, 11]. It is also locally consumed as a snack after it has been dried as thin layer or nugget, not to be ground [8]. There are four different types of tarhana, stated by Turkish Standardization Institute: flour tarhana, göce (cracked wheat) tarhana, semolina tarhana, and mixed tarhana. The difference between them is the usage of the wheat flour, cracked wheat, and semolina separately or as combinations in the recipe [5].

Tarhana-like products are known under different names in the other countries: kishk (sour milk-wheat mixture with boiled chicken stock) in Egypt, Syria, Lebanon, and Jordan, kushuk (milk-sour dough mixture with turnips) in Iraq, and tahonya/talkuna (fermented cereal mixture with vegetables) in Hungary and Finland [1, 2, 5, 10, 12].

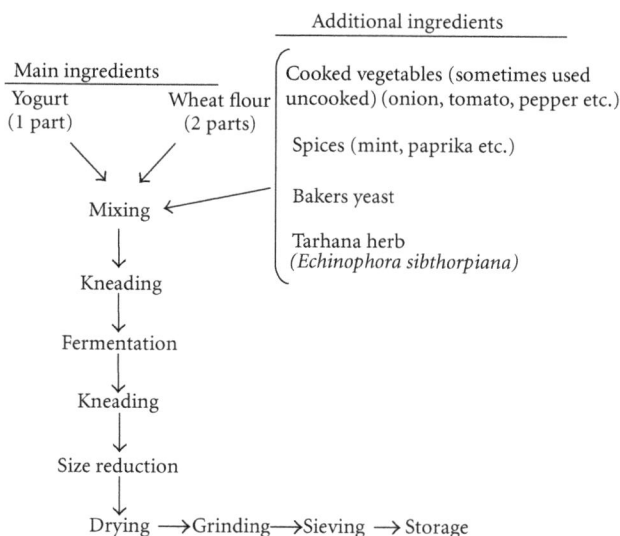

FIGURE 1: Flow diagram for traditional tarhana production [12].

TABLE 1: Number of samples collected according to month.

Season	Month	Year	Number of sample
Autumn	September-October-November	2010	42
Winter	December-January-February	2010-2011	56
Spring	March-April-May	2011	40

2. Materials and Methods

2.1. Samples. During the period September–May 2011, a total of 138 tarhana powder samples were collected randomly from bazaars located in different regions of Istanbul (Figure 2). Samples were transported under cold conditions from their place of collection to the laboratory. The number of samples gathered according to month bought was given in Table 1.

2.2. Aflatoxin Analysis

2.2.1. Sample Preparation. Sample preparation procedures were performed according to the instructions of the test kit (Rida Aflatoxin Column Art no.: R5001/5002, R-Biopharm, Darmstadt, Germany) manual [22]. 25 mL of methanol (70%) was added to 5 g of tarhana. Afterwards, the solution was extracted by mixing gently for 10 minutes at room temperature. The extract was filtered through a paper filter and 15 mL of distilled water were added to 5 mL of filtered solution. 0.25 mL Tween 20 were added and stirred for 2 minutes, followed by entire amount of the sample solution (20 mL) passing over the column. Clean up procedure was performed according to the kit's manual. Toxin containing eluate was diluted $1:10$ with the sample dilution buffer (supplied with the test kit) and used $50\,\mu L$ per well in the assay.

2.2.2. Test Procedure of Total Aflatoxins. According to Ridascreen Aflatoxin Total (Art no.: 4701) test kit manual [23], $50\,\mu L$ of the standard solutions or prepared sample in duplicate were added to the wells of microtiter plate. Then $50\,\mu L$ of the diluted enzyme conjugate and $50\,\mu L$ of the diluted antibody solution were added to each well. The solution was mixed gently, and incubated for 30 min at room temperature (20–25°C) in the dark. The unbound conjugate was removed during washing for three times (ELISA Washer ELX 50, Bio-tek Inst.). Afterwards, $100\,\mu L$ of substrate/chromogen solution was added to each well, mixed gently, and incubated for 30 min at room temperature (20–25°C) in the dark. Then, $100\,\mu L$ of the stop solution ($1\,M$ H_2SO_4) was added to each well and the absorbance was measured at 450 nm in ELISA plate reader (ELX 800, Bio-tek Inst.). The mean lower detection limit is $0.25\,\mu g/kg$.

2.2.3. Test Procedure of AFB1. According to Ridascreen Aflatoxin B1 30/15 (Art no.: 1211) test kit manual [24], $50\,\mu L$ of the standard solutions or prepared sample in duplicate

Lactic acid bacteria and yeast are responsible for acid formation during fermentation in Tarhana [1]. The low pH (3.8–4.5) and low moisture content (about 10%) of tarhana provide a bacteriostatic effect against pathogenic and spoilage microorganisms [1, 2, 12]. However, some certain mould species such as *Aspergillus, Penicillium,* and *Fusarium* can grow even at low moisture and pH values and produce mycotoxins in several food commodities [13, 14].

Among all mycotoxins, aflatoxins are a group of highly toxic secondary metabolic products named as aflatoxin B1(AFB1), aflatoxin B2 (AFB2), aflatoxin G1 (AFG1), and aflatoxin G2 (AFG2)[15, 16]. Aflatoxins are carcinogenic, mutagenic, teratogenic, and immunosuppressive to most animal species and humans [17]. AFB1 has the highest potency as a toxin and is classified as group I carcinogen by International Agency for Research on Cancer (IARC)[18]. The order of toxicity, AFB1 > AFG1 > AFB2 > AFG2, indicates that the terminal furan moiety of AFB1 is the critical point for determining the degree of biological activity of this group of mycotoxins [19]. Aflatoxins easily occur in feeds and foods during growth, harvest, or storage [20].

Due to their frequent occurrence and toxicity, guidelines and tolerance levels of aflatoxins have been set in several countries including Turkey. According to the Turkish Food Codex, the maximum residue limits for AFB1 and total aflatoxin in risky foods is 5 and $10\,\mu g/kg$, respectively [21].

Although several studies are available for aflatoxin levels in different food types which are consumed in Turkey, there is very little information on the presence of aflatoxins in tarhana. On the other hand, limited studies were conducted on mould contamination of tarhana. Therefore, this study was planned to determine aflatoxin levels and mould contamination in tarhana powder which are consumed to a great extend at Turkish kitchen and to compare the obtained results with maximum aflatoxin tolerance limits accepted by the Turkish Food Codex.

FIGURE 2: Map of samples collected in Istanbul, Turkey.

was added to the wells of microtiter plate. Then $50\,\mu L$ of the enzyme conjugate and $50\,\mu L$ of the anti-aflatoxin antibody solution were added to each well, mixed gently and incubated for 30 min at room temperature (20–25°C). The washing procedure was applied for three times (ELISA Washer ELX 50, Bio-tek Inst.). After the washing step, $100\,\mu L$ of substrate/chromogen solution were added to each well and mixed gently and incubated for 30 min at room temperature (20–25°C) in the dark. Finally, $100\,\mu L$ of the stop solution (1 M H_2SO_4) were added to each well and the absorbance was measured at 450 nm in ELISA plate reader (ELX 800, Bio-tek Inst.). The mean lower detection limit is $1.0\,\mu g/kg$.

2.2.4. Determination of Moisture and Water Activity.
Moisture contents of tarhana samples were determined by drying a homogeneous mixture of the sample in an oven (Heraeus, Germany) at 105 ± 2°C until a constant weight was obtained according to AOAC procedures [25]. The water activity analysis was determined by means of water activity meter (Decagon, AquaLab Lite, USA).

2.2.5. Determination of pH.
The pH was determined after mixing a 10 g sample with 90 mL distilled water (1/10 sample/water) and the pH value measurements were carried out using a Hanna pH meter (Hanna HI-9321, Woonsocket, Rhode Island, USA), equipped with a FC220B electrode (Hanna HI-9321, Woonsocket, Rhode Island, USA), after calibration with standard buffers of pH 4.0 and 7.0 [26].

2.2.6. Mould Analysis.
Mould was defined on Dichloran Rose Bengal Chloramphenicol agar with Chloramphenicol Selective supplement (DRBC, Oxoid, CM0727, and SR0078). Spread plates were incubated at 25°C for 5 days [27].

3. Results and Discussion

The distribution and evaluation of mould counts, aflatoxin amounts, pH, moisture, and water activity values of analyzed tarhana samples are given in Tables 2, 3, and 4, respectively.

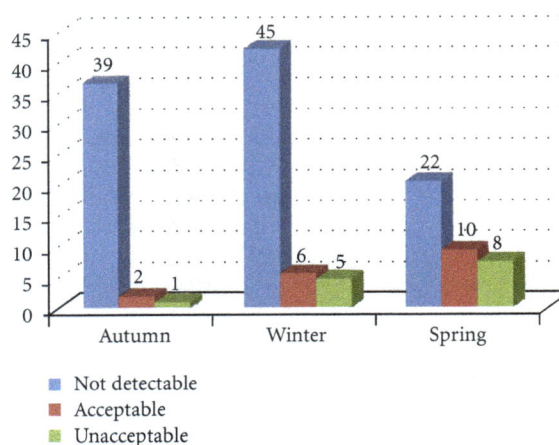

FIGURE 3: The number of acceptable and unacceptable samples according to Turkish Food Codex (TFC).

In this study 32 (3 in autumn, 11 in winter, 18 in spring) out of 138 tarhana samples (23.2%) were found to be contaminated with aflatoxins in the range of 0.7–$16.8\,\mu g/kg$, whereas 29 out of 138 (21.0%) tarhana samples contained AFB1 ranging from 0.2–$13.2\,\mu g/kg$ (Table 3). According to these results, 14 tarhana samples exceeded the maximum limits of AFB1 ($5\,\mu g/kg$) and total aflatoxin ($10\,\mu g/kg$) set in the Turkish Food Codex [21] (Figure 3).

Because the presence of aflatoxins in food is a hazard to human health, numerous studies have been conducted in different countries and also in Turkey to examine the presence and levels of aflatoxins in various food commodities. However, there is very little information on the presence of aflatoxins in tarhana, except for a report by Arici [28], who detected AFB1 in 4 out of 31 (12.9%) tarhana samples. Our results were higher than the results of the above-mentioned researcher.

The mould contamination rate on examined tarhana samples was fairly high. As can be seen from Table 2, all

TABLE 2: Distribution and evaluation of mould contamination in Tarhana samples ($n = 138$).

Range (cfu/g)	ND* (<10)	10 to <10^2	10^2 to <10^3	10^3 to <10^4	10^4 to <10^5	10^5 to <10^6	10^6 to <10^7	>10^7
Number	—	13	22	43	32	15	8	5
Percentage (%)	—	9.4	15.9	31.2	23.2	10.9	5.8	3.6

TABLE 3: Mould counts and Aflatoxin amounts of contaminated tarhana samples.

Season	Sample no.	Mould Count (cfu/g)	Total aflatoxins $\pm$ SE (μg/kg)	AFB1 $\pm$ SE (μg/kg)
Autmn	No: 4	3.8×10^5	4.1 ± 0.1	2.9 ± 0.3
	No: 7	1.3×10^6	10.5 ± 0.2	9.7 ± 0.6
	No: 11	4.1×10^5	1.2 ± 0.1	0.7 ± 0.4
Winter	No: 18	2.1×10^4	3.6 ± 0.7	ND
	No: 25	3.5×10^6	15.1 ± 0.8	8.8 ± 0.9
	No: 32	5.7×10^6	16.2 ± 0.9	10.4 ± 0.4
	No: 45	5.8×10^7	16.8 ± 0.5	13.2 ± 0.3
	No: 48	1.9×10^5	8.2 ± 0.6	4.4 ± 0.2
	No: 53	1.2×10^5	6.2 ± 0.9	3.4 ± 0.2
	No: 57	6.9×10^5	11.9 ± 0.3	10.2 ± 0.7
	No: 59	4.3×10^4	3.4 ± 0.1	ND
	No: 63	1.2×10^7	14.1 ± 0.2	10.4 ± 0.7
	No: 68	4.1×10^5	2.6 ± 1.1	2.1 ± 0.3
	No: 71	4.8×10^6	12.1 ± 0.3	7.5 ± 0.4
Spring	No: 74	5.2×10^4	7.6 ± 0.5	4.8 ± 0.8
	No: 76	3.5×10^4	3.1 ± 0.7	2.6 ± 0.4
	No: 80	2.2×10^6	11.6 ± 0.9	6.8 ± 0.7
	No: 82	2.8×10^5	1.7 ± 0.2	0.8 ± 0.3
	No: 87	6.1×10^3	0.9 ± 0.1	0.5 ± 0.1
	No: 95	1.1×10^7	10.8 ± 1.2	8.7 ± 0.4
	No: 99	4.5×10^3	2.7 ± 0.3	1.8 ± 0.2
	No: 101	1.7×10^6	14.3 ± 1.1	7.8 ± 0.9
	No: 107	5.8×10^5	11.1 ± 0.9	8.3 ± 0.6
	No: 108	2.4×10^7	12.5 ± 0.5	9.1 ± 0.3
	No: 111	1.8×10^4	0.7 ± 0.2	0.2 ± 0.1
	No: 113	6.2×10^6	12.3 ± 0.8	10.6 ± 0.4
	No: 115	6.4×10^5	5.2 ± 0.5	3.1 ± 0.2
	No: 119	3.2×10^7	15.1 ± 0.8	8.6 ± 0.5
	No: 120	1.8×10^3	2.9 ± 0.3	1.1 ± 0.2
	No: 125	7.2×10^5	4.5 ± 0.6	1.4 ± 0.5
	No: 127	3.2×10^4	6.2 ± 0.2	3.8 ± 0.7
	No: 134	3.1×10^3	0.8 ± 0.3	ND

ND: not detected, SE: standart error.

samples (100%) contaminated with moulds in the range of 1.4×10^1–5.8×10^7 cfu/g. The average count was detected as 4.6×10^3 cfu/g.

In Turkey, there are few studies on mould contamination in tarhana. In a study conducted by Soyyigit [29], the yeast-mould counts were detected as <10–3.3×10^7 cfu/g in 27 examined tarhana samples produced in Isparta city. Coskun [30] reported the mean yeast-mould counts in tarhana samples as 3.04×10^3, 3.52×10^3, and 3.37×10^1 cfu/g in Edirne, Kırklareli and Tekirdağ cities, respectively. Daglioglu et al. [31] found the mould/yeast contamination at a level of 1.5×10^3 cfu/g in traditional dried tarhana samples.

TABLE 4: pH, moisture, and water activity results of analyzed tarhana samples.

Parameters	Minimum	Maximum	Average
pH	3.25	4.50	3.82
Moisture (%)	10.35	17.85	12.71
(Dry matter %)	(89.65)	(82.15)	(87.29)
Water activity (a_w)	0.658	0.895	0.695

As expected, samples with high mould contamination contained high aflatoxin levels (Table 3). Source of mould and aflatoxin contamination in tarhana may result from wheat flour and spices used in the production. Wheat flour is an ingredient used in many foods in European and American culture and also is main ingredient of tarhana. Flour is generally regarded as a microbiologically safe product as it is a low water activity commodity [32]. However, toxigenic moulds may contaminate and grow in flour at different phases of production and processing, mainly in appropriate humidity and temperature conditions. Hence, there are several researches on mould contamination and aflatoxin levels in wheat and wheat flour [32–34]. Spices are exposed to a wide range of microbial contamination due to poor collection conditions, unpretentious production process, and extended drying times. In addition, spices can be contaminated through dust, waste water, and animal/human excreta in unpackaged spices which are sold in markets and bazaars. Several studies have demonstrated that spices are contaminated with various microorganisms including toxigenic moulds (especially *Aspergillus* spp.) and aflatoxins [35]. Therefore, spices pose health problems because they are often added to foods without further processing or are eaten raw. Therefore, to protect the consumer's health, it should not be used mould and aflatoxin-contaminated flour and spices in tarhana production.

As can be seen from Table 4 the average pH, moisture, and a_w results were 3.82, 12.71%, and 0.695, respectively. Similar results were also reported by other researchers [3, 9, 36, 37]. It is obvious that moisture has a great importance for the safe storage of food regarding microorganisms, particularly certain species of moulds. In addition to this, poor hygienic production conditions and absence of standard production method of tarhana may enhance of aflatoxin production by moulds.

In conclusion, aflatoxin producing mould species contaminate numerous food commodities, in warm climates where they may produce aflatoxins at different points of the food chain, such as preharvest, processing, transportation, or storage. The results of this study demonstrated that in spite of the low moisture and pH levels, moulds may grow and synthesized aflatoxins in tarhana. In order to prevent the health risk, a number of methods (storing in proper moisture and temperature, standardization of production method, improving the production conditions, microwave treatments, packing, etc.) can be applied to reduce/eliminate moulds from tarhana.

References

[1] N. Bilgiçli and S. Ibanoğlu, "Effect of wheat germ and wheat bran on the fermentation activity, phytic acid content and colour of tarhana, a wheat flour-yoghurt mixture," *Journal of Food Engineering*, vol. 78, no. 2, pp. 681–686, 2007.

[2] H. Erkan, S. Çelik, B. Bilgi, and H. Köksel, "A new approach for the utilization of barley in food products: barley tarhana," *Food Chemistry*, vol. 97, no. 1, pp. 12–18, 2006.

[3] S. Ibanoglu and E. Ibanoglu, "Rheological properties of cooked tarhana, a cereal-based soup," *Food Research International*, vol. 32, no. 1, pp. 29–33, 1999.

[4] M. Erbaş, M. Certel, and M. Kemal Uslu, "Microbiological and chemical properties of Tarhana during fermentation and storage as wet - Sensorial properties of Tarhana soup," *LWT*, vol. 38, no. 4, pp. 409–416, 2005.

[5] L. Settanni, H. Tanguler, G. Moschetti, S. Reale, V. Gargano, and H. Erten, "Evolution of fermenting microbiota in tarhana produced under controlled technological conditions," *Food Microbiology*, vol. 28, no. 7, pp. 1367–1373, 2011.

[6] N. Bilgiçli, "Effect of buckwheat flour on chemical and functional properties of tarhana," *LWT*, vol. 42, no. 2, pp. 514–518, 2009.

[7] S. Ibanoğlu and M. Maskan, "Effect of cooking on the drying behaviour of tarhana dough, a wheat flour-yoghurt mixture," *Journal of Food Engineering*, vol. 54, no. 2, pp. 119–123, 2002.

[8] M. Erbaş, M. Kemal Uslu, M. Ozgun Erbaş, and M. Certel, "Effects of fermentation and storage on the organic and fatty acid contents of tarhana, a Turkish fermented cereal food," *Journal of Food Composition and Analysis*, vol. 19, no. 4, pp. 294–301, 2006.

[9] N. Bilgiçli, A. Elgün, E. N. Herken, SelmanTürker, N. Ertaş, and S. Ibanoğlu, "Effect of wheat germ/bran addition on the chemical, nutritional and sensory quality of tarhana, a fermented wheat flour-yoghurt product," *Journal of Food Engineering*, vol. 77, no. 3, pp. 680–686, 2006.

[10] S. Ibanoğlu and P. Ainsworth, "Effect of canning on the starch gelatinization and protein in vitro digestibility of tarhana, a wheat flour-based mixture," *Journal of Food Engineering*, vol. 64, no. 2, pp. 243–247, 2004.

[11] O. Bozkurt and O. Gürbüz, "Comparison of lactic acid contents between dried and frozen tarhana," *Food Chemistry*, vol. 108, no. 1, pp. 198–204, 2008.

[12] I. Y. Sengun, D. S. Nielsen, M. Karapinar, and M. Jakobsen, "Identification of lactic acid bacteria isolated from Tarhana, a traditional Turkish fermented food," *International Journal of Food Microbiology*, vol. 135, no. 2, pp. 105–111, 2009.

[13] A. Zinedine, C. Brera, S. Elakhdari et al., "Natural occurrence of mycotoxins in cereals and spices commercialized in Morocco," *Food Control*, vol. 17, no. 11, pp. 868–874, 2006.

[14] H. Akiyama, Y. Goda, T. Tanaka, and M. Toyoda, "Determination of aflatoxins B1, B2, G1 and G2 in spices using a multifunctional column clean-up," *Journal of Chromatography A*, vol. 932, no. 1-2, pp. 153–157, 2001.

[15] A. Tarín, M. G. Rosell, and X. Guardino, "Use of high-performance liquid chromatography to assess airborne mycotoxins: aflatoxins and ochratoxin A," *Journal of Chromatography A*, vol. 1047, no. 2, pp. 235–240, 2004.

[16] A. Kamkar, "A study on the occurrence of aflatoxin M1 in Iranian Feta cheese," *Food Control*, vol. 17, no. 10, pp. 768–775, 2006.

[17] M. Sharma and C. Márquez, "Determination of aflatoxins in domestic pet foods (dog and cat) using immunoaffinity

column and HPLC," *Animal Feed Science and Technology*, vol. 93, no. 1-2, pp. 109–114, 2001.

[18] IARC Working Group on the Evaluation of Carcinogenic Risks to Humans, "Monographs on the evaluation of carcinogenic risk to humans: some naturally occuring substances," in *Food Items and Constituents Heterocyclic Aromatic Amines and Mycotoxins*, vol. 56, pp. 245–395, IARC, Şyon, France, 1993.

[19] E. Chiavaro, C. Dall'Asta, G. Galaverna et al., "New reversed-phase liquid chromatographic method to detect aflatoxins in food and feed with cyclodextrins as fluorescence enhancers added to the eluent," *Journal of Chromatography A*, vol. 937, no. 1-2, pp. 31–40, 2001.

[20] G. Piva, F. Galvano, A. Pietri, and A. Piva, "Detoxification methods of aflatoxins. A review," *Nutrition Research*, vol. 15, no. 5, pp. 767–776, 1995.

[21] "Turkish Food Codex. Legislation about Determination of Maximum Levels of Certain Contaminants in Foods," Basbakanlik Basimevi. Ankara, Turkey, 2008.

[22] "Immunoaffinity column for sample clean up prior to analysis of aflatoxins," R5001/R5002. R-Biopharm GmbH, Darmstadt, Germany, 2005.

[23] "Enzyme immunoassay for the quantitative analysis of aflatoxins," article R4701, R-Biopharm GmbH, Darmstadt, Germany, 2002.

[24] "Enzyme immunoassay for the quantitative analysis of aflatoxin B$_1$," R1211. R-Biopharm GmbH, Darmstadt, Germany, 2004.

[25] AOAC, *Official Methods of Analysis*, Association of Official Analytical Chemists, Washington, DC, USA, 15th edition, 1990.

[26] AOAC, *Official Methods of Analysis, (Centennial Edition)*, Association of Official Analytical Chemists, Washington, DC, USA, 1984.

[27] W. F. Harrigan, *Laboratory Methods in Food Microbiology*, Academic Press, London, UK, 1998.

[28] M. Arici, "Microbiological studies and mycotoxin analysis of a fermented cereal product (Tarhana)," *Ernahrungs-Umschau*, vol. 47, p. 477, 2000.

[29] H. Soyyigit, "Isparta ve yöresinde üretilen ev yapımı tarhanaların mikrobiyolojik ve teknolojik özellikleri," Süleyman Demirel Üniversitesi Fen Bilimleri Enstitüsü Yüksek Lisans Tezi, Isparta, 2004.

[30] F. Coskun, "Trakya'nın değişik yörelerinde üretilen ev tarhanalarının kimyasal, mikrobiyolojik ve duyusal özellikleri üzerine bir araştırma," *Gıda Mühendisliği Dergisi*, vol. 12, pp. 48–52, 2002.

[31] O. Daglioglu, M. Arıcı, M. Konyalı, and T. Gumus, "Effects of tarhana fermentation and drying methods on the fate of *Escherichia coli* 0157:H7 and *Staphylococcus aureus*," *European Food Research and Technology*, vol. 215, no. 6, pp. 515–519, 2002.

[32] L. K. Berghofer, A. D. Hocking, D. Miskelly, and E. Jansson, "Microbiology of wheat and flour milling in Australia," *International Journal of Food Microbiology*, vol. 85, no. 1-2, pp. 137–149, 2003.

[33] N. Karagozlu and M. Karapinar, "Bazı tahıl ve ürünlerinde okratoksin—a ve fungal kontaminasyon," *Turkish Journal of Biology*, vol. 24, pp. 561–572, 2000.

[34] B. Giray, G. Girgin, A. B. Engin, S. Aydin, and G. Sahin, "Aflatoxin levels in wheat samples consumed in some regions of Turkey," *Food Control*, vol. 18, no. 1, pp. 23–29, 2007.

[35] H. Colak, E. B. Bingol, H. Hampikyan, and B. Nazli, "Determination of aflatoxin contamination in red-scaled, red and black pepper by ELISA and HPLC," *Journal of Food and Drug Analysis*, vol. 14, no. 3, pp. 292–296, 2006.

[36] N. Karagozlu, B. Ergonul, and C. Karagozlu, "Microbiological attributes of instant tarhana during fermentation and drying," *Bulgarian Journal of Agricultural Science*, vol. 14, no. 6, pp. 535–541, 2008.

[37] O. Sagdıc, H. Soyyigit, S. Ozcelik, and H. Gul, "Viability of *Escherichia coli* O157:H7 during the fermentation of tarhana produced with different spices," *Annals of Microbiology*, vol. 55, no. 2, pp. 97–100, 2005.

18

Fluorescence-Based Rapid Detection of Microbiological Contaminants in Water Samples

Hervé Meder, Anne Baumstummler, Renaud Chollet, Sophie Barrier, Monika Kukuczka, Frédéric Olivieri, Esther Welterlin, Vincent Beguin, and Sébastien Ribault

Merck Millipore, Lab Solutions, BioMonitoring, Research & Development, Applications Group, 39, Route industrielle de la Hardt, 67120 Molsheim, France

Correspondence should be addressed to Hervé Meder, herve.meder@merckgroup.com

Academic Editors: A. P. Hudson and A. L. MacHado

Microbiological contamination of process waters is a current issue for pharmaceutical industries. Traditional methods require several days to obtain results; therefore, rapid microbiological methods are widely requested to shorten time-to-result. Milliflex Quantum was developed for the rapid detection and enumeration of microorganisms in filterable samples. It combines membrane filtration to universal fluorescent staining of viable microorganisms. This new alternative method was validated using European and United States Pharmacopeia definitions, with sterile water and/or sterile water artificially contaminated with microorganisms. The Milliflex Quantum method was demonstrated to be reliable, robust, specific, accurate, and linear over the whole range of assays following these guidelines. The Milliflex Quantum system was challenged to detect natural contaminants in different types of pharmaceutical purified process waters. Milliflex Quantum was demonstrated to detect accurately contaminants 3- to 7-fold faster than traditional membrane filtration method. The staining procedure is nondestructive allowing downstream identification following a positive result. The Milliflex Quantum offers a fast, sensitive, and robust alternative to the compendial membrane filtration method.

1. Introduction

Microbiological contamination of process waters is a key issue for pharmaceutical, biotechnology, and food and beverage industries. Traditional methods are the reference for the control of microbiological quality of water as they are reliable, easy to use and allow microorganisms identification. Nevertheless, these methods are time-consuming and labor-intensive. Moreover, they depend on the ability of microorganisms to yield visible colonies after an incubation period of typically 3 days that can go up to 14 days (European and United States Pharmacopeia). This long time-to-result is a concern for industries as improvement in processes and products requires faster methods to control microbiological quality. Therefore, over the past 25 years, many technologies have been developed to reduce the time-to-result. These new alternative rapid methods have to be sensitive, accurate, and cost-effective. The most studied and used technologies are polymerase chain reaction (PCR), impedimetry, bioluminescence, enzyme-linked immunosorbent assay (ELISA), flow cytometry (FCM), and solid-phase cytometry (SPC) [1–3].

The Milliflex Quantum (Millipore, Molsheim, France) is a new system developed for the rapid detection and enumeration of microorganisms in filterable samples. It combines membrane filtration and fluorescent staining, which are two proven and widely used technologies. The detection is based on a universal enzymatic fluorescent staining of viable microorganisms. The staining procedure is nondestructive allowing downstream identification following a positive result.

This paper describes the validation of the performances of the Milliflex Quantum method. The study was done according to definitions for the validation of a quantitative estimation of viable microorganisms in a sample from European (chapter 5.1.6.) and United States (chapter <1223>) Pharmacopeia. Robustness, ruggedness, accuracy, linearity, range, limit of quantification, limit of detection, and specificity of the method were assessed with sterile water and/or sterile water artificially spiked with microorganisms. The Milliflex Quantum method was also challenged to detect natural contaminants in different types of pharmaceutical process waters. All results obtained with Milliflex Quantum method were compared to the traditional membrane filtration method.

2. Materials and Methods

2.1. Media. Prefilled Tryptic Soy Agar plates (TSA, Millipore), R2A plates (Millipore), and Sabouraud Dextrose Agar plates (SDA, Millipore) were used to promote growth of microorganisms.

2.2. Microorganisms Strains. The following American-Type Culture Collection (ATCC) strains were used to validate the Milliflex Quantum method: *Candida albicans* ATCC 10231, *Aspergillus brasiliensis* ATCC 16404, *Staphylococcus epidermidis* ATCC 12228, *Ralstonia pickettii* ATCC 27511, *Brevundimonas diminuta* ATCC 19146, *Staphylococcus aureus* ATCC 6538, *Pseudomonas aeruginosa* ATCC 9027, *Bacillus subtilis* ATCC 6633, and *Escherichia coli* ATCC 8739. An environmental isolate of *Caulobacter sp.* was also tested with Milliflex Quantum. The cultures were maintained at $-80°C$ in Tryptic Soy Broth (TSB; BioMérieux, Craponne, France) with 5% (v/v) glycerol (Sigma-Aldrich, St. Quentin Fallavier, France) in 0.5 mmol L^{-1} HEPES buffer (Sigma-Aldrich).

2.3. Milliflex Quantum Method. The procedure used is as described by Baumstummler et al. [4]. Briefly, samples were filtered through mixed cellulose ester membranes and membranes were placed onto media agar plates and incubated at temperatures recommended by Pharmacopeia or at optimal growth temperatures. After incubation, membranes were stained for 30 min and placed in the Milliflex Quantum reader for counting fluorescent microcolonies. Membranes were reincubated onto media agar plates for visual counting of colony-forming unit (CFU), viability assessment, and contaminants identification. The compendial method was performed in parallel.

Fluorescence counts and CFU counts obtained after reincubation were compared to the compendial method. The fluorescence recovery and viability recovery were calculated as follows:

$$\text{Fluorescence recovery (\%)}$$
$$= \frac{\text{Fluorescence count}}{\text{Compendial method count}} \times 100,$$

$$\text{Viability recovery (\%)}$$
$$= \frac{\text{CFU count after reincubation}}{\text{Compendial method count}} \times 100.$$

$$(1)$$

Acceptance criterion for these parameters was set to equal to or higher than 70% (European Pharmacopeia chapter 5.1.6. and United States Pharmacopeia chapter <1223>).

2.4. Milliflex Quantum Method Validation

2.4.1. Statistical Analysis. An Anderson-Darling test or a Goodness-Of-Fit Chi2-test was used to determine if data obtained with the Milliflex Quantum method and with the compendial method follow a normal distribution. When data are normally distributed (Anderson-Darling P value $\geq$ 0.1; Chi2-value ≤ 4.61 for a 5 classes distribution; Chi2-value ≤ 2.71 for a 4 classes distribution), a one-way analysis of variance (ANOVA), a Student's two samples t-test or a Chi2-test was performed to compare results obtained with both methods. All statistical analysis were carried out using the Minitab Statistical Software (version 14; Minitab Inc., State College, PA, USA) except the Goodness-Of-Fit Chi2-test which was performed with Microsoft Office Excel (version 2003; Microsoft, Redmond, WA, USA).

2.4.2. Negative Controls. Negative controls were carried out in parallel of microorganisms testing. One hundred mL of 0.9% NaCl water (B. Braun Medical, Boulogne Billancourt, France) was filtered, incubated, and analyzed in the same conditions as samples containing microorganisms.

2.4.3. Incubation Time Robustness. Spiked samples were filtered, and membranes were incubated during various times before being stained following the Milliflex Quantum protocol as described previously. Membranes were reincubated to assess viability. Incubation conditions used were those required for the microbiological examination of nonsterile products in chapters of European (2.6.12. and 2.6.13.) and United States (<61> and <62>) Pharmacopeia. Fluorescence and viability recoveries were determined in comparison with the compendial method. An ANOVA was used to assess the time range for stable enumeration with Milliflex Quantum.

2.4.4. Ruggedness. The effect of using different media lots, membrane lots, reagent lots, analysts, and instruments was assessed. *Candida albicans* and *Ralstonia pickettii*, spiked separately in sterile water, were detected using 2 different lots of either membranes or reagents. Moreover, ruggedness of media was evaluated on TSA with *Bacillus subtilis* and *Escherichia coli*, on SDA with *Candida albicans* and on R2A with *Ralstonia pickettii*. For each ruggedness test, two different test runs were performed. Each run was tested by a different analyst, with a different set of instruments. Fluorescence and viability recoveries were calculated, and ANOVA was performed to check if recoveries corresponding to the different tested conditions were not statistically different (P value ≥ 0.05).

2.4.5. Accuracy, Linearity, Range, and Limit of Quantification. For each challenged microorganism spiked in sterile water, tests were performed at the following targeted spike levels per sample: 0 CFU, 5 CFU, 25 CFU, 50 CFU, 75 CFU, and 100 CFU. Milliflex Quantum method and traditional method were performed at the same time. Fluorescence and viability recoveries were calculated and a Student's two samples t-test was performed to check if the Milliflex Quantum counts were not statistically different from the traditional Milliflex counts (P value ≥ 0.05).

Linearity, range, and limit of quantification were established from the data generated during the accuracy test. Acceptance criteria for linearity include an R^2-value greater than 0.95 (United States Pharmacopeia) and a linear regression slope between 0.8 and 1.2.

2.4.6. Limit of Detection. The microorganisms used for accuracy testing were adjusted separately to approximately 3–5 CFU per 100 mL until at least 50% of the samples showed growth in the compendial method. Twenty replicate samples were assessed with each microorganism with each method. As the aim of the test was to demonstrate that the Milliflex Quantum method enabled to detect 1 CFU, the method had to detect at least one time 1 CFU during the experiment. Furthermore, a Student's two samples t-test was performed to check if the Milliflex Quantum counts were not statistically different from the traditional Milliflex counts (P value ≥ 0.05). Finally, the equivalence between the Milliflex Quantum proportion of growth and the traditional method proportion was assessed with a Chi2-test (P value ≥ 0.05).

2.4.7. Specificity. The specificity of Milliflex Quantum method was established from the data generated during the robustness and accuracy tests, where a panel of microorganisms was tested.

2.5. Detection of Microorganisms in Pharmaceutical Process Waters. In-process nonsterile water samples were taken at different steps of the water treatment process in 5 pharmaceutical plants. These various types of purified waters were diluted in 100 mL of 0.9% NaCl water (B. Braun Medical) and filtered. Membranes were placed onto R2A plates (Millipore) and incubated at 32.5°C (European Pharmacopeia General Monographs: Water For Injections; Water Highly Purified; Water Purified). Several incubation times were tested to assess the minimal incubation time required for the fluorescence detection. After incubation, membranes were stained following the Milliflex Quantum protocol. Membranes were reincubated onto R2A plates (Millipore) for visual counting of CFU and contaminants identification. The compendial method was performed in parallel. Fluorescence and viability recoveries were determined in comparison with the compendial method.

3. Results

3.1. Milliflex Quantum Method Validation

3.1.1. Incubation Time Robustness. The robust incubation time range required for detection with Milliflex Quantum was evaluated on 10 microorganisms. Table 1 summarizes results obtained. Conforming detection of microorganisms was achieved after 22 h of incubation for *Candida albicans* and after 28 h for *Aspergillus brasiliensis*. The minimal incubation time to detect *Escherichia coli* and *Bacillus subtilis* was 8 h and 9 h, respectively. The environmental isolate of

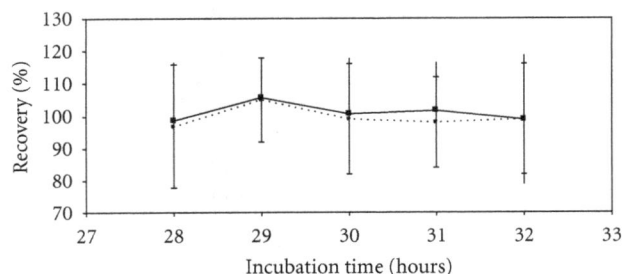

FIGURE 1: Determination of the robust incubation time range required to detect *Aspergillus brasiliensis* with the Milliflex Quantum method ($n = 10$). Fluorescence recovery (■, solid line); viability recovery (•, dashed line). Standard deviation is denoted by the vertical bars (delineated with—for viability standard deviation).

Caulobacter sp. was detected after 28 hours. The other 5 bacteria tested needed between 12 h and 22 h of incubation. The Milliflex Quantum method was demonstrated to be robust over several incubation times for each strain (fluorescence and viability recoveries $\geq 70\%$ and ANOVA P value ≥ 0.05, $n = 10$). These incubation time ranges were used during the further tests of the method validation. Recoveries results over the whole tested incubation time range obtained with *Aspergillus brasiliensis* are presented as example in Figure 1, which proves the results stability and method robustness over the 4-hour tested range.

3.1.2. Ruggedness. Using different media lots, membrane lots and reagent lots has no significant effect on Milliflex Quantum results as ANOVA results proved that fluorescence and viability recoveries were statistically equivalent in all tested conditions (data not shown). Reproducibility of tests results is guaranteed as well with different operators and equipment sets. Therefore, the method's performance is ensured in terms of ruggedness.

3.1.3. Accuracy, Linearity, Range, and Limit of Quantification. *Candida albicans* was accurately detected with the method at each tested contamination level (Table 2) as fluorescence recoveries ranged from 98% to 102% and viability recoveries from 97% to 102%. Moreover, no statistical differences between Milliflex Quantum and traditional method (Student's two samples t-test; $P \geq 0.05$) were found. Similar results were obtained with *Aspergillus brasiliensis*, *Escherichia coli*, and *Bacillus subtilis* (data not shown).

Linearity, range, and limit of quantification results obtained with microorganisms tested are showed in Table 3. A linear correlation between either fluorescence counts or viability counts and counts obtained with the compendial method were demonstrated since R^2-values varied from 0.95 to 0.98. The detection observed with Milliflex Quantum ranged from 0 CFU with each microorganism to 97 to 163 CFU, depending on the microorganism tested.

Limits of quantification of the Milliflex Quantum method were demonstrated to be between 4 and 10 CFU, depending on the microorganism tested. These levels are equal

TABLE 1: Determination of the robust incubation time range required to detect collection strains with the Milliflex Quantum method ($n = 10$).

Microorganism	Robust incubation time range	Fluorescence recovery ± SD (%)	ANOVA P value on fluorescence recoveries within robust incubation time range	Viability recovery ± SD (%)	ANOVA P value on viability recoveries within robust incubation time range
Candida albicans ATCC 10231	22 h–26 h	102 ± 38	0.66	101 ± 39	0.75
Aspergillus brasiliensis ATCC 16404	28 h–32 h	99 ± 18	0.71	97 ± 19	0.66
Staphylococcus epidermidis ATCC 12228	14 h–18 h	98 ± 18	0.20	101 ± 15	0.28
Ralstonia pickettii ATCC 27511	22 h–26 h	91 ± 26	0.33	92 ± 26	0.34
Brevundimonas diminuta ATCC 19146	22 h–26 h	112 ± 21	0.78	100 ± 18	0.13
Staphylococcus aureus ATCC 6538	12 h–16 h	128 ± 33	0.06	117 ± 27	0.33
Pseudomonas aeruginosa ATCC 9027	16 h–20 h	111 ± 28	0.91	103 ± 24	0.16
Bacillus subtilis ATCC 6633	9 h–10 h	111 ± 23	0.44	102 ± 21	0.40
Escherichia coli ATCC 8739	8 h–10 h	122 ± 29	0.06	111 ± 24	0.44
Caulobacter sp. (environmental strain)	28 h–32 h	90 ± 17	0.16	102 ± 17	0.10

ANOVA, One-way analysis of variance; ATCC, American Type Culture Collection; SD, Standard deviation.

TABLE 2: Accuracy results obtained with *Candida albicans*: comparison between fluorescence and viability results obtained with the Milliflex Quantum method and compendial method results ($n = 10$).

Microorganism	Incubation time	Target concentration (CFU)	Fluorescence count mean ± SD (CFU)	Fluorescence recovery ± SD (%)	Student's two samples t-test P value on fluorescence counts	Viability count mean ± SD (CFU)	Viability recovery ± SD (%)	Student's two samples t-test P value on viability counts
Candida albicans ATCC 10231	22 h	0	0.0 ± 0.0	NA	NA	0.0 ± 0.0	NA	NA
		5	6.2 ± 2.9	102 ± 61	0.93	6.2 ± 2.9	102 ± 60	0.93
		25	24.9 ± 5.4	101 ± 29	0.93	24.2 ± 5.6	98 ± 30	0.83
		50	62.7 ± 10.4	98 ± 22	0.73	62.2 ± 10.1	97 ± 21	0.64
		75	102.0 ± 6.3	101 ± 10	0.68	101.4 ± 5.5	101 ± 9	0.82
		100	163.9 ± 15.3	102 ± 15	0.65	161.7 ± 14.7	101 ± 14	0.87

ATCC, American Type Culture Collection; CFU, Colony-forming unit; NA, Not applicable; SD, Standard deviation.

TABLE 3: Linearity, range and limit of quantification results obtained with the Milliflex Quantum method: analysis done using data generated for the accuracy testing of the Milliflex Quantum method, comparison between fluorescence and viability results ($n = 10$).

| Microorganism | Linearity | | | | Range (CFU) | | Limit of Quantification (CFU) | | |
	Fluorescence R^2-value	Fluorescence linear regression slope	Viability R^2-value	Viability linear regression slope	Fluorescence	Viability	Fluorescence lowest count mean	Viability lowest count mean	Lowest mean count with compendial method
Candida albican ATCC 10231	0.98	1.02	0.98	1.01	0·to 163	0 to 161	7	7	7
Aspergillus brasiliensis ATCC 16404	0.95	1.09	0.96	1.10	0 to 124	0 to 120	10	10	9
Bacillus subtilis ATCC 6633	0.96	1.16	0.96	0.97	0 to 121	0 to 97	6	6	6
Escherichia coli ATCC 8739	0.95	1.06	0.96	1.02	0 to 120	0 to 114	4	4	5

ATCC, American Type Culture Collection; CFU, Colony-forming unit.

to or lower than the corresponding limit of quantification of the compendial method except with *Aspergillus brasiliensis*. With this strain, the limit of quantification of Milliflex Quantum method was found to be 10 CFU, while, 9 CFU was the level demonstrated with compendial method. This latter result is explained by the *Aspergillus* morphology. Due to filaments and colonies merging, the colonies are more difficult to count with the traditional method. The Milliflex Quantum allows earlier counting, with the microcolonies being smaller and separated.

3.1.4. Limit of Detection. The method was proved to detect 1 CFU of *Candida albicans*, *Aspergillus brasiliensis*, *Bacillus subtilis*, and *Escherichia coli* (data not shown). The Milliflex Quantum method gave equivalent result to the traditional method (Student's two samples t-test, $P \geq 0.05$). The Chi2-test demonstrated as well the equivalence of results since proportion of growth is similar for the 2 methods (Chi2-test, $P \geq 0.05$).

3.1.5. Specificity. The specificity of the Milliflex Quantum method was confirmed as all challenged microorganisms were successfully detected during robustness and accuracy tests.

3.2. Detection of Microorganisms in Pharmaceutical Process Waters. Different purified waters were sampled in 5 different pharmaceutical plants, at various stages of the pipes. Table 4 summarizes the results obtained at incubation times allowing conforming results with Milliflex Quantum and allowing a stable count with the traditional method.

The contaminants of tested pharmaceutical waters were detected with Milliflex Quantum within a time range of 24 to 40 hours. In comparison, 5 to 7 days were needed to visually count all colonies with naked eyes. The Milliflex Quantum detected the contaminants in waters from 3- to 7-fold faster than the compendial method.

After staining and reincubation, the viability rate conformed to acceptance criteria, proving the nondestructiveness of the Milliflex Quantum method. Identifications using the MicroSEQ platform (Applied Biosystems, Carlsbad, CA, USA) were performed on some colonies stained and reincubated. Common water contaminants as *Rhodococcus sp.* and *Delftia acidovorans* and very slow-grower strains as *Aquabacterium parvum* and *Pelomonas saccharophila* were identified, proving that the Milliflex Quantum method is fully compatible with identification technology.

3.3. Specific Case Study: Detection of Slow Grower Microorganisms in Pharmaceutical Process Water. In-process nonsterile pharmaceutical water samples were collected after double reverse osmosis, UV and ozone exposure. Microorganisms being highly stressed in these conditions, an incubation time range from 1 to 7 days was applied before detection with Milliflex Quantum and parallel compendial counting were performed until 21 days.

Figure 2 compares counts obtained with both methods. With the compendial method, all counts increased up to

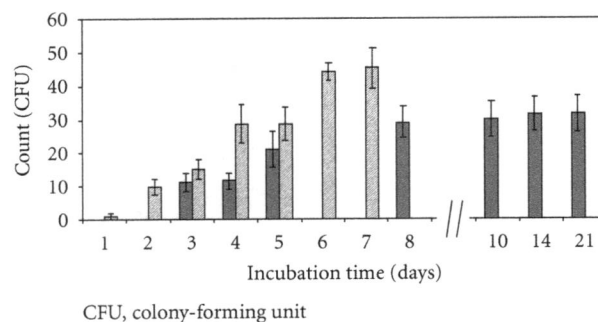

CFU, colony-forming unit

FIGURE 2: Detection of contaminants in pharmaceutical process water: comparison between fluorescence counts obtained with the Milliflex Quantum method (dashed) and counts obtained with the compendial method (solid grey) ($n = 5$). Standard deviation is denoted by the vertical bars.

14 days of incubation and remained stable between 14 and 21 days. The compendial count at 14 days was chosen as the reference for recoveries calculations. The fluorescence recoveries conformed to acceptance criteria from 4 days of incubation, compared to the compendial count obtained after a 14-day incubation (Figures 3(a) and 3(b)). Viability recoveries were calculated comparing CFU count after a 14-day incubation on membrane stained with Milliflex Quantum and CFU compendial count at 14 days (Figure 3(a)). Conforming viability detection was achieved on membranes stained after 4 days of incubation. The Milliflex Quantum allowed reducing the time-to-results 3.5-fold by detecting accurately contaminants after only 4 days. Identifications by sequencing were performed either on reincubated membranes after staining or on compendial membranes. Fourteen genii were identified: for example, *Variovorax paradoxus*, *Afipia broomeae*, and *Bradyrhizobium japonicum*. The latter was the slowest microorganism present in the samples as almost all colonies becoming visible after the 5th day of incubation or reincubation had the macroscopic aspect of *Bradyrhizobium japonicum*.

4. Discussion

We report the validation of the Milliflex Quantum method and the evaluation of its performances for the rapid detection of total microbial contaminants in pharmaceutical water samples. This new rapid method was compared to the traditional membrane filtration method for all tests carried out. The Milliflex Quantum method was demonstrated to be reliable, robust, specific, accurate, and linear over the whole range of assays following these guidelines.

Milliflex Quantum was demonstrated to detect accurately contaminants in 7 pharmaceutical process waters after 24 to 40 hours of incubation, in comparison with 5 to 7 days with the traditional method. An additional sample of highly purified water containing very high stressed microorganisms required 14 days to detect contaminants with the culture-based procedure, whereas Milliflex Quantum enabled to shorten this time-to-result to 4 days. Therefore, Milliflex

TABLE 4: Detection of microorganisms in pharmaceutical waters: comparison between fluorescence and viability results obtained with Milliflex Quantum and compendial method results ($n = 10$). "a" and "b" are 2 different types of waters, sampled in the same pharmaceutical plant.

Water	Tested volume	Incubation time	Fluorescence count mean ± SD (CFU)	Fluorescence recovery ± SD (%)	Viability count mean ± SD (CFU)	Viability recovery ± SD (%)	Incubation time	Count mean ± SD (CFU)
1	1 mL	40 h	248.3 ± 31.0	113 ± 19	196.3 ± 21.5	89 ± 14	5 days	248.3 ± 31.0
2	10 mL	24 h	105.4 ± 8.0	110 ± 11	98.8 ± 8.8	103 ± 11	5 days	95.8 ± 6.0
3a	10^{-2} mL	30 h	21.6 ± 5.0	72 ± 19	22.8 ± 6.4	76 ± 24	7 days	29.9 ± 3.9
3b	10^{-2} mL	24 h	30.6 ± 2.7	77 ± 15	30.6 ± 1.7	77 ± 14	7 days	39.9 ± 7.1
4a	10^{-1} mL	24 h	179.7 ± 5.5	95 ± 7	178.3 ± 11.5	95 ± 9	5 days	188.7 ± 12.9
4b	50 mL	30 h	24.0 ± 3.0	74 ± 19	27.7 ± 4.7	86 ± 24	5 days	32.3 ± 7.0
5	1 mL	26 h	21.8 ± 8.5	109 ± 55	17.0 ± 5.1	85 ± 37	5 days	20.0 ± 6.4

CFU, Colony-forming unit; SD, Standard deviation.

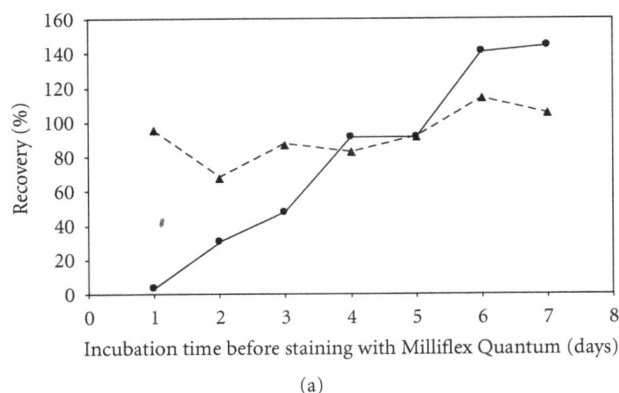

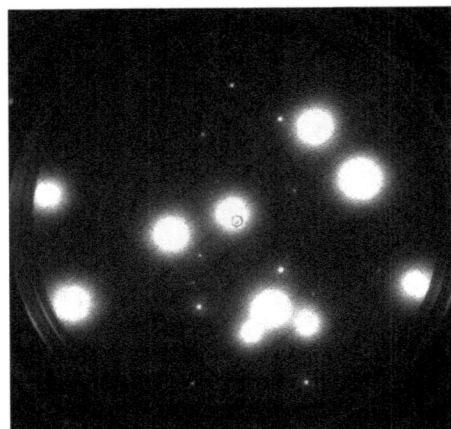

FIGURE 3: Detection of contaminants in pharmaceutical process water. (a) Fluorescence recovery (•, solid line); viability recovery (▲, dashed line). Viability recoveries were calculated comparing CFU count after a 14-day incubation with both methods. Each viability recovery is placed on the plot with the incubation time before staining with Milliflex Quantum as abscissa value. (b) Example of picture taken after staining of membrane incubated for 4 days on R2A plate at 32.5°C.

Quantum allows accurate enumeration of contaminants with time-to-results that are 3- to 7-fold shorter than traditional method. Gram-negative and Gram-positive bacteria, as well as yeasts and molds are universally detected with high sensitivity by the alternative method. Some additional tests proved the time-to-results can be shortened if microorganisms are incubated in optimal growing conditions, following microbiological literature instead of Pharmacopeia guidelines. For instance, *Pseudomonas aeruginosa* and *Staphylococcus aureus* were accurately detected after 9 hours on TSA at 37°C, instead of, respectively, 16 hours and 12 hours at the 32.5°C temperature recommended by the Pharmacopeia, and *Aspergillus brasiliensis* needed 17 hours at 37°C versus 28 hours at 22.5°C (data not shown). Validation of the Milliflex Quantum method, in optimal conditions, as an alternative method for control of microbiological quality is consistent and would enhance time-to-results performances of the Milliflex Quantum system.

The Milliflex Quantum method utilizes mixed cellulose ester membrane filtration for sample preparation, which ensures consistent and reliable results. Large sample volume up to several hundred milliliters can be processed to monitor microbiological quality of very low-contaminated samples. PCR allows testing only small volumes (generally 0.1 to 1 mL) after complex sample preparation methods and FCM available systems are limited to the treatment of a 1 mL volume of sample.

As fluorescent detection is based on metabolism activity of cells, needing integrity of membranes, Milliflex Quantum enables enumeration of all viable-culturable cells, after a short incubation step. Molecular-based methods are unable to distinguish between live and dead cells, which is a shortcoming of methods like PCR [5, 6]. On the other hand, FCM detection needs to be combined with differential staining techniques to enumerate specifically viable cells.

Viable-culturable cells final result can only be calculated by comparison with the number of plate count, thus requiring the growth and visualization of colonies on traditional agar plate after one or several days incubation [7].

The method was proved to detect 1 CFU while reaching this sensitivity level with molecular-based method requires development of complex and often laborious sample preparation methods [8]. The SPC's claimed sensitivity is 1 CFU with immediate detection but discrimination is often difficult to distinguish between microorganisms and dust and raw results can only be analyzed by high skilled operator after a long specific training. On the other hand, FCM is not adapted to the detection of rare event and enables direct counting of contamination levels down to 100–200 cells per mL [9, 10]. Moreover, this technology remains complex to be implemented routinely as data analysis is sophisticated and needs experienced operators [1, 11]. Therefore, FCM is perfectly suitable to monitoring of drinking water and wastewater treatments, whereas followup of microbiological quality of pharmaceutical or biopharmaceutical waters remains very limited as they may contain very low numbers of viable microorganisms [12].

The staining procedure was demonstrated to be nondestructive since colonies grow after staining and reincubation and can be collected for identification. Sequencing allowed identifying natural contaminants of tested pharmaceutical waters. Moreover, all standard identifications methods (Gram-staining, biochemical tests, PCR, etc.) were applied successfully on other samples after filtration, staining, and reincubation to prove their compatibility with the Quantum procedure (data not shown). The nondestructiveness of the method is an advantage over molecular-based and adenosine triphosphate-based bioluminescence methods, which do not enable to recover and identify contaminants in case of a positive result.

Implementation and use of new rapid method by industries are limited, among others, by regulatory acceptance [1, 10]. Milliflex Quantum method is a membrane-filtration and growth-based procedure, very close to the principle of the compendial method. Sample preparation and incubation conditions remain identical to traditional microbiology. These features have the advantage to facilitate validation and reassure regulation agencies for their acceptance of rapid method as routine procedure for water monitoring, replacing longer traditional methods.

Milliflex Quantum was demonstrated to be compatible and efficient in beverages and raw materials matrices (data not shown). The method has been also evaluated for its applicability in the detection of low concentrations of microorganisms in animal cell cultures and cell culture media. Combined to a fast pretreatment method to selectively lyse mammalian cells, the time-to-results were 2–5 times shorter than traditional method [13]. This feature is of great interest for monitoring biotechnology processes including high contaminated cells bioreactors.

We demonstrate in this paper that Milliflex Quantum is a validated and reliable tool enabling the rapid detection and enumeration of microbial contamination in pharmaceutical process waters. This easy-to-use protocol and simple system is totally suitable to be implemented in routine for the monitoring of purified water. The Milliflex Quantum system can be applied to a wide range of filterable products as it was proved to be effective as well with beverages, raw materials, and mammalian cells contaminated matrices.

References

[1] R. T. Noble and S. B. Weisberg, "A review of technologies for rapid detection of bacteria in recreational waters," *Journal of Water and Health*, vol. 3, no. 4, pp. 381–392, 2005.

[2] R. Chollet, M. Kukuczka, N. Halter et al., "Rapid detection and enumeration of contaminants by ATP bioluminescence using the milliflex® rapid microbiology detection and enumeration system," *Journal of Rapid Methods and Automation in Microbiology*, vol. 16, no. 3, pp. 256–272, 2008.

[3] L. Yang and R. Bashir, "Electrical/electrochemical impedance for rapid detection of foodborne pathogenic bacteria," *Biotechnology Advances*, vol. 26, no. 2, pp. 135–150, 2008.

[4] A. Baumstummler, R. Chollet, H. Meder et al., "Development of a nondestructive fluorescence-based enzymatic staining of microcolonies for enumerating bacterial contamination in filterable products," *Journal of Applied Microbiology*, vol. 110, no. 1, pp. 69–79, 2011.

[5] J. T. Keer and L. Birch, "Molecular methods for the assessment of bacterial viability," *Journal of Microbiological Methods*, vol. 53, no. 2, pp. 175–183, 2003.

[6] A. Nocker and A. K. Camper, "Novel approaches toward preferential detection of viable cells using nucleic acid amplification techniques," *FEMS Microbiology Letters*, vol. 291, no. 2, pp. 137–142, 2009.

[7] M. M. T. Khan, B. H. Pyle, and A. K. Camper, "Specific and rapid enumeration of viable but nonculturable and viable-culturable gram-negative bacteria by using flow cytometry," *Applied and Environmental Microbiology*, vol. 76, no. 15, pp. 5088–5096, 2010.

[8] B. S. Byron, Y. Charles, J. Lee-Ann, and M. L. Tortorello, "Sample preparation: the forgotten beginning," *Journal of Food Protection*, vol. 72, no. 8, pp. 1774–1789, 2009.

[9] F. Hammes, M. Berney, Y. Wang, M. Vital, O. Köster, and T. Egli, "Flow-cytometric total bacterial cell counts as a descriptive microbiological parameter for drinking water treatment processes," *Water Research*, vol. 42, no. 1-2, pp. 269–277, 2008.

[10] J. Moldenhauer, "Overview of rapid microbiological methods," in *Principles of Bacterial Detection: Biosensors, Recognition Receptors and Microsystems*, M. Zourob, S. Elwary, and A. Turner, Eds., pp. 49–79, Springer, New York, NY, USA, 2008.

[11] F. Hammes and T. Egli, "Cytometric methods for measuring bacteria in water: advantages, pitfalls and applications," *Analytical and Bioanalytical Chemistry*, vol. 397, no. 3, pp. 1083–1095, 2010.

[12] D. T. Reynolds and C. R. Fricker, "Application of laser scanning for the rapid and automated detection of bacteria in water samples," *Journal of Applied Microbiology*, vol. 86, no. 5, pp. 785–795, 1999.

[13] A. Baumstummler, R. Chollet, H. Meder, C. Rofel, A. Venchiarutti, and S. Ribault, "Detection of microbial contaminants in mammalian cell cultures using a new fluorescence-based staining method," *Letters in Applied Microbiology*, vol. 51, no. 6, pp. 671–677, 2010.

19

Effects of Irradiation Dose and O_2 and CO_2 Concentrations in Packages on Foodborne Pathogenic Bacteria and Quality of Ready-to-Cook Seasoned Ground Beef Product (Meatball) during Refrigerated Storage

Gurbuz Gunes, Neriman Yilmaz, and Aylin Ozturk

Food Engineering Department, Istanbul Technical University, Maslak, 34469 Istanbul, Turkey

Correspondence should be addressed to Gurbuz Gunes, gunesg@itu.edu.tr

Academic Editors: R. AbuSabha, S.-J. Ahn, H. Chen, and P. Janssen

Combined effects of gamma irradiation and concentrations of O_2 (0, 5, 21%) and CO_2 (0, 50%) on survival of *Escherichia coli* O157:H7, *Salmonella enteritidis*, *Listeria monocytogenes*, lipid oxidation, and color changes in ready-to-cook seasoned ground beef (meatball) during refrigerated storage were investigated. Ground beef seasoned with mixed spices was packaged in varying O_2 and CO_2 levels and irradiated at 2 and 4 kGy. Irradiation (4 kGy) caused about 6 Log inactivation of the inoculated pathogens. Inactivation of *Salmonella* was 0.9- and 0.4-Log lower in 0 and 5% O_2, respectively, compared to 21% O_2. Irradiation at 2 and 4 kGy increased thiobarbituric acid reactive substances in meatballs by 0.12 and 0.28 mg malondialdehyde kg^{-1}, respectively, compared to control. In reduced-O_2 packages, radiation-induced oxidation was lower, and the initial color of an irradiated sample was maintained. Packaging with 0% + 50% CO_2 or 5% O_2 + 50% CO_2 maintained the oxidative and the color quality of irradiated meatballs during 14-day refrigerated storage. MAP with 5%O_2 + 50% CO_2 combined with irradiation up to 4 kGy is suggested for refrigerated meatballs to reduce the foodborne pathogen risk and to maintain the quality.

1. Introduction

Consumer demand for minimally processed ready-to-cook and ready-to-eat food products has been increasing throughout the world. Ready-to-cook meat products such as ground beef seasoned with various spices and herbs in the form of meatballs or patties are very popular among consumers. However, potentially dangerous pathogens including Shiga toxin producing *Escherichia coli* O157:H7, *Salmonella,* and *Listeria monocytogenes* in these products can create a great risk of foodborne diseases [1]. In fact, there have been a number of foodborne-disease outbreaks associated with consumption of meat products contaminated with these pathogens in the last couple of decades [2–4].

Gamma irradiation is an effective cold decontamination process approved for beef in many countries including USA and Turkey. Effective inactivation of pathogens and shelf life extension of meat products by irradiation have been reported in a number of studies [5–9]. Irradiation below 3 kGy doses significantly decreased populations of *E. coli* O157:H7, *Salmonella*, and *L. monocytogenes* in meat and poultry [6, 10]. Irradiation increased 2-thiobarbituric acid-reactive substances (TBARSs) in different meat products under different conditions [11, 12]. Color is an important quality attribute that influences consumer acceptance of meat or meat products. Red color of beef products can be lost, turning into brown or gray upon irradiation [13]. Therefore, the irradiation dose applied to inactivate the target microorganisms is often limited by the induced degradations in the organoleptic qualities of the products.

Modified atmosphere packaging (MAP) involves enclosure of a product in a package atmosphere different from air [14]. Elevated CO_2 and reduced O_2 levels are commonly used to extend shelf-life of food products through inhibition of microbial growth and oxidative changes [14, 15]. Use of O_2-free atmospheres in packages has been suggested for different

Effects of Irradiation Dose and O_2 and CO_2 Concentrations in Packages on Foodborne Pathogenic Bacteria and Quality
of Ready-to-Cook Seasoned Ground Beef Product (Meatball) during Refrigerated Storage

143

meat products [15]. However, there is a risk of *C. botulinum* growth in anaerobic packages [10]. Thus, inclusion of a low level of O_2 in packages can be suggested. MAP containing 5% O_2 with elevated CO_2 did not adversely affect the quality of meatball compared to anaerobic MAP and was suggested for meatball [16].

The combination of preservation technologies can have synergistic effects on microbial inactivation and may allow the use of milder treatments to control their undesirable effects on the quality of food products. Combination of MAP and gamma irradiation can be a useful approach to inactivate pathogenic microorganisms and maintain the quality of ready-to-cook meatball. Benefits of the combined effects of MAP and irradiation have been shown in different products including turkey meat, minced pork, and ground beef [8, 17, 18]. The atmospheric composition around the product during irradiation can affect radiation-induced changes in the product, and thus it needs to be investigated [19]. Therefore, the effects of reduced O_2 levels on quality of irradiated meatball compared to O_2-free (anaerobic) and ambient atmosphere (air) would be useful for determination of optimum packaging atmosphere for the product.

The objective of this study was to investigate the combined effects of O_2 and CO_2 concentrations in packages and gamma irradiation on the survival of *E. coli* O157:H7, *S. enteritidis*, and *L. monocytogenes*, and lipid oxidation and color changes in ready-to-cook seasoned ground beef (meatball) during refrigerated storage.

2. Material and Methods

2.1. Materials. Beef, salt, black pepper, cumin, red pepper, onion powder, garlic powder, and bread powder were purchased from a local supermarket. MacConkey agar (SMAC) with 5-bromo-4-chloro-3-indoyl-b-d-glucuronide (SMAC-BCIG), Cefixime Tellurite (CT), peptone, and Dryspot *E. coli* O157:H7 latex agglutination test were purchased from Oxoid (Hampshire, UK). Xylose lysine desoxycholate (XLD) agar, Oxford Listeria selective agar, Listeria selective supplement, tryptic soy agar (TSA), and thiobarbituric acid (TBA) were purchased from Merck (Darmstadt, Germany); Butylated hydroxytoluene (BHT), 1,1,3,3-Tetraethoxypropane (TEP), and trichloroacetic acid (TCA) were purchased from Fluka (Buschs, Switzerland). Low-density polyethylene (LDPE) film (oxygen permeability: 3800 cc/m²·day·atm at 0°C 0% RH) and a multilayered polyethylene teraphylate/polyethylene-ethylene vinyl alcohol-polyethylene (PET/PE-EVOH-PE) film (oxygen permeability 1.2 cc/m².day.atm at 0°C 0% RH) were donated by Korozo Packaging Company GmbH (Istanbul, Turkey).

2.2. Preparation of Meatballs. The surface of beef was sterilized with ethanol and the residual ethanol was burned-off before the beef was ground through a plate with 4.7 mm pores using a meat grinder (Moulinex HV4, France) in our laboratory. Total fat content of the ground beef was measured as 14% on average. Meatballs were prepared with ground beef seasoned with the following dry ingredients: bread

powder (8%), onion powder (3%), red pepper (2%), cumin (2%), salt (2%), garlic powder (0.5%), and black pepper (0.1%). All the ingredients were mixed and kneaded by hand with sanitized gloves for about 30 minutes and left in a refrigerator at $3 \pm 1°C$ for one night. Then the seasoned ground beef was shaped into small pieces (18 ± 3 g) using a disc-shaped mould (3 cm diameter × 1 cm height) before packaging.

2.3. Preparation of Inoculums and Inoculation of Meatball Samples. *S. enteritidis* KUKEN 369, *L. monocytogenes* SLCC 2371, and *E. coli* O157:H7 (ATCC 700728) in TSA slants were suspended in peptone (0.1%) and spread on TSA plates and incubated at 35°C for 24 hours. A loop of the cultures was transferred to TSA again and incubated for another 24 hours. A loop of the cultures from the TSA plates was added to test tubes with 0.1% peptone water and vortexed to prepare separate stock cultures of each pathogen with approximately 10^8 CFU mL^{-1} concentration using McFarland standard turbidity tubes. A 0.2 mL from each stock culture was added to the center of the pieces of meatballs to get a final population of 10^6 CFU g^{-1} for each pathogen.

2.4. Packaging and Irradiation of Samples. Meatballs were packaged in a low barrier LDPE film for aerobic packaging (21% O_2 + 0% CO_2) and a high barrier PET/PE-EVOH-PE film for MAP. The packaging treatments were the following combinations of O_2 and CO_2: 0% O_2 + 0% CO_2, 0% O_2 + 50% CO_2, 5% O_2 + 0% CO_2, 5% O_2 + 50% CO_2, 21% O_2 + 50% CO_2, and 21% O_2 + 0% CO_2. The proportion of the gas components were completed to 100% with N_2 when needed. The gas mixtures were made by mixing O_2, CO_2, and N_2 in a gas mixer (PBI Dansensor Map Mix 9000, Denmark) connected to a packaging machine (Multivac C200, Germany). Packages were sealed after flushed with the gas mixtures using the packaging machine. Gas compositions in the gas mixtures and the packages were measured using a gas analyzer (PBI Dansensor CheckMate, Denmark).

The packaged samples were irradiated at a commercial food irradiation plant (Gamma-Pak GmbH., Cerkezkoy, Turkey) with a ^{60}Co source at an average sample temperature of $9 \pm 3°C$ in refrigerated boxes. The samples were exposed to gamma rays at a rate of 1.33 kGy h^{-1} for 1.5 and 3 hours amounting to a cumulative dose of 2 and 4 kGy, respectively. The dose rate was measured by Amber 3042 dosimeters (Harwell Dosimeters Ltd, Oxfordshire, UK). Unirradiated (0 kGy) control samples were kept outside the irradiation chamber during the treatments. The experiments were carried out in two replicates. All the samples were stored at $3 \pm 1°C$ for 14 days after the irradiation treatments. Gas compositions in packages, *E. coli* O157:H7, *S. enteritidis*, and *L. monocytogenes* counts, TBARS, and color were determined on day 1, day 7, and day 14 during storage.

2.5. Enumeration of the Pathogens. *E. coli* O157:H7, *S. enteritidis*, and *L. monocytogenes* were enumerated using Thin Agar Layer method [20]. Selective media used for these 3 organisms were SMAC-BCIG with CT supplement, XLD agar, and

TABLE 1: Changes in *S. enteritidis*, *L. monocytogenes*, and *E. coli* O157:H7 counts (Log CFU g^{-1}) in ready-to-cook meatballs during storage at 3°C as affected by irradiation dose (kGy) and O_2 levels in 0% CO_2 containing packages.

Storage (day)	O_2^* (%)	*S. enteritidis*			*L. monocytogenes*			*E. coli* O157:H7		
		0 kGy	2 kGy	4 kGy	0 kGy	2 kGy	4 kGy	0 kGy	2 kGy	4 kGy
1	0	6.61[d,x]	3.22[ab,y]	<2.0	6.60[e,x]	2.60[d,y]	<2.0	4.85[e,x]	<2.0	<2.0
	5	6.65[cd,x]	3.04[b,y]	<2.0	7.00[bc,x]	2.33[e,y]	<2.0	5.57[d,x]	<2.0	<2.0
	21	6.93[a,x]	2.28[f,y]	<2.0	6.97[cd,x]	2.54[d,y]	<2.0	5.95[cd,x]	<2.0	<2.0
7	0	6.70[bcd,x]	3.03[b,y]	<2.0	6.85[d,x]	2.97[b,y]	<2.0	6.13[c,x]	<2.0	<2.0
	5	6.87[ab,x]	2.74[cd,y]	<2.0	7.00[bc,x]	2.89[bc,y]	<2.0	5.98[c,x]	<2.0	<2.0
	21	6.64[d,x]	2.43[ef,y]	<2.0	7.11[ab,x]	2.07[f,y]	<2.0	6.32[bc,x]	<2.0	<2.0
14	0	6.86[abc,x]	3.25[a,y]	<2.0	7.07[bc,x]	3.22[a,y]	<2.0	6.82[a,x]	<2.0	<2.0
	5	6.63[d,x]	2.81[c,y]	<2.0	7.02[bc,x]	3.13[a,y]	<2.0	6.69[ab,x]	<2.0	<2.0
	21	6.86[abc,x]	2.58[de,y]	<2.0	7.21[a,x]	2.80[c,y]	<2.0	6.88[a,x]	<2.0	<2.0

Values with different superscript letters (a–f) within a column differ significantly ($P < 0.05$).
Values with different superscript letters (x–y) within a row for each pathogen differ significantly ($P < 0.05$).
<2.0: below detection limit.
*Target O_2 levels in the original packages.

Oxford Listeria selective agar with Listeria selective supplement, respectively. Petri plates with double agar layer were prepared 1 day before the day of analyses as follows: 14 mL of the selective media was poured into a petri plate. Then a 14 mL nonselective TSA was added onto the solidified selective media layer. Meatball samples (18 ± 3 g each) diluted in sterile peptone water (0.1%) were homogenized at medium speed for 2 minutes using a stomacher (Seward 400, Seward Ltd, London, UK). Homogenized samples were serially diluted and spread-plated onto the double layer agar plates. The plates were incubated at 35°C for 24–48 hours and typical colonies were counted. Representative *E. coli* O157:H7 colonies (one to two colonies from some plates in each experiment) were verified with the Dryspot latex agglutination test.

2.6. Measurement of Lipid Oxidation. Lipid oxidation was determined by the TBARS method [21]. A 5% trichloroacetic acid (TCA) solution in water and a 7.2% BHT solution in ethanol were prepared. A 10 g of sample was mixed with 34 mL of the TCA and 1 mL BHT solutions and homogenized at 11200 rpm for 2 min using a homogenizer (IKA-Ultra-Turrax T18, Germany). The homogenized sample was filtered through a filter paper (Whatman No. 4) into a 50 mL flask and diluted to the volume with the TCA solution. Next, 5 mL of the solution was transferred to a test tube, mixed with 5 mL of 0.02 M thiobarbituric acid (TBA) solution, and incubated at 80°C in a water bath for 20 min. Absorbance was measured at 532 nm against a blank containing 5 mL of TBA and 5 mL of 5% TCA using a spectrophotometer (T80 UV/VIS, PG Instrument, UK). Standard curve was prepared using 1,1,3,3-Tetraethoxypropane (TEP) as the malondialdehyde (MDA) standard at different concentrations. The TBARS values were expressed as mg MDA kg^{-1} of samples.

2.7. Color Measurements. Color (Hunter L, a, b values) of the samples was measured using a Chroma meter (CR-400, Konica Minolta, Japan). The instrument was calibrated with a standard white reflector plate prior to the measurements. The measurements were performed on each piece of meatball at five different locations and averaged.

2.8. Statistical Analyses. The data were subjected to analysis of variance and Tukey pairwise comparison using a statistical software (Minitab Inc., State College, PA, USA). Regression equation ($\text{Log CFU g}^{-1} = A \times \text{Dose} + B$) was fitted to the microbiological data, and D_{10}-value was calculated as the negative reciprocal of the slope ($D_{10} = -1/A$). All statistical analyses were evaluated at $\alpha = 0.05$ significance level.

3. Results

Gas compositions of the packages were determined periodically during storage. There were small reductions in O_2 and small increase in CO_2 levels in all the packages possibly due to microbial activities and permeation through the packages (data not shown). However, these changes can be considered negligible because the gas concentrations remained close to the initial target levels during storage, and there were no overlaps among different gas levels targeted in packaging treatments.

3.1. Growth/Survival of the Pathogens. The initial population of the pathogens in meatballs were significantly reduced by gamma irradiation ($P < 0.001$). Irradiation at 4 kGy resulted in about 6 Log inactivation of the pathogens, and their counts dropped below the detection limit ($<10^2 \text{ CFU g}^{-1}$) in all samples (Tables 1 and 2). *E. coli* O157:H7 was more sensitive to irradiation than *L. monocytogenes* and *S. enteritidis*. The *E. coli* O157:H7 count in the samples irradiated at 2 kGy was below the detection limit during 14-day storage. D_{10}-values for the pathogens were calculated using day 1 data presented in Tables 1 and 2 and reported for each O_2 level in the package headspace (Table 3). These values tend to decrease with increased O_2 levels, but the differences were not statistically

Effects of Irradiation Dose and O_2 and CO_2 Concentrations in Packages on Foodborne Pathogenic Bacteria and Quality
of Ready-to-Cook Seasoned Ground Beef Product (Meatball) during Refrigerated Storage

145

TABLE 2: Changes in *S. enteritidis*, *L. monocytogenes*, and *E. coli* O157:H7 counts (Log CFU g^{-1}) in ready-to-cook meatballs during storage at 3°C as affected irradiation dose (kGy) and O_2 level in 50% CO_2 containing packages.

Storage (day)	O_2^* (%)	*S. enteritidis*			*L. monocytogenes*			*E. coli* O157:H7		
		0 kGy	2 kGy	4 kGy	0 kGy	2 kGy	4 kGy	0 kGy	2 kGy	4 kGy
1	0	6.56^{d,x}	3.07^{b,y}	<2.0	6.92^{b,x}	2.57^{d,y}	<2.0	5.15^{c,x}	<2.0	<2.0
	5	6.59^{cd,x}	2.24^{e,y}	<2.0	6.94^{ab,x}	2.50^{d,y}	<2.0	5.51^{c,x}	<2.0	<2.0
	21	6.81^{ab,x}	2.19^{e,y}	<2.0	7.05^{ab,x}	2.54^{d,y}	<2.0	5.47^{c,x}	<2.0	<2.0
7	0	6.83^{ab,x}	3.05^{b,y}	<2.0	6.92^{b,x}	2.94^{b,y}	<2.0	6.19^{ab,x}	<2.0	<2.0
	5	6.83^{ab,x}	2.98^{bc,y}	<2.0	6.74^{c,x}	2.70^{cd,y}	<2.0	6.10^{b,x}	<2.0	<2.0
	21	6.74^{abc,x}	2.12^{e,y}	<2.0	6.93^{ab,x}	2.09^{e,y}	<2.0	6.13^{b,x}	<2.0	<2.0
14	0	6.89^{a,x}	3.32^{a,y}	<2.0	7.09^{a,x}	3.20^{a,y}	<2.0	6.58^{a,x}	<2.0	<2.0
	5	6.64^{bcd,x}	2.77^{c,y}	<2.0	6.89^{bc,x}	3.24^{a,y}	<2.0	6.50^{ab,x}	<2.0	<2.0
	21	6.86^{a,x}	2.47^{d,y}	<2.0	7.06^{ab,x}	2.89^{bc,y}	<2.0	6.50^{ab,x}	<2.0	<2.0

Values with different superscript letters (a–d) within a column differ significantly ($P < 0.05$).
Values with different superscript letters (x–y) within a row for each pathogen differ significantly ($P < 0.05$).
<2.0: below detection limit.
*Target O_2 levels in the original packages.

TABLE 3: D_{10}-values (kGy) of the pathogens in ready-to-cook meatballs at 9°C as affected by the O_2 levels.

O_2 Level (%)	*E. coli* O157:H7	*S. enteritidis*	*L. monocytogenes*
0	0.40*	0.58	0.48
5	0.36*	0.50	0.44
21	0.35*	0.43	0.45
Average	0.37*	0.51	0.46

*Calculated by assuming complete inactivation of the pathogen by 2 kGy as no viable cells were detected in the irradiated samples.

significant. The inactivation *S. enteritidis* by irradiation was 0.54 and 1.2 Log higher at 5% and 21% O_2, respectively, compared to 0% O_2 as assessed from day 1 data (Tables 1 and 2). On the other hand, no effects of O_2 on the radiation induced inactivation of *L. monocytogenes* were detected. The growth of the three pathogens in unirradiated samples during storage was slightly higher (0.07–0.25 Log difference) in 21% O_2 than 0 and 5% O_2 ($P < 0.05$), while no difference between the effect of 0% and 5% O_2 was detected.

Carbon dioxide did not affect the radiation-induced inactivation of pathogens. No inhibitory effects of CO_2 on number of *L. monocytogenes* and *S. enteritidis* in both the irradiated and the unirradiated samples were detected during 14-day storage. However, the 50% CO_2 resulted in 0.12 Log lower *E. coli* O157:H7 counts, on average, in unirradiated samples during storage ($P < 0.05$).

3.2. Lipid Oxidation.
TBARS values of the meatballs increased significantly as the irradiation dose increased (Figure 1). Irradiation at 2 and 4 kGy resulted in, respectively, 0.12 and 0.28 mg MDA kg^{-1} higher TBARS values on day 1 compared to the unirradiated (0 kGy) samples ($P < 0.01$). These effects became more pronounced with 0.36 and 0.42 mg MDA kg^{-1} higher TBARS in the samples irradiated at 2 and

4 kGy, respectively, than in the unirradiated samples on day 14 of storage. There was a significant interaction of O_2 and irradiation dose. Irradiation at 2 and 4 kGy resulted in 0.26 and 0.42 mg MDA kg^{-1} higher TBARS values, respectively, in the presence of 21% O_2 compared to 0% O_2 on day 1 ($P < 0.01$). The elevated CO_2 (50%) in the package headspace did not affect the radiation-induced TBARS values.

As the O_2 concentration in the packages increased, the TBARS values increased significantly ($P < 0.01$). Oxygen levels of 0% and 5% in the packages resulted in 1.15 and 0.9 mg MDA kg^{-1} lower TBARS values, respectively, compared to 21% O_2 during 14-day storage (Figure 1). The difference between the TBARS values due to 0 and 5% O_2 was relatively small (0.25 mg MDA kg^{-1}) during 14-day storage. The elevated CO_2 (50%) resulted in slightly higher (0.186 mg MDA kg^{-1}) TBARS value during 14-day storage, but its effect was not significant at the low O_2 levels (Figure 1).

3.3. Color.
In general, there were no changes in a-values in the first 7-day storage, but the a-values decreased significantly after 14-day storage. Irradiation at 2 and 4 kGy resulted in significant reduction ($P < 0.001$) in the a-values of the meatballs after 7-day storage (Figure 2). Average a-value of the irradiated (2 and 4 kGy) meatball was 9.9, while it was 13.1 in the unirradiated (0 kGy) samples after 7-day storage. No difference between the effect of 2 and 4 kGy was detected. As the O_2 concentration decreased, smaller reduction in the a-values was observed (Figure 2). Average a-value of the samples in the packages with 21% O_2 was 7.8, while it was 10.8 in the 5% O_2-containing packages after 7-day storage. The packages with no O_2 maintained the initial a-values of the meatballs during storage (Figure 2). Carbon dioxide had no effect on the a-values of the samples. L- and b-values of the samples were in the range of 29.1–34.7 and 9.1–13.1, respectively, and they were not affected by the irradiation dose and O_2 and CO_2 levels.

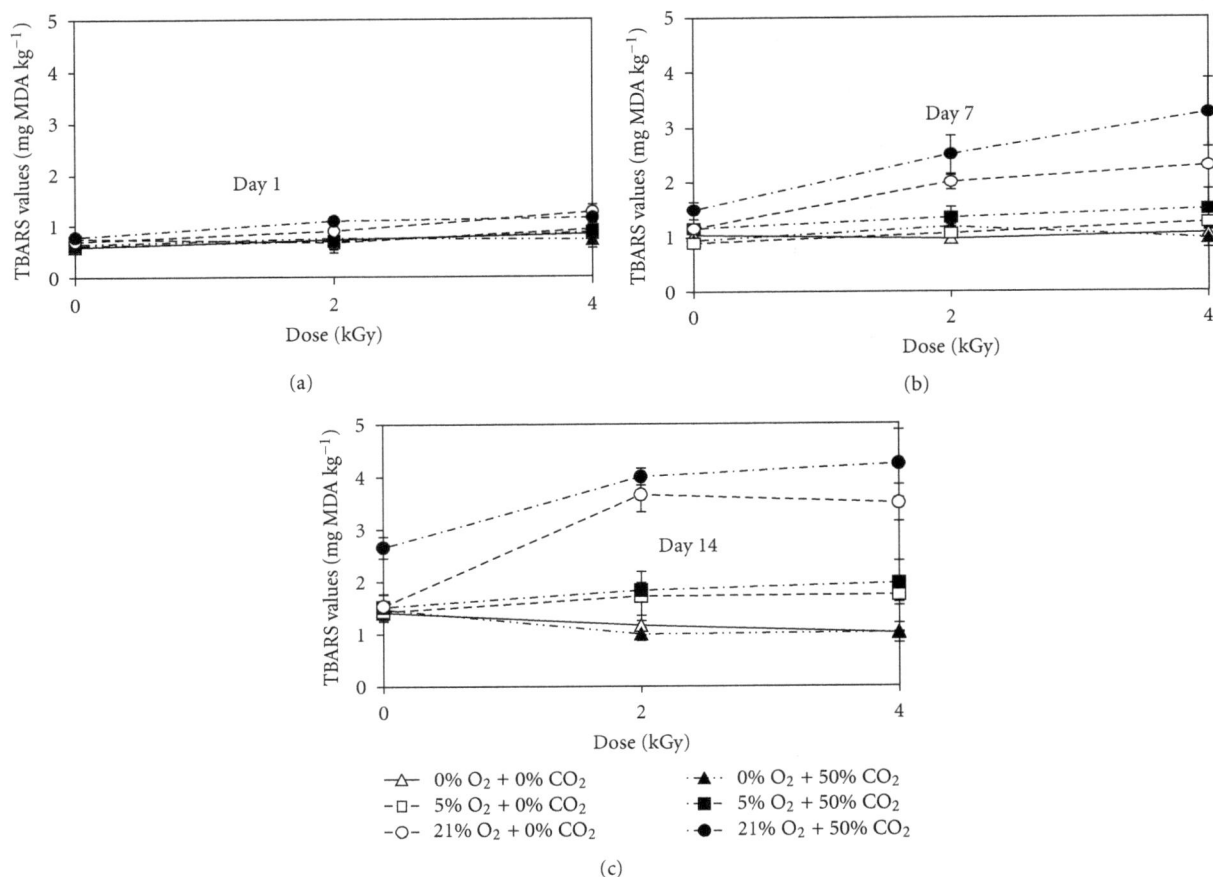

FIGURE 1: Effects of irradiation dose and O_2 and CO_2 levels in packages on TBARS values of ready-to-cook meatballs during storage at $3°C$. Error bars represent standard deviations.

4. Discussion

Irradiation effectively inactivated the pathogenic microorganisms in the meatball in all the packaging conditions tested. The inactivation was higher in the packages containing 21% O_2. The D_{10}-values for the pathogens in our study were in general agreement with the literature which reported D_{10}-values for *S. enteritidis*, *L. monocytogenes*, and *E. coli* O157:H7 in various conditions as 0.51–0.71, 0.45–0.50, and 0.24–0.41 kGy, respectively [6, 10, 18, 19]. Differences in the reported D_{10}-values in the literature for the same pathogens are probably due to the differences in the irradiation conditions (temperature, atmosphere, dose rate) and the product formulations, since these factors significantly affect the radiation-induced inactivation of microorganisms [5, 9, 10, 19].

Although 21% O_2 in the packages resulted in higher radiation-induced inactivation of the pathogens, the difference between the number of pathogens in the 21% O_2 and the 5% O_2 packages was relatively small (approximately 0.2 Log). On the other hand, the lower O_2 concentration in the packages maintained the oxidative and the color stability of the irradiated samples.

The oxidative and the color qualities of the irradiated meatballs were best maintained in the 0% O_2 packages.

However, the absence of O_2 in the packages can create a great risk of anaerobic pathogens such as *C. botulinum* spores which have a high inherent radiation resistance (D_{10} = 2–3 kGy) [22]. Thus, the irradiation dose applied up to 4 kGy in this study would cause only 1-2 Log reduction in the spore count. Growth and toxin production by *C. botulinum* in packaged meat products have been detected at O_2 levels below 2% [15]. The differences between the effects of 0 and 5% O_2 on the quality attributes of the meatballs were negligible in our study. Therefore, we suggest that the meatball packages should contain 2–5% O_2 during storage for the safety assurance of the product.

In conclusion, the safety risk due to the *L. monocytogenes*, *E. coli* O157:H7, and *S. enteritidis* in ready-to-cook meatballs can be significantly reduced by gamma irradiation. More than 6 Log reduction in *L. monocytogenes* and *S. enteritidis* by 4 kGy and 5 Log reduction in *E. coli* O157:H7 by 2 kGy were achieved in the meatballs. Lower irradiation doses would likely be sufficient for the products in commerce since the contamination levels would be lower than 5-6 Log. The low O_2 (0% and 5%) levels inhibited the irradiation-induced color change and the oxidation significantly during refrigerated storage. An inclusion of 5% O_2 in the packages is suggested as it did not adversely affect the quality of

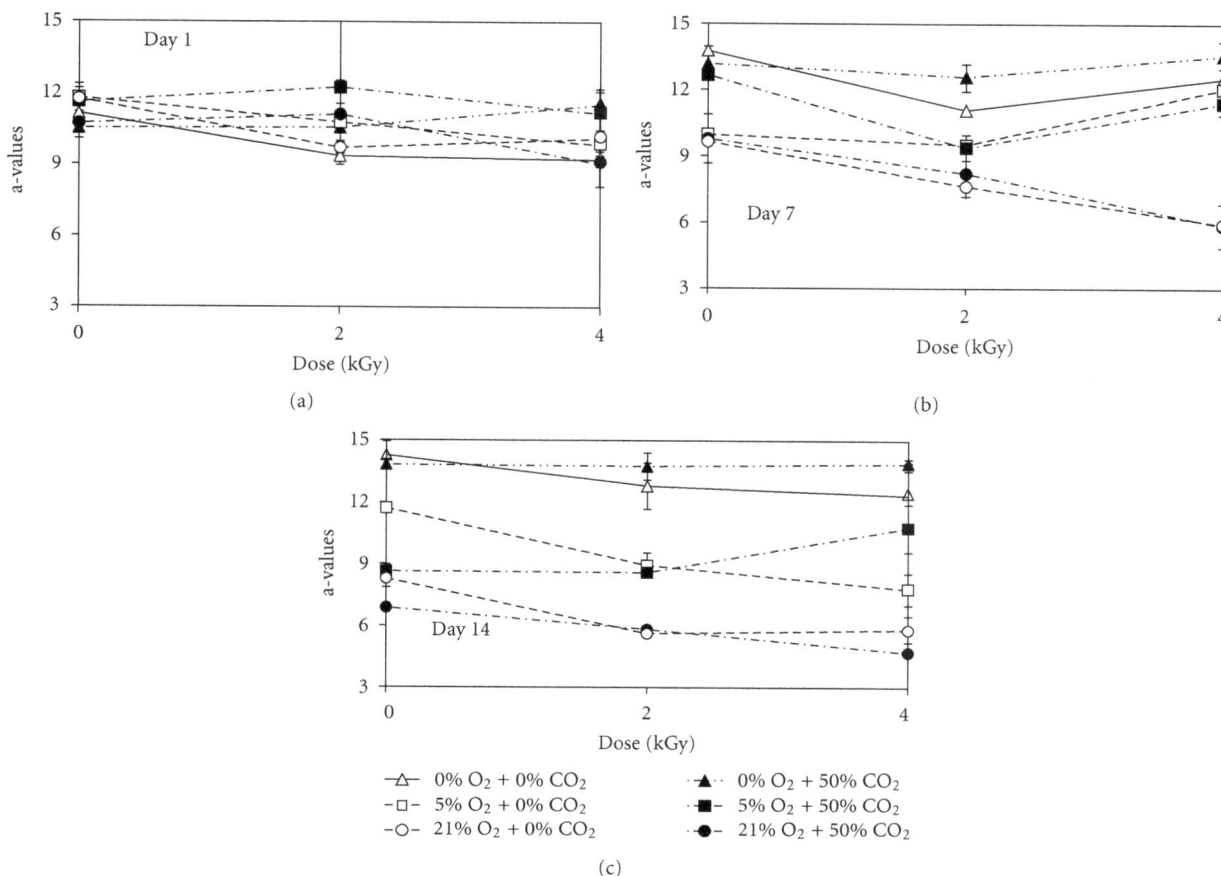

FIGURE 2: Effects of irradiation dose and O_2 and CO_2 levels in packages on a-values of ready-to-cook meatballs during storage at 3°C. Error bars represent standard deviations.

the irradiated meatballs compared to 0% O_2. Although 50% CO_2 did not affect the radiation-induced inactivation of the pathogens, it can inhibit growth of the surviving pathogens and other microorganisms in meatballs during refrigerated storage. Thus, MAP with 5% O_2 + 50% CO_2 combined with irradiation up to 4 kGy is suggested for refrigerated meatballs to reduce the foodborne pathogen risk and to maintain the quality.

Acknowledgments

This research was funded by TUBITAK and Research Fund of Istanbul Technical University. The authors greatly acknowledge Gamma-Pak GmbH and Dr. Hasan Alkan for irradiating samples. The packaging materials were provided by Korozo Packaging Company GmbH (Istanbul, Turkey). They also acknowledge Necla Aran for valuable discussions and suggestions on microbial analyses.

References

[1] B. Nørrung and S. Buncic, "Microbial safety of meat in the European Union," Meat Science, vol. 78, no. 1-2, pp. 14–24, 2008.

[2] N. V. V. Padhye and M. P. Doyle, "Escherichia coli O157:H7 epidemiology, pathogenesis and methods for detection in food," Journal of Food Protection, vol. 55, pp. 555–565, 1992.

[3] D. M. Frye, R. Zweig, J. Sturgeon et al., "An outbreak of febrile gastroenteritis associated with delicatessen meat contaminated with Listeria monocytogenes," Clinical Infectious Diseases, vol. 35, no. 8, pp. 943–949, 2002.

[4] M. Kivi, A. Hofhuis, D. W. Notermans et al., "A beef-associated outbreak of Salmonella Typhimurium DT104 in The Netherlands with implications for national and international policy," Epidemiology and Infection, vol. 135, no. 6, pp. 890–899, 2007.

[5] J. Farkas, "Irradiation as a method for decontaminating food: a review," International Journal of Food Microbiology, vol. 44, no. 3, pp. 189–204, 1998.

[6] Z. Gezgin and G. Gunes, "Influence of gamma irradiation on growth and survival of Escherichia coli O157:H7 and quality of cig kofte, a traditional raw meat product," International Journal of Food Science and Technology, vol. 42, no. 9, pp. 1067–1072, 2007.

[7] S. R. Kanatt, R. Chander, and A. Sharma, "Effect of radiation processing on the quality of chilled meat products," Meat Science, vol. 69, no. 2, pp. 269–275, 2005.

[8] D. W. Thayer and G. Boyd, "Irradiation and modified atmosphere packaging for the control of Listeria monocytogenes on turkey meat," Journal of Food Protection, vol. 62, no. 10, pp. 1136–1142, 1999.

[9] C. A. O'Bryan, P. G. Crandall, S. C. Ricke, and D. G. Olson, "Impact of irradiation on the safety and quality of poultry and meat products: a review," *Critical Reviews in Food Science and Nutrition*, vol. 48, no. 5, pp. 442–457, 2008.

[10] D. W. Thayer, "Use of irradiation to kill enteric pathogens on meat and poultry," *Journal of Food Safety*, vol. 15, pp. 181–192, 1995.

[11] D. U. Ahn, J. L. Sell, M. Jeffery et al., "Dietary vitamin E affects lipid oxidation and total volatiles of irradiated raw turkey meat," *Journal of Food Science*, vol. 62, no. 5, pp. 954–958, 1997.

[12] A. Karadag and G. Gunes, "The effects of gamma irradiation on the quality of ready-to-cook meatballs," *Turkish Journal of Veterinary and Animal Sciences*, vol. 32, no. 4, pp. 269–274, 2008.

[13] S. J. Millar, B. W. Moss, and M. H. Stevenson, "The effect of ionising radiation on the colour of beef, pork and lamb," *Meat Science*, vol. 55, no. 3, pp. 349–360, 2000.

[14] I. J. Church and A. L. Parsons, "Modified atmosphere packaging technology: a review," *Journal of the Science of Food and Agriculture*, vol. 67, pp. 143–152, 1995.

[15] D. Narasimha Rao and N. M. Sachindra, "Modified atmosphere and vacuum packaging of meat and poultry products," *Food Reviews International*, vol. 18, no. 4, pp. 263–293, 2002.

[16] A. Ozturk, N. Yilmaz, and G. Gunes, "Effect of different modified atmosphere packaging on microbial quality, oxidation and colour of a seasoned ground beef product (Meatball)," *Packaging Technology and Science*, vol. 23, no. 1, pp. 19–25, 2010.

[17] I. R. Grant and M. F. Patterson, "Effect of irradiation and modified atmosphere packaging on the microbial safety of minced pork stored under temperature abuse conditions," *International Journal of Food Science & Technology*, vol. 26, pp. 521–533, 1991.

[18] V. López-González, P. S. Murano, R. E. Brennan, and E. A. Murano, "Influence of various commercial packaging conditions on survival of escherichia coli O157:H7 to irradiation by electron beam versus gamma rays," *Journal of Food Protection*, vol. 62, no. 1, pp. 10–15, 1999.

[19] M. Patterson, "Sensitivity of bacteria to irradiation on poultry meat under various atmospheres," *Letters in Applied Microbiology*, vol. 7, no. 3, pp. 55–58, 1988.

[20] V. C. H. Wu and D. Y. C. Fung, "Evaluation of Thin Agar Layer method for recovery of heat-injured foodborne pathogens," *Journal of Food Science*, vol. 66, no. 4, pp. 580–583, 2001.

[21] J. Pikul, D. E. Leszczynski, and F. A. Kummerow, "Evaluation of three modified TBA methods for measuring lipid oxidation in chicken meat," *Journal of Agricultural and Food Chemistry*, vol. 37, no. 5, pp. 1309–1313, 1989.

[22] I. Kiss, C. O. Rhee, N. Grecz, T. A. Roberts, and J. Farkas, "Relation between radiation resistance and salt sensitivity of spores of five strains of Clostridium botulinum types A, B, and E," *Applied and Environmental Microbiology*, vol. 35, no. 3, pp. 533–539, 1978.

20

A Constitutively Mannose-Sensitive Agglutinating *Salmonella enterica* subsp. *enterica* Serovar Typhimurium Strain, Carrying a Transposon in the Fimbrial Usher Gene *stbC*, Exhibits Multidrug Resistance and Flagellated Phenotypes

Kuan-Hsun Wu,[1] Ke-Chuan Wang,[2] Lin-Wen Lee,[3] Yi-Ning Huang,[4] and Kuang-Sheng Yeh[4]

[1] *Department of Pediatrics, Wan Fang Hospital, Taipei Medical University, Taipei 116, Taiwan*
[2] *Graduate Institute of Medical Sciences, College of Medicine, Taipei Medical University, Taipei 110, Taiwan*
[3] *Department of Microbiology and Immunology, School of Medicine, College of Medicine, Taipei Medical University, Taipei 110, Taiwan*
[4] *Department of Veterinary Medicine, School of Veterinary Medicine, College of Bioresources and Agriculture, National Taiwan University, Taipei 106, Taiwan*

Correspondence should be addressed to Kuang-Sheng Yeh, ksyeh@ntu.edu.tw

Academic Editors: P. Mastroeni and R. Rivas

Static broth culture favors *Salmonella enterica* subsp. *enterica* serovar Typhimurium to produce type 1 fimbriae, while solid agar inhibits its expression. A transposon inserted in *stbC*, which would encode an usher for Stb fimbriae of a non-flagellar *Salmonella enterica* subsp. *enterica* serovar Typhimurium LB5010 strain, conferred it to agglutinate yeast cells on both cultures. RT-PCR revealed that the expression of the fimbrial subunit gene *fimA*, and *fimZ*, a regulatory gene of *fimA*, were both increased in the *stbC* mutant when grown on LB agar; *fimW*, a repressor gene of *fimA*, exhibited lower expression. Flagella were observed in the *stbC* mutant and this phenotype was correlated with the motile phenotype. Microarray data and RT-PCR indicated that the expression of three genes, *motA*, *motB*, and *cheM*, was enhanced in the *stbC* mutant. The *stbC* mutant was resistant to several antibiotics, consistent with the finding that expression of *yhcQ* and *ramA* was enhanced. A complementation test revealed that transforming a recombinant plasmid possessing the *stbC* restored the mannose-sensitive agglutination phenotype to the *stbC* mutant much as that in the parental *Salmonella enterica* subsp. *enterica* serovar Typhimurium LB5010 strain, indicating the possibility of an interplay of different fimbrial systems in coordinating their expression.

1. Introduction

Salmonella enterica subsp. *enterica* contains more than 2,300 serovars; one of these, Typhimurium, is an important cause of gastroenteritis [1]. The ability to adhere to the host epithelial cell is considered a prerequisite step for infection. Fimbriae, proteinaceous hair-like appendages on the outer membrane of bacteria, have been implicated in such adherence. Many types of fimbriae have been described for *Salmonella enterica* subsp. *enterica* serovar Typhimurium, among which type 1 fimbriae is the most common; therefore, type 1 fimbriae are also referred to as common fimbriae [2].

Type 1 fimbriae adhere to a variety of cells, including erythrocytes, leukocytes, intestinal cells, and even plant root hairs. More than 80% of *Salmonella* species isolates produce type 1 fimbriae, and this fimbrial type may play some role in the life cycle of bacteria [3].

Old et al. described the phenotypic variation of the expression of type 1 fimbriae in *Salmonella enterica* subsp. *enterica* serovar Typhimurium [4, 5]. In brief, strongly type 1 fimbriate-phase bacterial cells were obtained following successive passage every 48 hr in static broth culture medium, while nonfimbriated-phase bacteria were found when cultured on solid media [4, 5]. Subsequent investigations have indicated that phenotypic expression of type 1 fimbriae in *Salmonella enterica* subsp. *enterica* serovar Typhimurium involves the interaction of several genes in the *fim* gene cluster [6–9]. To explore whether there is another genetic determinant outside the *fim* genes that would also participate in regulation of type 1 fimbriae in *Salmonella enterica* subsp. *enterica* serovar Typhimurium, Chuang et al. constructed an insertional library to screen for mutant strains that showed different type 1 fimbrial phenotypes than the parental strain [10]. Yeast agglutination test and southern blot analysis were used to screen for those mutants and validated that the transpositional event occurred only once. A group of genes classified as having fimbrial biosynthesis and regulation were involved in the expression of type 1 fimbriae in response to different culture conditions [10]. One mutant *stbC* strain agglutinated yeast obtained either from static LB broth or on solid LB agar medium, and the agglutination was mannose-sensitive [10]. The gene *stbC* encoded the usher protein for Stb fimbriae in *Salmonella enterica* subsp. *enterica* serovar Typhimurium [11]. The usher protein helps anchor the developing fimbrial subunit structure to the outer membrane [12]. Cross-talk between different fimbrial systems in one strain has been documented [13–15]. Therefore, the interaction of *fim* and *stb* fimbrial systems is an intriguing topic. This paper describes the interesting characteristics of *Salmonella enterica* subsp. *enterica* serovar Typhimurium *stbC* mutant that we discovered.

2. Materials and Methods

2.1. Bacterial Strains and Culture Media. The *Salmonella enterica* subsp. *enterica* serovar Typhimurium strain used in this study was the LT2 derivative strain LB5010 [16]. The *stbC* mutant strain was obtained from the *Salmonella enterica* subsp. *enterica* serovar Typhimurium transposon library collection in our laboratory [10]. This strain has a transposon inserted in the position of *stbC*, encoding the usher protein of StbA fimbriae. *Salmonella* cells were cultured in Luria-Bertani (LB) broth (Difco/Becton Dickinson, Franklin Lakes, NJ) or on LB agar. Mueller-Hinton agar (Difco/Becton Dickinson) was used when performing the antimicrobial susceptibility test, and modified semisolid Rappaport-Vassiliadis (MSRV) (Difco/Becton Dickinson) was used to detect the motility of *Salmonella* strains.

2.2. Yeast Agglutination Test. The LB agar plates were incubated at 37°C for 18 h, while the broth preparations were incubated statically at 37°C for 48 h. Bacterial cells from the solid agar were collected by a sterile loop and resuspended in 100 μL of 1 × phosphate-buffered saline (PBS). Cells in the broth medium were collected by centrifugation, and the pellet was resuspended in 100 μL of 1 × PBS. Subsequently, 30 μL of a 3% (vol/vol) suspension of *Candida albicans* in PBS and an equal amount of bacterial cells to be tested were mixed together on a glass slide [4]. Visible agglutination after gentle agitation indicated a positive reaction for the presence of type 1 fimbriae. Any bacterial suspension that produced type 1 fimbriae was further mixed with *C. albicans*,

along with 3% (wt/vol) of a D-mannose solution (Sigma Chemical Company, St. Louis, MO). The mannose-sensitive agglutination conferred by type 1 fimbriae was inhibited in the presence of mannose.

2.3. Electron Microscopy. Bacterial strains were grown in static broth or on solid agar and resuspended in 1 × PBS. The bacterial cells were then negatively stained with 2% phosphotungstic acid and observed with a Hitachi H-600 transmission electron microscope (Hitachi Ltd., Tokyo, Japan).

2.4. Motility Test. *Salmonella enterica* subsp. *enterica* serovar Typhimurium LB5010 and the *stbC* mutant strains were grown in static broth for 48 h and on solid agar for 18 h. Cells were resuspended in 1 × PBS and adjusted to the same turbidity. A drop of cell suspension was then spotted on the modified semisolid Rappaport-Vassiliadis (MSRV) agar medium and incubated at 42°C for 16 h.

2.5. Antimicrobial Susceptibility Test. *Salmonella enterica* subsp. *enterica* serovar Typhimurium LB5010 and the *stbC* mutant strains were analyzed for drug resistance. Antimicrobial susceptibility was tested by a disc diffusion method using commercially available discs and Mueller-Hinton agar. The following antibiotics were used: ampicillin (10 μg), penicillin (10 μg), cephalothin (30 μg), tetracycline (30 μg), doxycycline (10 μg), chloramphenicol (50 μg), florfenicol (30 μg), streptomycin (50 μg), and ciprofloxacin (5 μg). *Escherichia coli* ATCC 25922 was used as the control strain. The results were interpreted according to criteria specified by the Clinical and Laboratory Standards Institute (CLSI) [17].

2.6. RNA Purification. Bacterial cells were harvested in TRIzole Reagent (Invitrogen, Carlsbad, CA) and disrupted using the MagNA Lyser System (Roche Diagnostics, Mannheim, Germany), with ceramic bead shaking at 5,000 rpm for 15 sec. After phenol-chloroform extraction, the aqueous layer was applied to an RNeasy column (Qiagen, Valencia, CA) for RNA purification according to the protocol provided by the manufacturer. The RNA was quantified by using a ND-1000 spectrophotometer (NanoDrop Technology, Wilmington, DE) and then examined with a Bioanalyzer 2100 (Agilent Technology, Palo Alto, CA) with an RNA 6000 Nano LabChip kit (Agilent). To enhance the sensitivity of microarray signal, the purified total RNA was subjected to a ribosomal RNA removal procedure by MICROB Express Bacterial mRNA Purification Kit (Ambion, Applied Biosystems, Foster City, CA).

2.7. Custom Array Design. Probes were designed by eArray of Agilent Technologies. In the design process, 4,718 *Salmonella enterica* subsp. *enterica* serovar Typhimurium LT2 transcripts were uploaded to Agilent eArray and designed by Tm matching methodology. Target sequences that were duplicated were removed. Probes with Tm at around 80°C, optimal base content, and low cross-hybridization were

A Constitutively Mannose-Sensitive Agglutinating Salmonella enterica subsp. enterica Serovar Typhimurium Strain, Carrying a Transposon in the Fimbrial Usher Gene stbC, Exhibits Multidrug Resistance and Flagellated Phenotypes

151

selected. The resulting 4,646 probes were then generated. The custom microarray was manufactured in $4 \times 44 \mathrm{k}$ format by in situ synthesis of oligonucleotide probes. Each array consisted of 4,646 *Salmonella enterica* subsp. *enterica* serovar Typhimurium LT2-specific probes and printed in 9 replicates.

2.8. Microarray Experiment. One microgram of enriched mRNA was reverse-transcribed to cDNA with a CyScribe 1st-strand cDNA labeling kit (GE Healthcare, Buckinghamshire, UK) and labeled with Cy3-CTP or Cy5-CTP (CyDye, PerkinElmer, Waltham, MA). Correspondingly labeled cDNA was then pooled and hybridized to microarrays at $60^{\circ} \mathrm{C}$ for 17 h. After washing and drying by nitrogen gun blowing, microarrays were scanned with an Agilent microarray scanner at 535 nm for Cy3-CTP and 625 nm for Cy5-CTP. Scanned images were then analyzed by Feature Extraction 9.5.3 software (Agilent). Image analysis and normalization software was used to quantify signal and background intensity for each feature and substantially normalized the data by rank-consistency-filtering LOWESS method.

2.9. Reverse Transcription Polymerase Chain Reaction (RT-PCR) Analysis. The bacteria were first stabilized by adding RNAProtect Bacteria Reagent (Qiagen) and total RNA was isolated by using an RNeasy Mini Kit (Qiagen) and RNase-free DNase I (1 unit/1 μg RNA) (Promega, Madison, WI), according to the protocol provided by the manufacturer. RT-PCR was performed by a SuperScript III One-Step RT-PCR System (Invitrogen). RNA was denatured at $58^{\circ} \mathrm{C}$ for 5 min and followed by cDNA synthesis at $42^{\circ} \mathrm{C}$ for 30 min. The reaction was stopped by heating the bacteria at $94^{\circ} \mathrm{C}$ for 2 min. The following PCR programming was used: 35 cycles of denaturing at $94^{\circ} \mathrm{C}$ for 30 sec, annealing at $54^{\circ} \mathrm{C}$ for 30 sec, and extension at $72^{\circ} \mathrm{C}$ for 30 sec. An additional extension was performed at $72^{\circ} \mathrm{C}$ for 10 min. Primers used in the present study are listed in Table 1.

2.10. Complementation Test. The primer set, stbC-F and stbC-R, was used to amplify the *stbC* coding sequence from the genomic DNA of *Salmonella enterica* subsp. *enterica* serovar Typhimurium LB 5010 with the Epicentre FailSafe PCR PreMix Selection kit (Epicentre, Madison, WI). The resulting DNA fragment was cleaved with *Bam*HI and ligated into the pACYC 184 vector. *Hind*III and *Sal*I were then used to cleave a 2.9 kb DNA fragment possessing the *stbC* coding sequence from the aforementioned recombinant plasmid and ligated into the pBBR1MCS-5 vector that carries a gentamycin-resistant gene [18]. The resulting recombinant plasmid was transformed into the *stbC* mutant strain, and transformants were selected on gentamycin-containing (100 μg/mL) LB agar medium. A transformant labeled *stbC* (pStbC) was selected for further study. For the control, the pBBR1MCS-5 vector was transformed into the *stbC* mutant, and a resulting transformant labeled *stbC* (vector) was selected.

3. Results

3.1. Yeast Agglutination Test. A *stbC* mutant prepared on both static LB broth and LB agar medium both exhibited agglutination when mixed with yeast cells on a glass slide. The addition of mannose inhibited agglutination, indicating that agglutination was mannose-sensitive. *Salmonella enterica* subsp. *enterica* serovar Typhimurium LB5010 from static broth agglutinated yeast cells, while that from agar medium did not. The *stbC* (pStbC) harboring the plasmid that contains the coding sequence of *stbC* exhibited the same agglutination pattern as the parental strain, while *stbC* (vector), containing the vector alone, prepared on both agar and broth medium, agglutinated yeast cells as did the *stbC* mutant strain. Table 2 compares the yeast agglutination capabilities conferred by these strains.

3.2. Electron Microscopy. *Salmonella enterica* subsp. *enterica* serovar Typhimurium LB5010 prepared in static LB broth culture showed fimbrial appendages on the outside of the cell (Figure 1(a)). Besides fimbriae, additional flagella-like structures were present on the *stbC* mutant strain cultured in static broth (Figure 1(b)). On the contrary, *Salmonella enterica* subsp. *enterica* serovar Typhimurium LB5010 grown on agar medium did not produce type1 fimbriae (Figure 2(a)). The *stbC* mutant prepared from agar medium did not produce fimbriae; either, however, flagella-like structures were still observed around the *stbC* mutant prepared from agar medium (Figure 2(b)).

3.3. Motility Test. The semisolid characteristic of MSRV allows mobility to be detected as "halos" of growth around the point of inoculation. *Salmonella enterica* subsp. *enterica* serovar Typhimurium LB5010 did not exhibit a "halo" effect on the MSRV medium, and the medium remained blue around the inoculated drop (Figure 3(a)). In contrast, a gray-white zone was observed extending from the inoculated drop of the *stbC* mutant, prepared either from static LB broth or LB agar (Figure 3(b)). The *stbC* was a motile *Salmonella* strain. We also used polyvalent *Salmonella* H antiserum (Difco/Becton Dickinson) to confirm that the flagellar antigen was present on the *stbC* mutant but not on the *Salmonella enterica* subsp. *enterica* serovar Typhimurium LB5010 by slide agglutination test.

3.4. Antimicrobial Susceptibility Test. Since we had difficulty complementing the *stbC* with a recombinant plasmid containing the coding sequence of *stbC*, we tested whether *stbC* mutant exhibited resistance to the antibiotic marker carried in the cloning vector. We found that *stbC* was resistant to ampicillin and tetracycline carried in TA cloning and pACYC 184 vectors, respectively. Therefore, a battery of antibiotics was tested on the *stbC* mutant and the parental strain LB5010. Both strains were resistant to penicillin and streptomycin. In addition, the *stbC* mutant also exhibited resistance to ampicillin, cephalothin, tetracycline, doxycycline, chloramphenicol, florfenicol, and ciprofloxacin.

TABLE 1: Oligonucleotide primers used in the present study.

Primer	Sequence (5′-3′)	
gvcH-RT-F	TAAAGATCCAGCCACCG	
gvcH-RT-R	GGCATTACTGAACACGC	
ramA-RT-F	CGATTGTCGAGTGGATT	
ramA-RT-R	GCGTAAAGGTTTGCTGC	
csgA-RT-F	CGACCATTACCCAGAGC	
csgA-RT-R	TTGCCAAAACCAACCTG	
cheM.-RT-F	GCCAGATTACGCACCTC	
cheM-RT-R	TGCCAGCATGGAACAAC	
yhcQ-RT-F	ATTCCCCTGCTGCTCGT	
yhcQ-RT-R	ATGTCGTCGCTATTGCC	
motA-RT-F	GTCTGCTGCTGGTTTGG	
motA-RT-R	ATAGGGGCGTTCATTGT	
motB-RT-F	CTTTTGGCGATGTGGGT	
motB-RT-R	CAGCAGGGTGAAGTGGA	
STM0347-RT-F	GGCATCGCTTCACTCTT	
STM0347-RT-R	TCACCGACCGCTACATC	
rpsS-RT-F	ATAAGTACGAGTCGGTGCG	
rpsS-RT-R	CACTTGCTGAAGAAGGTAGA	
16S-F	TTCCTCCAGATCTCTCTACGCA	
16S-R	GTGGCTAATACCGCATAACG	
fimA-RT-F	ACTATTGCGAGTCTGATGTTTG	
fimA-RT-R	CGTATTTCATGATAAAGGTGGC	
fimZ-RT-F	ATTCGTGTGATTTGGCGT	
fimZ-RT-R	ACTTATCCTGTTGACCTT	
fimY-RT-F	GAGTTACTGAACCAACAGCT	
fimY-RT-R	GCCGGTAAACTACACGATGA	
fimW-RT-F	AAAGTGAAAGTAAAGCGG	
fimW-RT-R	AAGAGATAGATAATGCCG	
stbC-F	ATACGGGATCCCG-GCTGACAAACAGGCTGGTGATAAACAAT 3′	The underlined sequence denotes the BamHI restriction site
stbC-R	CTACGGGATCCCG-TGACGGGCTAGGTAAACCTGATAATCTG 3	The underlined sequence denotes the BamHI restriction site

TABLE 2: Yeast agglutination test of *Salmonella enterica* subsp. *enterica* serovar Typhimurium LB5010 and *stbC* mutant strains.

	Agglutination of yeast cells mixed with different concentrations of bacterial cells with a 2-fold dilution[a]					
Strain	1×	2×	4×	8×	16×	32×
LB5010/agar	−	−	−	−	−	−
LB5010/broth	+++	+++	++	+	+	−
stbC mutant/agar	++	++	+	−	−	−
stbC mutant/broth	+++	+++	+++	++	++	+
stbC (pStbC)/agar	−	−	−	−	−	−
stbC (pStbC)/broth	+++	++	++	+	+	−
stbC (vector)/agar	++	++	+	−	−	−
stbC (vector)/broth	+++	+++	+++	++	++	+

[a] Strong agglutination is indicated by (+++) and a negative result by (−).

A Constitutively Mannose-Sensitive Agglutinating Salmonella enterica subsp. enterica Serovar Typhimurium Strain, Carrying a Transposon in the Fimbrial Usher Gene stbC, Exhibits Multidrug Resistance and Flagellated Phenotypes

153

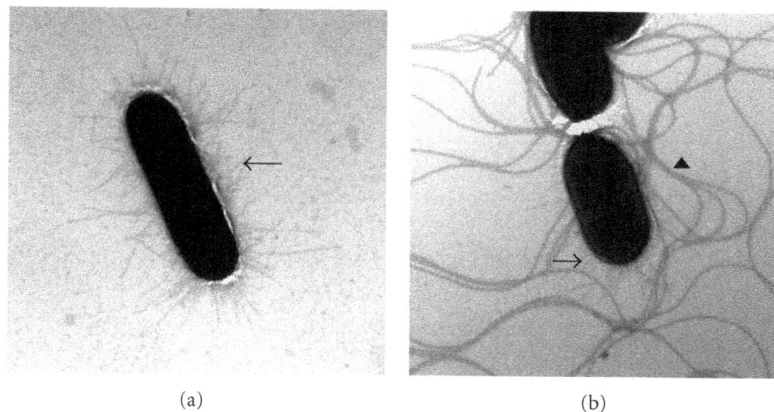

(a)

(b)

FIGURE 1: Observation of *Salmonella enterica* subsp. *enterica* serovar Typhimurium LB5010 and the *stbC* mutant strain grown in static broth culture. (a) *Salmonella enterica* subsp. *enterica* serovar Typhimurium LB5010 strain grown in static LB broth condition at 37°C for 48 h produced fimbrial appendages (arrow). No flagella structures were observed. (b) *Salmonella enterica* subsp. *enterica* serovar Typhimurium *stbC* mutant strain grown in static LB broth condition at 37°C for 48 h produced fimbrial appendages (arrow) and flagella structures (arrowhead). Bacterial cells were negatively stained with 2% of phosphotungstic acid (20,000x).

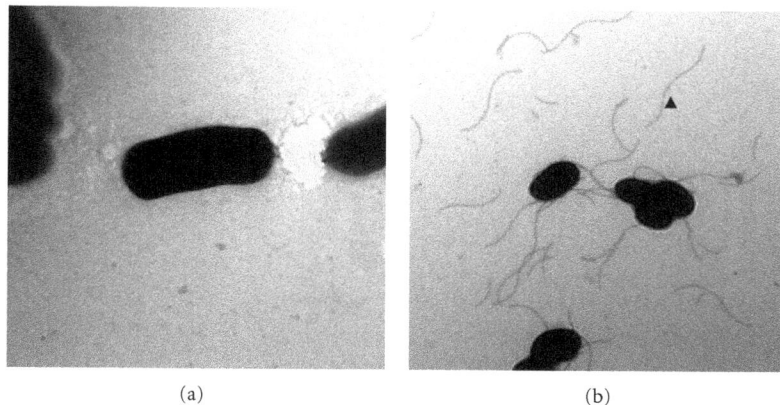

(a)

(b)

FIGURE 2: Observation of *Salmonella enterica* subsp. *enterica* serovar Typhimurium LB5010 and the *stbC* mutant strain grown on solid agar. (a) *Salmonella enterica* subsp. *enterica* serovar Typhimurium LB5010 grown on LB agar at 37°C for 16 h did not produce fimbrial appendages. (b) The *Salmonella enterica* subsp. *enterica* serovar Typhimurium *stbC* mutant grown on LB agar at 37°C for 16 h exhibited flagella structures (arrowhead) but no fimbrial appendages were observed (3,000x). Bacterial cells were negatively stained with 2% of phosphotungstic acid (20,000x).

3.5. Microarray Analysis. Since the *stbC* mutant strain prepared from solid agar produced different results on yeast agglutination testing and exhibited multidrug resistant characteristics, it is tempting to identify the genes that would express differently between the *stbC* mutant strain and its parental strain *Salmonella enterica* subsp. *enterica* serovar Typhimurium LB5010 when grown on solid LB agar. Total RNA was isolated from both strains cultured on solid LB agar and analyzed by hybridization to a *Salmonella enterica* subsp. *enterica* serovar Typhimurium LT2 DNA microarray. The median hybridization results of 9 arrays showed that about 50 genes from the *stbC* mutant grown on agar were upregulated more than 8-fold (Table 3). Analysis of the microarray data revealed that these genes can be classified according to their functions, including fimbrial structure, motility, drug resistance, gene regulation,

transportation, outer membrane porin structure, prophage, ribosomal protein, inner membrane protein, periplasmic protein, and enzymes of different functions. Twenty-two genes are enzymes with a variety of functions (44%), and these comprised the major part of the genes upregulated in the *stbC* mutant strain grown on solid agar. The detail results of the microarray assay in the present study can be accessed at GEO accession number GSE34685.

3.6. Reverse Transcription Polymerase Chain Reaction (RT-PCR) Analysis. To validate our microarray finding, several genes were selected and assayed by RT-PCR. The transcriptional levels of *gcvH* (5.9), *ramA* (3.9), *csgA* (2.4), *cheM* (7.2), *yhcQ* (3.0), *motA* (2.1), *motB* (3.7), STM0347 (1.5), and *rpsS* (1.4) in *stbC* were higher than those in LB5010 when both strains were grown on LB agar. (The vales in the parenthesis

TABLE 3: Identification of selected genes of *Salmonella enterica* subsp. *enterica* serovar Typhimurium *stbC* mutant strain grown on LB agar by microarray analysis.

Group	Function	Ratio of expression in *stbC*/Lb5010 on agar
Fimbriae		
csgA	Curlin major subunit	24.3
Motility		
cheM	Methyl-accepting sensory transducer	19.7
motA	Proton conductor component of motor	10.3
motB	Enables flagellar motor rotation	8.8
Drug resistance		
yhcQ	Putative membrane located multidrug resistance protein	16.9
ramA	Putative regulatory protein of efflux pump	33.1
Porin		
STM0346	Homologue of Ail and OmpX, putative outer membrane protein	18.4
nmpC	Outer membrane protein, porin	8.9
cirA	Pori, receptor for colicin, requires TonB	8.3
Prophage		
STM2706	Fels-2 prophage	77.7
STM2595	Gifsy-1 prophage	35.8
Gene regulation		
STM0347	LuxR family putative response regulator	16.7
Ribosomal protein		
rpmC	50S ribosomal subunit protein L29	8.5
rplD	50S ribosomal subunit protein L4	8.4
rpsS	30S ribosomal subunit protein S19	8.3
Transportation		
ynfM	Putative MFS family transport protein	32.7
cysU	ABC superfamily thiosulfate transport protein	11.7
sbp	ABC superfamily sulfate transport protein	8.9
Inner membrane protein		
STM3350	Putative inner membrane protein	11.3
yjcB	Putative inner membrane protein	9.7
Periplasmic protein		
STM3650	Putative periplasmic protein	10.9
ybfA	Putative periplasmic protein	33.1
Enzymes		
gcvH	Glycine cleavage complex protein H	93.7
gcvT	Glycine cleavage complex protein H	77.7
aceK	Isocitrate dehydrogenase kinase/phosphatase	27.7
prpE	Putative acetyl-CoA synthetase	9.6

represent the ratio of *stbC*/LB5010.) This was correlated with the microarray data. Expression of type 1 fimbriae involved several Fim regulatory proteins and other gene products outside the *fim* gene cluster. *FimZ* and *FimY* are positive regulators for type 1 fimbriae, while *FimW* is a repressor for fimbrial expression [6–9].

We also used RT-PCR to investigate the transcription level of the major fimbrial subunit gene *fimA* and those of three regulatory genes *fimZ*, *fimY*, and *fimW*. Total RNAs from the parental strain *Salmonella enterica* subsp. *enterica*

serovar Typhimurium LB5010, *stbC* mutant, *stbC* (pStbC), and *stbC* (vector) were prepared and analyzed for *fimA*, *fimZ*, *fimY*, *fimW* mRNA, and 16S ribosomal (r)RNA expression by RT-PCR. Figure 4 shows the RT-PCR results. Expression levels of *fim* genes of LB5010 obtained on LB agar were used as the reference. When LB 5010 was cultured in static LB broth, *fimA*, *fimZ*, *fimY*, and *fimW* had higher levels of expression than when cultured on LB agar. The *stbC* strain demonstrated the same tendency, except for *fimW*. The expression of *fimA* obtained from the *stbC* mutant strain was

A Constitutively Mannose-Sensitive Agglutinating Salmonella enterica subsp. enterica Serovar Typhimurium Strain, Carrying a Transposon in the Fimbrial Usher Gene stbC, Exhibits Multidrug Resistance and Flagellated Phenotypes

155

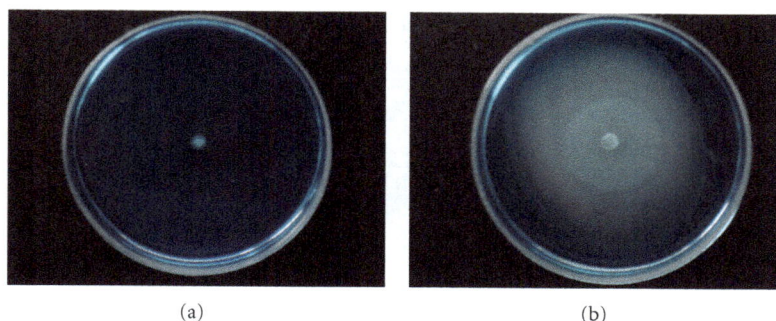

FIGURE 3: *Salmonella enterica* subsp. *enterica* serovar Typhimurium LB 5010 and the *stbC* mutant grown on MSRV agar medium. (a) *Salmonella enterica* subsp. *enterica* serovar Typhimurium LB5010 did not exhibit any "halo" appearance on the agar surface. (b) a gray-white zone was observed extending from the inoculated drop of the *stbC* mutant.

approximately 4.6-fold higher than that of the LB5010 strain when both strains were grown on LB agar. The expression of *fimZ* was also higher than that of LB5010 when grown on LB agar. As a control, 16S rRNA was consistently expressed in all of the strains tested. Transforming the recombinant plasmid harboring the *stbC* coding sequence to the *stbC* strain conferred on it the ability to express a level of *fimA* similar to that of LB5010. Gene expression of *fimZ*, *fimY*, and *fimW* was higher in *stbC* (StbC) grown in broth than on agar medium. The *stbC* (vector), possessing the cloning vector alone, exhibited similar gene expression patterns as the *stbC* strain.

4. Discussions

Salmonella enterica subsp. *enterica* serovar Typhimurium LB5010, a LT2 strain derivative, exhibits mannose-sensitive phenotype when grown in static broth culture but not on the solid agar medium culture. This strain is nonflagellated and not multidrug resistant. However, the present study revealed that a transposon inserted in the *stbC* gene of LB5010 strain conferred it to become a constitutively mannose-sensitive agglutinating, multidrug resistance, and flagellated strain.

Whole-genome sequence analysis of *Salmonella enterica* subsp. *enterica* serovar Typhimurium LT2 strain has revealed the presence of 13 gene clusters that contain open reading frames (ORFs) to encode putative fimbrial subunit and fimbrial-accessory proteins [11]. Laboratory-grown cultures of *Salmonella enterica* subsp. *enterica* serovar Typhimurium commonly produce only type 1 fimbriae and thin aggregative fimbriae (curli fimbriae) [19, 20]. Although the laboratory culture condition could not induce bacteria to produce Stb fimbriae, such fimbriae did express *in vivo* by flow cytometry using bovine ligated ileal loops [21]. The fact that mice infected with *Salmonella enterica* subsp. *enterica* serovar Typhimurium seroconvert to StbA, a major fimbrial subunit of Stb fimbriae, also provides evidence for *in vivo* Std fimbrial expression [21]. The amino acid sequence of the StbC subunit shares similarities to those of the usher proteins of the chaperone/usher assembly pathway. Usher protein is an integral outer membrane protein that interacts with the chaperone/fimbrial subunit complex, facilitating the release

of the fimbrial subunits and their secretion through the usher channel [12, 22]. This finding leads to an interesting question: Does cross-talk occur between different fimbrial systems in one bacterial strain? For example, expression of pyelonephritis-associated pili [23] represses type 1 fimbrial expression in the same *E. coli* strain [14]. A transposon inserted in the *stbC* gene of *Salmonella enterica* subsp. *enterica* serovar Typhimurium conferred it to exhibit mannose-sensitive agglutination constitutively when cultured in static broth or agar medium. Nonetheless the *stbC* mutant collected from agar medium exhibited less agglutinating power than those cultured in static broth medium. Titration of the original *stbC* mutant/agar suspension to 8-fold abolished the agglutination power. One reason could be due to the decreased number of fimbriae present on the *stbC* mutant.

Previously we have encountered difficulties when attempting to transform the recombinant plasmid containing the coding sequence of *stbC* to the *stbC* mutant to perform a complementation test. The results of our study revealed that one major reason for this could be the multidrug-resistant (MDR) phenotype of the *stbC* strain. The antimicrobial susceptibility test indicated that the *stbC* mutant resisted a battery of antibiotics. Since these antimicrobial agents have different mechanisms of antibiotic action, it is speculated that the transposon inserted in *stbC* had changed gene(s) expression that is associated with MDR. Concurrently, the microarray result indicated that *ramA* and *yhcQ* were enhanced in the *stbC* mutant compared to its parental strain LB5010. Our RT-PCR analysis also confirmed this result. RamA is a member of the AraC-XylS family of transcriptional regulators that controls one of the efflux pump genes *acrB* [24], and the product of *yhcQ* is a putative MDR pump [25]. Deletion of *yhcQ* in *E. coli* reduced the penicillin G resistance [26]. The MDR phenotype of the *stbC* mutant could be reasonably explained, at least partially, by the increased expression of *ramA* and *yhcQ*.

Salmonella enterica subsp. *enterica* serovar Typhimurium LB5010 was derived from LT2 strain, which is routinely used in laboratories for molecular genetics and as a representative of the wild type of *Salmonella enterica* subsp. *enterica* serovar Typhimurium [16]. This strain has defects in its flagellar synthesis gene, making it suitable for observing fimbrial

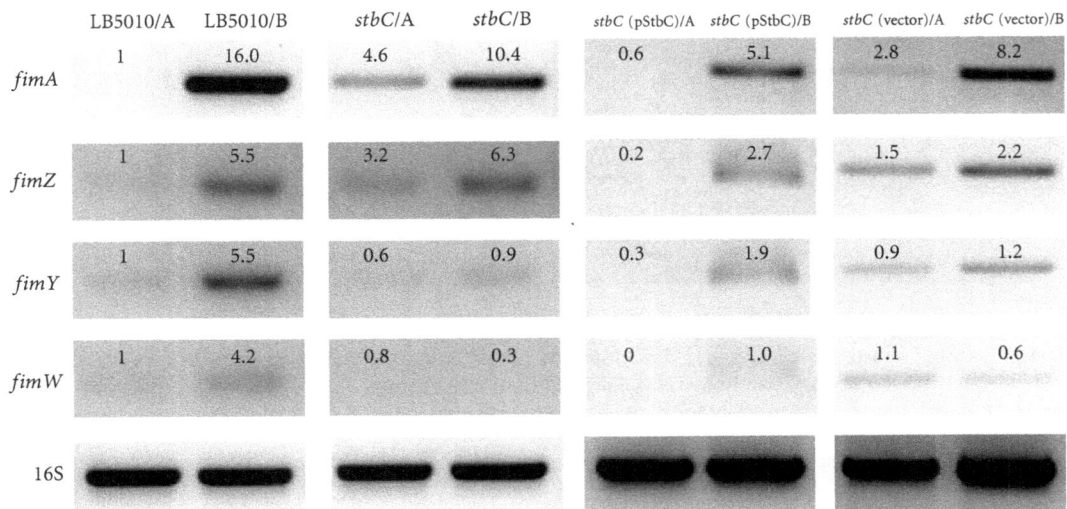

FIGURE 4: Effect of a transposon inserted in *stbC* on transcription within the *fim* gene cluster. RT-PCR assays were used to monitor *fim* gene transcription in *Salmonella enterica* subsp. *enterica* serovar Typhimurium LB5010, the *stbC* mutant, *stbC* (pStbC), and *stbC* (vector) cultured on static LB broth and solid LB agar. The intensities of the bands on the gel were determined by densitometry and are expressed relative to the value for *Salmonella enterica* subsp. *enterica* serovar Typhimurium LB5010 grown on LB agar.

appendages without flagella background [27]. Accordingly, *Salmonella enterica* subsp. *enterica* serovar Typhimurium LB5010 did not demonstrate motility on MSRV, a semisolid agar medium for isolating motile *Salmonella*. There were no flagellar structures on *Salmonella enterica* subsp. *enterica* serovar Typhimurium LB5010, prepared either from static broth or on agar medium, while the *stbC* mutant exhibited flagellar structures under electron microscopy, which was unexpected. When tested in MSRV, the *stbC* mutant exhibited extended growth from the inoculated center on MSRV, indicating the motile capability of the tested strain.

A transposon inserted in *stbC* could cause a suppression effect that alleviated the original phenotype exhibited by LB5010. However, some flagella of the *stbC* mutant observed under electron microscopy were not anchored on the bacterial cells. This could be due to the method used to prepare the samples or to the fact that some flagella were actually secreted and without function. Interestingly, microarray data and RT-PCR also indicated that the expression of three genes, *motA*, *motB*, and *cheM*, was enhanced in the *stbC* mutant compared to the LB5010. MotA and MotB are integral to the cell membrane and are required for motor rotation [28]. The product of the *cheM* gene is a methyl-accepting chemotaxis protein II, a sensory transducer. All three genes are components of a complicated chemotaxis/flagella mechanism [29]. These results were correlated with the motile characteristic of the *stbC* mutant.

One dilemma of the *stbC* mutant strain we observed was the absence of type 1 fimbriae structures under electron microscopy. Agglutination of yeast cells did appear, and it was mannose-sensitive, indicating that the agglutination was mediated by type 1 fimbriae whose receptor contains mannose residue. RT-PCR also indicated that expression of the *fimA* was higher in the *stbC* mutant than that in the parental

LB5010 when these bacteria were grown on LB agar. This finding was correlated with the fact that *stbC* had higher *fimZ* expression and lower *fimW* expression than LB5010 when grown on agar, which correlated with the previous findings. These results demonstrate that the *fimZ* gene encodes a positive regulator for *fimA*, while FimW is a repressor for *fimA* [6, 7]. FimA protein may secrete and attach on the outer membrane but did not assemble into an intact fimbrial appendage. Use of immune-gold electron microscopy would test this hypothesis. Another possibility is that *stbC* mutant grown on LB agar induced a mannose-sensitive adhesion protein that has not previously been characterized and that is present on the outer membrane. The expression of the fimbrial major subunit gene of thin aggregative fimbriae *csgA* was increased in *stbC* mutant grown on LB agar. However, this laboratory condition is not suitable for *Salmonella enterica* subsp. *enterica* serovar Typhimurium to produce thin aggregative fimbriae. Collinson et al. demonstrated that static colonization factor antigen (CFA) broth at 30°C could induce *Salmonella enterica* subsp. *enterica* serovar Enteritidis to produce thin aggregative fimbriae [30].

To better understand whether the *stbC* gene really does account for the mannose-sensitive agglutination, flagellar formation, and MDR characteristics of this interesting strain, a complementation test was performed. When a recombinant plasmid carrying the *stbC* coding sequence and a gentamycin-resistant cassette were constructed and transformed into the *stbC* strain, mannose-sensitive agglutination phenotypes and gene expression levels of the type 1 fimbrial subunit gene, *fimA*, and 3 fimbrial regulatory genes, *fimZ*, *fimY*, and *fimW*, exhibited the same tendency toward the LB5010 parental strain by the *stbC* strain. This evidence suggests that the *stbC* gene product, an usher of Stb fimbriae, may play some role in the regulatory network

A Constitutively Mannose-Sensitive Agglutinating Salmonella enterica subsp. enterica Serovar Typhimurium Strain, Carrying a Transposon in the Fimbrial Usher Gene stbC, Exhibits Multidrug Resistance and Flagellated Phenotypes

157

of type 1 fimbrial expression. The interaction of different fimbrial systems in a single strain was reported [13–15], but in the present study, we could not identify 1 or more specific *fim* genes with which StbC directly interacts, nor could we determine if other non-*fim* genes were involved in connecting these 2 different fimbrial system. However, these interesting topics warrant further investigation. Protein-protein interaction using yeast two hybrid system is being investigated in our laboratory. Complementation tests did not restore the MDR and flagellar formation characteristics of the LB5010 parental to *stbC* (pStbC). Whether this was due to polar effect or other reasons was under exploration.

Acknowledgment

This work was supported by the research funds from the Taipei Medical University/Wan Fang Hospital, Taipei, Taiwan, under the contract no. 98TMU-WFH-12. The plasmid pBBR1MCS-5 was kindly provided by Dr. Wen-Ling Deng, Department of Plant Pathology, National Chung-Hsing University, Taichung, Taiwan. The authors would like to thank Ms. H.-M. Chen, Department of Anatomy, Taipei Medical University, for assistance with electron microscopy and Ms. D.-W. Lin for molecular cloning technique assistance. The authors do not have direct financial relation with the commercial identity mentioned in the paper.

References

[1] P. S. Mead, L. Slutsker, V. Dietz et al., "Food-related illness and death in the United States," *Emerging Infectious Diseases*, vol. 5, no. 5, pp. 607–625, 1999.

[2] S. Clegg and G. F. Gerlach, "Enterobacterial fimbriae," *Journal of Bacteriology*, vol. 169, no. 3, pp. 934–938, 1987.

[3] S. Clegg and D. L. Swenson, "Salmonella fimbriae," in *Fimbriae: Adhesion, Genetics, Biogenesis, and Vaccines*, P. Klemm, Ed., pp. 105–114, CRC Press, Boca Raton, Fla, USA, 1994.

[4] D. C. Old, I. Corneil, L. F. Gibson, A. D. Thomson, and J. P. Duguid, "Fimbriation, pellicle formation and the amount of growth of salmonellas in broth," *Journal of General Microbiology*, vol. 51, no. 1, pp. 1–16, 1968.

[5] D. C. Old and J. P. Duguid, "Selective outgrowth of fimbriate bacteria in static liquid medium," *Journal of Bacteriology*, vol. 103, no. 2, pp. 447–456, 1970.

[6] K. S. Yeh, L. S. Hancox, and S. Clegg, "Construction and characterization of a fimZ mutant of *Salmonella typhimurium*," *Journal of Bacteriology*, vol. 177, no. 23, pp. 6861–6865, 1995.

[7] J. K. Tinker, L. S. Hancox, and S. Clegg, "FimW is a negative regulator affecting type 1 fimbrial expression in *Salmonella enterica* serovar Typhimurium," *Journal of Bacteriology*, vol. 183, no. 2, pp. 435–442, 2001.

[8] J. K. Tinker and S. Clegg, "Control of fimY translation and type 1 fimbrial production by the arginine tRNA encoded by *fimU* in *Salmonella enterica* serovar Typhimurium," *Molecular Microbiology*, vol. 40, no. 3, pp. 757–768, 2001.

[9] D. L. Swenson, K. J. Kim, E. W. Six, and S. Clegg, "The gene fimU affects expression of *Salmonella typhimurium* type 1 fimbriae and is related to the *Escherichia coli* tRNA gene argU," *Molecular and General Genetics*, vol. 244, no. 2, pp. 216–218, 1994.

[10] Y. C. Chuang, K. C. Wang, Y. T. Chen et al., "Identification of the genetic determinants of *Salmonella enterica* serotype Typhimurium that may regulate the expression of the type 1 fimbriae in response to solid agar and static broth culture conditions," *BMC Microbiology*, vol. 8, article 126, 2008.

[11] M. McClelland, K. E. Sanderson, J. Spieth et al., "Complete genome sequence of *Salmonella enterica* serovar Typhimurium LT2," *Nature*, vol. 413, no. 6858, pp. 852–856, 2001.

[12] D. G. Thanassi, E. T. Saulino, and S. J. Hultgren, "The chaperone/usher pathway: a major terminal branch of the general secretory pathway," *Current Opinion in Microbiology*, vol. 1, no. 2, pp. 223–231, 1998.

[13] P. Klemm, G. Christiansen, B. Kreft, R. Marre, and H. Bergmans, "Reciprocal exchange of minor components of type 1 and F1C fimbriae results in hybrid organelles with changed receptor specificities," *Journal of Bacteriology*, vol. 176, no. 8, pp. 2227–2234, 1994.

[14] N. J. Holden, M. Totsika, E. Mahler et al., "Demonstration of regulatory cross-talk between P fimbriae and type 1 fimbriae in uropathogenic *Escherichia coli*," *Microbiology*, vol. 152, no. 4, pp. 1143–1153, 2006.

[15] S. P. Nuccio, D. Chessa, E. H. Weening, M. Raffatellu, S. Clegg, and A. J. Bäumler, "SIMPLE approach for isolating mutants expressing fimbriae," *Applied and Environmental Microbiology*, vol. 73, no. 14, pp. 4455–4462, 2007.

[16] L. R. Bullas and J. I. Ryu, "Salmonella typhimurium LT2 strains which are r⁻ m⁺ for all three chromosomally located systems of DNA restriction and modification," *Journal of Bacteriology*, vol. 156, no. 1, pp. 471–474, 1983.

[17] N. C. F. C. L. Standards, *Performance Standards for Antimicrobial Susceptibility Testing; Seventeenth Informational Supplement*, National Committee for Clinical Laboratory Standards, Wayne, Pa, USA, 2007.

[18] M. E. Kovach, P. H. Elzer, D. S. Hill et al., "Four new derivatives of the broad-host-range cloning vector pBBR1MCS, carrying different antibiotic-resistance cassettes," *Gene*, vol. 166, no. 1, pp. 175–176, 1995.

[19] J. P. Duguid, E. S. Anderson, and I. Campbell, "Fimbriae and adhesive properties in *Salmonellae*," *The Journal of Pathology and Bacteriology*, vol. 92, no. 1, pp. 107–138, 1966.

[20] S. Grund and A. Weber, "A new type of fimbriae on Salmonella typhimurium," *Zentralblatt fur Veterinarmedizin*, vol. 35, no. 10, pp. 779–782, 1988.

[21] A. Humphries, S. Deridder, and A. J. Bäumler, "*Salmonella enterica* serotype typhimurium fimbrial proteins serve as antigens during infection of mice," *Infection and Immunity*, vol. 73, no. 9, pp. 5329–5338, 2005.

[22] S. P. Nuccio and A. J. Bäumler, "Evolution of the chaperone/usher assembly pathway: fimbrial classification goes Greek," *Microbiology and Molecular Biology Reviews*, vol. 71, no. 4, pp. 551–575, 2007.

[23] E. A. Duffy, K. E. Belk, J. N. Sofos, G. R. Bellinger, A. Pape, and G. C. Smith, "Extent of microbial contamination in United States pork retail products," *Journal of Food Protection*, vol. 64, no. 2, pp. 172–178, 2001.

[24] A. M. Bailey, I. T. Paulsen, and L. J. V. Piddock, "RamA confers multidrug resistance in *Salmonella enterica* via increased expression of acrB, which is inhibited by chlorpromazine," *Antimicrobial Agents and Chemotherapy*, vol. 52, no. 10, pp. 3604–3611, 2008.

[25] A. Marchler-Bauer, J. B. Anderson, M. K. Derbyshire et al., "CDD: a conserved domain database for interactive domain family analysis," *Nucleic Acids Research*, vol. 35, no. 1, pp. D237–D240, 2007.

[26] S. V. Lynch, L. Dixon, M. R. Benoit et al., "Role of the *rapA* gene in controlling antibiotic resistance of *Escherichia coli* biofilms," *Antimicrobial Agents and Chemotherapy*, vol. 51, no. 10, pp. 3650–3658, 2007.

[27] K. E. Sanderson and J. R. Roth, "Linkage map of *Salmonella typhimurium*, edition VI," *Microbiological Reviews*, vol. 47, no. 3, pp. 410–453, 1983.

[28] M. Silverman and M. Simon, "Operon controlling motility and chemotaxis in *E. coli*," *Nature*, vol. 264, no. 5586, pp. 577–580, 1976.

[29] H. Szurmant and G. W. Ordal, "Diversity in chemotaxis mechanisms among the bacteria and archaea," *Microbiology and Molecular Biology Reviews*, vol. 68, no. 2, pp. 301–319, 2004.

[30] S. K. Collinson, L. Emody, K. H. Muller, T. J. Trust, and W. W. Kay, "Purification and characterization of thin, aggregative fimbriae from *Salmonella enteritidis*," *Journal of Bacteriology*, vol. 173, no. 15, pp. 4773–4781, 1991.

21

Detection of *Helicobacter pylori* in City Water, Dental Units' Water, and Bottled Mineral Water in Isfahan, Iran

Ahmad Reza Bahrami,[1] Ebrahim Rahimi,[2] and Hajieh Ghasemian Safaei[3]

[1] *Faculty of Veterinary Medicine, Islamic Azad University, Shahrekord Branch, Shahrekord, Iran*
[2] *Department of Food Hygiene, Faculty of Veterinary Medicine, Islamic Azad University, Shahrekord Branch, Shahrekord, Iran*
[3] *Department of Microbiology, Faculty of Medical Sciences, Isfahan University of Medical Sciences, Isfahan, Iran*

Correspondence should be addressed to Hajieh Ghasemian Safaei; ghasemian@med.mui.ac.ir

Academic Editors: C. Lu and H. Marcotte

Helicobacter pylori infection in human is one of the most common infections worldwide. However, the origin and transmission of this bacterium has not been clearly explained. One of the suggested theories is transmission via water. This study was conducted to determine the prevalence rate of *H. pylori* in tap water, dental units' water, and bottled mineral water in Iran. In the present study, totally 200 water samples were collected in Isfahan province and tested for *H. pylori* by cultural method and polymerase chain reaction (PCR) by the detection of the *ureC (glmM)* gene. Using cultural method totally 5 cultures were positive. Two out of 50 tap water samples (4%), 2 out of 35 dental units' water (5.8%) samples, and 1 out of 40 (2.5%) from water cooler in public places were found to be contaminated with *H. pylori*. *H. pylori ureC* gene was detected in 14 (7%) of water samples including 5 tap water (10%), 4 dental units' water (11.4%), 1 refrigerated water with filtration, and 4 (10%) water cooler in public places samples. This may be due to the coccoid form of bacteria which is detected by PCR method.

1. Introduction

Helicobacter pylori is a gram-negative microaerophilic rod found in the human gastric mucosa and is associated with different digestive diseases, such as peptic ulcer, gastritis, and mucosa-associated lymphoid tissue lymphoma [1], and it is considered a risk factor in the development of gastric cancer [2].

H. pylori infection is frequently acquired during childhood, and symptoms such as vomiting and epigastric or recurrent abdominal pain are associated with *H. pylori* infection [3]. In developing countries, it is estimated that 70–90% of the population carries *H. pylori*, contrasting to 25–50% of infection among the inhabitants of developed countries [4]. It has been demonstrated that people living in developing countries acquire the infection earlier in life, when compared with individuals of the same age group in developed countries [5].

Despite the high incidence of the infection, the reservoirs and the transmission pathways of *H. pylori* to humans are still unclear, although multiple routes of transmission have been suggested [6]. The current literature suggests that the transmission of *H. pylori* occurs by person to person both via the oral-oral and fecal-oral routes [7]. Furthermore, many authors suggested that the human infection may occur by contaminated foods [8, 9]. Indeed, *H. pylori* has been detected from drinking water [10–13], sea water [14], and foods of animal origin, such as sheep and cow milk [15–17]. Epidemiological studies have shown that infection with *H. pylori* is associated with the level of sanitation, particularly water sanitation.

This study was conducted to determine the occurrence of *H. pylori* in tap water, dental units' water, water cooler in public places, refrigerated water with filtration, and bottled mineral water in Isfahan provinces, by means of a conventional bacteriological procedures and polymerase chain reaction (PCR).

2. Materials and Methods

2.1. Sample Collection. Isfahan province—with a population of 4,800,000 and area of 291,107,044 square kilometers—is the second biggest province of Iran and is located in the central part of Iran among Iran's central mountains eastern hillside of Zagros at the margin of the Zayande-Rood River. The drinking water of this province is supplied from the Zayande-Rood River which is considered to be surface water. It is probable that this water is contaminated with industrial and urban sewerage at the margin of this river. Other than refinement, water receives no treatment such as radiation. Taking all this, and given that this river is the only water source for companies producing bottled mineral water in this province, it is likely that one of the sources of microbial contamination in this area is water. In this study, a total of 200 samples including 50 samples of tape water, 35 samples of dental units, 30 samples of home refrigerator with filtration system, 25 samples tape water equipped with filtration system, 40 samples from water cooler in public places, and 20 samples of mineral bottled water were examined over a period of 6 months, from July to December 2011 from four different geographical regions of Isfahan province. For each region, 10–15 samples were collected in 1,000 mL glass bottles containing 0.5 g of sodium thiosulphate for dechlorination of the water. The 20 bottled mineral water samples were purchased from five different companies (using the same water system) on the day that the experiment was conducted.

2.2. Isolation of Helicobacter pylori. Samples of 1000 mL water collected in sterile glass flasks and transferred to laboratory within 2 hours. Samples were filtered through 0.045 μm filter membrane (Albet Co.). Each membrane was then immersed into 2 mL of tryptic soy broth (TSB) for 1 h. After that each 2 mL TSB was taken and cultured for *H. pylori* and DNA extraction. Samples were cultured on Brucella agar (Merck, Germany) containing campylobacter selective supplement (5 mg/L, Merck), trimethoprim (0.25 mg/L), amphotericin B, sheep blood (5%), and 7% fetal calf serum (Sigma). After 72 h incubation at 37°C in microaerophilic condition (5% O_2, 85% N_2, 10% CO_2) using MART system (Anoxamat, Lichtenvoorde, The Netherlands), the bacterial growth was tested and confirmed as *H. pylori* by gram staining, urease, and oxidase tests [1]. The isolates were identified as *H. pylori* were also positive, using the PCR assay. For comparison, a reference strain of *H. pylori* (ATCC 43504) was employed.

2.3. Detection of Helicobacter pylori Using PCR Method. DNA was extracted by a DNA isolation kit from mentioned TSB (Roche Applied Science, Germany) according to the manufacturer's instructions, and its density was assessed by optic densitometry. Extracted genomic DNA was amplified for the *ureC* (*glmM*) gene and detected with the specific primers HP-F: 5′-GAATAAGCTTTTAGGGGTGTTAGGGG-3′ and HP-R: 5′-AAGCTTACTTTCTAACACTAACGCGC-3′. The gene product was 294 bp. PCR reactions were performed in a final volume of 50 μL containing 5 μL 10 × buffer + MgCl$_2$,

2 mM dNTP, 2 unit Taq DNA polymerase, 100 ng genomic DNA as a template, and 25 picomoles of each primer. PCR was performed using a thermal cycler (Eppendorf Co., Germany) under the following conditions: an initial denaturation for 10 minutes at 94°C; and 35 cycles for 1 minute at 94°C, 1 minute at 55°C, 1 minute at 72°C, and a final extension at 72°C for 10 minutes. The PCR products were electrophoresed through a 1.5% agarose gels (Fermentas, Germany) containing Ethidium bromide. A DNA ladder (Fermentas Co., Germany) was used to detect the molecular weight of observed bands under a UV lamp. All tests were performed in triplicate. Samples inoculated with *H. pylori* were used as positive controls.

2.4. Statistical Analysis. Data were transferred to Microsoft Excel spreadsheet (Microsoft Corp., Redmond, WA, USA) for analysis. Using SPSS 16.0 statistical software (SPSS Inc., Chicago, IL, USA), Chi-square test and Fisher's exact two-tailed test analysis was performed, and differences were considered significant at values of $P < 0.05$.

3. Results and Discussion

Using traditional bacteriologic methods, totally 5 cultures were positive. Two of 50 tap water samples (4%), 2 out of 35 dental units' water (5.8%) samples, and 1 of 40 (2.5%) from water cooler in public places were found to be contaminated with *H. pylori*. *H. pylori ureC* gene was detected in 14 (7%) of water samples including 5 tap water (10%), 4 dental units' water (11.4%), and 4 (10%) water cooler in public places samples (Table 1). Statistically significant differences ($P > 0.05$) were not observed in the prevalence of *H. pylori ureC* gene in water samples collected from different geographical regions of Isfahan province.

The association of serum antibodies against *H. pylori* with serum antibodies against two known waterborne pathogens hepatitis A virus [18] and Giardia [19] suggests that the infection may be waterborne or related to poor sanitary practices [20]. Klein et al. [21] studied the prevalence of *H. pylori* infection in 407 children (two months to 12 years old), in Lima, Peru. *H. pylori* infection rate was 56% among children from low-income families and 32% among those from high-income families. However, children from high-income families whose homes were supplied with municipal water were 12 times more likely to be infected than those from the same socioeconomic status whose water supply came from community wells. These results showed that the acquisition of *H. pylori* infection by Peruvian children was correlated with socioeconomic status, but additionally the municipal water supply seemed to be involved in the spread of infection among them. Indirect evidence that the transmission of *H. pylori* is waterborne is based upon four sets of data: (i) presence of DNA in water samples, (ii) observation of coccoid forms in water samples, (iii) survival of *H. pylori* in artificially contaminated water, and (iv) growth of *H. pylori* from water samples [20].

In the present study, only two tap water samples (4%) were found to be contaminated with *H. pylori* using traditional bacteriologic methods. *H. pylori* has rarely been

TABLE 1: Frequency of *Helicobacter pylori* detected in different water samples in Iran by PCR.

Water sample	No. of samples	No. of *H. pylori*-positive by culture*	No. of *H. pylori*-positive by PCR
Tap water	50	2 (4.0%)	5 (10.0%)
Dental units' water	35	2 (5.8%)	4 (11.4%)
Bottled mineral water	20	0	0
Refrigerated water with filtration	30	0	1 (3.3%)
Tape water equipped with filtration system	25	0	0
Water cooler in public places	40	1 (2.5%)	4 (10.0%)
Total	200	5 (2.5 %)	14 (7.0%)

*Results expressed as the number of *H. pylori*-positive samples (percent positive samples analyzed).

isolated from water samples [22, 23]. In several studies no *H. pylori* was found in water samples [8, 24, 25]. This could be attributed to the fact that *H. pylori* can survive for short period of time in water [8, 20]. Moreover, the method employed for *H. pylori* isolation may lack sufficient sensitivity to recover very low numbers of *H. pylori* [2, 23, 26, 27].

Two out of 35 dental units' water samples were found to be contaminated with *H. pylori* using traditional bacteriologic methods. The presence of *H. pylori* associated with biofilms from wells, rivers, and water distribution systems has been reported by different investigators [28–32]. Biofilms are slimy films of bacteria, other microbes, and organic materials that cover underwater surfaces, particularly inside plumbing. This makes them rather inaccessible and provides a matrix difficult to be reached by disinfectants. The detachment of biofilms is the principal form of contamination of treated water [33, 34]. Taken together, these results suggest that biofilms in water distribution systems are responsible for the contamination of water.

In this study, *ureC* gene of *H. pylori* was detected in tap water, dental units' water, refrigerated water with filtration, and public cooler water samples. *H. pylori* DNA has been identified in several water sources using diverse gene targets. Drinking, river, sea, ground, and wastewater have provided positive results by PCR analysis [9, 14, 23, 24, 35–37]. The *H. pylori* DNA present in water samples could be from dead *H. pylori* cells or from VBNC forms, since culture is usually not possible. Water spiked with viable *H. pylori* cells rapidly led to the observation of coccoid forms [23, 27, 38, 39]. Whether the coccoid form of *H. pylori* is viable in the dormant state or is degenerative and undergoing apoptosis is still an unanswered question. Coccoid *H. pylori* appears to conserve the capacity to produce proteins for at least 100 days when stored at 4°C, in either phosphate-buffered saline (PBS) or distilled water [40]. It has been suggested that although the virulence of coccoid *H. pylori* induced by water decreases, the coccoid forms still retain a considerable urease activity and preserve adhering ability to epithelial cells. These coccoid forms induced by water have been capable of colonizing the gastric mucosa, causing gastritis in mice [41].

The PCR assay employed in this work specifically targets a region of the *ureC* (*glmM*) gene which has been shown to be unique and essential for the growth of *H. pylori*. It has been previously reported that detecting this gene improves sensitivity and specificity of recognition of *H. pylori* in

samples containing prokaryotic cells as well as many organic impurities [9, 17, 37]. However, because the PCR assay detects *ureC* (*glmM*) gene of *H. pylori*, we are unable to speculate on the viability of organisms in water samples.

The high prevalence of *H. pylori* isolated from healthy human carrier [42, 43] suggests that water contamination is due to poor hygiene management. Therefore, the consumption of tap water and dental units' water would be a potential risk of *H. pylori* infection for the consumer. To the author's knowledge, the present study is the first report of the isolation of *H. pylori* from water in Iran and the first demonstration of *H. pylori* DNA in tap water, dental units' water, refrigerated water with filtration, and public water cooler samples. Further studies will be necessary to determine the prevalence of *H. pylori* in water and other foods in Iran and to explore the potential risk of human infection with *H. pylori* via consumption of water and foods.

Acknowledgment

This work was supported by Vice Chancellor of Research of Isfahan University of Medical Sciences, Isfahan, Iran.

References

[1] B. E. Dunn, H. Cohen, and M. J. Blaser, "*Helicobacter pylori*," *Clinical Microbiology Reviews*, vol. 10, no. 4, pp. 720–741, 1997.

[2] World Health Organization, "Infection with *Helicobacter pylori*," *IARC Monographs on the Evaluation of Carcinogenic Risks To Humans*, vol. 61, pp. 177–240, 1994.

[3] B. Drumm, S. Koletzko, and G. Oderda, "*Helicobacter pylori* infection in children: a consensus statement," *Journal of Pediatric Gastroenterology and Nutrition*, vol. 30, no. 2, pp. 207–213, 2000.

[4] L. M. Brown, "*Helicobacter pylori*: epidemiology and routes of transmission," *Epidemiologic Reviews*, vol. 22, no. 2, pp. 283–297, 2000.

[5] A. Lee, "The microbiology and epidemiology of *Helicobacter pylori* infection," *Scandinavian Journal of Gastroenterology*, vol. 29, no. S201, pp. 2–6, 1994.

[6] R. P. Allaker, K. A. Young, J. M. Hardie, P. Domizio, and N. J. Meadows, "Prevalence of *Helicobacter pylori* at oral and gastrointestinal sites in children: evidence for possible oral-to-oral transmission," *Journal of Medical Microbiology*, vol. 51, no. 4, pp. 312–317, 2002.

[7] L. Cellini, L. Marzio, G. Ferrero et al., "Transmission of *Helicobacter pylori* in an animal model," *Digestive Diseases and Sciences*, vol. 46, no. 1, pp. 62–68, 2001.

[8] B. C. Gomes and E. C. P. De Martinis, "The significance of *Helicobacter pylori* in water, food and environmental samples," *Food Control*, vol. 15, no. 5, pp. 397–403, 2004.

[9] N. C. Quaglia, A. Dambrosio, G. Normanno et al., "High occurrence of *Helicobacter pylori* in raw goat, sheep and cow milk inferred by glmM gene: a risk of food-borne infection?" *International Journal of Food Microbiology*, vol. 124, no. 1, pp. 43–47, 2008.

[10] M. K. Glynn, C. R. Friedman, B. D. Gold et al., "Seroincidence of *Helicobacter pylori* infection in a cohort of rural Bolivian children: acquisition and analysis of possible risk factors," *Clinical Infectious Diseases*, vol. 35, no. 9, pp. 1059–1065, 2002.

[11] J. P. Hegarty, M. T. Dowd, and K. H. Baker, "Occurrence of *Helicobacter pylori* in surface water in the United States," *Journal of Applied Microbiology*, vol. 87, no. 5, pp. 697–701, 1999.

[12] Y. Lu, T. E. Redlinger, R. Avitia, A. Galindo, and K. Goodman, "Isolation and genotyping of *Helicobacter pylori* from untreated municipal wastewater," *Applied and Environmental Microbiology*, vol. 68, no. 3, pp. 1436–1439, 2002.

[13] N. Queralt, R. Bartolomé, and R. Araujo, "Detection of *Helicobacter pylori* DNA in human faeces and water with different levels of faecal pollution in the north-east of Spain," *Journal of Applied Microbiology*, vol. 98, no. 4, pp. 889–895, 2005.

[14] L. Cellini, A. Del Vecchio, M. Di Candia, E. Di Campli, M. Favaro, and G. Donelli, "Detection of free and plankton-associated *Helicobacter pylori* in seawater," *Journal of Applied Microbiology*, vol. 97, no. 2, pp. 285–292, 2004.

[15] M. P. Dore, A. R. Sepulveda, and H. El-Zimaty, "Isolation of *Helicobacter pylori* from milk sheep-implications for transmission to humans," *American Journal of Gastroenterology*, vol. 96, pp. 1396–1401, 2001.

[16] S. Fujimura, T. Kawamura, S. Kato, H. Tateno, and A. Watanabe, "Detection of *Helicobacter pylori* in cow's milk," *Letters in Applied Microbiology*, vol. 35, no. 6, pp. 504–507, 2002.

[17] H. G. Safaei, E. Rahimi, A. Zandi, and A. Rashidipour, "*Helicobacter pylori* as a zoonotic infection: the detection of *H. pylori* antigens in the milk and faeces of cows," *Journal of Research in Medical Sciences*, vol. 16, no. 2, pp. 184–187, 2011.

[18] A. R. Bizri, I. A. Nuwayhid, G. N. Hamadeh, S. W. Steitieh, A. M. Choukair, and U. M. Musharrafieh, "Association between hepatitis A virus and *Helicobacter pylori* in a developing country: the saga continues," *Journal of Gastroenterology and Hematology*, vol. 21, no. 10, pp. 1615–1621, 2006.

[19] E. D. Moreira, V. B. Nassri, R. S. Santos et al., "Association of *Helicobacter pylori* infection and giardiasis: results from a study of surrogate markers for fecal exposure among children," *World Journal of Gastroenterology*, vol. 11, no. 18, pp. 2759–2763, 2005.

[20] F. F. Vale and J. M. B. Vítor, "Transmission pathway of *Helicobacter pylori*: does food play a role in rural and urban areas?" *International Journal of Food Microbiology*, vol. 138, no. 1-2, pp. 1–12, 2010.

[21] P. D. Klein, R. Gilman, R. Leon-Barua et al., "Water source as risk factor for *Helicobacter pylori* infection in Peruvian children," *The Lancet*, vol. 337, no. 8756, pp. 1503–1506, 1991.

[22] N. F. Azevedo, C. Almeida, I. Fernandes et al., "Survival of gastric and enterohepatic *Helicobacter* spp. in water: Implications for transmission," *Applied and Environmental Microbiology*, vol. 74, no. 6, pp. 1805–1811, 2008.

[23] N. Queralt and R. Araujo, "Analysis of the survival of *H. pylori* within a laboratory-based aquatic model system using molecular and classical techniques," *Microbial Ecology*, vol. 54, no. 4, pp. 771–777, 2007.

[24] K. Hulten, S. W. Han, H. Enroth et al., "*Helicobacter pylori* in the drinking water in Peru," *Gastroenterology*, vol. 110, no. 4, pp. 1031–1035, 1996.

[25] S. L. Percival and J. G. Thomas, "Transmission of *Helicobacter pylori* and the role of water and biofilms," *Journal of Water and Health*, vol. 7, no. 3, pp. 469–477, 2009.

[26] A. S. Angelidis, I. Tirodimos, and M. Bobos, "Detection of *Helicobacter pylori* in raw bovine milk by fluorescence in situ hybridization (FISH)," *International Journal of Food Microbiology*, vol. 151, no. 2, pp. 252–256, 2011.

[27] Y. Moreno, P. Piqueres, J. L. Alonso, A. Jimenez, A. Gonzalez, and M. A. Ferrus, "Survival and viability of *Helicobacter pylori* after inoculation into chlorinated drinking water," *Water Research*, vol. 41, no. 15, pp. 3490–3496, 2007.

[28] J. E. G. Bunn, W. G. MacKay, J. E. Thomas, D. C. Reid, and L. T. Weaver, "Detection of *Helicobacter pylori* DNA in drinking water biofilms: implications for transmission in early life," *Letters in Applied Microbiology*, vol. 34, no. 6, pp. 450–454, 2002.

[29] S. M. Bragança, N. F. Azevedo, L. C. Simões, C. W. Keevil, and M. J. Vieira, "Use of fluorescent in situ hybridisation for the visualisation of *Helicobacter pylori* in real drinking water biofilms," *Water Science and Technology*, vol. 55, no. 8-9, pp. 387–393, 2007.

[30] M. S. Giao, N. F. Azevedo, S. A. Wilks, M. J. Vieira, and C. W. Keevil, "Persistence of *Helicobacter pylori* in heterotrophic drinking-water biofilms," *Applied and Environmental Microbiology*, vol. 74, no. 19, pp. 5898–5904, 2008.

[31] S. R. Park, W. G. Mackay, and D. C. Reid, "*Helicobacter* sp. recovered from drinking water biofilm sampled from a water distribution system," *Water Research*, vol. 35, no. 6, pp. 1624–1626, 2001.

[32] C. L. Watson, R. J. Owen, B. Said et al., "Detection of *Helicobacter pylori* by PCR but not culture in water and biofilm samples from drinking water distribution systems in England," *Journal of Applied Microbiology*, vol. 97, no. 4, pp. 690–698, 2004.

[33] M. Gouider, J. Bouzid, S. Sayadi, and A. Montiel, "Impact of orthophosphate addition on biofilm development in drinking water distribution systems," *Journal of Hazardous Materials*, vol. 167, no. 1–3, pp. 1198–1202, 2009.

[34] P. Stoodley, S. Wilson, L. Hall-Stoodley, J. D. Boyle, H. M. Lappin-Scott, and J. W. Costerton, "Growth and detachment of cell clusters from mature mixed-species biofilms," *Applied and Environmental Microbiology*, vol. 67, no. 12, pp. 5608–5613, 2001.

[35] T. Horiuchi, T. Ohkusa, M. Watanabe, D. Kobayashi, H. Miwa, and Y. Eishi, "*Helicobacter pylori* DNA in drinking water in Japan," *Microbiology and Immunology*, vol. 45, no. 7, pp. 515–519, 2001.

[36] M. Mazari-Hiriart, Y. López-Vidal, and J. J. Calva, "*Helicobacter pylori* in water systems for human use in Mexico City," *Water Science and Technology*, vol. 43, no. 12, pp. 93–98, 2001.

[37] N. C. Quaglia, A. Dambrosio, G. Normanno, and G. V. Celano, "Evaluation of a Nested-PCR assay based on the phosphoglucosamine mutase gene (glmM) for the detection of *Helicobacter pylori* from raw milk," *Food Control*, vol. 20, no. 2, pp. 119–123, 2009.

[38] B. L. Adams, T. C. Bates, and J. D. Oliver, "Survival of *Helicobacter pylori* in a natural fresh water environment," *Applied*

and Environmental Microbiology, vol. 69, no. 12, pp. 7462–7466, 2003.

[39] A. K. Nayak and J. B. Rose, "Detection of *Helicobacter pylori* in sewage and water using a new quantitative PCR method with SYBR green," *Journal of Applied Microbiology*, vol. 103, no. 5, pp. 1931–1941, 2007.

[40] K. Mizoguchi, S. D. Meyers, S. Basu, and J. J. O'Brien, "Multi- and quasidecadal variations of sea surface temperature in the North Atlantic," *Journal of Physical Oceanography*, vol. 29, no. 12, pp. 3133–3144, 1999.

[41] F. F. She, J. Y. Lin, J. Y. Liu, C. Huang, and D. H. Su, "Virulence of water-induced coccoid *Helicobacter pylori* and its experimental infection in mice," *World Journal of Gastroenterology*, vol. 9, no. 3, pp. 516–520, 2003.

[42] G. I. Perez-Perez, D. Rothenbacher, and H. Brenner, "Epidemiology of *Helicobacter pylori* infection," *Helicobacter*, vol. 9, supplement 1, pp. 1–6, 2004.

[43] D. Olivares and J. P. Gisbert, "Factors involved in the pathogenesis of *Helicobacter pylori* infection," *Revista Espanola de Enfermedades Digestivas*, vol. 98, no. 5, pp. 374–386, 2006.

22

In Vitro Antibacterial and Antifungal Activity of Salicylanilide Benzoates

Martin Krátký,[1] Jarmila Vinšová,[1] and Vladimír Buchta[2, 3]

[1] Department of Inorganic and Organic Chemistry, Faculty of Pharmacy, Charles University, Heyrovského 1203,
  500 05 Hradec Králové, Czech Republic
[2] Department of Clinical Microbiology, Faculty of Medicine and University Hospital, Charles University, Sokolská 581,
  500 12 Hradec Králové, Czech Republic
[3] Department of Biological and Medical Sciences, Faculty of Pharmacy, Charles University, Heyrovského 1203,
  500 05 Hradec Králové, Czech Republic

Correspondence should be addressed to Jarmila Vinšová, vinsova@faf.cuni.cz

Academic Editor: Adam Shih-Yuan Lee

The resistance to antimicrobial agents brings a need of novel antimicrobial agents. We have synthesized and found the *in vitro* antibacterial activity of salicylanilide esters with benzoic acid (2-(phenylcarbamoyl)phenyl benzoates) in micromolar range. They were evaluated *in vitro* for the activity against eight fungal and eight bacterial species. All derivatives showed a significant antibacterial activity against Gram-positive strains with minimum inhibitory concentrations $\geq 0.98\,\mu$mol/L including methicillin-resistant *Staphylococcus aureus* strain. The most active compounds were 5-chloro-2-(3,4-dichlorophenylcarbamoyl)phenyl benzoate and 4-chloro-2-(4-(trifluoromethyl)phenylcarbamoyl)phenyl benzoate. The antifungal activity is significantly lower.

1. Introduction

The worldwide epidemic of antibiotic resistance is in danger of ending the "golden age" of antibiotic therapy and therefore is touching all people [1]. Major current problems arise from the spread of nosocomial antibiotic-resistant bacteria such as methicillin-resistant *Staphylococcus aureus* (MRSA), extended-spectrum β-lactamases-producing (ESBL) *Escherichia coli* or *Klebsiella* spp., multiresistant *Pseudomonas*, or *Acinetobacter* sp. as well as *Clostridium difficile* [2].

The situation in the fungal kingdom is a bit different. Over the past decades there has been a growing number of immunocompromised patients (e.g., patients with AIDS or after transplantations) who can develop opportunistic mycoses caused by expanding spectrum of fungal pathogens, including those with problematic susceptibility to current antifungal drugs. Pathogenic fungi can use different mechanisms of resistance to diverse drugs with unrelated modes of action [3].

The searching for potential antimicrobial agents is still challenging and new groups of compounds are desired [4].

Various salicylanilide (2-hydroxy-*N*-phenylbenzamide) esters have displayed good antibacterial and antifungal activities, especially against Gram-positive strains [5–7]. Recently salicylanilides were described besides an excellent antibacterial acting against both drug-sensitive and methicillin-resistant *S. aureus* inhibition activity towards bacterial transglycosylase, an enzyme necessary for the formation of the cell wall [8].

Benzoic acid alone is known as a nonspecific antimicrobial agent with the wide spectrum of the activities against human pathogenic fungi and bacteria with different minimum inhibitory concentration (MIC) values [9–14]; moreover it was being evaluated as an inhibitor of β-carbonic anhydrase, a new molecular target occurring in *C. albicans* and *Cryptococcus neoformans* [15]. The review of the benzoic acid as preservative agent, the mechanisms of action, and resistance were published [16].

Based on these facts, we designed and evaluated new salicylanilide benzoates as potential antibacterial and antifungal agents.

2. Material and Methods

2.1. Chemistry. Salicylanilides were prepared by the procedure described previously [7]. The esters were prepared from salicylanilides by using benzoic acid and *N,N′*-dicyclohexylcarbodiimide as dehydrating and condensation agent (e.g., [7]). The general structure is presented in the head of Table 1.

All used chemicals were purchased from commercial sources (Sigma-Aldrich) and they were used without a further purification. Reactions were monitored by thin-layer chromatography plates coated with 0.2 mm silica gel 60 F_{254} (Merck) visualized by UV irradiation (254 nm). All synthesized compounds were characterized. Elemental analysis (C, H, N) was performed on an automatic microanalyser CHNS-O CE instrument (FISONS EA 1110, Italy). Melting points were determined on a Melting Point machine B-540 (Büchi) apparatus using open capillaries and they are uncorrected. Infrared spectra (ATR) were recorded on FT-IR spectrometer Nicolet 6700 FT-IR in the range of 400–4000 cm^{-1}. The NMR spectra were recorded on a Varian VNMR S500 (500 MHz for ^{1}H and 125 MHz for ^{13}C; Varian, Inc., Palo Alto, USA) at ambient temperature using deuterated dimethyl sulfoxide (DMSO-*d6*) solutions of the samples. The chemical shifts δ are given in ppm, with respect to tetramethylsilane as an internal standard. The coupling constants (J) are reported in Hz.

2.2. Biology

2.2.1. Antibacterial Evaluation. The *in vitro* antibacterial activity was assayed against next Gram-positive and Gram-negative strains: *Staphylococcus aureus* CCM 4516/08, methicillin-resistant *Staphylococcus aureus* H 5996/08 (MRSA), *Staphylococcus epidermidis* H 6966/08, *Enterococcus* sp. J 14365/08, *Escherichia coli* CCM4517, *Klebsiella pneumoniae* D 11750/08, ESBL-positive *Klebsiella pneumoniae* J 14368/08, and *Pseudomonas aeruginosa* CCM 1961.

The microdilution broth method modified according to standard M07-A07 [17] in Mueller-Hinton broth (HiMedia Laboratories, India) was adjusted to pH 7.4 (±0.2). The investigated compounds were dissolved in DMSO to the final concentrations ranging from 500 to 0.49 μmol/L. Penicillin G (benzylpenicillin) and benzoic acid were used as comparative standard drugs. Bacterial inoculum in sterile water was prepared to match 0.5 McFarland scale (1.5 × 10^8 CFU/mL). The minimum inhibitory concentrations (MICs) were assayed as 80% (IC$_{80}$) or higher reduction of growth in comparison to the control. The determination of results was performed visually and spectrophotometrically (at 540 nm). The values of MICs were determined after 24 and 48 h of incubation in the darkness at 35°C (±0.1) in a humid atmosphere.

2.2.2. Antifungal Evaluation. The inhibitory activity was determined *in vitro* against four yeast strains (*Candida albicans* ATCC 44859, *Candida tropicalis* 156, *Candida krusei* E28, and *Candida glabrata* 20/I) and four moulds (*Trichosporon asahii* 1188, *Aspergillus fumigatus* 231, *Absidia corymbifera* 272, and *Trichophyton mentagrophytes* 445).

The method used was microdilution broth method in the format of the CLSI M27-A3 and M38 A2 guidelines for yeasts and moulds [18, 19] in RPMI 1640 with glutamine (KlinLab, the Czech Republic) buffered to pH 7.0 with 0.165 M of 3-morpholino-propane-1-sulphonic acid (Sigma-Aldrich, Germany). DMSO served as a diluent for all compounds. Fungal inoculum was prepared to give a final concentration of 5 × 10^3 ± 0.2 CFU/mL. Fluconazole was used as a reference drug. Other conditions were the same as for antibacterial assay; only for *T. mentagrophytes* the final MIC were determined after 72 and 120 h of incubation. MICs were determined twice and in duplicate.

3. Results and Discussion

3.1. Chemistry. Eighteen new salicylanilide benzoates were synthesized. The yields ranged from 44 to 88%.

4-Chloro-2-(3-chlorophenylcarbamoyl)phenyl Benzoate (1). White solid; yield 82%; mp 146.5–149°C. IR (ATR): 3325 (NH amide; m), 3081, 2932, 2853, 1716 (CO ester; s), 1672 (CO amide; s), 1590, 1525, 1483, 1451, 1424, 1309, 1286, 1267, 1251, 1208, 1181, 1104, 1085, 1067, 1023, 902, 873, 786, 735, 703, 681. ^{1}H NMR (500 MHz, DMSO): δ 10.69 (1H, bs, NH), 8.07 (2H, d, J = 7.5 Hz, H2″, H6″), 7.83 (1H, d, J = 2.5 Hz, H3), 7.76–7.69 (3H, m, H5, H6, H2′), 7.56 (1H, t, J = 7.7 Hz, H4″), 7.53–7.49 (3H, m, H6′, H3″, H5″), 7.31 (1H, t, J = 8.1 Hz, H5′), 7.11 (1H, dd, J = 1.9 Hz, J = 7.9 Hz, H4′). ^{13}C NMR (125 MHz, DMSO): δ 164.3, 163.0, 147.0, 140.3, 134.4, 133.1, 131.8, 131.2, 130.6, 130.4, 130.0, 129.1, 128.6, 128.0, 125.7, 123.8, 119.4, 118.4. Anal. Calcd. for $C_{20}H_{13}Cl_2NO_3$ (386.23): C, 62.19; H, 3.39; N, 3.63. Found: C, 61.89; H, 3.50; N, 3.87.

5-Chloro-2-(3-chlorophenylcarbamoyl)phenyl Benzoate (2). White solid; yield 68%; mp 166–168°C. IR (ATR): 3282 (NH amide; m), 3072, 1739 (CO ester; s), 1647 (CO amide; s), 1600, 1589, 1548, 1481, 1450, 1410, 1320, 1255, 1241, 1192, 1075, 1051, 1021, 915, 896, 873, 854, 829, 782, 702, 676, 660. ^{1}H NMR (500 MHz, DMSO): δ 10.65 (1H, bs, NH), 8.08 (2H, d, J = 7.9 Hz, H2″, H6″), 7.79 (1H, d, J = 8.3 Hz, H3), 7.76–7.67 (3H, m, H4, H6, H2′), 7.58–7.48 (4H, m, H6′, H3″, H4″, H5″) 7.30 (1H, t, J = 8.1 Hz, H5′), 7.10 (1H, dd, J = 1.8 Hz, J = 7.9 Hz, H4′). ^{13}C NMR (125 MHz, DMSO): δ 164.2, 163.5, 148.9, 140.3, 135.8, 134.4, 133.1, 131.0, 130.6, 130.1, 129.1, 128.6, 128.5, 126.5, 124.0, 123.7, 119.4, 118.4. Anal. Calcd. for $C_{20}H_{13}Cl_2NO_3$ (386.23): C, 62.19; H, 3.39; N, 3.63. Found: C, 62.34; H, 3.22; N, 3.79.

4-Chloro-2-(4-chlorophenylcarbamoyl)phenyl Benzoate (3). White solid; yield 80%; mp 185–187°C. IR (ATR): 3309 (NH amide; m), 3072, 2928, 2850, 1741 (CO ester; s), 1649 (CO

TABLE 1: Antibacterial activity of benzoates **1-18**.

			MIC/IC_{80} [μmol/L]									
	R^1	R^2	*Staphylococcus aureus*		MRSA *S. aureus*		*Staphylococcus epidermidis*		*Enterococcus* sp.		*Pseudomonas aeruginosa*	
			24 h	48 h	24 h	48 h	24 h	48 h	24 h	48 h	24 h	48 h
(1)	4-Cl	3-Cl	**0.98**	3.9	**0.98**	3.9	1.95	7.81	31.25	62.5	**3.9**	7.81
(2)	5-Cl	3-Cl	3.9	3.9	3.9	3.9	3.9	62.5	>125	>125	>125	>125
(3)	4-Cl	4-Cl	3.9	3.9	3.9	3.9	>125	>125	>125	>125	>125	>125
(4)	5-Cl	4-Cl	15.62	31.25	15.62	31.25	31.25	31.25	>125	>125	>125	>125
(5)	4-Cl	3,4-diCl	**1.95**	3.9	**1.95**	7.81	1.95	1.95	1.95	1.95	>125	>125
(6)	5-Cl	3,4-diCl	**1.95**	1.95	**1.95**	1.95	1.95	1.95	1.95	1.95	>125	>125
(7)	4-Cl	3-Br	3.9	7.81	3.9	7.81	1.95	3.9	7.81	>125	>125	>125
(8)	5-Cl	3-Br	**1.95**	3.9	3.9	7.81	**1.95**	3.9	7.81	>125	>125	>125
(9)	4-Cl	4-Br	**1.95**	7.81	7.81	15.62	3.9	15.62	125	>125	>125	>125
(10)	5-Cl	4-Br	62.5	>125	125	>125	15.62	>125	>125	>125	>125	>125
(11)	4-Cl	3-F	3.9	7.81	**3.9**	7.81	1.95	7.81	31.25	>125	>125	>125
(12)	5-Cl	3-F	3.9	7.81	**3.9**	15.62	1.95	7.81	7.81	>125	>125	>125
(13)	4-Cl	4-F	7.81	7.81	15.62	15.62	1.95	3.9	62.5	>125	>125	>125
(14)	5-Cl	4-F	3.9	7.81	7.81	15.62	1.95	3.9	>125	>125	>125	>125
(15)	4-Cl	4-CF_3	**1.95**	1.95	**1.95**	1.95	1.95	1.95	1.95	1.95	>125	>125
(16)	5-Cl	4-CF_3	3.9	3.9	3.9	3.9	3.9	3.9	>125	>125	>125	>125
(17)	4-Cl	3-CF_3	**1.95**	3.9	**1.95**	7.81	1.95	3.9	7.81	>125	>125	>125
(18)	4-Br	4-CF_3	**1.95**	1.95	**1.95**	1.95	1.95	1.95	1.95	>125	>125	>125
PNC	—	—	**0.98**	0.98	62.5	125	250	250	7.81	15.62	>500	>500
BA	—	—	>500	>500	>500	>500	>500	>500	>500	>500	>500	>500

PNC: penicillin G; BA: benzoic acid. The lowest MIC value(s) for each strain are bolded.

amide; s), 1593, 1543, 1537, 1490, 1451, 1405, 1314, 1257, 1245, 1197, 1099, 1053, 1023, 875, 836, 814, 724, 706, 669. ^{1}H NMR (500 MHz, DMSO): δ 10.65 (1H, bs, NH), 8.06 (2H, d, J = 7.2 Hz, H2″, H6″), 7.82 (1H, d, J = 2.6 Hz, H3), 7.73–7.68 (2H, m, H5, H6), 7.63 (2H, d, J = 8.9 Hz, H2′, H6′), 7.55 (1H, t, J = 7.8 Hz, H4″), 7.50 (2H, t, J = 8.7 Hz, H3″, H5″), 7.33 (2H, d, J = 8.9 Hz, H3′, H5′). ^{13}C NMR (125 MHz, DMSO): δ 164.3, 162.8, 146.9, 137.8, 134.3, 131.6, 131.3, 130.4, 130.0, 129.1, 128.7, 128.0, 127.7, 127.1, 125.7, 121.5. Anal. Calcd. for $C_{20}H_{13}Cl_2NO_3$ (386.23): C, 62.19; H, 3.39; N, 3.63. Found: C, 62.00; H, 3.45; N, 3.87.

5-Chloro-2-(4-chlorophenylcarbamoyl)phenyl Benzoate (**4**). White solid; yield 81%; mp 192–194°C. IR (ATR): 3344 (NH amide; m), 3070, 1746 (CO ester; s), 1650 (CO amide; s), 1592, 1533, 1491, 1453, 1401, 1307, 1259, 1243, 1191, 1176, 1077, 1052, 1021, 915, 893, 827, 762, 703. ^{1}H NMR (500 MHz, DMSO): δ 10.61 (1H, bs, NH), 8.07 (2H, d, J = 7.4 Hz, H2″, H6″), 7.78 (1H, d, J = 8.3 Hz, H3), 7.73–7.67 (2H, m, H4, H6), 7.63 (2H, d, J = 8.8 Hz, H2′, H6′), 7.58–7.53 (3H, m, H3″, H4″, H5″), 7.32 (2H, d, J = 8.8 Hz, H3′, H5′). ^{13}C NMR (125 MHz, DMSO): δ 164.2, 163.3, 148.9, 137.9, 135.7, 134.4, 130.9, 130.1, 129.1, 128.8, 128.7, 128.6,

127.6, 126.5, 123.9, 121.4. Anal. Calcd. for $C_{20}H_{13}Cl_2NO_3$ (386.23): C, 62.19; H, 3.39; N, 3.63. Found: C, 61.87; H, 3.54; N, 3.90.

4-Chloro-2-(3,4-dichlorophenylcarbamoyl)phenyl Benzoate (5). White solid; yield 52%; mp 169.5–172°C. IR (ATR): 3304 (NH amide; m), 3074, 2928, 2850, 1713 (CO ester; s), 1668 (CO amide; s), 1578, 1516, 1478, 1469, 1449, 1384, 1298, 1275, 1247, 1208, 1101, 1087, 1066, 1026, 889, 866, 704. ^{1}H NMR (500 MHz, DMSO): δ 10.79 (1H, bs, NH), 8.06 (2H, d, J = 7.4 Hz, H2″, H6″), 7.92 (1H, s, H2′), 7.84 (1H, d, J = 2.6 Hz, H3), 7.74–7.69 (2H, m, H5, H6), 7.58–7.50 (5H, m, H5′, H6′, H3″, H4″, H5″). ^{13}C NMR (125 MHz, DMSO): δ 164.3, 163.0, 147.0, 138.9, 134.4, 131.9, 131.1, 130.9, 130.4, 130.0, 129.1, 128.6, 128.0, 127.9, 127.0, 125.7, 121.1, 120.0. Anal. Calcd. for $C_{20}H_{12}Cl_3NO_3$ (420.67): C, 57.10; H, 2.88; N, 3.33. Found: C, 57.40; H, 2.99; N, 3.41.

5-Chloro-2-(3,4-dichlorophenylcarbamoyl)phenyl Benzoate (6). White solid; yield 82%; mp 155–157°C. IR (ATR): 3412 (NH amide; m), 3093, 2930, 2851, 1754 (CO ester; s), 1682 (CO amide; s), 1593, 1527, 1475, 1450, 1400, 1375, 1303, 1244, 1180, 1135, 1042, 1020, 914, 881, 823, 759, 700. ^{1}H NMR (500 MHz, DMSO): δ 10.75 (1H, bs, NH), 8.07 (2H, d, J = 7.9 Hz, H2″, H6″), 7.92 (1H, s, H2′), 7.79 (1H, d, J = 8.3 Hz, H3), 7.74–7.69 (2H, m, H4, H6), 7.59–7.53 (5H, m, H5′, H6′, H3″, H4″, H5″). ^{13}C NMR (125 MHz, DMSO): δ 164.2, 163.6, 148.9, 139.0, 136.0, 134.4, 131.1, 130.9, 130.1, 129.1, 128.8, 128.0, 127.1, 126.5, 125.6, 124.0, 121.1, 119.9. Anal. Calcd. for $C_{20}H_{12}Cl_3NO_3$ (420.67): C, 57.10; H, 2.88; N, 3.33. Found: C, 56.95; H, 2.82; N, 3.59.

2-(3-Bromophenylcarbamoyl)-4-chlorophenyl Benzoate (7). White solid; yield 84%; mp 144–146°C. IR (ATR): 3325 (NH amide; m), 3079, 2930, 2852, 1716 (CO ester; s), 1671 (CO amide; s), 1586, 1520, 1479, 1450, 1419, 1305, 1284, 1266, 1249, 1208, 1103, 1084, 1065, 1023, 893, 872, 783, 733, 702, 684, 672. ^{1}H NMR (500 MHz, DMSO): δ 10.67 (1H, bs, NH), 8.07 (2H, d, J = 7.8 Hz, H2″, H6″), 7.89 (1H, s, H2′), 7.83 (1H, d, J = 2.5 Hz, H3), 7.73–7.69 (2H, m, H5, H6), 7.58–7.50 (3H, m, H3″, H4″, H5″), 7.45 (1H, d, J = 7.4 Hz, H6′), 7.36 (1H, t, J = 7.7 Hz, H5′), 7.24 (1H, d, J = 5.2 Hz, H4′). ^{13}C NMR (125 MHz, DMSO): δ 164.3, 163.0, 147.0, 140.4, 134.4, 131.7, 130.9, 130.4, 130.0, 129.1, 128.6, 128.0, 127.1, 126.7, 125.7, 122.3, 121.6, 118.7. Anal. Calcd. for $C_{20}H_{13}BrClNO_3$ (430.68): C, 55.78; H, 3.04; N, 3.25. Found: C, 55.49; H, 3.20; N, 3.48.

2-(3-Bromophenylcarbamoyl)-5-chlorophenyl Benzoate (8). White solid; yield 73%; mp 153.5–156°C. IR (ATR): 3279 (NH amide; m), 1739 (CO ester; s), 1647 (CO amide; s), 1599, 1585, 1541, 1476, 1452, 1407, 1320, 1254, 1241, 1190, 1075, 1050, 1020, 914, 894, 854, 781, 702, 659. ^{1}H NMR (500 MHz, DMSO): δ 10.63 (1H, bs, NH), 8.08 (2H, d, J = 7.5 Hz, H2″, H6″), 7.89 (1H, s, H2′), 7.79 (1H, d, J = 8.3 Hz, H3), 7.73–7.68 (2H, m, H4, H6), 7.58–7.53 (3H, m, H3″, H4″, H5″), 7.45 (1H, d, J = 7.0 Hz, H6′), 7.36 (1H, t, J = 7.6 Hz, H5′), 7.23 (1H, d, J = 5.2 Hz, H4′). ^{13}C

NMR (125 MHz, DMSO): δ 164.1, 163.5, 148.9, 140.5, 135.8, 134.4, 130.9, 130.1, 129.1, 128.6, 128.0, 127.1, 126.6, 126.5, 124.0, 122.2, 121.6, 118.7. Anal. Calcd. for $C_{20}H_{13}BrClNO_3$ (430.68): C, 55.78; H, 3.04; N, 3.25. Found: C, 55.87; H, 3.31; N, 3.46.

2-(4-Bromophenylcarbamoyl)-4-chlorophenyl Benzoate (9). White solid; yield 79%; mp 199–201°C. IR (ATR): 3308 (NH amide; m), 2932, 1739 (CO ester; s), 1668 (CO amide; s), 1597, 1541, 1487, 1449, 1403, 1314, 1258, 1246, 1197, 1099, 1053, 1023, 875, 833, 813, 706. ^{1}H NMR (500 MHz, DMSO): δ 10.64 (1H, bs, NH), 8.06 (2H, d, J = 7.3 Hz, H2″, H6″), 7.82 (1H, d, J = 2.5 Hz, H3), 7.73–7.68 (2H, m, H5, H6), 7.59–7.45 (7H, m, H2′, H3′, H5′, H6′, H3″, H4″, H5″). ^{13}C NMR (125 MHz, DMSO): δ 164.3, 162.8, 147.0, 138.3, 134.4, 131.8, 131.7, 131.3, 130.4, 130.0, 129.1, 129.0, 128.7, 125.6, 121.9, 115.7. Anal. Calcd. for $C_{20}H_{13}BrClNO_3$ (430.68): C, 55.78; H, 3.04; N, 3.25. Found: C, 55.55; H, 3.01; N, 3.51.

2-(4-Bromophenylcarbamoyl)-5-chlorophenyl Benzoate (10). White solid; yield 73%; mp 201–203°C. IR (ATR): 3326 (NH amide; m), 2929, 1745 (CO ester; s), 1651 (CO amide; s), 1601, 1587, 1531, 1487, 1450, 1397, 1260, 1241, 1189, 1176, 1073, 1051, 1020, 914, 893, 823, 810, 762, 703. ^{1}H NMR (500 MHz, DMSO): δ 10.61 (1H, bs, NH), 8.07 (2H, d, J = 7.5 Hz, H2″, H6″), 7.78 (1H, d, J = 8.3 Hz, H3), 7.73–7.67 (2H, m, H4, H6), 7.59–7.53 (5H, m, H2′, H6′, H3″, H4″, H5″), 7.45 (2H, d, J = 8.8 Hz, H3′, H5′). ^{13}C NMR (125 MHz, DMSO): δ 164.2, 163.3, 148.9, 138.3, 135.7, 134.4, 131.7, 130.9, 130.0, 129.1, 128.7, 128.6, 126.5, 123.9, 121.8, 115.7. Anal. Calcd. for $C_{20}H_{13}BrClNO_3$ (430.68): C, 55.78; H, 3.04; N, 3.25. Found: C, 55.67; H, 2.93; N, 3.29.

4-Chloro-2-(3-fluorophenylcarbamoyl)phenyl Benzoate (11). White solid; yield 61%; mp 143–144.5°C. IR (ATR): 3324 (NH amide; m), 2930, 2852, 1716 (CO ester; s), 1673 (CO amide; s), 1601, 1531, 1485, 1452, 1437, 1316, 1268, 1208, 1176, 1104, 1086, 1067, 1022, 965, 858, 784, 733, 704, 681. ^{1}H NMR (500 MHz, DMSO): δ 10.71 (1H, bs, NH), 8.07 (2H, d, J = 7.9 Hz, H2″, H6″), 7.83 (1H, d, J = 2.5 Hz, H3), 7.73–7.69 (2H, m, H5, H6), 7.57–7.46 (4H, m, H2′, H3″, H4″, H5″), 7.40–7.29 (2H, m, H5′, H6′), 6.92–6.86 (1H, m, H4′). ^{13}C NMR (125 MHz, DMSO): δ 164.3, 162.9, 163.1 and 161.2 (J = 241.5 Hz), 147.0, 140.6 and 140.5 (J = 10.9 Hz), 131.7, 131.2, 130.6 and 130.5 (J = 9.5 Hz), 130.4, 130.0, 129.1, 128.6, 128.0, 127.1, 125.7, 115.7 and 115.7 (J = 2.5 Hz), 110.6 and 110.5 (J = 21.0 Hz), 106.8 and 106.6 (J = 26.0 Hz). Anal. Calcd. for $C_{20}H_{13}ClFNO_3$ (369.77): C, 64.96; H, 3.54; N, 3.79. Found: C, 64.80; H, 3.29; N, 3.58.

5-Chloro-2-(3-fluorophenylcarbamoyl)phenyl Benzoate (12). White solid; yield 78%; mp 149–151°C. IR (ATR): 3295 (NH amide; m), 3073, 2931, 1739 (CO ester; s), 1650 (CO amide; s), 1596, 1550, 1489, 1450, 1423, 1324, 1256, 1241, 1196, 1171, 1149, 1075, 1053, 1022, 911, 845, 779, 704, 662. ^{1}H NMR (500 MHz, DMSO): δ 10.68 (1H, bs, NH), 8.08 (2H, d, J = 7.9 Hz, H2″, H6″), 7.79 (1H, d, J = 8.3 Hz, H3), 7.73–7.68 (2H, m, H4, H6), 7.58–7.51 (4H, m, H2′, H3″, H4″,

H5″), 7.39–7.27 (2H, m, H5′, H6′), 6.90–6.85 (1H, m, H4′). ^{13}C NMR (125 MHz, DMSO): δ 164.2, 163.5, 163.1 and 161.2 ($J = 241.4$ Hz), 148.9, 140.7 and 140.6 ($J = 11.0$ Hz), 135.8, 134.4, 130.9, 130.6 and 130.5 ($J = 9.4$ Hz), 130.1, 129.1, 128.6, 128.5, 126.5, 124.0, 115.7 and 115.6 ($J = 2.6$ Hz), 110.5 and 110.4 ($J = 21.1$ Hz), 106.7 and 106.5 ($J = 26.1$ Hz). Anal. Calcd. for $C_{20}H_{13}ClFNO_3$ (369.77): C, 64.96; H, 3.54; N, 3.79. Found: C, 64.90; H, 3.24; N, 3.99.

4-Chloro-2-(4-fluorophenylcarbamoyl)phenyl Benzoate (13). White solid; yield 88%; mp 149–151°C. IR (ATR): 3318 (NH amide; m), 3076, 2929, 2852, 1715 (CO ester; s), 1665 (CO amide; s), 1622, 1571, 1527, 1505, 1479, 1450, 1406, 1310, 1275, 1250, 1207, 1177, 1151, 1098, 1086, 1064, 1023, 892, 822, 778, 733, 697. ^{1}H NMR (500 MHz, DMSO): δ 10.56 (1H, bs, NH), 8.07 (2H, d, $J = 7.2$ Hz, H2″, H6″), 7.81 (1H, d, $J = 2.6$ Hz, H3), 7.73–7.68 (2H, m, H5, H6), 7.63–7.59 (2H, m, H2′, H6′), 7.55 (1H, t, $J = 7.8$ Hz, H4″), 7.51–7.48 (2H, m, H3″, H5″), 7.14–7.09 (2H, m, H3′, H5′). ^{13}C NMR (125 MHz, DMSO): δ 164.3, 162.6, 159.5 and 157.5 ($J = 240.5$ Hz), 147.0, 135.2 and 135.2 ($J = 2.6$ Hz), 134.3, 131.5, 130.4, 130.0, 129.1, 128.7, 128.0, 127.1, 125.6, 121.8 and 121.8 ($J = 7.9$ Hz), 115.5 and 115.4 ($J = 22.2$ Hz). Anal. Calcd. for $C_{20}H_{13}ClFNO_3$ (369.77): C, 64.96; H, 3.54; N, 3.79. Found: C, 65.15; H, 3.47; N, 3.70.

5-Chloro-2-(4-fluorophenylcarbamoyl)phenyl Benzoate (14). White solid; yield 84%; mp 142–144°C. IR (ATR): 3295 (NH amide; m), 2930, 2852, 1739 (CO ester; s), 1647 (CO amide; s), 1601, 1547, 1505, 1452, 1411, 1317, 1258, 1245, 1194, 1177, 1155, 1072, 1054, 1023, 893, 835, 826, 706. ^{1}H NMR (500 MHz, DMSO): δ 10.52 (1H, bs, NH), 8.07 (2H, d, $J = 7.9$ Hz, H2″, H6″), 7.78 (1H, d, $J = 8.3$ Hz, H3), 7.73–7.66 (2H, m, H4, H6), 7.63–7.59 (2H, m, H2′, H6′), 7.57–7.53 (3H, m, H3″, H4″, H5″), 7.13–7.08 (2H, m, H3′, H5′). ^{13}C NMR (125 MHz, DMSO): δ 164.2, 163.1, 159.4 and 157.5 ($J = 240.5$ Hz), 148.9, 135.6, 135.3 and 135.3 ($J = 2.5$ Hz), 134.4, 130.9, 130.0, 129.1, 128.6, 128.0, 126.4, 123.9, 121.8 and 121.7 ($J = 7.9$ Hz), 115.5 and 115.3 ($J = 22.3$ Hz). Anal. Calcd. for $C_{20}H_{13}ClFNO_3$ (369.77): C, 64.96; H, 3.54; N, 3.79. Found: C, 64.74; H, 3.61; N, 3.94.

4-Chloro-2-(4-(trifluoromethyl)phenylcarbamoyl)phenyl Benzoate (15). White solid; yield 44%; mp 179.5–181.5°C. IR (ATR): 3310 (NH amide; m), 1715 (CO ester; s), 1672 (CO amide; s), 1601, 1526, 1476, 1452, 1407, 1317, 1278, 1253, 1207, 1168, 1110, 1086, 1063, 1015, 891, 867, 844, 820, 775, 703. ^{1}H NMR (500 MHz, DMSO): δ 10.87 (1H, bs, NH), 8.06 (2H, d, $J = 7.8$ Hz, H2″, H6″), 7.86 (1H, d, $J = 2.6$ Hz, H3), 7.82 (2H, d, $J = 8.5$ Hz, H2′, H6′), 7.75–7.68 (3H, m, H5, H6), 7.65 (2H, d, $J = 8.6$ Hz, H3′, H5′), 7.57–7.51 (3H, m, H3″, H4″, H5″). ^{13}C NMR (125 MHz, DMSO): δ 164.3, 163.2, 147.0, 142.5, 134.4, 131.9, 131.1, 130.4, 130.0, 129.2, 129.1, 128.6, 126.8 (q, $J = 30.5$ Hz), 126.2 (d, $J = 3.7$ Hz), 125.7, 124.4 (q, $J = 271.4$ Hz), 119.9. Anal. Calcd. for $C_{21}H_{13}ClF_3NO_3$ (419.78): C, 60.08; H, 3.12; N, 3.34. Found: C, 60.31; H, 2.87; N, 3.09.

5-Chloro-2-(4-(trifluoromethyl)phenylcarbamoyl)phenyl Benzoate (16). White solid; yield 80%; mp 177.5–179.5°C. IR (ATR): 3325 (NH amide; m), 2927, 1746 (CO ester; s), 1654 (CO amide; s), 1598, 1534, 1481, 1450, 1408, 1323, 1243, 1192, 1160, 1104, 1067, 1048, 1020, 916, 894, 845, 825, 763, 702. ^{1}H NMR (500 MHz, DMSO): δ 10.84 (1H, bs, NH), 8.07 (2H, d, $J = 7.2$ Hz, H2″, H6″), 7.83–7.79 (3H, m, H3, H2′, H6′), 7.71–7.67 (2H, m, H4, H6), 7.64 (2H, d, $J = 8.1$ Hz, H3′, H5′), 7.60–7.54 (3H, m, H3″, H4″, H5″). ^{13}C NMR (125 MHz, DMSO): δ 164.2, 163.7, 149.0, 142.5, 135.9, 134.4, 131.0, 130.0, 129.1, 128.5, 128.5, 126.5, 126.2 (q, $J = 3.8$ Hz), 124.6 (q, $J = 285.3$ Hz), 124.0, 123.9 (q, $J = 32.0$ Hz), 119.8. Anal. Calcd. for $C_{21}H_{13}ClF_3NO_3$ (419.78): C, 60.08; H, 3.12; N, 3.34. Found: C, 60.22; H, 3.21; N, 3.43.

4-Chloro-2-(3-(trifluoromethyl)phenylcarbamoyl)phenyl Benzoate (17). White solid; yield 72%; mp 159.5–162°C. IR (ATR): 3314 (NH amide; m), 2931, 2853, 1714 (CO ester; s), 1670 (CO amide; s), 1597, 1540, 1443, 1326, 1274, 1206, 1166, 1119, 1097, 1084, 1064, 1023, 906, 800, 735, 695. ^{1}H NMR (500 MHz, DMSO): δ 10.83 (1H, bs, NH), 8.07 (2H, d, $J = 7.2$ Hz, H2″, H6″), 8.00 (1H, s, H2′), 7.87 (1H, d, $J = 2.6$ Hz, H3), 7.74–7.68 (2H, m, H5, H6), 7.56–7.50 (4H, m, H6′, H3″, H4″, H5″), 7.46–7.35 (2H, m, H4′, H5′). ^{13}C NMR (125 MHz, DMSO): δ 164.3, 163.1, 147.0, 139.6, 134.4, 131.8, 131.1, 130.4, 130.2, 130.0, 129.6 (q, $J = 31.7$ Hz), 129.1, 128.6, 128.0, 125.7, 124.2 (q, $J = 272.6$ Hz), 123.6, 120.4 (q, $J = 3.6$ Hz), 116.1 (q, $J = 3.9$ Hz). Anal. Calcd. for $C_{21}H_{13}ClF_3NO_3$ (419.78): C, 60.08; H, 3.12; N, 3.34. Found: C, 60.19; H, 3.35; N, 2.99.

4-Bromo-2-(4-(trifluoromethyl)phenylcarbamoyl)phenyl Benzoate (18). White solid; yield 79%; mp 177.5–180°C. IR (ATR): 3309 (NH amide; m), 1714 (CO ester; s), 1672 (CO amide; s), 1600, 1526, 1474, 1451, 1406, 1317, 1280, 1252, 1208, 1169, 1111, 1086, 1064, 1041, 862, 843, 818, 720, 703. ^{1}H NMR (500 MHz, DMSO): δ 10.87 (1H, bs, NH), 8.06 (2H, d, $J = 7.2$ Hz, H2″, H6″), 7.97 (1H, d, $J = 2.4$ Hz, H3), 7.85 (1H, dd, $J = 2.5$ Hz, $J = 8.6$ Hz, H5), 7.82 (2H, d, $J = 8.5$ Hz, H2′, H6′), 7.70 (1H, t, $J = 7.5$ Hz, H4″), 7.65 (2H, d, $J = 8.2$ Hz, H3′, H5′), 7.54 (2H, t, $J = 7.9$ Hz, H3″, H5″), 7.45 (1H, d, $J = 8.6$ Hz, H6). ^{13}C NMR (125 MHz, DMSO): δ 164.2, 163.1, 147.5, 142.5, 134.8, 134.4, 132.0, 131.4, 130.0, 129.1, 128.6, 126.7 (q, $J = 37.3$ Hz), 126.2 (d, $J = 3.7$ Hz), 126.0, 124.4 (q, $J = 271.5$ Hz), 119.9, 118.5. Anal. Calcd. for $C_{21}H_{13}BrF_3NO_3$ (464.23): C, 54.33; H, 2.82; N, 3.02. Found: C, 54.16; H, 3.03; N, 3.35.

3.2. Biology

3.2.1. Antibacterial Evaluation. Salicylanilide benzoates were assayed for their *in vitro* antibacterial activity towards eight strains. Benzoic acid expressed no activity up to 500 μmol/L. Table 1 summarizes the results with respect to structure of the esters.

Almost all benzoates exhibited a very good activity against both strains of *S. aureus* with MIC from 0.98 to 31.25 μmol/L (**10** being an exception). Importantly, MIC for

TABLE 2: *In vitro* antifungal activity of salicylanilide benzoates.

| | \multicolumn{10}{c}{MIC/IC$_{80}$ [μmol/L]} | | | | | | | | | |
| | Candida albicans | | Candida krusei | | Trichosporon asahii | | Absidia corymbifera | | Trichophyton mentagrophytes | |
	24 h	48 h	24 h	48 h	24 h	48 h	24 h	48 h	72 h	120 h
(1)	250	>500	125	>500	>500	>500	500	>500	250	>500
(6)	**125**	>125	**15.62**	>125	**3.9**	**15.62**	125	125	**31.25**	**31.25**
(8)	>250	>250	>250	>250	>250	>250	250	>250	>250	>250
(9)	>125	>125	>125	>125	>125	>125	125	125	62.5	62.5
(13)	>125	>125	>125	>125	>125	>125	**62.5**	>125	125	125
(18)	>125	>125	>125	>125	>125	>125	>125	>125	62.5	125
BA	>500	>500	>500	>500	>500	>500	>500	>500	>500	>500
FLU	1.00	2.00	>50.0	>50.0	4.00	9.00	>50.0	>50.0	17.0	26.0

FLU: fluconazole; BA: benzoic acid. The best MIC value for each strain is bolded.

drug-sensitive and MRSA strain did not differ practically, which indicates none cross resistance. Also *S. epidermidis* was affected by the majority of evaluated compounds (without **3**) at only slightly higher concentrations from 1.95 μmol/L. *Enterococcus* was the most insensitive Gram-positive strain, although it was inhibited by four esters (**5, 6, 15, 18**) at 1.95 μmol/L; six esters (**2, 3, 4, 10, 14, 16**) were inactive at 125 μmol/L. Molecules **6** and **15** were found having the highest *in vitro* inhibitory potency. MIC values against three *Staphylococci* support the hypothesis that these derivatives act as bactericidal agents.

Some salicylanilide benzoates MIC values are comparable to benzylpenicillin towards *S. aureus*, but almost all esters are favorable against MRSA and *S. epidermidis*. The situation for *Enterococcus* is quite more complex; some benzoates exhibited better MIC value (**5, 6, 15, 18**), some comparable after 24 h (**7, 8, 12, 17**) and other worse than benzylpenicillin.

When concentrated on the structure-activity relationships, the position of the substituents on the salicylic ring is ambiguous—in some cases are superior 4-chloro derivatives (e.g., **1** versus **2**, **9** versus **10**, **15** versus **16**), in other 5-chloro ones (**5** versus **6** or **7** versus **8**). For the substitution of the aniline ring, 3,4-dichloro (**5, 6**) and CF$_3$- (**15–18**) moieties improved the antibacterial activity the most significantly. In general, it seems that 3-substituted anilines produced a higher antibacterial activity than 4-substituted anilines.

Gram-negative species (*E. coli*, two strains of *Klebsiella pneumoniae*, *P. aeruginosa*) were almost completely resistant to benzoates up to 125 μmol/L at the testing conditions with one notable exception; *P. aeruginosa* was inhibited by **1** at low concentration of 3.9/7.81 μmol/L.

3.2.2. Antifungal Evaluation. Synthesized esters were tested for their *in vitro* activity against eight human pathogenic fungi. MIC values are presented in Table 2. Benzoic acid alone was completely inactive at the concentration of 500 μmol/L at both neutral and slightly acidic (pH $\sim$ 5) environment.

Unforeseen, in contrast to the antibacterial activity, salicylanilide benzoates expressed only mild antifungal potency. From eight strains, *C. tropicalis*, *C. glabrata*, and *A. fumigatus* showed a complete insensitivity to all tested derivatives at the value of 125 μmol/L. When concentrated on the esters, benzoates **2** and **11** did not affect the growth of any fungal species at the concentration of 500 μmol/L and lower, derivatives **12, 14,** and **17** up to 250 μmol/L and **3, 4, 5, 7, 10, 15, 16** expressed activity levels >125 μmol/L. Only six derivatives (**1, 6, 8, 9, 13, 18**) displayed certain *in vitro* efficacy. The most active compound was assayed trichlorinated ester **6** (MIC $\geq$ 3.9 μmol/L), which surpassed standard fluconazole against *C. krusei*. In general, *Candida* is more resistant to salicylanilide esters than moulds and moreover, it seems that the mechanism of the action is only fungistatic. On the other side, the growth of *T. mentagrophytes* was affected by the highest number of the derivatives; based on the MIC values, activity against filamentous fungi is probably fungicidal.

Because pH effect on the efficiency of benzoic acid derivatives (generally of weak organic acids) was described (e.g., [16]), we measured MIC values of the esters **6** and **15** not only at approximately neutral environment, but additionally at slightly acidic pH ($\sim$5; without buffering of the testing medium). Unfortunately, the change of pH did not result in the improvement of activity, even it led to a bit worse values.

Although we expected that the introduction of benzoyl fragment into salicylanilide molecules resulted in the significantly increased antifungal potency, this modification failed in this point.

4. Conclusion

In sum, we have designed and synthesized new esters of halogenated salicylanilides with benzoic acid. This series of compounds was evaluated to be a new group with promising *in vitro* antibacterial (against Gram-positive strains) activity. Unfortunately these derivatives disappointed expectation about them as potential antifungal agents.

Acknowledgments

This work was financially supported by GAUK 27610/2010, the Research Project MSM 0021620822, IGA NS 10367-3, and SVV 2012-265-001.

References

[1] I. M. Gould, "Antibiotic resistance: the perfect storm," *International Journal of Antimicrobial Agents*, vol. 34, supplement 3, pp. S2–S5, 2009.

[2] M. Mielke, "Prevention and control of nosocomial infections and resistance to antibiotics in Europe—primum non-nocere: elements of successful prevention and control of healthcare-associated infections," *International Journal of Medical Microbiology*, vol. 300, no. 6, pp. 346–350, 2010.

[3] J. Morschhäuser, "Regulation of multidrug resistance in pathogenic fungi," *Fungal Genetics and Biology*, vol. 47, no. 2, pp. 94–106, 2010.

[4] A. R. Duval, P. H. Carvalho, M. C. Soares et al., "7-Chloroquinolin-4-yl arylhydrazone derivatives: synthesis and antifungal activity," *TheScientificWorldJOURNAL*, vol. 11, pp. 1489–1495, 2011.

[5] J. Vinsova, A. Imramovsky, V. Buchta et al., "Salicylanilide acetates: synthesis and antibacterial evaluation," *Molecules*, vol. 12, no. 1, pp. 1–12, 2007.

[6] A. Imramovský, J. Vinšová, J. M. Férriz, V. Buchta, and J. Jampílek, "Salicylanilide esters of N-protected amino acids as novel antimicrobial agents," *Bioorganic and Medicinal Chemistry Letters*, vol. 19, no. 2, pp. 348–351, 2009.

[7] M. Krátký, J. Vinšová, V. Buchta, K. Horvati, S. Bösze, and J. Stolaříková, "New amino acid esters of salicylanilides active against MDR-TB and other microbes," *European Journal of Medicinal Chemistry*, vol. 45, no. 12, pp. 6106–6113, 2010.

[8] T. J. R. Cheng, Y. T. Wu, S. T. Yang et al., "High-throughput identification of antibacterials against methicillin-resistant *Staphylococcus aureus* (MRSA) and the transglycosylase," *Bioorganic and Medicinal Chemistry*, vol. 18, no. 24, pp. 8512–8529, 2010.

[9] E. Melliou and I. Chinou, "Chemical analysis and antimicrobial activity of Greek propolis," *Planta Medica*, vol. 70, no. 6, pp. 515–519, 2004.

[10] W. K. Kong, Y. L. Zhao, L. M. Shan, X. H. Xiao, and W. Y. Guo, "Thermochemical studies on the quantity—antibacterial effect relationship of four organic acids from *Radix isatidis* on *Escherichia coli* growth," *Biological and Pharmaceutical Bulletin*, vol. 31, no. 7, pp. 1301–1305, 2008.

[11] J. Alvesalo, H. Vuorela, P. Tammela, M. Leinonen, P. Saikku, and P. Vuorela, "Inhibitory effect of dietary phenolic compounds on *Chlamydia pneumoniae* in cell cultures," *Biochemical Pharmacology*, vol. 71, no. 6, pp. 735–741, 2006.

[12] I. Kubo, P. Xiao, K. I. Nihei, K. I. Fujita, Y. Yamagiwa, and T. Kamikawa, "Molecular design of antifungal agents," *Journal of Agricultural and Food Chemistry*, vol. 50, no. 14, pp. 3992–3998, 2002.

[13] J. Y. Cho, J. H. Moon, K. Y. Seong, and K. H. Park, "Antimicrobial activity of 4-hydroxybenzoic acid and trans 4-hydroxycinnamic acid isolated and identified from rice hull," *Bioscience, Biotechnology, and Biochemistry*, vol. 62, no. 11, pp. 2273–2276, 1998.

[14] A. Nishina, F. Kajishima, M. Matsunaga, H. Tezuka, H. Inatomi, and T. Osawa, "Antimicrobial substance, 3′,4′-dihydroxyacetophenone, in coffee residue," *Bioscience, Biotechnology, and Biochemistry*, vol. 58, no. 2, pp. 293–296, 1994.

[15] A. Innocenti, R. A. Hall, C. Schlicker, F. A. Muhlschlegel, and C. T. Supuran, "Carbonic anhydrase inhibitors. Inhibition of the β-class enzymes from the fungal pathogens *Candida albicans* and *Cryptococcus neoformans* with aliphatic and aromatic carboxylates," *Bioorganic and Medicinal Chemistry*, vol. 17, no. 7, pp. 2654–2657, 2009.

[16] S. Brul and P. Coote, "Preservative agents in foods: mode of action and microbial resistance mechanisms," *International Journal of Food Microbiology*, vol. 50, no. 1-2, pp. 1–17, 1999.

[17] Clinical and Laboratory Standards Institute (CLSI), "Methods for dilution antimicrobial susceptibility test for bacteria that growth aerobically," Approved Standard—7th edition, 2011, http://www.clsi.org/source/orders/free/m11a7.pdf.

[18] Clinical and Laboratory Standards Institute (CLSI), "Reference method for brothdilution antifungal susceptibility testing of filamentous fungi," Approved Standard—2nd edition, 2011, http://www.clsi.org/source/orders/free/M38-a2.pdf.

[19] Clinical and Laboratory Standards Institute (CLSI), "Reference method for broth dilution antifungal susceptibility testing of yeasts," Approved Standard—3rd edition, 2011, http://www.clsi.org/source/orders/free/M27-A3.pdf.

23

Bacillus cereus in Infant Foods: Prevalence Study and Distribution of Enterotoxigenic Virulence Factors in Isfahan Province, Iran

Ebrahim Rahimi,[1] **Fahimeh Abdos,**[2] **Hassan Momtaz,**[3] **Zienab Torki Baghbadorani,**[1] **and Mohammad Jalali**[4]

[1] *Department of Food Hygiene and Public Health, College of Veterinary Medicine, Islamic Azad University, Shahrekord Branch, P.O. Box 166, Shahrekord, Iran*
[2] *Department of Food Science and Technology, College of Agriculture, Islamic Azad University, Shahrekord Branch, Shahrekord, Iran*
[3] *Department of Microbiology, College of Veterinary Medicine, Islamic Azad University, Shahrekord Branch, Shahrekord, Iran*
[4] *Infectious Disease and Tropical Medicine Research Center and School of Food Science and Nutrition, Isfahan University of Medical Sciences, Isfahan, Iran*

Correspondence should be addressed to Ebrahim Rahimi; ebrahimrahimi55@yahoo.com

Academic Editors: S.-J. Ahn and D. Zhou

This study was carried out in order to investigate the presences of *Bacillus cereus* and its enterotoxigenic genes in infant foods in Isfahan, Iran. Overall 200 infant foods with various based were collected and immediately transferred to the laboratory. All samples were culture and the genomic DNA was extracted from colonies with typical characters of *Bacillus cereus*. The presences of enterotoxigenic genes were investigated using the PCR technique. Eighty-four of two hundred samples (42%) were found to be contaminated with *B. cereus* with a ranges of 3×10^1–9.3×10^1 spore per gram sample. Totally, *entFM* had the highest (61.90%) incidences of enterotoxigenic genes while *hblA* had the lowest (13.09%) incidences of enterotoxigenic genes. Overall, 6.7% of *B. cereus* isolates had all studied enetrotoxigenic genes while 25.5% of *B. cereus* strains had all studied enetrotoxigenic genes expectance *bceT* gene. Thisstudyisthe first prevalence report of *B. cereus* and its enterotoxigenic genes in infant foods in Iran. Results showed that the infant food is one of the main sources of enterotoxigenic genes of *B. cereus* in Iran. Therefore, the accurate food inspection causes to reducing outbreak of diseases.

1. Introduction

Baby foods are the primary source of nutrition for kids before they are able to digest other types of food. Their high values for proteins, minerals, fats, and vitamins are undeniable. In a day, millions of babies use these foods in the world. Babies have the weak immune system and any infection in their foods causes their illness. Therefore, the hygienic quality of baby foods is very important but sometimes it will be changed and several infections and illness occur.

Foodborne diseases are a worldwide growing health problem involving a wide spectrum of illnesses caused by bacterial, viral, parasitic, or chemical contamination of food.

A previous report of the World Health Organization (WHO) showed that the *Bacillus cereus* (*B. cereus*) is the most common foodborne bacterium in pasteurized food products [1–3].

The *B. cereus* is a Gram-positive and rod-shaped bacterium which is responsible for causing diarrhea, emesis, fatal meningitis, and spoilage of different food products [4, 5]. The *B. cereus* is a spore former organism. Therefore, there is a risk in its transmission through processed, pasteurized, sterilized, and heat-treated food products. The spore of this bacterium can survive in low and high temperatures. The most common source of this bacterium is liquid food products, milk powder, mixed food products, and, with particular concerns, the baby formula industry [6, 7]. The occurrence of *B. cereus* as a contaminant of baby food was previously reported [8, 9].

Food spoilage, diarrhea, emesis, and other complications of *B. cereus* are caused by several virulence genes. Virulence genes of *B. cereus* have been ascribed to different extracellular factors. Two of these virulence factors are protein complexes, that is, the hemolysin BL (HBL) [10] and the nonhemolytic enterotoxin (NHE) [11]. Other factors are single-gene products encoded by *entFM* (enterotoxin FM) and *bceT* (*B. cereus* enterotoxin) [12].

It is important to know which genes of *B. cereus* are endemic in various regions and samples. Besides, the epidemiology and prevalence of *B. cereus* are indeed unknown in Iranian baby foods. Therefore, the present study was carried out in order to study the prevalence of *B. cereus* and its enterotoxigenic virulence factors in infant foods in Isfahan province, Iran.

2. Materials and Methods

2.1. Samples and Identification of B. cereus. Overall 200 baby food samples including baby food with rice and milk based ($n = 50$), baby food with wheat and milk based ($n = 50$), baby food with wheat, honey, and milk based ($n = 50$), and baby food with wheat, banana, and milk based ($n = 50$) were purchased from the supermarkets of Isfahan, Iran. All of these products were pasteurized and after collection were kept under refrigeration in plastic bags; information about dates of production and of assigned shelf lives was not presented.

First, tenfold 10 g of each sample was added into 90 mL 0.1% (wv^{-1}) peptone water. The samples were well mixed and homogenized by vigorous vortexing at room temperature for 3 min. Tenfold dilution was prepared in 20% (vv^{-1}) glycerol-peptone water. A 50 μL aliquot from this dilution was inoculated into 5 mL Nutrient Broth (NB) (Applichem) and incubated at 37°C for 18 h with shaking at 150 rpm. The tubes were pasteurized at 80°C for 10 min to eliminate nonsporulating bacteria. The suspension was streaked onto chromogenic *B. cereus* agar (BCA) supplemented with chromogenic *B. cereus* selective supplement (Oxoid). The plates were incubated at 37°C overnight and blue/green colonies were subcultured on chromogenic BCA until obtaining a pure culture. After identification by biochemical tests (Gram staining and catalase test), the isolated strains were stored in sterile NB containing 20% (vv^{-1}) glycerol at −80°C. The colonies with the typical characters of *B. cereus* were tested using the polymerase chain reaction (PCR) method [13, 14].

2.2. DNA Extraction. Genomic DNA was extracted from the culture positive colonies. Genomic DNA was extracted by freezing first and then boiling the cells. Strains were grown at 37°C for 16 h. A loopful of cells was scraped off from NA plate and resuspended in 150 μL sterile water. After 20 min of freezing at −80°C, the samples were boiled for 10 min to lyse the cells completely. Cell debris was removed by centrifugation (11000 g, 10 s). The supernatant containing genomic DNA was stored at −20°C.

2.3. PCR for Detection of Virulence Genes. The strains were tested for the presence of enterotoxin genes. The primer sets used in this study are presented in Table 1. Each amplification process was performed in a 50 μL reaction mixture containing 100 ng of genomic DNA as the template, 5 μL of 10x reaction buffer (100 mMTris-HCl (pH 8.8), 500 mMKCl, 0.8% (vv^{-1}) Nonidet P-40, and 1.5 mM MgCl$_2$), 10 μM of each of the primers, 0.2 mM of each of the four dNTPs (Fermentas), and 2 U Taq DNA polymerase (Fermentas). Reactions were initiated at 95°C for 5 min, followed by 30 cycles of 95°C for 1 min, 58°C for 1 min, 72°C for 1 min, and a final elongation step at 72°C for 10 min, with a final hold at 4°C in a DNA thermal cycler (Mastercycler Gradiant, Eppendorf, Germany). The diarrheagenic strain of *B. cereus* (ATCC 14579) was used as a positive control and the sterile water was used as a negative control. PCR products were analyzed in 1.5% (wv^{-1}) TAE agarose gels and all PCR experiments were performed twice for each strain.

2.4. Statistical Analysis. Statistical analysis was performed using SPSS/18.0 software for significant relationship between hot and cold seasons for occurrence of bacteria in water. Chi-square test was performed and differences were considered significant at P value < 0.05.

3. Results and Discussion

The presence of *B. cereus* in typical colonies has been confirmed by the PCR techniques. In this study, 84 of 200 samples (42%) were found to be contaminated with *B. cereus* between 3×10^1 and 9.3×10^1 (Table 2). Statistical analysis showed significant differences ($P < 0.05$) for the presence and spore ranges of *B. cereus* between baby food with rice and milk based and baby food with wheat, banana, and milk based which was in agreement with the results of Reyes et al. [20]. Reyes et al. [20] showed that 35 out of 56 baby foods with rice and milk based (62.5%) were contaminated with *B. cereus* with a range of 3 to over than 1000 spores per gram sample. Despite the similar results of our study with the study of Reyes et al. [20], the high differences in percentage and range of contamination may be due to the different formulas of the samples, seasons of sample production, methods of sampling, and methods of testing.

The higher contamination of infant foods with cereal based including wheat and rice is maybe due to the higher presence of *B. cereus* in cereal [21]. Various studies showed the high percentages (40%–100%) and ranges (10 to 1000 spores per gram sample) of *B. cereus* in rice [22]. Other studies showed that the *B. cereus* is a flora of rice [23]. In addition, the bacterium has been isolated from another type of cereals (prevalence rate of 56% and contamination range of 10^2 to 10^5 spores per gram sample) [24]. It seems that infant food additives have the high role in contamination of baby foods.

Totally, 42% of samples were found to be contaminated with *B. cereus*. Despite the high percentage of *B. cereus* in baby foods in this study, the numbers of spore per gram of each sample were very low and in all samples were lower than 100 spores per gram. The results of Becker et al. [9] were in

Bacillus cereus in Infant Foods: Prevalence Study and Distribution of Enterotoxigenic Virulence Factors in Isfahan
Province, Iran

173

TABLE 1: The primer sequences and amplicon sizes used in PCR analysis.

Gene	Primer sequences (50–30) F: forward; R: reverse	Amplicon size (bp)	Reference
bceT	F-TTACATTACCAGGACGTGCTT R-TGTTTGTGATTGTAATTCAGG	428	[15]
entFM	F-ATGAAAAAAGTAATTTGCAGG R-TTAGTATGCTTTTGTGTAACC	1269	[16]
hblC	F-GATACTAATGTGGCAACTGC R-TTGAGACTGCTGTCTAGTTG	740	[17]
nheA	F-TACGCTAAGGAGGGGCA R-GTTTTTATTGCTTCATCGGCT	499	[18]
nheB	F-CTATCAGCACTTATGGCAG R-ACTCCTAGCGGTGTTCC	769	[18]
nheC	F-CGGTAGTGATTGCTGGG R-CAGCATTCGTACTTGCCAA	581	[19]
hblA	F-AAGCAATGGAATACAATGGG R-AGAATCTAAATCATGCCACTGC	1154	[17]
hblB	F-AAGCAATGGAATACAATGGG R-AATATGTCCCAGTACACCCG	2684	[17]

TABLE 2: Occurrence and distribution of B. cereus in various baby food samples.

Samples	Number of samples	B. cereus positive (%)	B. cereus (spore in gram)	
			Average	Range
Baby food with rice and milk based	50	31 (62)	2.09×10^1	$3–9.3 \times 10^1$
Baby food with wheat and milk based	50	25 (50)	1.77×10^1	$3–9.3 \times 10^1$
Baby food with wheat, honey, and milk based	50	18 (36)	2.51×10^1	$3–7.5 \times 10^1$
Baby food with wheat, banana, and milk based	50	10 (20)	1.61×10^1	$2.8–7 \times 10^1$
Total	200	84 (42)	1.69×10^1	$3–9.3 \times 10^1$

agreement with our study. They reported that 54% of infant foods were contaminated with *B. cereus* between 3 and 600 spores per gram samples.

Our results showed that *entFM, nheC, bheA, hblC,* and *nheB* with the frequencies of 61.90%, 51.19%, 44.04%, 34.52%, and 33.33% were the highest enterotoxigenic virulence genes isolated from infant foods, respectively (Table 3). This result is in agreement with others [21, 25, 26]. Hansen and Hendriksen [27] showed that 52% of *B. cereus* strains which were isolated from foods had *hblA* genes which were higher than our results. The presence of enterotoxigenic virulence genes in baby food with rice and milk based was the most common while their presence in baby food with wheat, banana, and milk based was rare. Statistical analysis showed significant differences ($P < 0.05$) for the presence of enterotoxigenic virulence genes between baby food with rice and milk based and baby food with wheat, banana, and milk based. Besides, statistical analyses were significant ($P < 0.05$) between the presence of *entFM* with *hblA, hblB,* and *bceT* enterotoxigenic virulence genes in all baby food samples.

In the present study, 14 various patterns have been obtained for detection of enterotoxigenic genes of *B. cereus* in Iranian infant foods (Table 4). These patterns showed that 6.7% of *B. cereus* strains had all studied enterotoxigenic genes while 25.5% of *B. cereus* strains had all studied enterotoxigenic genes except *bceT* gene. This finding is similar to

the report of Samapundo et al. [3]. Samapundo et al. [3] showed that 16 various profiles for the presence of *nhe* and *hble* enterotoxigenic genes were obtained from a total of 324 *B. cereus* strains which were isolated from food products in Belgium. They showed that 52.5% of strains had all *hblA, hblB, hblC, nheA, nhrB,* and *nheC* enterotoxigenic genes which were in harmony with our results.

The results of the present study showed that despite the high presence of *B. cereus* in infant food samples, the numbers of spore in gram sample were low. Therefore, more studies with higher numbers of samples are needed to opine about the baby food contamination with *B. cereus*. It seems that some food safety and quality standards (good agricultural practices (GAPs), good manufacturing practices (GMPs), and the hazard analysis and critical control point (HACCP)) systems need to be applied and performed in most of the Iranian infant food units to control growth of *B. cereus* and production of its enterotoxigenic virulence genes during harvesting, distribution, and storage periods. In addition, the hygienic quality of raw materials and even additives are important in reducing the microbial load of infant foods. Also, disinfection of room air, fridge halls, and handling systems is essential. The moisture content of infant foods should be monitored, and storage and transport should be modified according to what has been observed in the models. We suggested that many studies should have been performed

TABLE 3: Pattern of enterotoxigenic genes of *B. cereus* isolated from various baby food samples in Isfahan, Iran.

Pattern	Hemolytic enterotoxins			Non-hemolytic enterotoxins			*entFM*	*bceT*
	hblA	*hblB*	*hblC*	*nheA*	*nheB*	*nheC*		
1	+	+	+	+	+	+	+	+
2	+	+	+	+	+	+	+	−
3	−	−	+	+	+	+	+	−
4	−	−	−	+	+	+	+	−
5	−	+	+	+	+	+	+	−
6	+	+	+	+	−	+	+	+
7	−	+	+	+	+	+	+	+
8	−	−	+	+	−	+	+	−
9	−	−	−	+	−	+	+	−
10	−	+	+	−	−	−	+	+
11	−	+	+	−	−	−	+	−
12	−	+	−	+	−	−	+	+
13	−	−	−	−	+	−	−	+
14	−	−	−	−	−	−	−	+

TABLE 4: Distribution of various enterotoxigenic virulence genes of *B. cereus* isolated from baby foods in Isfahan, Iran.

Samples	*B. cereus* positive	Enterotoxigenic virulence genes (%)							
		hblA	*hblB*	*hblC*	*nheA*	*nheB*	*nheC*	*entFM*	*bceT*
Baby food with rice and milk based	31	6 (19.35)	7 (22.57)	10 (32.25)	13 (41.93)	11 (35.48)	17 (54.83)	25 (80.64)	7 (22.57)
Baby food with wheat and milk based	25	3 (12)	4 (16)	8 (32)	9 (36)	8 (32)	11 (44)	15 (60)	5 (20)
Baby food with wheat, honey, and milk based	18	1 (5.55)	3 (16.66)	7 (38.88)	9 (50)	6 (33.33)	10 (55.55)	7 (38.88)	3 (16.66)
Baby food with wheat, banana, and milk based	10	—	1 (10)	4 (40)	6 (60)	4 (40)	5 (50)	5 (50)	1 (10)
Total	84	11 (13.09)	15 (17.85)	29 (34.52)	37 (44.04)	28 (33.33)	43 (51.19)	52 (61.90)	16 (19.04)

on different Iranian foods to study the presence of *B. cereus* and its virulence factors.

Acknowledgments

The authors would like to thank Dr. Amir Shakerian and Dr. Farhad Safarpour at the Biotechnology Research Centre of the Islamic Azad University of Shahrekord for their important technical and clinical support. This work was supported by the Islamic Azad University, Shahrekord Branch, Iran, Grant 90/8965.

References

[1] G. Zhou, H. Liu, J. He, Y. Yuan, and Z. Yuan, "The occurrence of *Bacillus cereus*, *B. thuringiensis* and *B. mycoides* in Chinese pasteurized full fat milk," *International Journal of Food Microbiology*, vol. 121, no. 2, pp. 195–200, 2008.

[2] WHO, "Food Safety & Food-borne Illness," Fact Sheet No. 237, World Health Organization, Geneva, Switzerland, 2007.

[3] S. Samapundo, M. Heyndrickx, R. Xhaferi, and F. Devlieghere, "Incidence, diversity and toxin gene characteristics of *Bacillus cereus* group strains isolated from food products marketed in Belgium," *International Journal of Food Microbiology*, vol. 150, no. 1, pp. 34–41, 2011.

[4] K. Dierick, E. Van Coillie, I. Swiecicka et al., "Fatal family outbreak of *Bacillus cereus*-associated food poisoning," *Journal of Clinical Microbiology*, vol. 43, no. 8, pp. 4277–4279, 2005.

[5] F. Evreux, B. Delaporte, N. Leret, C. Buffet-Janvresse, and A. Morel, "A case of fatal neonatal Bacillus cereus meningitis," *Archives de Pediatrie*, vol. 14, no. 4, pp. 365–368, 2007.

[6] K. Arshak, C. Adley, E. Moore, C. Cunniffe, M. Campion, and J. Harris, "Characterisation of polymer nanocomposite sensors for quantification of bacterial cultures," *Sensors and Actuators B*, vol. 126, no. 1, pp. 226–231, 2007.

[7] F. Carlin, J. Brillard, V. Broussolle et al., "Adaptation of *Bacillus cereus*, an ubiquitous worldwide-distributed foodborne pathogen, to a changing environment," *Food Research International*, vol. 43, no. 7, pp. 1885–1894, 2010.

Bacillus cereus in Infant Foods: Prevalence Study and Distribution of Enterotoxigenic Virulence Factors in Isfahan
Province, Iran

175

[8] M. Fricker, U. Messelhäußer, U. Busch, S. Scherer, and M. Ehling-Schulz, "Diagnostic real-time PCR assays for the detection of emetic Bacillus cereus strains in foods and recent foodborne outbreaks," Applied and Environmental Microbiology, vol. 73, no. 6, pp. 1892–1898, 2007.

[9] H. Becker, G. Schaller, W. von Wiese, and G. Terplan, "Bacillus cereus in infant foods and dried milk products," International Journal of Food Microbiology, vol. 23, no. 1, pp. 1–15, 1994.

[10] D. J. Beecher and A. C. L. Wong, "Tripartite haemolysin BL: isolation and characterization of two distinct homologous sets of components from a single Bacillus cereus isolate," Microbiology, vol. 146, no. 6, pp. 1371–1380, 2000.

[11] T. Lund, M. L. de Buyser, and P. E. Granum, "A new cytotoxin from Bacillus cereus that may cause necrotic enteritis," Molecular Microbiology, vol. 38, no. 2, pp. 254–261, 2000.

[12] A. Fagerlund, O. Ween, T. Lund, S. P. Hardy, and P. E. Granum, "Genetic and functional analysis of the cytK family of genes in Bacillus cereus," Microbiology, vol. 150, no. 8, pp. 2689–2697, 2004.

[13] S. Das, P. K. Surendran, and N. Thampuran, "PCR-based detection of enterotoxigenic isolates of Bacillus cereus from tropical seafood," Indian Journal of Medical Research, vol. 129, no. 3, pp. 316–320, 2009.

[14] M. Ehling-Schulz, M. Fricker, and S. Scherer, "Identification of emetic toxin producing Bacillus cereus strains by a novel molecular assay," FEMS Microbiology Letters, vol. 232, no. 2, pp. 189–195, 2004.

[15] N. Agata, M. Ohta, Y. Arakawa, and M. Mori, "The bceT gene of Bacillus cereus encodes an enterotoxic protein," Microbiology, vol. 141, no. 4, pp. 983–988, 1995.

[16] S. I. Asano, Y. Nukumizu, H. Bando, T. Iizuka, and T. Yamamoto, "Cloning of novel enterotoxin genes from Bacillus cereus and Bacillus thuringiensis," Applied and Environmental Microbiology, vol. 63, no. 3, pp. 1054–1057, 1997.

[17] J. Minnaard, L. Delfederico, V. Vasseur et al., "Virulence of Bacillus cereus: a multivariate analysis," International Journal of Food Microbiology, vol. 116, no. 2, pp. 197–206, 2007.

[18] P. E. Granum, K. O'Sullivan, and T. Lund, "The sequence of the non-haemolytic enterotoxin operon from Bacillus cereus," FEMS Microbiology Letters, vol. 177, no. 2, pp. 225–229, 1999.

[19] M. H. Guinebretiere, V. Broussolle, and C. Nguyen-The, "Enterotoxigenic profiles of food—poisoning and food-borne Bacillus cereus strains," Journal of Clinical Microbiology, vol. 40, no. 8, pp. 3053–3056, 2002.

[20] J. E. Reyes, J. M. Bastías, M. R. Gutiérrez, and M. Rodríguez, "Prevalence of Bacillus cereus in dried milk products used by Chilean School Feeding Program," Food Microbiology, vol. 24, no. 1, pp. 1–6, 2007.

[21] M. H. Guinebretiere, H. Girardin, C. Dargaignaratz, F. Carlin, and C. Nguyen-The, "Contamination flows of Bacillus cereus and spore-forming aerobic bacteria in a cooked, pasteurized and chilled zucchini purée processing line," International Journal of Food Microbiology, vol. 82, no. 3, pp. 223–232, 2003.

[22] J. A. Sarrías, M. Valero, and M. C. Salmerón, "Enumeration, isolation and characterization of Bacillus cereus strains from Spanish raw rice," Food Microbiology, vol. 19, no. 6, pp. 589–595, 2002.

[23] J. M. Kramer, R. T. Gilbert, and M. D. Doyle, "Bacillus cereus and other Bacillus species," Foodborne Bacterial Pathogens, vol. 19, pp. 21–70, 1989.

[24] M. C. Te Giffel, R. R. Beumer, P. E. Granum, and F. M. Rombouts, "Isolation and characterisation of Bacillus cereus from pasteurised milk in household refrigerators in the Netherlands," International Journal of Food Microbiology, vol. 34, no. 3, pp. 307–318, 1997.

[25] C. Molva, M. Sudagidan, and B. Okuklu, "Extracellular enzyme production and enterotoxigenic gene profiles of Bacillus cereus and Bacillus thuringiensis strains isolated from cheese in Turkey," Food Control, vol. 20, no. 9, pp. 829–834, 2009.

[26] J. B. Kim, J. M. Kim, C. H. Kim et al., "Emetic toxin producing Bacillus cereus Korean isolates contain genes encoding diarrheal-related enterotoxins," International Journal of Food Microbiology, vol. 144, no. 1, pp. 182–186, 2010.

[27] B. M. Hansen and N. B. Hendriksen, "Detection of enterotoxic Bacillus cereus and Bacillus thuringiensis: strains by PCR analysis," Applied and Environmental Microbiology, vol. 67, no. 1, pp. 185–189, 2001.

24

Isolation and Characterization of 89K Pathogenicity Island-Positive ST-7 Strains of *Streptococcus suis* Serotype 2 from Healthy Pigs, Northeast China

Shujie Wang,[1] Peng Liu,[2] Chunyu Li,[1] Yafang Tan,[2] Xuehui Cai,[1] Dongsheng Zhou,[2] and Yongqiang Jiang[2]

[1] *National Key Laboratory of Veterinary Biotechnology, Harbin Veterinary Research Institute, Chinese Academy of Agricultural Science, Harbin 150001, China*
[2] *State Key Laboratory of Pathogen and Biosecurity, Beijing Institute of Microbiology and Epidemiology, Beijing 100071, China*

Correspondence should be addressed to Dongsheng Zhou, dongshengzhou1977@gmail.com and Yongqiang Jiang, jiangyq@nic.bmi.ac.cn

Academic Editors: M. Dunn, Q. He, and B. Skoldenberg

Streptococcus suis is a swine pathogen which can also cause severe infection, such as meningitis, and streptococcal-like toxic shock syndrome (STSS), in humans. In China, most of the *S. suis* infections in humans were reported in the southern areas with warm and humid climates, but little attention had been paid to the northern areas. Data presented here showed that the virulent serotypes 1, 2, 7, and 9 of *S. suis* could be steadily isolated from the healthy pigs in the pig farms in all the three provinces of Northeast China. Notably, a majority of the serotype 2 isolates belonged to the 89K pathogenicity island-positive ST-7 clone that had historically caused the human STSS outbreaks in the Sichuan and Jiangsu provinces of China, although the human STSS case caused by *S. suis* had never been reported in northern areas of China. Data presented here indicated that the survey of *S. suis* should be expanded to or reinforced in the northern areas of China.

1. Introduction

Streptococcus suis is an important pathogen that can cause the severe systemic infection in the pigs reported worldwide [1]. *S. suis* can also be frequently isolated from other animals such as cats, dogs, and horses, and thus it is believed to be a commensal in the animal intestinal flora [2]. A total of 35 serotypes have been characterized for the *S. suis* isolates from the healthy pigs, but only a limited portion (serotypes 1 to 9, and 14) of them are responsible for the infections in pigs.

Being a causative agent of a zoonotic disease, *S. suis* can be transmitted from pigs to humans. Since the first human case caused by *S. suis* was reported in Denmark, increasing numbers of human cases have been reported in many countries especially including South Asia [3]. Human infections generally can be manifested as meningitis, septicaemia, endocarditis, and deafness. Nearly all the human cases characterized can be ascribed to the handling/consumption of unprocessed pork meat, or to the close contact with pigs [4]. Therefore, most of the infected people are pig farmers, abattoir workers, meat inspectors, butchers, or veterinarian practitioners.

S. suis serotype 2 is the most highly pathogenic one of the 35 serotypes for both pigs and humans [5], and it has caused the two recent outbreaks of human infection in China, which are characterized by a streptococcal toxic shock-like syndrome (STSS), with higher-than-usual human morbidity and mortality [6]. The STSS-causing *S. suis* serotype 2 strains have acquired a 89K pathogenicity island (89KPaI) harboring multiple virulence determinants [6–8]. The acquisition of 89KPaI through gene horizontal transfer plays important roles in the rapid adaptation and increased virulence of *S. suis* serotype 2 [6–8].

In the present work, a total of 155 *S. suis* strains were isolated from the 2204 nose swabs collected from the healthy pigs distributed in all three provinces in Northeast China. At

Isolation and Characterization of 89K Pathogenicity Island-Positive ST-7 Strains of Streptococcus suis Serotype 2 from
Healthy Pigs, Northeast China

177

least four virulent serotypes, 1, 2, 7, and 9 were discriminated from these strains. Notably, a collection of 89KPaI-positive ST-7 strains of serotype 2 were identified for the first time in northern areas of China.

2. Material and Methods

2.1. Specimens Collection. A total of 2204 nose swabs were collected with the aseptic procedures from 2204 different healthy pigs from March to November, 2007. The pigs included growing, nursery, and finishing ones, and sows from 23 pig farms in Northeast China (Heilongjiang, Liaoning and Jilin provinces).

2.2. Isolation of Bacterial Strains and Genomic DNA. The selective Todd-Hewitt broth containing polymyxin B (10 mg/mL), nalidixic acid (15 mg/mL), and crystal violet (0.1 mg/mL) was used for the primary isolation of *S. suis* from the nose swab specimens [9]. Bacteria were cultured overnight on the Columbia sheep agar plates at 37°C for isolation of the genomic DNA by using a genomic DNA isolation kit (Tigan, Beijing, China).

2.3. PCR Detection. *S. suis* was identified by detecting a specific *gdh*-amplification product through PCR [10]. The serotypes 1, 2, 7, and 9 were identified by PCR with the previously characterized serotype-specific primers [11, 12]. PCR was also done to detect the virulence genes [13, 14] encoding muramidase-released protein (MRP), extracellular protein factor (EF), and suilysin (SLY) and to detect the SSU05_0943 gene in the 89KPaI [7].

2.4. Multilocus Sequence Typing Analysis (MLST). A previously described MLST scheme [15] was applied to the indicated *S. suis* strains. This MLST scheme involved seven housing-keeping gene loci, that is, *cpn60*, *dpr*, *recA*, *aroA*, *thrA*, *gki*, and *mutS*, which encoded 60-kDa chaperonin, putative peroxide resistance protein, homologous recombination factor A, 5-enolpyruvylshikimate 3-phosphate synthase, aspartokinase/homoserine dehydrogenase, glucose kinase, and DNA mismatch repair enzyme, respectively. PCR products were purified by using the QIAquick PCR product purification columns and then sequenced from both ends with an ABI Prism 3700 DNA analyzer system. The obtained sequences were compared with the previously defined allelic sequences in the *S. suis* MLST database (http://ssuis.mlst.net/), to identify the allelic profile or sequence type (ST) of each isolate tested herein.

2.5. Microarray-Based Comparative Genomic Hybridization (M-CGH). Gene contents were compared between each paired test and reference (05ZYH33) DNAs using a whole-genome DNA microarray [16, 17] imprinted with 98% of the 2194 annotated ORFs of the *S. suis* serotype 2 strain 05ZYH33 (each ORF was printed in duplicate on a single glass slide). Each paired test and reference (05ZYH33) DNAs were labeled with difference fluorescent dyes (Cy3 or Cy5 dye) and then cohybridized to a microarray slide. Experiments were repeated in duplicate (two biological DNA

probes replicates, and accordingly two microarray slides for each strain), for which the incorporated dye was reversed. The hybridized slides were scanned by using a GenePix Personal 4100A Microarray Scanner (Axon Instruments). The scanning images were processed and the data were further analyzed by using GenePix Pro 5.0 software (Axon Instruments) combined with Microsoft Excel software. Spots with signal intensity (median) in the channel of Reference DNA less than two folds of local background intensity (median) were rejected from further analysis. Spots with bad data because of slide abnormalities were discarded as well. Data normalization was performed on the remaining spots by total intensity normalization methods. A ratio of intensity (test DNA normalized intensity/reference DNA normalized intensity) was recorded for each spot and then was converted to $\log_2$. Genes with fewer than three data points were considered unreliable and were accordingly removed. The averaged $\log_2$ ratio for each remaining gene on the two replicate slides was ultimately calculated. If 20% of the strains had a gene with missing data, the gene was removed. A $\log_2$ value equal to or lower than -1 was taken as defining the absence of a gene in given strain. The binary dataset of absent (0) or present (1) sign genes among strains was displayed by the TreeView software vesion1.60 [18].

3. Results and Discussion

3.1. Identification and Characterization of Total S. suis Strains. The Northeast China region consists of three provinces Jilin, Heilongjiang, and Liaoning. As shown in Figure 1(a), the 2204 nose swab specimens from 2204 different healthy pigs tested could approximately equally assigned into these three provinces. From the 2204 specimens, a total of 155 *S. suis* strains were isolated (Figure 1(b)), and these isolates were identified as *S. suis* by the positive PCR detection of the *S. suis*-specific *gdh* gene [10]. The three provinces Jilin, Heilongjiang, and Liaoning accounted for 83%, 12%, and 5% of the strains isolated, respectively. Accordingly, the *S. suis* isolation rate (no. of *S. suis* strain isolated/no. of specimens tested) for Jilin, Heilongjiang, and Liaoning were 18.43%, 0.95%, and 2.64%, respectively. The much higher isolation rate for Jilin might be due to the fact that the specimens from Liaoning and Heilongjiang were collected during colder months (in March, April, and October) whereas those from Jilin during warmer months (in June and July).

The serotypes 1, 2, 7, and 9 were screened by PCR for these 155 *S. suis* isolates with the previously characterized serotype-specific primers of *S. suis* [11, 12]. Accordingly, the 155 isolates were composed of 39 (25%) serotype-2 strains, 11 (7%) serotype-9 ones, 7 (4%) serotype-1 ones, 4 (3%) serotype-7 ones, and 94 (61%) ones of unknown serotypes (Figure 1(c)). The PCR assays showed that all the 39 serotype 2 strains harbored the three virulence genes encoding MRP, EF, and SLY. These results indicated the highly virulent serotype 2 strains could be frequently isolated from the healthy pigs in Northeast China.

3.2. Identification and Characterization of 89KPaI-Positive S. suis Serotype 2. The SSU05_0943 gene in the 89KPaI was

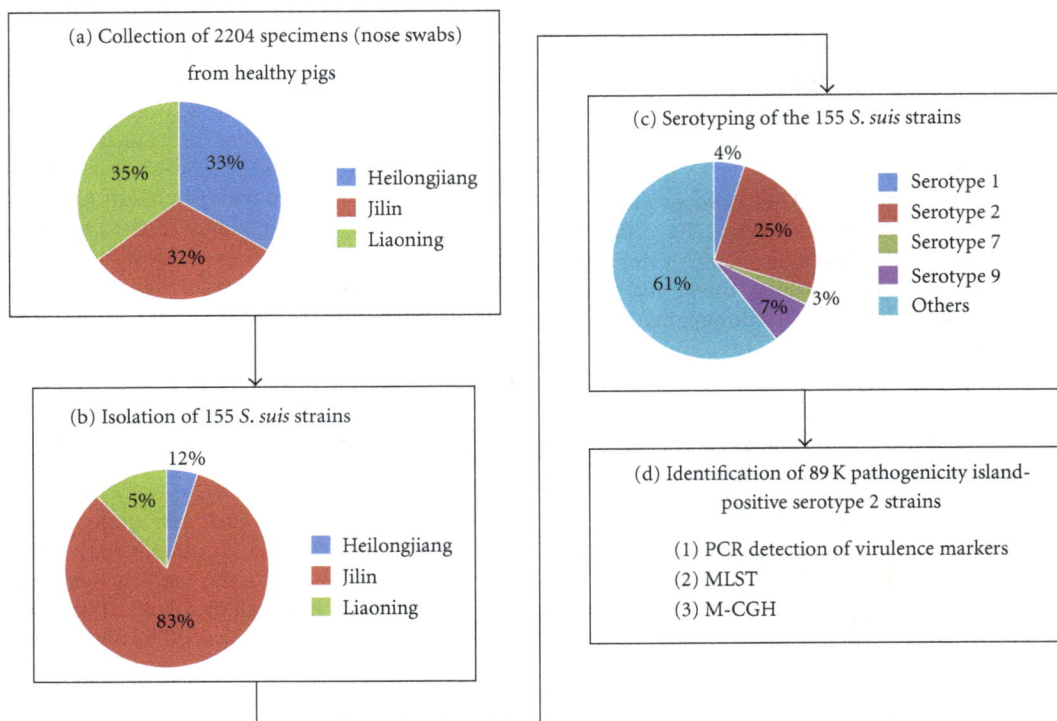

FIGURE 1: Flow charts of the analyses in this study.

chosen for the PCR-based screening for this island in the 39 serotype 2 strains; 32 (82%) of these 39 strains gave positive PCR reaction, indicating that the corresponding strains potentially harbored this genomic island.

Of the above 32 strains potentially harboring the 89KPaI, 15 were arbitrarily selected for the MLST assay. All these 15 strains were identified as ST-7 with an allelic profile 1,1,1,1,1,1,3, which was the same as that of the reference strain 05ZYH33 (an 89KPaI-positive strain, with the determined genome sequence, isolated from the human STSS case [7]). ST-7, represented by the STSS-causing, 89KPaI-positive strains of *S. suis* serotype 2, emerged first in Hong Kong in 1996, and caused 28 cases of human *S. suis* infection in Jiangsu Province, China, in 1998, and another large outbreak of human infection in Sichuan Province, China, in 2005 [6–8, 19]. As a member of the ST-1 (allelic profile 1,1,1,1,1,1,1) complex, ST-7 is a single-locus variant of ST-1 with increased virulence [19], as demonstrated by the fact that the toxicity of ST-7 to peripheral blood mononuclear cells is greater than that of ST-1 [19].

Of the above 15 ST-7 strains determined by MLST, eight were arbitrarily chosen for the further M-CGH analysis (Figure 2). M-CGH has been established as a standard method for the bacterial comparative genome analysis in our laboratory [20, 21]. In the present work, a total of *S. suis* 1918 genes were included in the final microarray dataset, and each gene was categorized as either present (1), absent (0), or missing data for each strain. The eight ST-7 strains tested harbored all the 89KPaI genes imprinted on the microarray, and they gave the gene profiles almost the

same as that of the reference 89KPaI-positive ST-7 strain 05ZYH33, indicating the high clonal genomic content of 05ZYH33 and the above eight ST-7 strains. Included also in the M-CGH analysis was S735 [22] that was a ST-1 strain (MRP+, EF+, SLY+, and 89KPaI-; serotype 2) isolated from a diseased pig in Netherlands in 1963. For S735, the absence of various genome loci (including 89KPaI) of 05ZYH33 was detected by M-CGH. It could be solidly concluded that the above eight strains, characterized by both MLST and M-CGH, belonged to the epidemic 89KPaI-positive ST-7 clone of *S. suis* serotype 2 that has historically cause the human STSS outbreaks in China.

3.3. Concluding Remarks. In China, most of the *S. suis* infections in humans were reported in the southern areas (such as Sichuan, Jiangsu, and Guangdong provinces, and Hongkong) with higher environmental humidity and temperature [23]. However, little attention had been paid to the northern areas of China. Our results showed that the virulent serotypes 1, 2, 7, and 9 could be steadily isolated from the healthy pigs in the pig farms in Northeast China. To the best of our knowledge, this is the first report of the isolation of the 89KPaI-positive ST-7 strains of *S. suis* serotype 2 in Northeast China. Although the human STSS case caused by *S. suis* has never been reported in northern areas of China, a routine survey of *S. suis* in these geographic regions is needed.

Authors' Contribution

S. Wang and P. Liu contributed equally to this work.

Isolation and Characterization of 89K Pathogenicity Island-Positive ST-7 Strains of Streptococcus suis Serotype 2 from Healthy Pigs, Northeast China

179

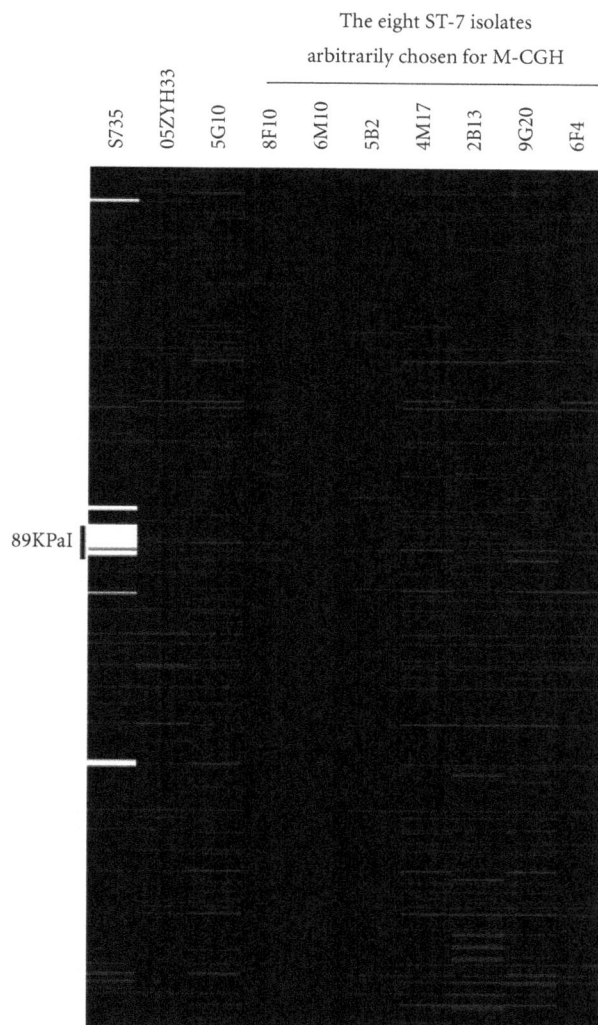

FIGURE 2: Schematic presentation of M-CGH data. Each column represented a strain, while each row standed for a different gene. The strain names were presented on the top. Genes were arranged according to the genomic location of the strain 05ZYH33. For each individual strain, the presence of a gene was represented by a black box, whereas the absence of a gene corresponded to a white box, and the grey area indicated the missing data.

Acknowledgments

This work was supported by the National Basic Research Program of China (2012CB518804), the National Natural Science Foundation of China (81171528), and the Fund of the Chinese Ministry of Agriculture Programs (2009ZX08009-141B).

References

[1] M. Segura, "Streptococcus suis an emerging human threat," Journal of Infectious Diseases, vol. 199, no. 1, pp. 4–6, 2009.

[2] J. J. Staats, I. Feder, O. Okwumabua, and M. M. Chengappa, "Streptococcus suis past and present," Veterinary Research Communications, vol. 21, no. 6, pp. 381–407, 1997.

[3] M. Gottschalk, M. Segura, and J. Xu, "Streptococcus suis infections in humans: the Chinese experience and the situation in North America," Animal Health Research Reviews, vol. 8, no. 1, pp. 29–45, 2007.

[4] M. Gottschalk, J. Xu, C. Calzas, and M. Segura, "Streptococcus suis: a new emerging or an old neglected zoonotic pathogen?" Future Microbiology, vol. 5, no. 3, pp. 371–391, 2010.

[5] H. F. L. Wertheim, H. D. T. Nghia, W. Taylor, and C. Schultsz, "Streptococcus suis an emerging human pathogen," Clinical Infectious Diseases, vol. 48, no. 5, pp. 617–625, 2009.

[6] J. Tang, C. Wang, Y. Feng et al., "Streptococcal toxic shock syndrome caused by Streptococcus suis serotype 2," PLoS Medicine, vol. 3, no. 5, article e151, 2006.

[7] C. Chen, J. Tang, W. Dong et al., "A glimpse of streptococcal toxic shock syndrome from comparative genomics of S. suis 2 Chinese isolates," Plos ONE, vol. 2, no. 3, article e315, 2007.

[8] M. Li, C. Wang, Y. Feng et al., "SalK/SalR, a two-component signal transduction system, is essential for full virulence of highly invasive Streptococcus suis serotype 2," PLoS ONE, vol. 3, no. 5, Article ID e2080, 2008.

[9] B. M. Gray, M. A. Pass, and H. C. Dillon, "Laboratory and field evaluation of selective media for isolation of group B streptococci," Journal of Clinical Microbiology, vol. 9, no. 4, pp. 466–470, 1979.

[10] O. Okwumabua, M. O'Connor, and E. Shull, "A polymerase chain reaction (PCR) assay specific for *Streptococcus suis* based on the gene encoding the glutamate dehydrogenase," *FEMS Microbiology Letters*, vol. 218, no. 1, pp. 79–84, 2003.

[11] H. E. Smith, V. Veenbergen, J. van der Velde, M. Damman, H. J. Wisselink, and M. A. Smits, "The cps genes of *Streptococcus suis* serotypes 1, 2, and 9: development of rapid serotype-specific PCR assays," *Journal of Clinical Microbiology*, vol. 37, no. 10, pp. 3146–3152, 1999.

[12] H. E. Smith, L. van Bruijnsvoort, H. Buijs, H. J. Wisselink, and M. A. Smits, "Rapid PCR test for *Streptococcus suis* serotype 7," *FEMS Microbiology Letters*, vol. 178, no. 2, pp. 265–270, 1999.

[13] H. J. Wisselink, J. J. Joosten, and H. E. Smith, "Multiplex PCR assays for simultaneous detection of six major serotypes and two virulence-associated phenotypes of *Streptococcus suis* in tonsillar specimens from pigs," *Journal of Clinical Microbiology*, vol. 40, no. 8, pp. 2922–2929, 2002.

[14] A. G. Allen, S. Bolitho, H. Lindsay et al., "Generation and characterization of a defined mutant of *streptococcus suis* lacking suilysin," *Infection and Immunity*, vol. 69, no. 4, pp. 2732–2735, 2001.

[15] S. J. King, J. A. Leigh, P. J. Heath et al., "Development of a multilocus sequence typing scheme for the pig pathogen *Streptococcus suis* Identification of virulent clones and potential capsular serotype exchange," *Journal of Clinical Microbiology*, vol. 40, no. 10, pp. 3671–3680, 2002.

[16] X. Zeng, Y. Yuan, Y. Wei et al., "Microarray analysis of temperature-induced transcriptome of Streptococcus suis serotype 2," *Vector-Borne and Zoonotic Diseases*, vol. 11, no. 3, pp. 215–221, 2011.

[17] Y. Wei, X. Zeng, Y. Yuan et al., "DNA microarray analysis of acid-responsive genes of Streptococcus suis serotype 2," *Annals of Microbiology*, vol. 61, pp. 505–510, 2010.

[18] M. B. Eisen, P. T. Spellman, P. O. Brown, and D. Botstein, "Cluster analysis and display of genome-wide expression patterns," *Proceedings of the National Academy of Sciences of the United States of America*, vol. 95, no. 25, pp. 14863–14868, 1998.

[19] C. Ye, X. Zhu, H. Jing et al., "Streptococcus suis sequence type 7 outbreak, Sichuan, China," *Emerging Infectious Diseases*, vol. 12, no. 8, pp. 1203–1208, 2006.

[20] H. Han, H. C. Wong, B. Kan et al., "Genome plasticity of Vibrio parahaemolyticus: microevolution of the "pandemic group"," *BMC Genomics*, vol. 9, article 570, 2008.

[21] D. Zhou, Y. Han, Y. Song et al., "DNA microarray analysis of genome dynamics in Yersinia pestis: insights into bacterial genome microevolution and niche adaptation," *Journal of Bacteriology*, vol. 186, no. 15, pp. 5138–5146, 2004.

[22] N. Fittipaldi, J. Harel, B. D'Amours, S. Lacouture, M. Kobisch, and M. Gottschalk, "Potential use of an unencapsulated and aromatic amino acid-auxotrophic *Streptococcus suis* mutant as a live attenuated vaccine in swine," *Vaccine*, vol. 25, no. 18, pp. 3524–3535, 2007.

[23] Y. Feng, H. Zhang, Y. Ma, and G. F. Gao, "Uncovering newly emerging variants of Streptococcus suis, an important zoonotic agent," *Trends in Microbiology*, vol. 18, no. 3, pp. 124–131, 2010.

25

Setup and Validation of Flow Cell Systems for Biofouling Simulation in Industrial Settings

Joana S. Teodósio,[1] Manuel Simões,[1] Manuel A. Alves,[2] Luís F. Melo,[1] and Filipe J. Mergulhão[1]

[1] LEPAE-Department of Chemical Engineering, Faculty of Engineering, University of Porto,
Rua Dr. Roberto Frias s/n, 4200-465 Porto, Portugal
[2] CEFT-Department of Chemical Engineering, Faculty of Engineering, University of Porto,
Rua Dr. Roberto Frias s/n, 4200-465 Porto, Portugal

Correspondence should be addressed to Filipe J. Mergulhão, filipem@fe.up.pt

Academic Editors: S. Fuchs and L. Ramirez

A biofouling simulation system consisting of a flow cell and a recirculation tank was used. The fluid circulates at a flow rate of $350\,L\cdot h^{-1}$ in a semicircular flow cell with hydraulic diameter of 18.3 mm, corresponding to an average velocity of $0.275\,m\cdot s^{-1}$. Using computational fluid dynamics for flow simulation, an average wall shear stress of 0.4 Pa was predicted. The validity of the numerical simulations was visually confirmed by inorganic deposit formation (using kaolin particles) and also by direct observation of pathlines of tracer PVC particles using streak photography. Furthermore, the validity of chemostat assumptions was verified by residence time analysis. The system was used to assess the influence of the dilution rate on biofilm formation by *Escherichia coli* JM109(DE3). Two dilution rates of 0.013 and $0.0043\,h^{-1}$ were tested and the results show that the planktonic cell concentration is increased at the lower dilution rate and that no significant changes were detected on the amount of biofilm formed in both conditions.

1. Introduction

Biofilms are complex communities of surface-attached microorganisms, comprised either of a single or multiple species [1–3]. Although adhesion to a surface may look unattractive at first glance due to a more difficult access to nutrients, the fact is that in nature probably 99% or more of all bacteria exist in biofilms [4].

Despite some beneficial applications of biofilms, for instance, in industrial production of various chemicals [5–7], wastewater treatment [8], removal of volatile compounds from waste streams [9], or even energy production [10], the fact is that detrimental biofilms are far more notorious than beneficial biofilms. Negative effects of biofilm formation are often found in industrial settings where biofouling costs can represent up to 30% of the plant operating costs [11], and in the overall expenditure of industrialized countries where estimates for fouling costs (of which biofouling possibly accounts for one third) represent 0.25% of the gross national product [11].

Since beneficial and detrimental biofilms exist, it is useful to develop strategies for biofilm control that promote the formation of beneficial biofilms and delay the formation of detrimental biofilms or promote their destruction. This requires intensive studies on the mechanisms of biofilm formation and resistance and prompted the need to develop *in-vitro* platforms for biofilm studies. These platforms are artificial biofilm model systems that are easy to control and reproducible enabling a more detailed study of this phenomena.

Flow cells have been used for more than 30 years for the study of dynamic biofilms. Although different configurations have been proposed for the biofilm forming system, flow cells are often placed downstream of a biological reactor that is the source of microorganisms. In order to study biofilm formation using equilibrium conditions, it is convenient to use the effluent from a chemostat to feed the flow cell. This system provides a constant concentration of exponentially growing cells with the added advantage that the composition of the effluent stream is also constant. This system has the drawback of being limited to relatively low flow rates due to the time and expense entailed with media preparation. Furthermore, the flow rate in the flow cell is fixed by the dilution rate used on the chemostat which reduces the range of hydrodynamic conditions that can be investigated. An alternative configuration can be used where the chemostat is part of a recirculating loop that is continuously fed by a nutrient stream. This configuration can be used to achieve the high flow rates that are common in industrial processes [12]. In this case, the flow velocity in the flow cell is independent of the dilution rate of the system and higher shear rates can be achieved. In essence these systems are "chemostats with irregular geometries" [13]. In order to prevent gradient formation along the system, the recirculation flow rates must be high and the volume of the recycle loops must be minimized thus decreasing residence times. Ideally, the residence time in the entire recycle loop and flow cell should not exceed a few minutes so that the whole system can be assumed to be completely mixed [13] given the time frame of the experiments that last for several days. In industrial settings, molecules and microorganisms are transported in process streams usually under turbulent flow and, therefore, the velocity field of the fluid in contact with the microbial layer will affect biofilm structure and behaviour [14–16]. Thus, hydrodynamic conditions will determine the rate of transport of cells, oxygen, and nutrients to the surface, as well as the magnitude of the shear stress acting on a developing biofilm [17]. The importance of hydrodynamic conditions in microbial adhesion to surfaces has prompted the development of flow cell types with different designs for biofilm studies [13]. These flow cells are constructed so that a large surface area is available on which the hydrodynamic conditions remain constant for a wide range of flow velocities. One of the key design issues concerns the geometry of the flow cell and the inlet conditions which dictate the length required for flow development [18]. The hydrodynamics in flow cells can be simulated numerically using computational fluid dynamics (CFD). For turbulent flow conditions, one of the most reliable two-equation-based models to describe turbulence is the k-ω model. This model is based on two transport equations, one for the turbulent kinetic energy and another for the specific dissipation rate. The shear-stress transport (SST) version of the k-ω model [19] blends effectively the k-ε model [20] and the standard k-ω model [21] and usually leads to accurate simulations both in free stream and wall bounded flows.

The use of CFD tools is particularly useful to obtain detailed information about the flow field in complex geometries, allowing the prediction of different flow variables, such as shear and normal stresses exerted on the flow cell walls, which are highly relevant to understand their influence on the development and growth of biofilms. On the other hand, the ability to accurately simulate the entire flow field in the flow cell allows us to determine the required flow-path lengths to achieve fully developed hydrodynamic conditions,

and to assess how uniform is the stress field in the flow cell walls.

In this work, we have validated results obtained using numerical simulation, by inorganic deposit formation studies and also by observation of fluid pathlines of polyvinyl chloride (PVC) particles using streak photography. The validity of chemostat assumptions was verified by residence time analysis. We have also assessed the effect of two dilution rates on *E. coli* biofilm formation under a constant turbulent flow rate.

2. Materials and Methods

2.1. Numerical Simulations. The Fluent CFD code (version 6.2.16, Fluent Inc.) was used in the numerical simulation of the flow field in the semicircular flow cell. The computational mesh used in the simulations was created using Gambit 2.2.30 mesh generator (Fluent Inc.) and has a total number of 1 250 472 hexahedral cells. As shown in Figure 1(a), the mesh resolution is very high, especially near the walls, hence numerical accuracy is high. The mesh generated includes the semicircular flow cell (110.0 cm length; 3.0 cm diameter), and the connecting circular tubes, both with a length of 30.0 cm and 8.0 mm internal diameter, as sketched in Figure 1(b).

To increase the numerical accuracy, the convective terms were discretized with the QUICK scheme, which is third-order accurate [22], while the pressure-velocity coupling was enforced using the PISO algorithm (Pressure-Implicit with Splitting of Operators) proposed by Issa [23]. The time integration used a fixed time step, $\delta t = 0.005$ s, and a second-order implicit Euler method was selected to enhance accuracy in time. Preliminary numerical tests, using different time-step values, showed that such time-step value is small enough to achieve high numerical accuracy along the oscillation cycles.

2.2. Flow Visualization. To complement the numerical simulations, a study of the flow patterns for a range of Reynolds numbers was performed using a long-exposure streak photography technique.

This method consists in the placement of reflective agents in the fluid. These agents were PVC spherical particles with a diameter of 50 μm at a weight concentration of 100 ppm. The illumination was provided by a 635 nm laser diode equipped with a cylindrical lens (Vector, model 5200-20, 5 mW), creating a laser light sheet which illuminates a specific plane of the flow cell. The flow pattern images representing the pathlines were obtained by a digital camera (Canon EOS 30D) equipped with a macro lens (Canon EF100 mm), which was positioned perpendicularly to the light sheet.

2.3. Bacterial Strain and Culture Conditions. Escherichia coli JM109(DE3) was used in this work because this particular strain is capable of producing significant amounts of biofilm [24]. A starter culture was prepared by inoculation of 300 μL of a glycerol stock (kept at −80°C) to a total volume of 0.5 L of inoculation media, as previously described by Teodósio et al. [24]. Optical density (O.D.) measurement at 610 nm was

(a)

(b)

(c)

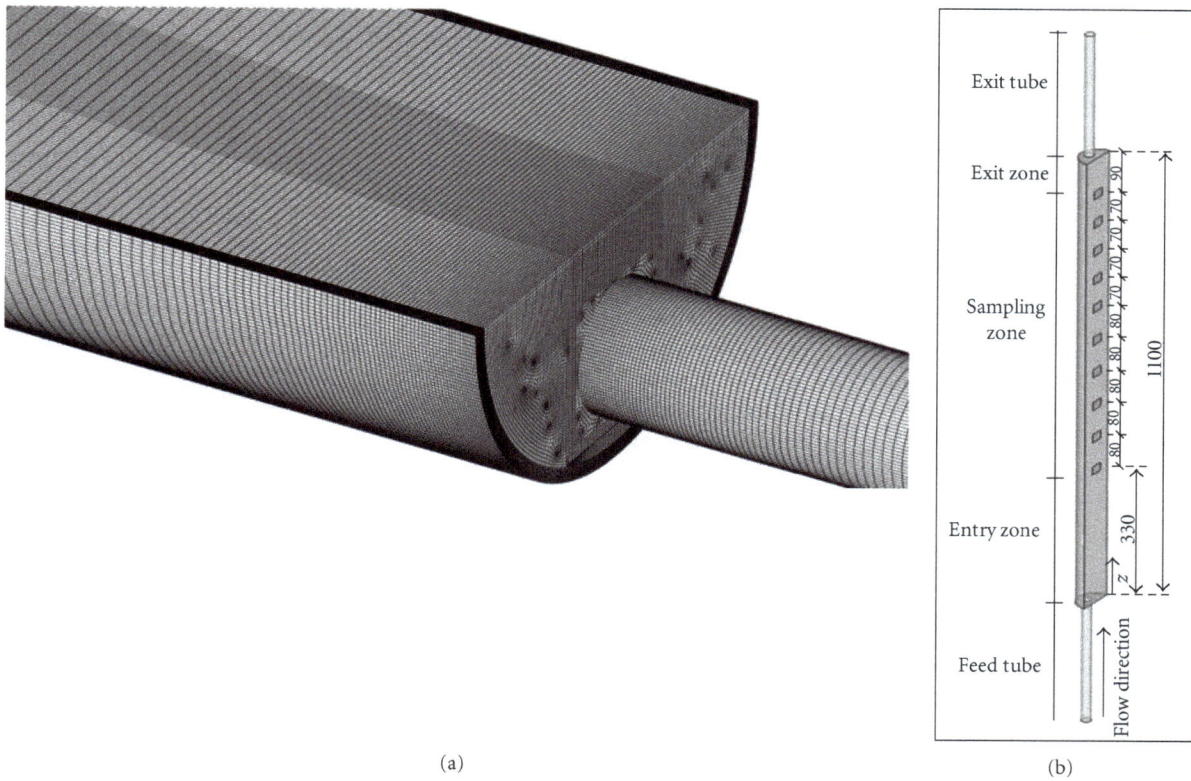

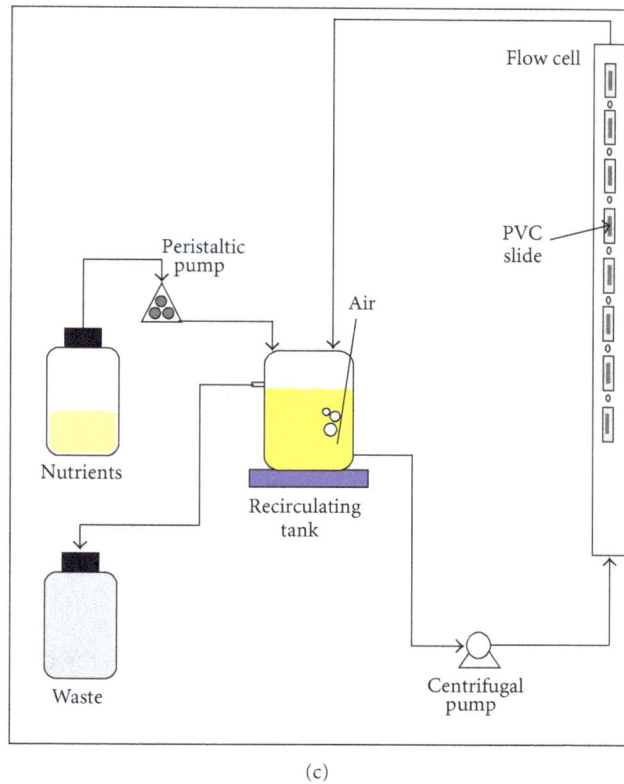

FIGURE 1: (a) Zoomed view near the entrance of the computational mesh used; (b) illustration of the flow cell and connecting tubes. Identification of the axial location of the center of the coupons (dimensions are reported in mm); (c) schematic representation of the biofilm producing flow cell system.

used to monitor cell growth and when the O.D. reached 1, cultures were used to inoculate the recirculating tank.

In order to obtain two different dilution rates, the same feed flow rate of 0.0252 L·h^{-1} was used in two systems where the only difference was the volume of the recirculating tank (all the remaining system components were maintained). Thus, a recirculating tank of 5 L (corresponding to a total system volume of 5.92 L–Table 1(a)) was used to obtain a dilution rate of 0.0043 h^{-1}, and a recirculating tank of 1 L (corresponding to a total volume of 1.92 L–Table 1(b)) was used to obtain a dilution rate of 0.013 h^{-1}. For inoculation, the 1 L recirculating tank contained 0.5 L of sterile water and the 5 L recirculating tank contained 2.5 L of sterile water. Recirculating tank feeding started 5 h after inoculation at a flow rate of 0.0252 L·h^{-1} with the nutrient media consisting of 0.55 g·L^{-1} glucose, 0.25 g·L^{-1} peptone, 0.125 g·L^{-1} yeast extract, and phosphate buffer (0.188 g·L^{-1} KH$_2$PO$_4$ and 0.26 g·L^{-1} Na$_2$HPO$_4$), pH 7.0.

2.4. Biofilm Producing System. A reactor system consisting of a recirculating tank, one vertical flow cell built in Perspex, peristaltic and centrifuge pumps, was used as described by Teodósio et al. [24] (Figure 1(c)). PVC slides (2 cm × 1 cm) are glued onto the removable coupons of the flow cell and are in contact with the bacterial suspension circulating in the system. PVC was chosen because it is commonly found in piping systems, although the slide material can be changed to simulate other surfaces. The recirculating tank, for planktonic cell growth, was built with a cooling jacket to enable temperature control. Temperature was kept constant at 30°C using a recirculating water bath in order to simulate conditions commonly found on industrial settings. *E. coli* cells were grown by recirculating the bacterial suspension during 13 days at a flow rate of 350 L·h^{-1}. Turbulent flow with Reynolds number of 6 290 was used for biofilm formation which is also typical for many industrial processes.

2.5. Sampling and Analysis. For biofilm sampling, the system was stopped to allow coupon removal and carefully started again maintaining the same flow conditions, as described by Teodósio et al. [24]. Biofilm wet weight was determined by weighing the coupon immediately after retrieval from the flow cell and subtracting the weight of the empty coupon that had been determined during system assembly. Biofilm thickness was determined using a digital micrometer (VS-30H, Mitsubishi Kasei Corporation) [24].

The optical density and glucose concentration were determined in the recirculating tank. Optical density was measured using a Spectrophotometer at 610 nm (T80 UV/VIS Spectrometer/PG Instrument, Ltd.). Glucose quantification was performed by dinitrosalicylic colorimetric method (DNS) adapted to a microtiter plate format as described by Teodósio et al. [24]. Glucose consumption values were obtained from a mass balance, by multiplying the glucose concentration difference (feed minus concentration on the tank) by the feed flow rate (0.0252 L·h^{-1}).

TABLE 1: (a) Residence times on the various components of the system (5.92 L system). (b) Residence times on the various components of the system (1.92 L system).

(a)

Component	Volume (mL)	Residence time (s)
Recirculating tank	5000	51.4
Flow cell	300	3.1
Recirculating tubing	620	6.4
Whole system	5920	60.9

(b)

Component	Volume (mL)	Residence time (s)
Recirculating tank	1000	10.3
Flow cell	300	3.1
Recirculating tubing	620	6.4
Whole system	1920	19.7

2.6. Deposit Formation by Inorganic Material. Kaolin powdered particles, with a distribution of diameters ranging between 5 and 10 μm [25], were used in an experiment to verify the formation of inorganic deposits on the biofilm producing system. The experimental setup was properly cleaned (with bleach) and sterile water was introduced in the system. A kaolin suspension with the final concentration of 2.0 g·L^{-1} was prepared and introduced in the system after the total removal of the sterile water. The system operated continuously at a flow rate of 350 L·h^{-1} for one week.

2.7. Statistical Analysis. Biofilm formation results originated from three independent experiments for each dilution rate condition. Paired *t*-test analyses were performed to estimate whether or not there was a significant difference between the results. Each time point was evaluated individually using the three independent results obtained in one condition and the three individual results obtained on the other condition. When a confidence level greater than 95% ($P < 0.05$) was obtained, these time points were marked with an *.

3. Results

3.1. Numerical Simulation of the Flow. In the numerical simulations, a recirculation flow rate of 350 L·h^{-1} was considered, at 30°C, corresponding to a Reynolds number of 6 290. The Reynolds number is here defined as

$$\text{Re} = \frac{\rho U D_h}{\mu}, \tag{1}$$

where ρ and μ are the density and dynamic viscosity of the fluid, respectively, U is the average velocity in the flow cell, and D_h is the hydraulic diameter of the semicircular flow cell ($D_h = \pi D/(2+\pi) = 1.83$ cm) of diameter D. In the inlet and outflow circular tubes connecting the flow cell (of diameter 8 mm), the Reynolds number was higher (Re = 19 300), therefore, the flow was turbulent in the full flow domain.

Figure 2 illustrates the time evolution of the average wall shear stresses acting on coupons 1, 2, 3, 6, and 10.

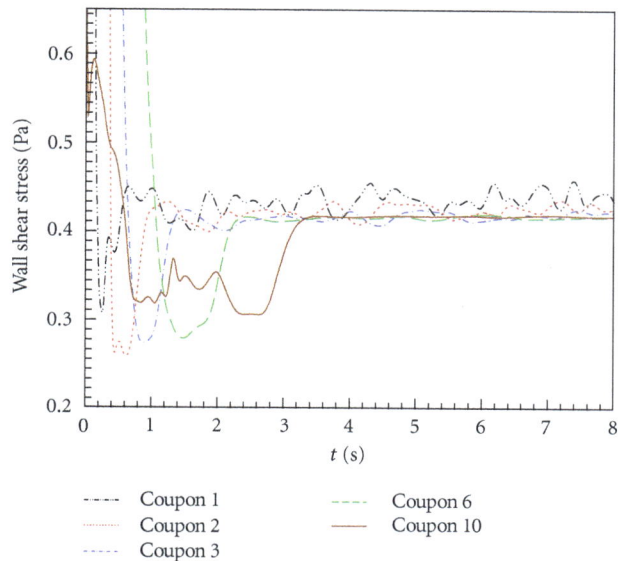

FIGURE 2: Time evolution of the magnitude of the wall shear stress acting on coupons 1, 2, 3, 6, and 10.

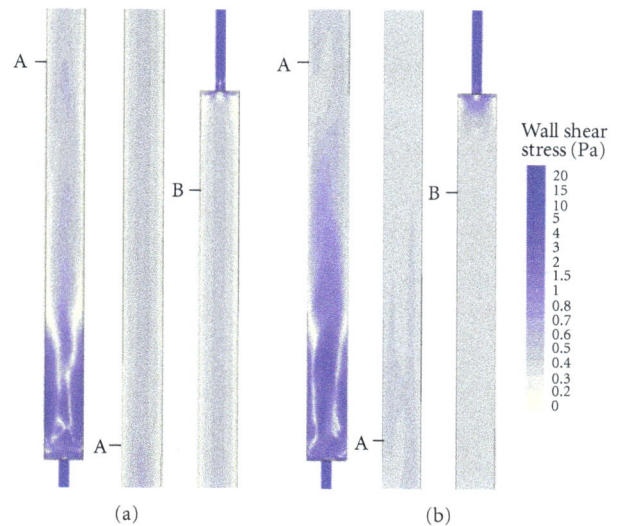

FIGURE 3: Contour plots of the instantaneous magnitude of the wall shear stress field on the (a) planar and (b) circular walls. In both cases the entry zone, fully developed flow region and exit zones are shown. The center of the first and last coupons is identified by letters A and B, respectively. The flow direction is from bottom to top.

Coupons 1–3 were selected to represent locations where flow is still spatially developing, while coupon 10 (last) was selected to show that, despite being close to the exit of the flow cell, the local flow was not influenced by the abrupt contraction to the downstream exit tube. Coupon 6 was representative of fully developed flow conditions and shows clearly that coupons 3–10 were also representative of fully developed flow conditions, to within a variation below 5%. In coupons 1 and 2, there were still some entrance effects, and the transients were more intense, nonetheless the variation of the average wall shear stress was still below 5%, an acceptable variation to still consider those first two coupons as representative of the fully developed flow conditions and accept them as reliable for monitoring biofilm growth in the semicircular flow cell.

Figure 2 also illustrates that after a transient period of about 1–3 s, depending on the coupon location, the flow patterns oscillated periodically with a dominant frequency of about 3 Hz. The average values of the wall shear stress on each coupon were 0.440, 0.425, 0.417, 0.417, and 0.417 Pa for coupons 1, 2, 3, 6, and 10, respectively. The amplitude of wall shear stress oscillations was higher in coupons 1 and 2 ($\pm$0.02 and 0.015 Pa, resp.), but overall the average shear stresses were very similar on all coupons. It is important to bear in mind that these small oscillations along time are not important, and the average values should be considered in the analysis given the time scale of the biofilm experiments, of the order of several days.

Figure 3 shows instantaneous contour plots of the wall shear stress acting on both the planar and round walls of the flow cell, and on the entry and outflow circular tubes. The wall shear stresses were similar in all coupons, except for the first ones, where an increase below 5% was observed. Also, it is relevant to emphasize that despite the transverse variation of the shear stress observed in the flat and rounded

walls, with minimum values on the intersection of both walls, the shear stresses predicted in the central part of the flat wall, where the coupons are located, were very similar to the predicted shear stresses on the curved wall, thus demonstrating that the average wall shear stresses acting on the coupons on the planar walls are representative of the shear stresses acting on the curved wall. This is an indication that the semicircular cell is a good representation of the wall shear stresses observed in circular tubes, which are typically used in industrial piping systems.

Figure 4 displays a frontal view through the planar wall of the instantaneous streamlines predicted numerically for the same instant illustrated in Figure 3. The results illustrate the long inlet developing length, and the negligible influence of the outflow boundary, justifying the use of a longer entry region to guarantee that in the first coupon the flow is nearly fully developed.

Figure 5 presents streamwise plots of the instantaneous magnitude of the wall shear stress along the center of the planar and circular walls. To better assess the entry length necessary to achieve fully developed flow conditions, the streamwise variation of the velocity magnitude along the center of the semicircular duct is also included. The results confirm unequivocally that even the first coupon was already under quasi-fully developed flow conditions. The difference in the wall shear stresses between the center of the planar and circular walls is small, showing that the wall shear stress on the planar wall is representative of the wall shear stress on the circular wall.

3.2. System Validation. After numerical simulation of the hydrodynamic conditions by CFD, the flow was visualized using tracer particles illuminated with a laser diode.

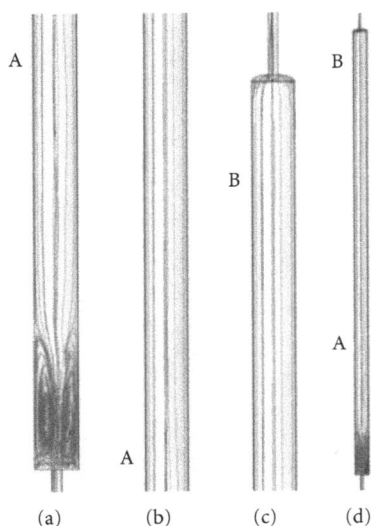

FIGURE 4: Instantaneous streamlines on the (a) entrance, (b) middle, and (c) exit regions of the semicircular flow cell. In (d) a global view is presented. The flow direction is from bottom to top, and the views illustrated are through the planar wall. Point A represents the position of the center of coupon 1 and point B the position of the center of coupon 10.

Photographs taken during this experiment are shown on Figure 6(a) where it is possible to verify that particles follow a quasilinear trajectory on the sampling area confirming numerical simulation results. This flow behavior is very different from the entry zone where significant vortices can be observed. Perturbations on the exit zone are not very clear from the tracer particles experiment and also confirm the results obtained by numerical simulation.

In order to further assess the hydrodynamic effects during biofilm formation, a suspension of kaolin particles (diameter 5–10 μm; density 2.6 g·cm^{-3}) was used and inorganic deposits were formed on the flow cell system (using the same flow rate conditions used during biofilm formation). Although the physical properties of kaolin particles are different from bacterial cells, these particles are highly adherent to Perspex surfaces as it can be seen on Figure 6(b). This figure shows that there are regions in the entry and exit zones where adhesion is prevented possibly due to the higher shear stresses generated in those regions and the more intense velocity fluctuations in those locations. The lengths of these deposit free zones are consistent with the CFD predictions presented on Figures 3, 4, and 5 which is an additional indication that the output of the numerical simulation is valid for the real life conditions used on this experiment. Indeed, by observing deposit formation (Figure 6(b)), it was possible to measure the flow cell entry and exit zone lengths: 18 cm and 0.5 cm, respectively. Deposit formation was observed on the entire flow cell, except for the entry zone (up to 18 cm from the inlet) and very close to the outlet (0.5 cm). These observations agree with the results presented in Figure 4, where a long entry zone region is required to guarantee fully developed flow in the first coupon

area. The negligible influence of the outflow boundary was also confirmed. One of the most significant advantages in using a recirculating system (where the flow cell is fed by the effluent of a chemostat and the flow is recirculated after passage back to the chemostat) is the possibility of adjusting the flow rate on the flow cell independently of the nutrient flow rate in order to achieve higher fluid velocities [13]. In order to guarantee that estimates of activity or the microbiology of the attached cells and the planktonic cell population in the effluent truthfully reflect the processes occurring on the system, chemostat assumptions are used and, therefore, the system is assumed to be completely mixed [13]. If the flow rate in the recirculation tube is low, significant gradients may build up along the length of the recycling tube, in the recirculating tank or even in the flow cell and, therefore chemostat kinetics cannot be used. In order to have a well-mixed system, the residence time in each of the components of the system must be minimized (e.g., by using the minimum length of connecting tube) and the total residence time in the system should not exceed a few minutes [13].

In order to obtain two different dilution rates, two different reactors of 1 L and 5 L were used in independent experiments and the feed flow rate and the recirculation rates were maintained (feed flow rate 0.0252 L·h^{-1} and recirculating rate 350 L·h^{-1}). The residence time analysis in both systems shows that the total residence time is equal to or less than one minute (Tables I(a) and I(b)). This indicates that the mass transfer processes controlled by the hydrodynamics of the system are much faster than the metabolic response by the cells and that the system can be considered a well-mixed chemostat with an irregular geometry [13].

3.3. Biofilm Formation. Biofilm formation by *E. coli* JM109(DE3) was assessed on a flow cell operated at a Reynolds number of 6 290 during 13 days. An experimental apparatus [24] including a recirculating tank system allowed biofilm growth under well-defined and controlled conditions and biofilm sampling for analysis. Two dilution rates (0.013 and 0.0043 h^{-1}) were tested using a glucose concentration of 0.55 g·L^{-1} on the feed stream.

Figure 7 shows the results obtained for the planktonic and biofilm parameters analyzed during the experiment. In order to evaluate planktonic cell growth, optical density (O.D.) measurements were performed in the recirculating tank (Figure 7(a)). Glucose consumption profiles (Figure 7(b)) were determined by sampling the recirculating tank and establishing a mass balance. Biofilm samples were retrieved from the coupons for quantification of wet weight (Figure 7(c)) and thickness (Figure 7(d)). The analysis of planktonic optical density (Figure 7(a)) shows a distinct behavior ($P < 0.05$) between cells grown under different dilution rates. With the exception of days 3 and 4, higher concentrations of planktonic cells were obtained for the lower dilution rate. For the higher dilution rate, planktonic cell concentration remained constant throughout the experiment whereas for the lower dilution rate cell, concentration

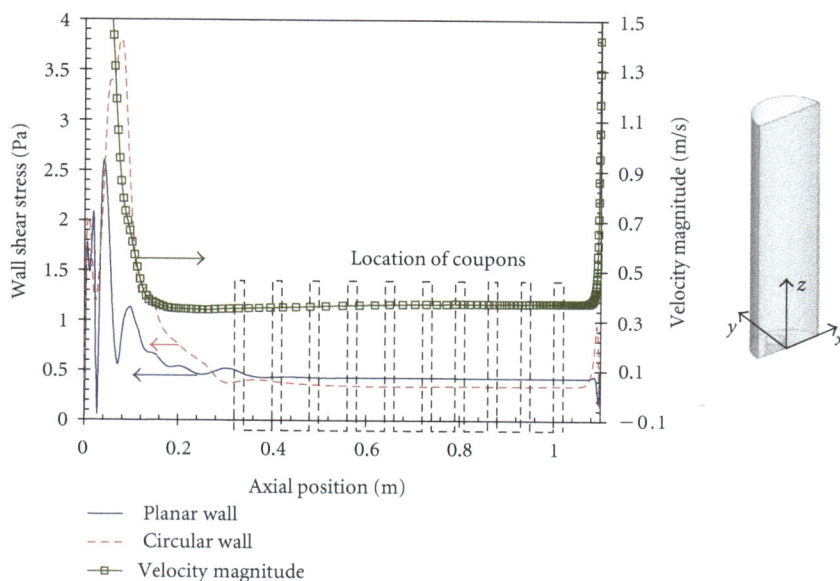

FIGURE 5: Streamwise variation of the magnitude of the wall shear stress along the center of the planar ($x = 0$, $y = 0$) and round walls ($x = 0$, $y = 0.015$ m). The velocity magnitude profile is also included, but along the center of the semicircular flow cell ($x = 0$, $y = 0.0075$ m).

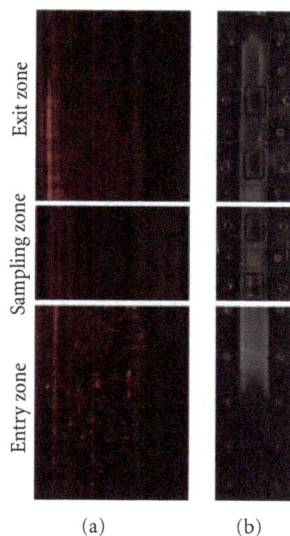

FIGURE 6: Flow visualization using tracer particles illuminated with a laser diode (a) and inorganic deposit formation (b) at three different zones of the flow cell: entry, sampling, and exit.

increased between days 4 and 5, then equilibrium was reached.

Glucose consumption increased during the experimental time for both dilution rates (Figure 7(b)). The profiles are very similar ($P > 0.05$) until day 9 and then glucose consumption in the lower dilution rate was higher ($P < 0.05$) which is consistent with an increased planktonic cell concentration in these conditions. Concerning biofilm formation results, similar profiles were observed for wet weight and thickness (Figure 7(c) and 7(d)) with thicker biofilms obtained in the lower dilution rate between days 6 and 11. Previous studies by Wijeyekoon et al. [26], using microscopic methods, showed that an increase in nutrient loading led to the formation of compact and thinner biofilms which is consistent with our results. The wet weight of the biofilms obtained in both conditions increases until days 10-11 and then decreases probably due a predominance of sloughing/detachment phenomena when compared to the attachment of new biofilm cells. The biofilm thickness results (Figure 7(d)) are consistent with the wet weight analysis and similar values are obtained for both parameters at the end of the experiment ($P > 0.05$).

Although it has been shown that high glucose concentrations can inhibit biofilm formation [27, 28], contradictory results were also reported [29] indicating that higher glucose concentrations may also be beneficial. Additionally, we have previously used higher dilution rates (of $0.165\,h^{-1}$) for biofilm formation with this same strain (not shown) and, therefore, we know that we were not exceeding the critical dilution rate in this situation. In the present conditions, an increase in dilution rate is not favorable for planktonic cell development and has no significant impact on the amount of biofilm formed. Apparently, the higher dilution rate may exceed the maximum growth rate of planktonic cells in these conditions. Arguably, washout is not observed due to the dynamic process of biofilm detachment and sloughing events which may be a source for planktonic cells. However, since the nutrient loading is increased at higher dilution rates but the hydrodynamic conditions were maintained (Re = 6 290), we hypothesize that in these conditions the hydrodynamics are more important than nutrient availability in controlling the amount of formed biofilm as observed by independent groups [30, 31] working with mixed species biofilms.

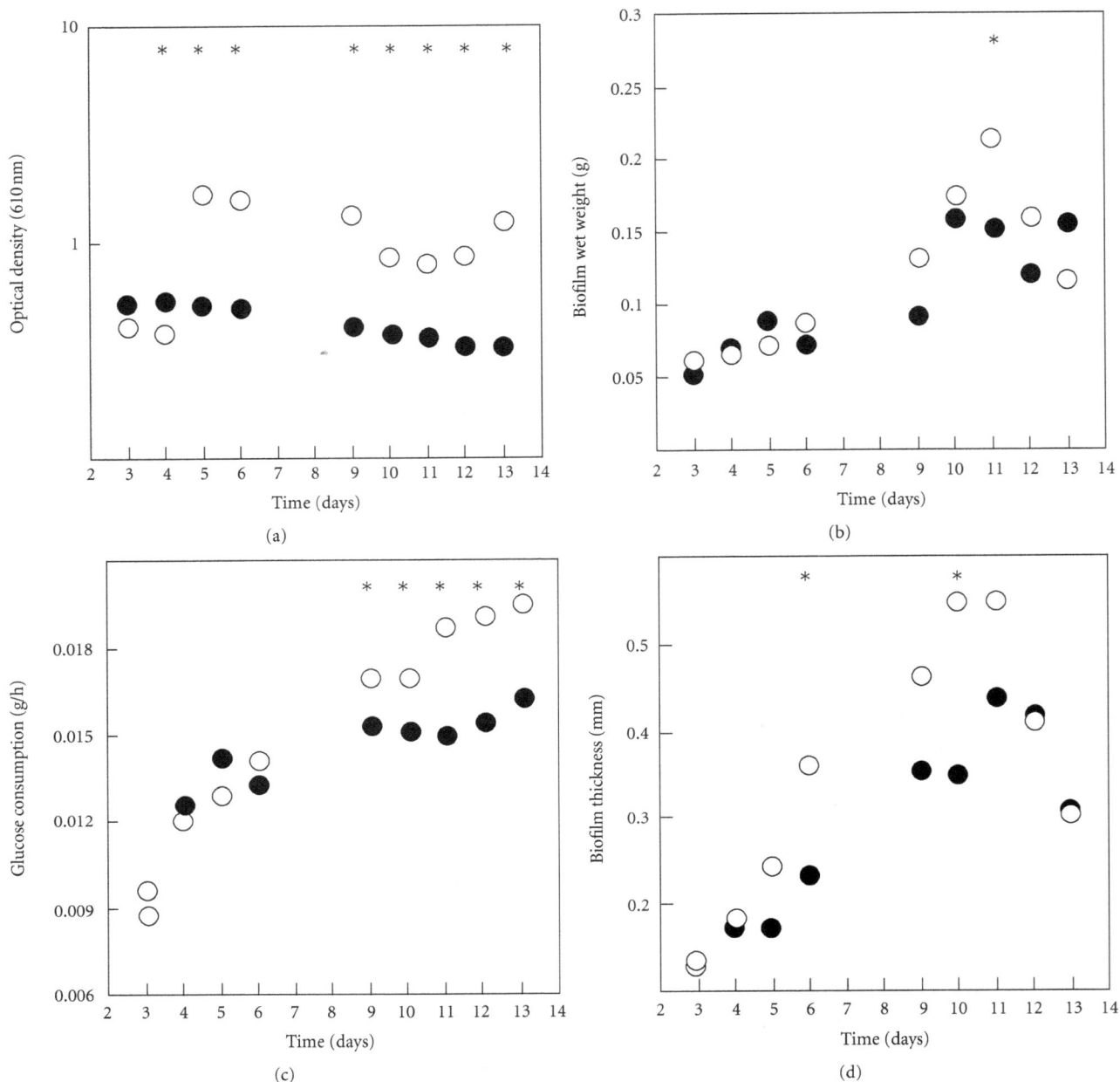

FIGURE 7: Time-course evolution of planktonic and biofilm assayed parameters for *E. coli* JM109(DE3). Closed circles: dilution rate of 0.013 h^{-1}, open circles: dilution rate of 0.0043 h^{-1}. (a) Planktonic optical density (610 nm), (b) glucose consumption in the system (g·h^{-1}), (c) biofilm wet weight (g) and (d) Biofilm thickness (mm). Results are an average of those obtained in three independent experiments for each condition. Time points marked with an * are those for which a statistically significant difference was obtained (for a confidence level greater than 95%).

4. Conclusions

A flow cell system was used to study the effects of the dilution rate on biofilm formation using *E. coli* JM109(DE3). The hydrodynamic conditions on the flow cell were simulated using CFD and it was shown that under the flow rate conditions that were used, a fully developed flow was obtained on the sampling section. It was also demonstrated that the entry zone had a long enough distance to allow flow development and that the effect of the sudden contraction on the exit zone was negligible. The validity of the models used for the simulation was confirmed by flow visualization and particle deposition experiments. Besides the hydrodynamic validation, the validity of chemostat assumptions was also verified by residence time analysis. Altogether these results show that this biofilm forming system comprising the flow cell and the recirculation tank is valid for biofilm formation studies at these flow rates. Biofilm formation assays showed that the lower dilution rate favored planktonic growth and biofilm thickness although the mass of biofilm formed was similar in both conditions. Although these biofilm experiments were performed using a particular *E. coli* strain, we have shown by time-residence analysis, particle deposition, and flow visualization experiments that this experimental setup and flow simulation by CFD are robust enough to be used with other biofilm producing bacteria.

Nomenclature

D: Diameter, m
D_h: Hydraulic diameter, $\pi D/(2 + \pi)$, m
μ: Fluid dynamic viscosity, Pa·s
Re: Reynolds number, Re $= \rho U D_h/\mu$, dimensionless
ρ: Fluid density, kg·m^{-3}
U: Average velocity, m·s^{-1}.

Acknowledgments

The authors would like to acknowledge the financial support provided by the Portuguese Foundation for Science and Technology (FCT) and FEDER, through projects PDT/BIO/69092/2007 and PTDC/EME-MFE/114322/2009. Preliminary numerical simulations by Mr. F. Silva are also acknowledged.

References

[1] J. D. Bryers, "Biofilms, microbial," in *Encyclopedia of Bioprocess Technology—Fermentation, Biocatalysis, and Bioseparation*, M. C. Flickinger and S. W. Drew, Eds., John Wiley & Sons, 1999.

[2] P. Stoodley, K. Sauer, D. G. Davies, and J. W. Costerton, "Biofilms as complex differentiated communities," *Annual Review of Microbiology*, vol. 56, pp. 187–209, 2002.

[3] P. Watnick and R. Kolter, "Biofilm, city of microbes," *Journal of Bacteriology*, vol. 182, no. 10, pp. 2675–2679, 2000.

[4] T. Shunmugaperumal, *Introduction and Overview of Biofilm, Biofilm Eradication and Prevention*, John Wiley & Sons, 2010.

[5] N. Qureshi, B. A. Annous, T. C. Ezeji, P. Karcher, and I. S. Maddox, "Biofilm reactors for industrial bioconversion process: employing potential of enhanced reaction rates," *Microbial Cell Factories*, vol. 4, article 24, pp. 1–21, 2005.

[6] N. Qureshi, P. Karcher, M. Cotta, and H. P. Blaschek, "High-productivity continuous biofilm reactor for butanol production: effect of acetate, butyrate, and corn steep liquor on bioreactor performance," *Applied Biochemistry and Biotechnology Part A*, vol. 114, no. 1–3, pp. 713–721, 2004.

[7] A. Tay and S. T. Yang, "Production of L(+)-lactic acid from glucose and starch by immobilized cells of *Rhizopus oryzae* in a rotating fibrous bed bioreactor," *Biotechnology and Bioengineering*, vol. 80, no. 1, pp. 1–12, 2002.

[8] F. K. J. Rabah and M. F. Dahab, "Nitrate removal characteristics of high performance fluidized-bed biofilm reactors," *Water Research*, vol. 38, no. 17, pp. 3719–3728, 2004.

[9] T. Manolov, H. Kristina, and G. Benoit, "Continuous acetonitrile degradation in a packed-bed bioreactor," *Applied Microbiology and Biotechnology*, vol. 66, no. 5, pp. 567–574, 2005.

[10] Z. W. Wang and S. Chen, "Potential of biofilm-based biofuel production," *Applied Microbiology and Biotechnology*, vol. 83, pp. 1–18, 2009.

[11] L. F. Melo and H. C. Flemming, "Mechanistic aspects of heat exchanger and membrane biofouling and prevention," in *The Science and Technology of Industrial Water Treatment*, Z. Amjad, Ed., pp. 365–380, Taylor and Francis Group, Boca Raton, Fla, USA, 2010.

[12] L. F. Melo and M. J. Vieira, "Physical stability and biological activity of biofilms under turbulent flow and low substrate concentration," *Bioprocess Engineering*, vol. 20, no. 4, pp. 363–368, 1999.

[13] P. Stoodley and B. K. Warwood, "Use of flow cells an annular reactors to study biofilms," in *Biofilms in Medicine, Industry and Environmental Biotechnology: Characteristics, Analysis and Control*, P. Lens, V. O'Flaherty, A. P. Moran, P. Stoodley, and T. Mahony, Eds., pp. 197–213, IWA Publishing, Cornwall, UK, 2003.

[14] M. O. Pereira, M. Kuehn, S. Wuertz, T. Neu, and L. F. Melo, "Effect of flow regime on the architecture of a Pseudomonas fluorescens biofilm," *Biotechnology and Bioengineering*, vol. 78, no. 2, pp. 164–171, 2002.

[15] Y. P. Tsai, "Impact of flow velocity on the dynamic behaviour of biofilm bacteria," *Biofouling*, vol. 21, no. 5-6, pp. 267–277, 2005.

[16] M. J. Vieira, L. F. Melo, and M. M. Pinheiro, "Biofilm formation: hydrodynamic effects on internal diffusion and structure," *Biofouling*, vol. 7, pp. 67–80, 1993.

[17] M. Simões, M. O. Pereira, S. Sillankorva, J. Azeredo, and M. J. Vieira, "The effect of hydrodynamic conditions on the phenotype of *Pseudomonas fluorescens* biofilms," *Biofouling*, vol. 23, no. 4, pp. 249–258, 2007.

[18] D. P. Bakker, A. Van der Plaats, G. J. Verkerke, H. J. Busscher, and H. C. Van der Mei, "Comparison of velocity profiles for different flow chamber designs used in studies of microbial adhesion to surfaces," *Applied and Environmental Microbiology*, vol. 69, no. 10, pp. 6280–6287, 2003.

[19] F. R. Menter, "Two-equation eddy-viscosity turbulence models for engineering applications," *AIAA Journal*, vol. 32, no. 8, pp. 1598–1605, 1994.

[20] B. E. Launder and D. B. Spalding, *Lectures in Mathematical Models of Turbulence*, Academic Press, London, UK, 1972.

[21] D. C. Wilcox, *Turbulence Modeling for CFD*, DCW Industries Inc., Canada, 1998.

[22] B. P. Leonard, "A stable and accurate convective modelling procedure based on quadratic upstream interpolation," *Computer Methods in Applied Mechanics and Engineering*, vol. 19, no. 1, pp. 59–98, 1979.

[23] R. I. Issa, "Solution of the implicitly discretised fluid flow equations by operator-splitting," *Journal of Computational Physics*, vol. 62, no. 1, pp. 40–65, 1986.

[24] J. S. Teodósio, M. Simões, L. F. Melo, and F. J. Mergulhão, "Flow cell hydrodynamics and their effects on *E. coli* biofilm formation under different nutrient conditions and turbulent flow," *Biofouling*, vol. 27, no. 1, pp. 1–11, 2011.

[25] M. O. Pereira, M. J. Vieira, and L. F. Melo, "The effect of clay particles on the efficacy of a biocide," *Water Science and Technology*, vol. 41, no. 4-5, pp. 61–64, 2000.

[26] S. Wijeyekoon, T. Mino, H. Satoh, and T. Matsuo, "Effects of substrate loading rate on biofilm structure," *Water Research*, vol. 38, no. 10, pp. 2479–2488, 2004.

[27] J. Domka, J. Lee, and T. K. Wood, "YliH (BssR) and YceP (BssS) regulate *Escherichia coli* K-12 biofilm formation by influencing cell signaling," *Applied and Environmental Microbiology*, vol. 72, no. 4, pp. 2449–2459, 2006.

[28] D. W. Jackson, J. W. Simecka, and T. Romeo, "Catabolite repression of *Escherichia coli* biofilm formation," *Journal of Bacteriology*, vol. 184, no. 12, pp. 3406–3410, 2002.

[29] T. Bühler, S. Ballestero, M. Desai, and M. R. W. Brown, "Generation of a reproducible nutrient-depleted biofilm of *Escherichia coli* and *Burkholderia cepacia*," *Journal of Applied Microbiology*, vol. 85, no. 3, pp. 457–462, 1998.

[30] K. Garny, H. Horn, and T. R. Neu, "Interaction between biofilm development, structure and detachment in rotating annular reactors," *Bioprocess and Biosystems Engineering*, vol. 31, no. 6, pp. 619–629, 2008.

[31] H. Beyenal and Z. Lewandowski, "Internal and external mass transfer in biofilms grown at various flow velocities," *Biotechnology Progress*, vol. 18, no. 1, pp. 55–61, 2002.

26

Plankton Microorganisms Coinciding with Two Consecutive Mass Fish Kills in a Newly Reconstructed Lake

Andreas Oikonomou,[1] **Matina Katsiapi,**[2] **Hera Karayanni,**[3]
Maria Moustaka-Gouni,[2] **and Konstantinos Ar. Kormas**[1]

[1] *Department of Ichthyology and Aquatic Environment, School of Agricultural Sciences, University of Thessaly, 384 46 Volos, Greece*
[2] *Department of Botany, School of Biology, Aristotle University of Thessaloniki, 541 24 Thessaloniki, Greece*
[3] *Department of Biological Applications and Technology, University of Ioannina, 451 10 Ioannina, Greece*

Correspondence should be addressed to Konstantinos Ar. Kormas, kkormas@uth.gr

Academic Editors: B. A. P. da Gama, J. B. Gurtler, and G. Hobbs

Lake Karla, Greece, was dried up in 1962 and its refilling started in 2009. We examined the Cyanobacteria and unicellular eukaryotes found during two fish kill incidents, in March and April 2010, in order to detect possible causative agents. Both microscopic and molecular (16S/18S rRNA gene diversity) identification were applied. Potentially toxic Cyanobacteria included representatives of the *Planktothrix* and *Anabaena* groups. Known toxic eukaryotes or parasites related to fish kill events were *Prymnesium parvum* and *Pfiesteria* cf. *piscicida*, the latter being reported in an inland lake for the second time. Other potentially harmful microorganisms, for fish and other aquatic life, included representatives of Fungi, Mesomycetozoa, Alveolata, and Heterokontophyta (stramenopiles). In addition, Euglenophyta, Chlorophyta, and diatoms were represented by species indicative of hypertrophic conditions. The pioneers of L. Karla's plankton during the first months of its water refilling process included species that could cause the two observed fish kill events.

1. Introduction

Planktonic Cyanobacteria and unicellular eukaryotes belonging to different functional groups constitute key components of aquatic ecosystems [1]. Among the unicellular plankton there are species that negatively influence the ecosystem [2, 3]. Several of these microorganisms lack distinct morphological features. Even if taxonomically useful morphological features are present, they may get lost throughout sampling, preservation, and examination procedures [4] making identification by traditional microscopic methods difficult. Molecular techniques have spawned new ways to access the diversity of the microbial world. Yet, molecular techniques have limitations [5]. Therefore, a combination of molecular techniques and microscopy methods is required in order to uncover the diversity of the microbial world [6].

Mass fish kills are known to occur in eutrophic lakes. They have been attributed mostly to hypoxic/anoxic conditions or uncommonly high/low temperatures. Other factors, related or not to the eutrophication, include floods, droughts, cyclonic storms, habitat loss, low water flow, and abrupt water level fluctuations [7]. Due to the changes of the grazing pressure, fish kills may lead to considerable changes in the food web structure of the lake ecosystem, with diminishing consequences for the possibilities of using the lake for recreation, fishing, or as a source of drinking water. Although such mass mortality events are well documented in the literature, to the best of our knowledge, there is no such data on newly reconstructed lakes.

In freshwater, the haptophyte *Prymnesium parvum* is considered one of the most dangerous microorganisms and is responsible for adverse effects on aquatic organisms [8] and in particular for several fish kill incidents [9]. It poses a serious threat to several ecosystems since it survives in a wide range of salinities and blooms in coastal and brackish inland waters worldwide [10, 11]. In Lake Koronia, Greece, *P. parvum* coincided with a mass death of birds and fish [2, 12]. The dinoflagellate *Pfiesteria* species can harm fish in coastal waters [13, 14] and has caused fish kills under certain circumstances in North Carolina, USA [13].

No *Pfiesteria*-induced fish kills have ever been reported in Mediterranean coastal waters, while the only, and most unusual, inland ecosystem where *Pfiesteria* has been reported is Ace Lake, Antarctica [15].

While acute fish kills due to toxic algae are well studied, another less obvious impact of toxic/parasitic unicellular eukaryotes is that exposure of aquatic animals to their toxins or parasitism might induce serious sublethal effects, including predisposing these populations to various infectious diseases resulting in, for example, reduction of growth and reproduction [8, 16]. This situation might be even more severe if one considers that we know only a few of the toxic/parasitic eukaryotes that can cause fish kills, while on the other hand our concept on the existing species diversity of the microscopic eukaryotes is still expanding [17]. This led us to investigate the planktonic Cyanobacteria and microeukaryotes of a newly reconstructed lake (Lake Karla, central Greece) during two consecutive fish kill events which occurred in less than six weeks. The aims of this study were to supplement the limited knowledge on the plankton Cyanobacterial and microeukaryotic diversity of newly reconstructed lakes and to identify potentially toxin-producing and parasitic taxa which coincided with the fish kill events and might have deleterious effects on the ecosystem.

2. Materials and Methods

2.1. Study Area. Lake Karla (Figure 1) is located in central Greece (39°29′02″ N, 22°51′41″ E). It formerly covered an area of ca. 180 km^2 but in the beginning of the 1960s it was drained through a tunnel leading the lake's drainage to the nearby Pagasitikos Gulf. A small permanent marsh remained at the area that once covered the lake. The structure and function of L. Karla was correlated with River Pinios, as the flooding events of the river supplied the lake with water rich in nutrients [18]. Several biological and physical-chemical criteria characterized the lake as a eutrophic but with high stability before its drainage [19]. It was not until the 1990s that the refilling of the lake was decided by inflowing water from the nearby River Pinios. Its actual filling started in September 2009, after building a peripheral dam which covers 38 km^2. We sampled in L. Karla in March and April 2010, during two fish kill events. As reported in local newspapers, the dead fish floated in the lake and lined along the shores of a 3.5 to 5 km stretch.

Water samples for microscopic analysis were collected on 17 March and 20 April 2010 at ca. 0.5 m depth from the water level pier at the southeast end of the lake (Figure 1). Three replicates of 500 mL each were collected in polyethylene bottles. Two of them were fixed with Lugol's solution and formaldehyde, while one was retained fresh for direct microscopic analysis. Water temperature, dissolved oxygen, salinity, and pH were measured in situ using a WTW sensor (Weilheim, Germany).

For each sampling date, at least three replicates of live and preserved samples were examined in sedimentation chambers using an inverted microscope with phase contrast (Nikon SE 2000). Cyanobacteria and microscopic eukaryotes were identified using classical taxonomic keys and previous works [20–23]. Phytoplankton counts (cells, colonies, and coenobia) were performed using the Utermöhl's sedimentation method [24]. For biomass (mg L^{-1}) estimation, the dimensions of 30 individuals (cells, filaments, or colonies) of each species were measured using tools of a digital microscope camera (Nikon DS-L1), while mean cell or filament volume estimates were calculated using appropriate geometric formulae, as described previously [25, 26]. Species and taxonomical groups comprising more than 10% (w/w) of the total phytoplankton biomass were considered to be dominant.

Water samples for DNA extraction were transported to the laboratory in 4-L collapsible plastic bottles (Nalgene, Rochester NY, USA) and processed within 1 h of collection. After screening through a 180 μm mesh net to exclude larger eukaryotes and particles, 200–250 mL of water was filtered through a 0.2 μm pore size Polycarbonate Isopore filter (Sartorius, Goettingen, Germany). The filtration was conducted under reduced pressure (≤100 mmHg) to prevent cell damage. Filters were stored immediately at −80°C until further analysis.

DNA was extracted using the UltraClean Soil DNA isolation kit (MoBio Laboratories, Carlsbad CA, USA) according to the manufacturer's protocol after slicing the filter with a sterile scalpel. The concentration of bulk DNA was estimated by spectrophotometry (NanoDrop ND-1000, NanoDrop Technologies, Wilmington DE, USA) and ranged between 11.9 and 15.4 ng μL^{-1} for the March and April samples, respectively. For PCR amplification, approximately 12 ng of environmental DNA was used as template for both samples. The 18S rRNA gene was amplified using the eukaryote specific primers EukA (5′-AACCTGGTTGATCCTGCCAGT-3′) and EukB (5′-GATCCTTCTGCAGGTTCACCTAC-3′) [27] for the March sample, while the primers EukA and Euk1633rE (5′-GGGCGGTGTGTACAARGRG-3′) [28] were used for amplification of the 18S rRNA gene for April sample.

PCR for the amplification of the March sample included an initial denaturation step at 95°C for 15 min, which was followed by 40 cycles consisting of denaturation at 95°C for 45 s, annealing at 55°C for 1 min, and elongation at 72°C for 2 min and 30 s; a final 7 min elongation step at 72°C was included. The PCR protocol for the April sample included an initial denaturation step at 95°C for 2 min followed by 40 cycles of denaturating at 95°C for 40 s, annealing at 50°C for 40 sec, and elongation at 72°C for 2 min and 15 s, with an additional step of final elongation at 72°C for 1 min. Each PCR from the two samples was repeated with different cycle numbers (between 20 and 37). The lowest number of cycles that gave a positive signal, that is, 26 and 28 cycles for the March and April sample, respectively, was further used in order to eliminate some of the major PCR innate limitations [29, 30] and to avoid differential representation of 18S rRNA genes with low and high copy numbers.

For PCR amplification of the Cyanobacterial 16S rDNAs, we used the Cyanobacteria-specific primers CYA106f (5′-CGGACGGGTGAGTAACGCGTGA-3′), CYA781r(a) (5′-GACTACTGGGGTATCTAATCCCATT-3′), and CYA781r(b) (5′-GACTACAGGGGTATCTAATCCCTTT-3′) [31]. PCR

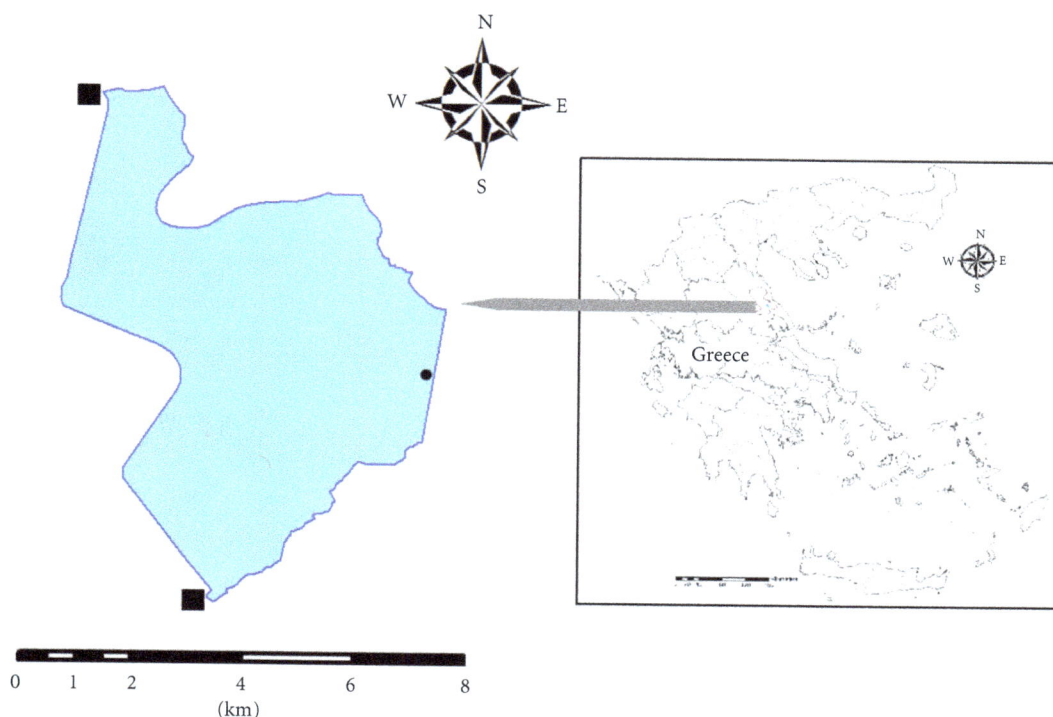

FIGURE 1: Map of Lake Karla, Greece, and sampling point (black dot). Black squares show points of inflowing water for reconstruction purposes. Centre of the lake is at 39°29′00″ N, 22°49′00″ E.

included an initial denaturation step at 94°C for 5 min, which was followed by 40 cycles consisting of denaturation at 94°C for 30 s, annealing at 57°C for 30 s, and elongation at 72°C for 3; a final 5 min elongation step at 72°C was included. Cycle optimization was performed as above which resulted in 26 cycles for the March sample. In April 2010, no sample was analysed for 16S rRNA gene diversity since the vast majority of the observed morphospecies was observed microscopically.

The PCR products from both the Eukarya- and Cyanobacteria-specific amplifications were visualized on a 1% agarose gel under UV light, purified using the Montage purification kit (Millipore Inc, Molsheim, France). The purified PCR products were ligated into the PCR XL TOPO Vector (Invitrogen-Life Technologies, Carlsbad CA, USA) and transformed in electrocompetent *Escherichia coli* cells according to the manufacturer's specifications. For each clone library a maximum of 151 clones were sequenced, each containing an insert of ca. 1800/1600 or 680 bp for the Eukarya and Cyanobacteria, respectively. These clones were grown in liquid Luria-Bertani medium with kanamycin and their plasmids were purified using the Nucleospin Plasmid Quick-Pure kit (Macherey-Nagel GmbH and Co. KG, Düren, Germany) for DNA sequencing. Sequence data were obtained by capillary electrophoresis (Macrogen Inc., Seoul, Korea) using the BigDye Terminator kit (Applied Biosystems-Life Technologies, Carlsbad, CA, USA) with the set of primers M13F (5′-GTAAAACGACGGCCAG-3′) and M13R (5′-CAGGAAACAGCTATGAC-3′). For the eukaryotic clones, intermediate sequencing was performed using the primer 1179rE

(5′-CCCGTGTTGAGTCAAATT-3′) [32]. Each sequence read was approximately 850 bp. For each individual clone, forward, reverse, and intermediate—for the Eukarya—reads were assembled, and then the assembled sequences were checked for chimeras. The Pintail program (http://www.bioinformatics-toolkit.org/Web-Pintail/, [33]) was used for the detection of putative chimeric sequences. Chimeras were discarded from the dataset. Using the multiple alignment program CLUSTALW2 (http://www.ebi.ac.uk/Tools/clustalw2/index.html/) and based on 98% gene similarity as a phylotype cutoff [17, 34], clones were grouped together and considered members of the same phylotype. All sequences were compared with the BLAST function (http://www.ncbi.nlm.nih.gov/BLAST/) for the detection of closest relatives. Sequence data were compiled using the MEGA4 software [35] and aligned with sequences obtained from the GenBank (http://www.ncbi.nlm.nih.gov/) database, using the ClustalX aligning utility. Phylogenetic analyses were performed using the MEGA version 4 software [35] and the topology of the tree was based on neighbour-joining according to Jukes-Cantor. Bootstrapping under parsimony criteria was performed with 1,000 replicates. Sequences of unique phylotypes found in this study have GenBank accession numbers JN090861-JN090912 and JN090913-JN090923 for the eukaryotes and Cyanobacteria, respectively.

Library clone coverage was calculated by the formula of the Good's C estimator $[1 - (n_i/N)]$ [36], where n_i is the number of phylotypes represented by only one clone and N is the total number of clones examined in each library. The number of predicted phylotypes for each clone library

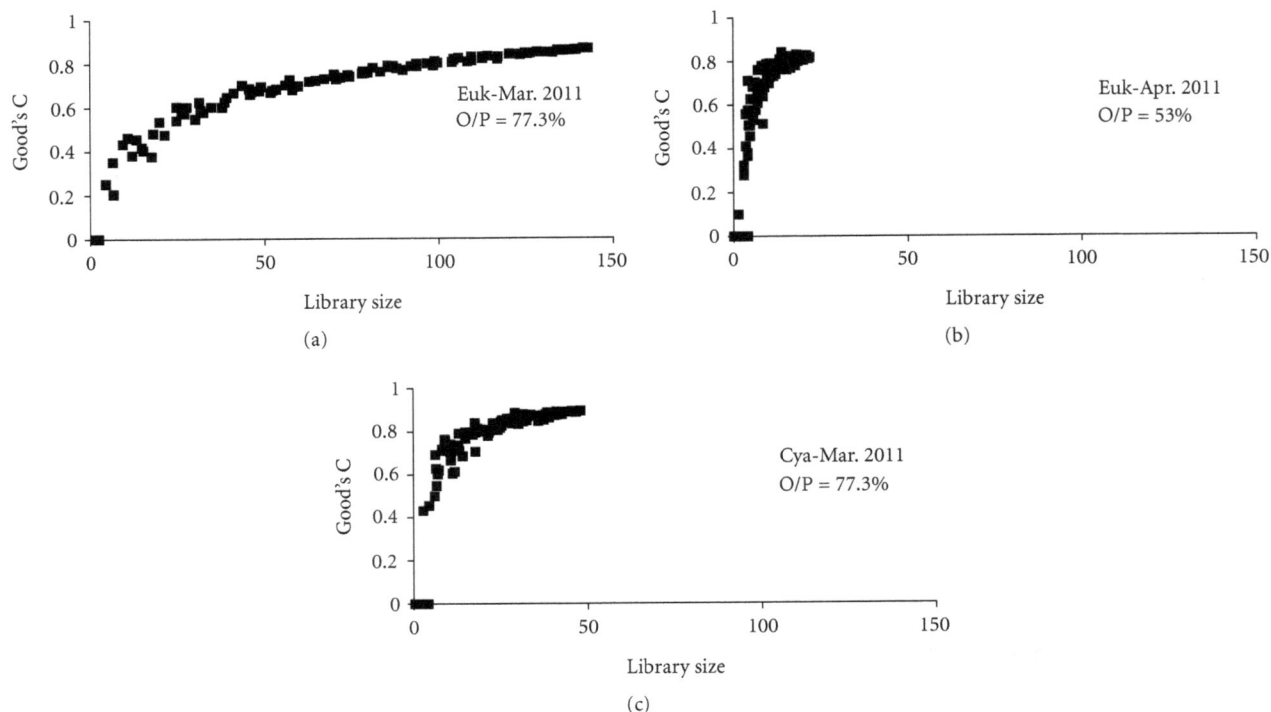

FIGURE 2: rRNA gene clone library coverage based on Good's C estimator of the unicellular eukaryotes (Euk) and Cyanobacteria (Cya) from Lake Karla, Greece. O/P = ratio of observed-to-predicted number of phylotypes.

was estimated after the abundance-based richness formula S_{Chao1} [37, 38].

3. Results and Discussion

We investigated the composition of plankton Cyanobacteria and unicellular eukaryotes by combing molecular, 18S/16S rRNA gene diversity, and microscopic analysis in Lake Karla during two fish kill events which happened within the first year of the lake's partial reconstruction. The prevailing abiotic factors (Table 1) indicated that dissolved oxygen (5.6–5.8 mg L^{-1}) was not limited, while the elevated salinity (7.6–8.1 psu) was possibly attributed to the drainage of the previous lake as well as the result of intensive agricultural and livestock use for four decades. Irrigation in the absence of leaching can increase soil salinity [39] and continued application of livestock manure to agricultural land may result in an accumulation of salt in soil [40].

The two eukaryotic clone libraries revealed that 45 phylotypes occurred in March and only seven in April 2010. However, in both cases, rarefaction curves (Figure 2) reached saturation levels for both clone libraries according to the Good's C estimator, indicating that the majority of the existing phylotypes were revealed. Based on the 18S rRNA gene diversity (Figures 3 and 4), members of the Chlorophyta, Cercozoa, Heterokontophyta (stramenopiles), Alveolata, Fungi, Euglenophyta, Choanoflagellata, Haptophyta, Mesomycetozoea, Katablepharidophyta, and Cryptophyta (Figures 2 and 3) were found. Chlorophyta was the most phylotype-rich group in both samplings, while the next most

TABLE 1: Prevailing physical and chemical parameters in L. Karla.

	Temperature (°C)	Salinity (PSU)	Dissolved oxygen (mg/L)	pH
17/03/2010	15.6	7.6	5.8	8.3
20/04/2010	17.2	8.1	5.6	8.0

abundant phylotypes belonged to the Cercozoa, Alveolata, and stramenopiles.

The Cyanobacteria 16S rRNA gene clone library coverage was satisfactory (Figure 2) and showed (Figure 5) that Cyanobacteria were represented by phylotypes related to the *Planktothrix* group, the Chroococcales, and several algal plastids. Along with these phylotypes, three Verrucomicro- bia-like phylotypes were also retrieved, reinforcing the notion that some Verrucomicrobia are associated with Cyanobacteria-dominated waters [41, 42].

Microscopic analysis (Figure 6) of phytoplankton gave a slightly different picture of the phytoplankton dominance. In March 2010, the diatom *Cyclotella* sp. dominated followed by *Prymnesium parvum* (Haptophyta), *Planktothrix* cf. *agardhii* (Cyanobacteria), *Euglena* sp. (Euglenophyta) and *Anabaena* sp. (Cyanobacteria) and from Alveolata *Pfiesteria* cf. *piscicida* (the latter consisted 0.4% of the high 46.5 mg L^{-1} total biomass and for this it is not included in Figure 6). Most of these microorganisms have been also found in April 2010 but in lower biomass. Nevertheless, the phylotypes of these organisms have been found in the respective clone libraries from both dates.

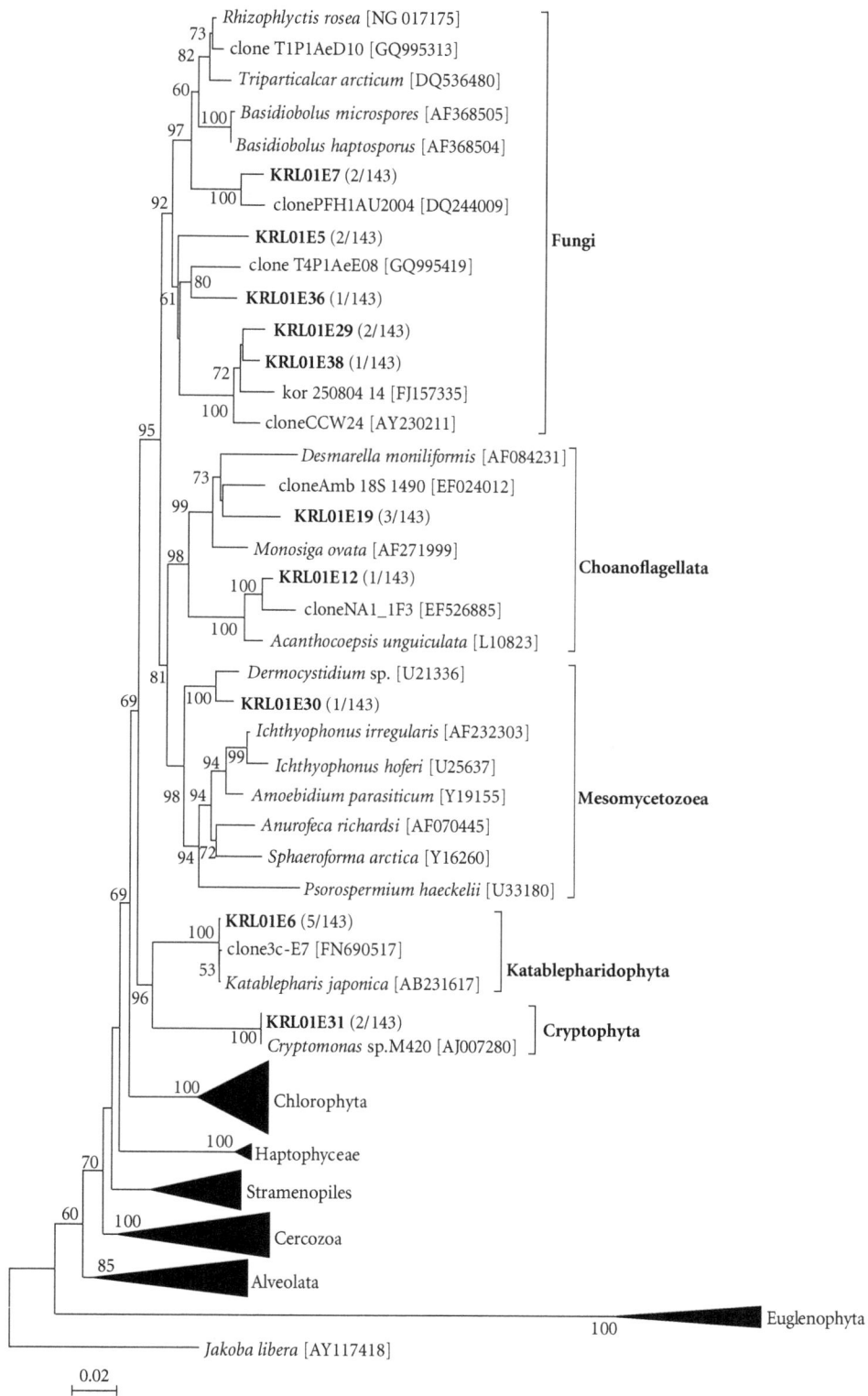

FIGURE 3: Continued.

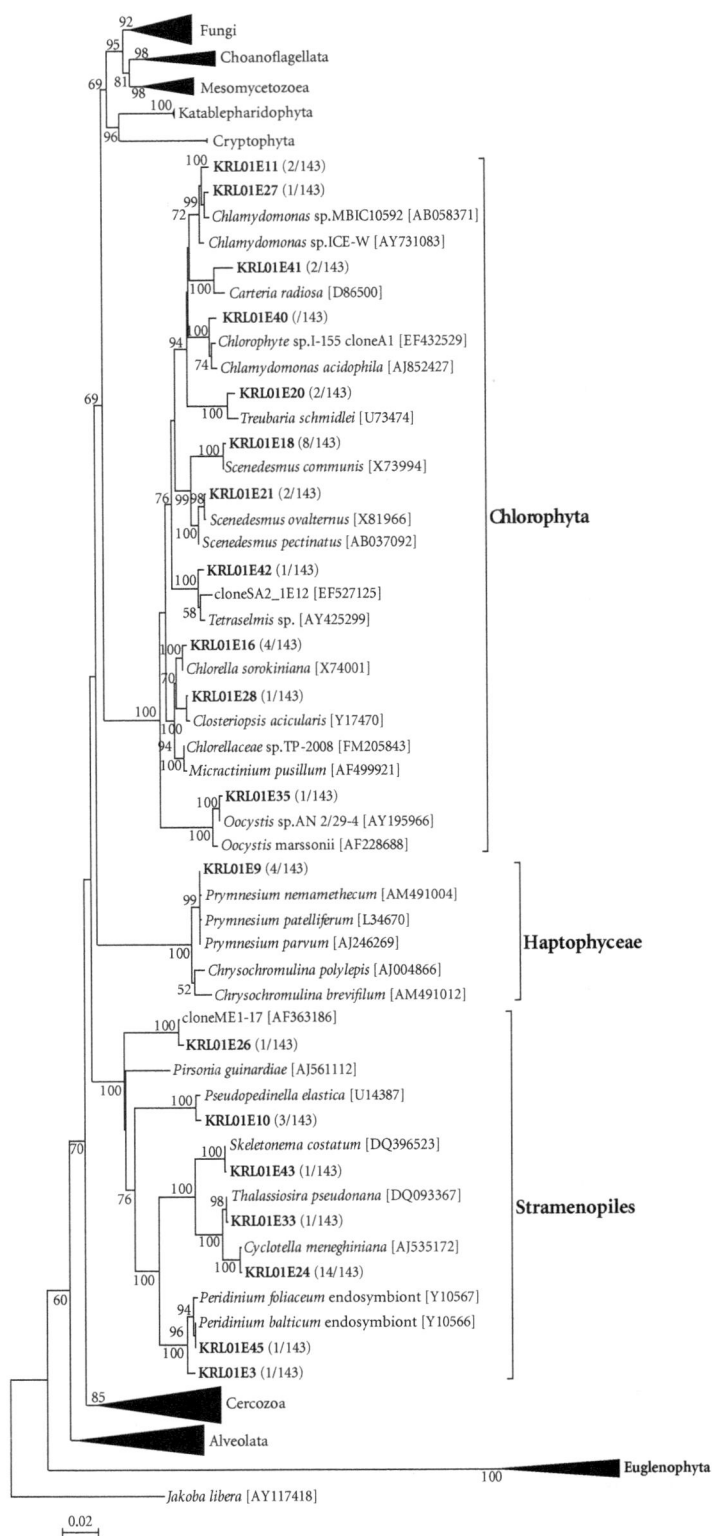

FIGURE 3: Continued.

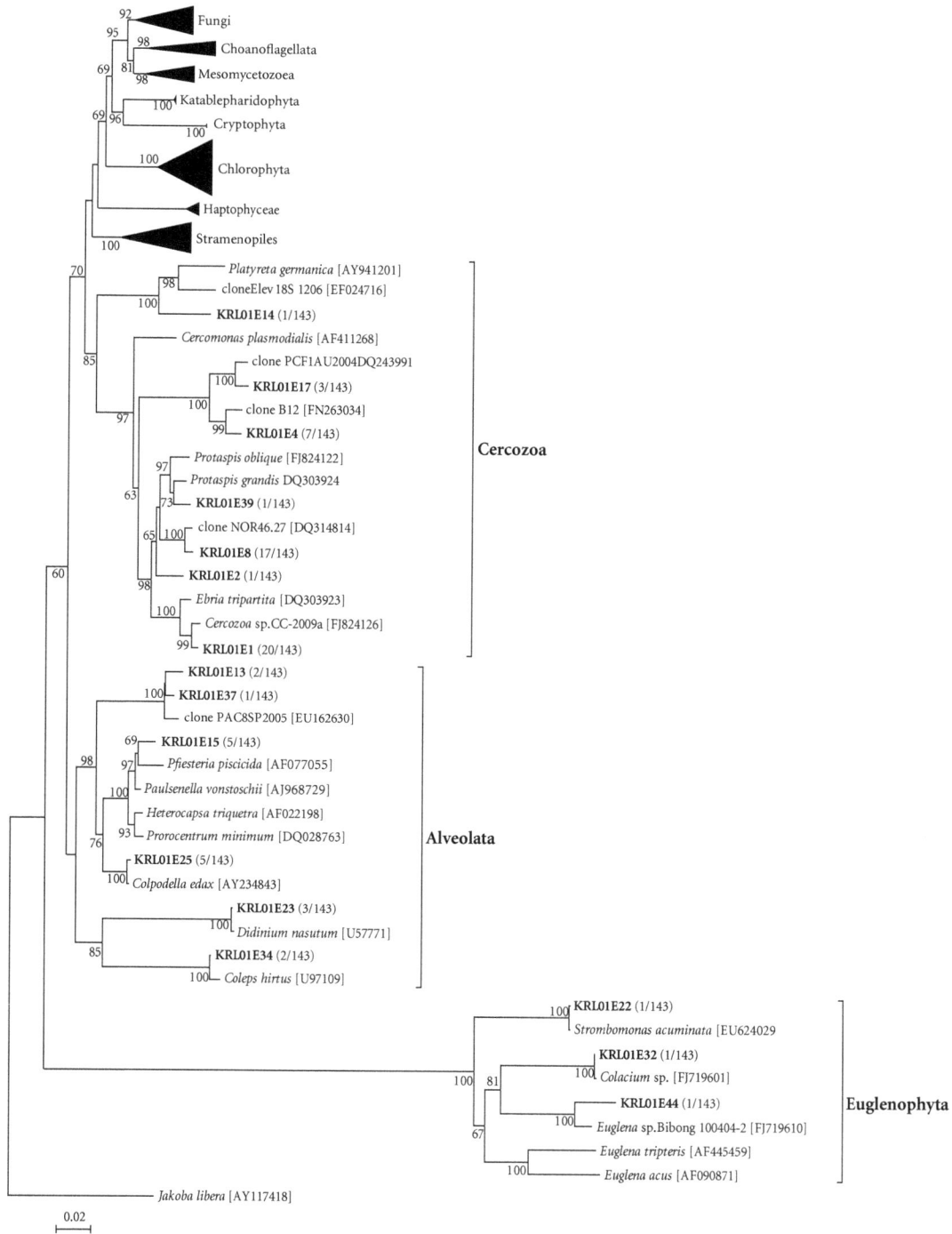

(c)

FIGURE 3: (a) Phylogenetic tree of relationships of 18S rDNA (ca. 1800 bp) of the representative unique (grouped on ≥98% similarity) eukaryotic clones (in bold) of the taxa Fungi, Choanoflagellata, Mesomycetozoea, Katablepharidophyta, and Cryptophyta, found in the Lake Karla water column, March 2010, based on the neighbour-joining method as determined by distance Jukes-Cantor analysis. One thousand bootstrap analyses (distance) were conducted. GenBank numbers are shown in parentheses. Numbers in parentheses indicate the relative abundance in the clone library. Scale bar represents 2% estimated. (b) Phylogenetic tree of relationships of 18S rDNA (ca. 1800 bp) of the representative unique (grouped on ≥98% similarity) eukaryotic clones (in bold) of the taxa Chlorophyta, Haptophyta, and Heterokontophyta (stramenopiles), found in the Lake Karla water column, March 2010, based on the neighbour-joining method as determined by distance Jukes-Cantor analysis. One thousand bootstrap analyses (distance) were conducted. GenBank numbers are shown in parentheses. Numbers in parentheses indicate the relative abundance in the clone library. Scale bar represents 2% estimated. (c) Phylogenetic tree of relationships of 18S rDNA (ca. 1800 bp) of the representative unique (grouped on ≥98% similarity) eukaryotic clones (in bold) of the taxa Cercozoa, Alveolata and Euglenophyta, found in the Lake Karla water column, March 2010, based on the neighbour-joining method as determined by distance Jukes-Cantor analysis. One thousand bootstrap analyses (distance) were conducted. GenBank numbers are shown in parentheses. Numbers in parentheses indicate the relative abundance in the clone library. Scale bar represents 2% estimated.

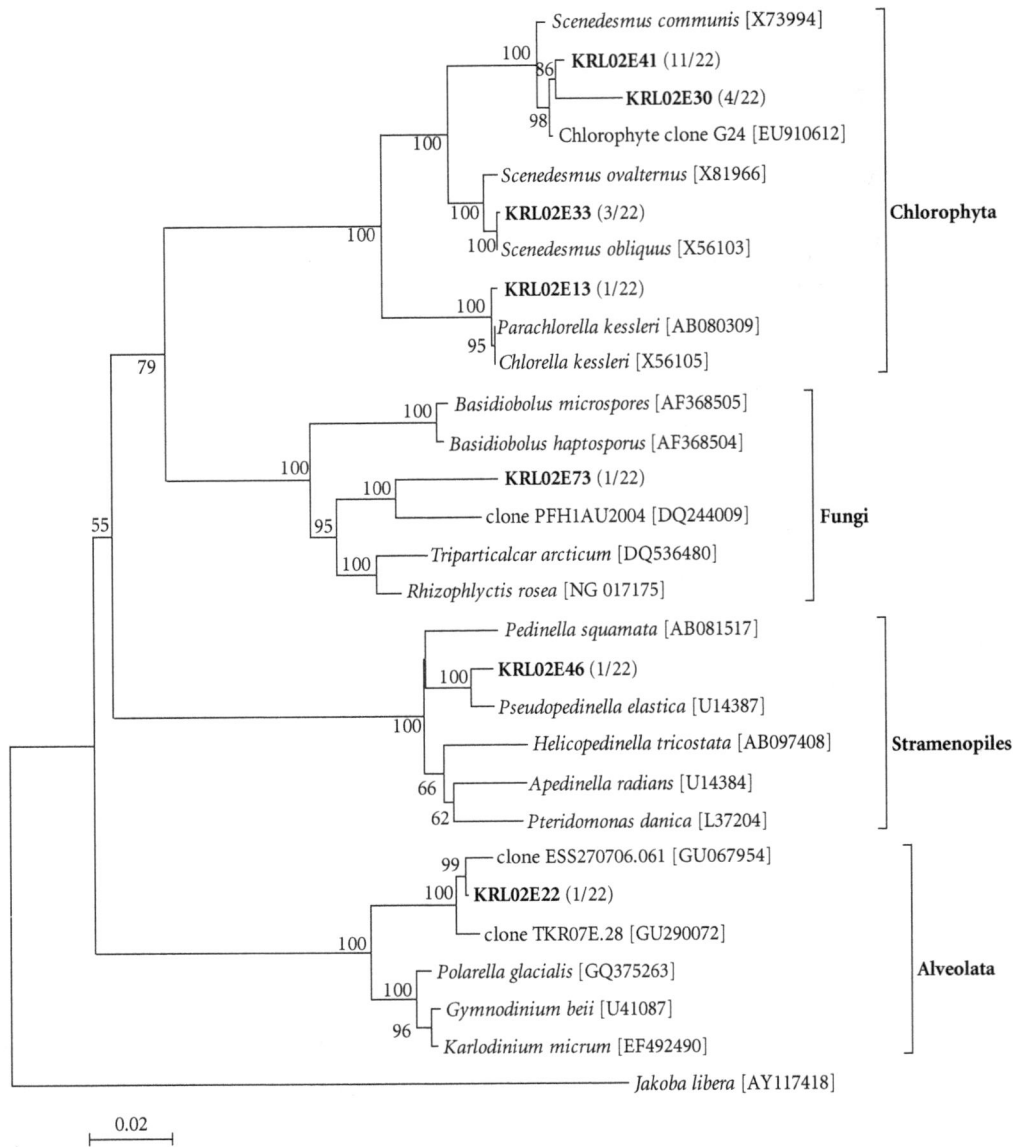

FIGURE 4: Phylogenetic tree of relationships of 18S rDNA (ca. 1600 bp) of the representative unique (grouped on ≥98% similarity) eukaryotic clones (in bold) found in the Lake Karla water column, April 2010, based on the neighbour-joining method as determined by distance Jukes-Cantor analysis. One thousand bootstrap analyses (distance) were conducted. GenBank numbers are shown in parentheses. Numbers in parentheses indicate the relative abundance in the clone library. Scale bar represents 2% estimated.

The slight discrepancy between the two approaches is expected (e.g., [43]) as PCR-based phylotype abundance is not quantitative but rather shows relative differences and can also be biased towards some groups. On the other hand, microscopic identification of unicellular phytoplankton can be problematic for some organisms, especially for these with complex/uncertain life cycles (e.g., [3, 25]). Thus, both approaches provide complementary rather redundant information. The gains of using both methods have already been depicted in limnological analysis (e.g., [42]) and especially for the unicellular eukaryotes [4, 6, 43].

The occurrence of diverse Chlorophyta phylotypes in both samplings (Figures 3 and 4), most of which were affiliated with well-characterized species, is related to the hypertrophic conditions prevailing in L. Karla. Chlorophyta are indicative of ecosystems receiving high nutrient loadings [1]. They have been found to dominate the clone library of a hypertrophic, polluted and heavily modified lake in Greece [3]. Some of these phylotypes, for example, *Scenedesmus* species, may constitute an important fraction of the freshwater total phytoplankton biomass, particularly in nutrient-rich ecosystems [44]. *Scenedesmus* species have capabilities

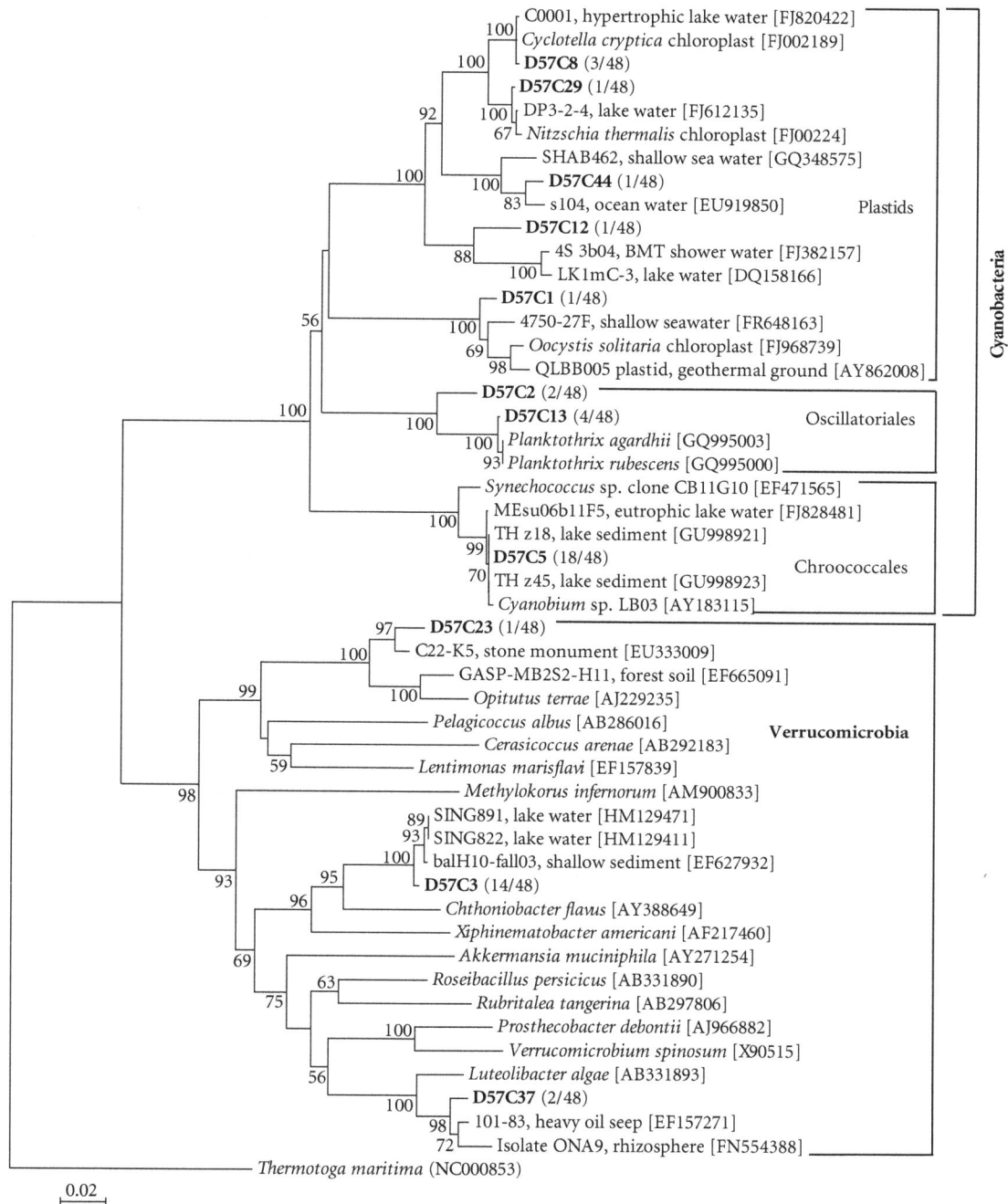

FIGURE 5: Phylogenetic tree of relationships of 16S rDNA (ca. 660 bp) of the representative unique (grouped on ≥98% similarity) Cyanobacterial clones (in bold), March 2010, based on the neighbour-joining method as determined by distance Jukes-Cantor analysis. One thousand bootstrap analyses (distance) were conducted. GenBank numbers are shown in parentheses. Numbers in parentheses indicate the relative abundance in the clone library. Scale bar represents 2% estimated.

of successful air dispersal and colonization of new aquatic habitats [45]. The hypertrophic conditions of the newly reconstructed L. Karla render its future rather erratic, since the prediction of community and ecosystem dynamics is decreased in eutrophic systems [46].

Apart from the Chlorophyta, other microorganisms in this study are associated with eutrophic/hypertrophic conditions. The found Euglenophyta-related phylotypes (Figure 3(c)) were affiliated with the genera *Colacium*, *Euglena* and *Strombomonas*. Members of the Euglenophyta are known to be abundant in highly eutrophic environments and on sediments polluted with organic matter [47]. Euglenophyta are considered biological indicators of organic pollution in seawater [48]. Cryptophyta (Figure 3(a)) are also a group forming blooms in eutrophic environments, yet their abundance are low due to high grazing rates of their protozoan predators [49]. Katablepharidophyta (Figure 3(a)) which were formerly classified as a subgroup of Cryptophyta, are

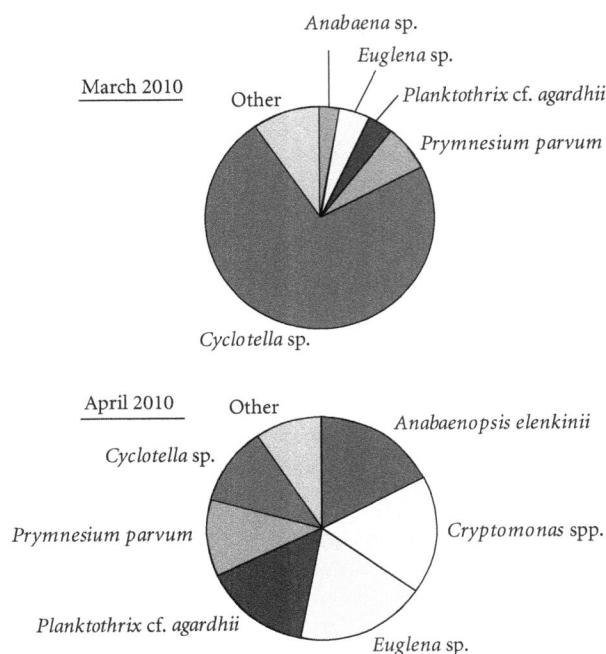

FIGURE 6: Relative biomass of the major taxa (90% dominance) recognized with light microscopy in the Lake Karla water column.

now considered to be a sister group of Cryptophyta [50] and could have similar environmental preferences. Choanoflagellata (Figure 3(a)) are epiphytic microorganisms depending on the quality of available organic matter, and many members of this group are adapted to using dissolved organic matter and colloidal organic particles [51].

The Cercozoa-related phylotypes (Figure 3(c)) were related to uncultivated environmental clones. Some well-characterized species such as *Ebria tripartita*, *Cercomonas plasmodialis*, and species of the genus *Protaspis* were affiliated with our retrieved sequences and fell in the Cercozoa taxonomic group. These taxa were also identified microscopically. Phylotypes KRL01E17 and KRL01E4 formed a novel clade in the Cercozoa, highly supported by the bootstrap test. Cercozoa phylotypes have been recovered from many different environments [52] but most of them are defined by molecular data and display huge morphological and ecological diversity [53]. They are mainly heterotrophs, including bacterivorous and predaceous species that phagocytize the cytoplasm of diatoms in marine ecosystems [54]. Cyst formation is a widespread characteristic among the Cercozoa [55], which probably allows their presence in anoxic sediments [56]. Members of the genus *Protaspis*, which was also recognized microscopically, comprise common predators in benthic marine ecosystems [55].

It is difficult to infer the trophic role of an organism by its phylogenetic position; however, the fact that most of the prementioned species/taxonomic groups have been detected with light microscopy of fixed and fresh samples in high numbers enforces the notion that these microorganisms are metabolically active in L. Karla. Based on the basic principle of ecology that the function of an ecosystem is defined by its dominant taxa, it is reasonable to characterize L. Karla

on the basis of its plankton as a hypertrophic system. Such systems tend to host various parasites as well as known toxin producers. Increased nutrient loadings are known to be associated with outbreaks of microparasitic species and blooms of harmful microalgae can also be indirectly promoted by nutrients inputs [57]. In the current study, such harmful eukaryotes belonging to the Alveolata, Fungi, Mesomycetozoea, and Haptophyta (Figures 3 and 4) along with some toxin-producing Cyanobacteria (Figure 5), have been identified by both molecular and microscopic analysis representing a very interesting but not previously described taxonomic and functional association [58].

Strict parasites are grouped in the Alveolata (Figures 3(c) and 4), as suggested by [59]. *Colpodella edax* can parasitize on Chlorophyta or Cryptophyta and can predate on protozoans smaller in size sucking out their cell contents by means of a rostrum [60]. Reference [59] associated this trophic strategy (myzocytosis) with parasitism. The Fungi (Figures 3(a) and 4) are exclusively composed of saprotrophs, known parasites of the phytoplankton community. Members of the Chytridiomycota can regulate the population of diatoms [59, 61]. Infection of certain phytoplankton species may suppress its development, thus Fungi parasitism can be an important factor controlling seasonal succession [61].

The taxonomic group of Mesomycetozoea (Figure 3(a)) includes facultative or obligate parasites [62]. Two orders have been described in Mesomycetozoea whereof Dermocystida consists exclusively of pathogenic microorganisms infecting fish (*Dermocystidium* sp.) as well as mammals and birds [62]. Members of this group have been found in another degraded lake ecosystem [3].

Known toxin producers such as *Prymnesium parvum* (Haptophyta) and *Pfiesteria* cf. *piscicida* (Alveolata) were also observed both in the clone libraries and by microscopic observations (Figures 3(b) and 6). To the best of our knowledge, it is the first time that these species occur simultaneously in the same ecosystem. *P. parvum* may form extensive blooms with major biogeochemical and ecological impact in brackish or inland waters [9, 63]. Massive kills of fish and birds have been attributed to blooms of *Prymnesium* [9, 25, 64]. *Pfiesteria piscicida* and *P. vonstochii* are parasites with similar feeding strategy and life cycle [65]. Temperature and salinity were suitable for the presence of *Pfiesteria* cf. *piscicida* in the lake as the species is detected in salinity ranging from 0.1–17.8 psu and temperature ranging from 3.2 to 25.5°C [66]. Toxin production of the *Pfiesteria* species increases in high nutrient loadings [13, 67, 68]. The genus *Peridinium* belonging to Alveolata (Figures 3(b) and 4) also includes species apparently related with toxin production [69]. Finally, the harmful organisms community of L. Karla hosts well-known toxin-producing Cyanobacteria [70] like *Planktothrix* cf. *agardhii*, *Anabaena* sp., and *Anabaenopsis elenkinii* (Figures 5 and 6).

During our samplings, salinity of L. Karla was elevated, generating the hypothesis that in L. Karla the occurrence of brackish or marine protists is feasible. Indeed, in both samplings we found phylotypes that were closely related to marine stramenopiles [71, 72]. *Cyclotella meneghiniana* present in the clone library of March, which was found by microscopy dominant in March 2010 and was identified as *Cyclotella* sp., is a common diatom species but tends to become abundant in organic, inorganic, heavy metal, or toxin-polluted environments [73]. *C. meneghiniana* has been recorded as being predominant or in remarkable occurrences in five polluted rivers and in four hypertrophic lakes [73]. *Thalassiosira* genus constitutes primarily of marine species (about 180 described species), while at least 12 species have been observed in freshwater ecosystems [74, 75]. The genus *Skeletonema* significantly contributes to phytoplankton blooms in many regions (e.g., [76–78]). In particular, *Skeletonema costatum* is a species that flourishes in nutrient-rich coastal waters throughout the world [79].

The presence of marine species within the stramenopiles (Figure 3(b)) poses the issue of the origin of these species in our study site. Karla is a newly reconstructed lake which is still under constant change and new microscopic eukaryotes colonize that ecosystem. Cyst formation is known for most of the groups observed like Cercozoa [54], Haptophyta [80], and Alveolata [81], so some microorganisms could have remained in the marsh and in the soil that formerly was the lakebed. The origin of the dominant freshwater microscopic eukaryotes (*C. meneghiniana*, *Scenedesmus* species) can be attributed to the inflow of River Pinios (the species were observed in the River's plankton, Moustaka-Gouni et al. unpublished data). Air dispersal is another possible vector for microorganisms. Chlorophyta have been found to be dominant in aerobiological studies [82] and are successful colonists in new aquatic habitats [45, 83].

In conclusion, our study showed that during two consecutive fish kill incidents which occurred in the recently reconstructed Lake Karla, Greece, in a six-week interval, the lake's water represented a cocktail of potentially toxic, *Planktothrix* cf. *agardhii*, *Prymnesium parvum*, and *Pfiesteria* cf. *piscicida* and parasitic species including *Dermocystidium* sp. Since the water temperature was far from the freezing point and the dissolved oxygen concentration was not even close to hypoxia, it is possible that the fish kills were caused by some of the microorganisms we observed. Apart from this risk, another problem for the ecosystem during the filling process of Lake Karla is the occurrence of other plankton, both freshwater and marine species, which are typical of eutrophic-hypertrophic conditions.

Acknowledgments

Pantelis Sidiropoulos is acknowledged for providing an earlier version of the Lake Karla map. Part of this work was supported by the John S. Latsis Public Benefit Foundation, Research Programs 2011. The three anonymous reviewers are fully acknowledged for their comments on the paper.

References

[1] C. Lepère, I. Domaizon, and D. Debroas, "Community composition of lacustrine small eukaryotes in hyper-eutrophic conditions in relation to top-down and bottom-up factors," *FEMS Microbiology Ecology*, vol. 61, no. 3, pp. 483–495, 2007.

[2] E. Michaloudi, M. Moustaka-Gouni, S. Gkelis, and K. Pantelidakis, "Plankton community structure during an ecosystem disruptive algal bloom of *Prymnesium parvum*," *Journal of Plankton Research*, vol. 31, no. 3, pp. 301–309, 2009.

[3] S. Genitsaris, K. A. Kormas, and M. Moustaka-Gouni, "Microscopic eukaryotes living in a dying lake (Lake Koronia, Greece)," *FEMS Microbiology Ecology*, vol. 69, no. 1, pp. 75–83, 2009.

[4] D. A. Caron, P. D. Countway, and M. V. Brown, "The growing contributions of molecular biology and immunology to protistan ecology: molecular signatures as ecological tools," *Journal of Eukaryotic Microbiology*, vol. 51, no. 1, pp. 38–48, 2004.

[5] F. Zhu, R. Massana, F. Not, D. Marie, and D. Vaulot, "Mapping of picoeucaryotes in marine ecosystems with quantitative PCR of the 18S rRNA gene," *FEMS Microbiology Ecology*, vol. 52, no. 1, pp. 79–92, 2005.

[6] T. Stoeck, W. H. Fowle, and S. S. Epstein, "Methodology of protistan discovery: from rRNA detection to quality scanning electron microscope images," *Applied and Environmental Microbiology*, vol. 69, no. 11, pp. 6856–6863, 2003.

[7] C. Brönmark and L.-A. Hansson, *The Biology of Lakes and Ponds*, Oxford University Press, Oxford, UK, 2nd edition, 2005.

[8] B. W. Brooks, S. V. James, T. W. Valenti et al., "Comparative toxicity of *Prymnesium parvum* in Inland waters," *Journal of the American Water Resources Association*, vol. 46, no. 1, pp. 45–62, 2010.

[9] D. L. Roelke, J. P. Grover, B. W. Brooks et al., "A decade of fish-killing *Prymnesium parvum* blooms in Texas: roles of in flow and salinity," *Journal of Plankton Research*, vol. 33, no. 2, pp. 243–253, 2011.

[10] B. Edvardsen and E. Paasche, "Bloom dynamics and physiology of *Prymnesium* and *Chrysochromulina*," in *Physiological Ecology of Harmful Algae Blooms*, D. M. Anderson, A. D. Cembella, and G. M. Hallegraeff, Eds., pp. 193–208, Springer, Berlin, Germany, 1998.

[11] J. W. Baker, J. P. Grover, B. W. Brooks et al., "Growth and toxicity of *Prymnesium parvum* (Haptophyta) as a function of salinity, light, and temperature," *Journal of Phycology*, vol. 43, no. 2, pp. 219–227, 2007.

[12] M. Moustaka-Gouni, C. M. Cook, S. Gkelis et al., "The coincidence of a *Prymnesium parvum* bloom and the mass kill of birds and fish in Lake Koronia," *Harmful Algae News*, vol. 26, pp. 1–2, 2004.

[13] J. M. Burkholder, H. B. Glasgow, and C. W. Hobbs, "Fish kills linked to a toxic ambush-predator dinoflagellate: distribution and environmental conditions," *Marine Ecology Progress Series*, vol. 124, no. 1–3, pp. 43–61, 1995.

[14] K. S. Jakobsen, T. Tengs, A. Vatne et al., "Discovery of the toxic dinoflagellate *Pfiesteria* in northern European waters," *Proceedings of the Royal Society B*, vol. 269, no. 1487, pp. 211–214, 2002.

[15] T. G. Park, E. M. Bell, I. Pearce, P. A. Rublee, C. J. S. Bolch, and G. M. Hallegraeff, "Detection of a novel ecotype of *Pfiesteria piscicida* (Dinophyceae) in an Antarctic saline lake by real-time PCR," *Polar Biology*, vol. 30, no. 7, pp. 843–848, 2007.

[16] E. J. Noga, "Toxic algae, fish kills and fish disease," *Fish Pathology*, vol. 33, no. 4, pp. 337–342, 1998.

[17] D. A. Caron, P. D. Countway, P. Savai et al., "Defining DNA-based operational taxonomic units for microbial-eukaryote ecology," *Applied and Environmental Microbiology*, vol. 75, no. 18, pp. 5797–5808, 2009.

[18] Y. Chatzinikolaou, A. Ioannou, and M. Lazaridou, "Intra-basin spatial approach on pollution load estimation in a large Mediterranean river," *Desalination*, vol. 250, no. 1, pp. 118–129, 2010.

[19] C. I. Ananiadis, "Limnological study of Lake Karla," *Bulletin de l'Institut Océanographique*, vol. 1083, pp. 1–19, 1956.

[20] N. Carter, "New or interesting algae from brackish water," *Archiv fuer Protistenkunde*, vol. 90, pp. 1–68, 1937.

[21] J. C. Green, D. J. Hibberd, and R. N. Pienaar, "The taxonomy of *Prymnesium* (Prymnesiophyceae) including a description of a new cosmpolitan species P. patellifera sp. nov. and further observations on *P. parvum* N. Carter," *British Phycological Journal*, vol. 17, pp. 363–382, 1982.

[22] K. Steidinger, J. Landsberg, R. W. Richardson et al., "Classification and identification of *Pfiesteria* and *Pfiesteria*-like species," *Environmental Health Perspectives*, vol. 109, no. 5, pp. 661–665, 2001.

[23] R. W. Litaker, M. W. Vandersea, S. R. Kibler, V. J. Madden, E. J. Noga, and P. A. Tester, "Life cycle of the heterotrophic dinoflagellate *Pfiesteria piscicida* (Dinophyceae)," *Journal of Phycology*, vol. 38, no. 3, pp. 442–463, 2002.

[24] H. Utermöhl, "Zur Vervollkommung der quantitativen Phyto-plankton-Methodik," *Verhandlungen der Internationalen Vereinigung für Theoretische und Angewandte Limnologie*, vol. 9, pp. 1–38, 1958.

[25] M. Moustaka-Gouni, K. A. Kormas, E. Vardaka, M. Katsiapi, and S. Gkelis, "Raphidiopsis mediterranea Skuja represents non-heterocytous life-cycle stages of Cylindrospermopsis raciborskii (Woloszynska) Seenayya et Subba Raju in Lake Kastoria (Greece), its type locality: evidence by morphological and phylogenetic analysis," *Harmful Algae*, vol. 8, no. 6, pp. 864–872, 2009.

[26] M. Katsiapi, M. Moustaka-Gouni, E. Michaloudi, and K. A. Kormas, "Phytoplankton and water quality in a Mediterranean drinking-water reservoir (Marathonas Reservoir, Greece)," *Environmental Monitoring and Assessment*, vol. 181, no. 1–4, pp. 563–575, 2011.

[27] L. Medlin, H. J. Elwood, S. Stickel, and M. L. Sogin, "The characterization of enzymatically amplified eukaryotic 16S-like rRNA-coding regions," *Gene*, vol. 71, no. 2, pp. 491–499, 1988.

[28] S. C. Dawson and N. R. Pace, "Novel kingdom-level eukaryotic diversity in anoxic environments," *Proceedings of the National Academy of Sciences of the United States of America*, vol. 99, no. 12, pp. 8324–8329, 2002.

[29] F. V. Wintzingerode, U. B. Göbel, and E. Stackebrandt, "Determination of microbial diversity in environmental samples: pitfalls of PCR-based rRNA analysis," *FEMS Microbiology Reviews*, vol. 21, no. 3, pp. 213–229, 1997.

[30] D. Spiegelman, G. Whissell, and C. W. Greer, "A survey of the methods for the characterization of microbial consortia and communities," *Canadian Journal of Microbiology*, vol. 51, no. 5, pp. 355–386, 2005.

[31] U. Nübel, F. Garcia-Pichel, and G. Muyzer, "PCR primers to amplify 16S rRNA genes from cyanobacteria," *Applied and Environmental Microbiology*, vol. 63, no. 8, pp. 3327–3332, 1997.

[32] P. B. Brown and G. V. Wolfe, "Protist genetic diversity in the acidic hydrothermal environments of Lassen Volcanic National Park, USA," *Journal of Eukaryotic Microbiology*, vol. 53, no. 6, pp. 420–431, 2006.

[33] K. E. Ashelford, N. A. Chuzhanova, J. C. Fry, A. J. Jones, and A. J. Weightman, "At least 1 in 20 16S rRNA sequence records currently held in public repositories is estimated to contain substantial anomalies," *Applied and Environmental Microbiology*, vol. 71, no. 12, pp. 7724–7736, 2005.

[34] M. Nebel, C. Pfabel, A. Stock, M. Dunthorn, and T. Stoeck, "Delimiting operational taxonomic units for assessing ciliate environmental diversity using small-subunit rRNA gene sequences," *Environmental Microbiology Reports*, vol. 3, no. 2, pp. 154–158, 2011.

[35] K. Tamura, J. Dudley, M. Nei, and S. Kumar, "MEGA4: Molecular Evolutionary Genetics Analysis (MEGA) software version 4.0," *Molecular Biology and Evolution*, vol. 24, no. 8, pp. 1596–1599, 2007.

[36] I. J. Good, "The population frequencies of species and the estimation of population parameters," *Biometrika*, vol. 40, pp. 237–264, 1953.

[37] A. Chao, "Non-parametric estimation of the number of classes in a population," *Scandinavian Journal of Statistics*, vol. 11, pp. 265–270, 1984.

[38] A. Chao, "Estimating the population size for capture—recapture data with unequal catchability," *Biometrics*, vol. 43, no. 4, pp. 783–791, 1987.

[39] ILRI, "Effectiveness and social/environmental impacts of irrigation projects: a review," in *Annual Report 1988 of the International Institute for Land Reclamation and Improvement (ILRI)*, pp. 18–34, Wageningen, The Netherlands, 1989.

[40] X. Hao and C. Chang, "Does long-term heavy cattle manure application increase salinity of a clay loam soil in semi-arid southern Alberta?" *Agriculture, Ecosystems and Environment*, vol. 94, no. 1, pp. 89–103, 2003.

[41] K. A. Kormas, E. Vardaka, M. Moustaka-Gouni et al., "Molecular detection of potentially toxic cyanobacteria and their associated bacteria in lake water column and sediment," *World Journal of Microbiology and Biotechnology*, vol. 26, no. 8, pp. 1473–1482, 2010.

[42] K. A. Kormas, S. Gkelis, E. Vardaka, and M. Moustaka-Gouni, "Morphological and molecular analysis of bloom-forming Cyanobacteria in two eutrophic, shallow Mediterranean lakes," *Limnologica*, vol. 41, pp. 167–173, 2011.

[43] W. Luo, C. Bock, H. R. Li, J. Padisák, and L. Krienitz, "Molecular and microscopic diversity of planktonic eukaryotes in the oligotrophic Lake Stechlin (Germany)," *Hydrobiologia*, vol. 661, pp. 133–143, 2011.

[44] S. S. An, T. Friedl, and E. Hegewald, "Phylogenetic relationships of *Scenedesmus* and *Scenedesmus*-like coccoid green algae as inferred from ITS-2 rDNA sequence comparisons," *Plant Biology*, vol. 1, no. 4, pp. 418–428, 1999.

[45] S. Genitsaris, M. Moustaka-Gouni, and K. A. Kormas, "Airborne microeukaryote colonists in experimental water containers: diversity, succession, life histories and established food webs," *Aquatic Microbial Ecology*, vol. 62, no. 2, pp. 139–152, 2011.

[46] K. L. Cottingham, J. A. Rusak, and P. R. Leavitt, "Increased ecosystem variability and reduced predictability following fertilisation: evidence from palaeolimnology," *Ecology Letters*, vol. 3, no. 4, pp. 340–348, 2000.

[47] F. E. Round, *The Ecology of Algae*, Cambridge University Press, Cambridge, UK, 1984.

[48] I. V. Stonik and M. S. Selina, "Species composition and seasonal dynamics of density and biomass of euglenoids in Peter the Great Bay, Sea of Japan," *Russian Journal of Marine Biology*, vol. 27, no. 3, pp. 174–176, 2001.

[49] M. Latasa, R. Scharek, M. Vidal et al., "Preferences of phytoplankton groups for waters of different trophic status in the northwestern Mediterranean sea," *Marine Ecology Progress Series*, vol. 407, pp. 27–42, 2010.

[50] N. Okamoto and I. Inouye, "The katablepharids are a distant sister group of the Cryptophyta: a proposal for Katablepharidophyta divisio nova/Kathablepharida phylum novum based on SSU rDNA and beta-tubulin phylogeny," *Protist*, vol. 156, no. 2, pp. 163–179, 2005.

[51] E. B. Sherr, B. F. Sherr, and L. Fessenden, "Heterotrophic protists in the Central Arctic Ocean," *Deep-Sea Research Part II*, vol. 44, no. 8, pp. 1665–1682, 1997.

[52] K. Romari and D. Vaulot, "Composition and temporal variability of picoeukaryote communities at a coastal site of the English Channel from 18S rDNA sequences," *Limnology and Oceanography*, vol. 49, no. 3, pp. 784–798, 2004.

[53] P. J. Keeling, "Foraminifera and Cercozoa are related in actin phylogeny: two orphans find a home?" *Molecular Biology and Evolution*, vol. 18, no. 8, pp. 1551–1557, 2001.

[54] E. Schnepf and S. F. Kühn, "Food uptake and fine structure of *Cryothecomonas longipes* sp. nov., a marine nanoflagellate incertae sedis feeding phagotrophically on large diatoms," *Helgoland Marine Research*, vol. 54, no. 1, pp. 18–32, 2000.

[55] M. Hoppenrath and B. S. Leander, "Dinoflagellate, euglenid, or cercomonad? The ultrastructure and molecular phylogenetic position of Protaspis grandis n. sp," *Journal of Eukaryotic Microbiology*, vol. 53, no. 5, pp. 327–342, 2006.

[56] K. Piwosz and J. Pernthaler, "Seasonal population dynamics and trophic role of planktonic nanoflagellates in coastal surface waters of the Southern Baltic Sea," *Environmental Microbiology*, vol. 12, no. 2, pp. 364–377, 2010.

[57] P. T. Johnson and S. T. Carpenter, "Influence of eutrophication on disease in aquatic ecosystems: patterns, processes and predictions," in *Infectious Disease Ecology. The Effects of Ecosystems on Disease and of Disease on Ecosystems*, R. S. Ostfeld, F. Keesing, and V. T. Eviner, Eds., pp. 71–101, Princeton University Press, Princeton, NJ, USA, 2008.

[58] J. Padisák, L. O. Crossetti, and L. Naselli-Flores, "Use and misuse in the application of the phytoplankton functional classification: a critical review with updates," *Hydrobiologia*, vol. 621, no. 1, pp. 1–19, 2009.

[59] E. Lefèvre, B. Roussel, C. Amblard, and T. Sime-Ngando, "The molecular diversity of freshwater picoeukaryotes reveals high occurrence of putative parasitoids in the plankton," *PLoS ONE*, vol. 3, no. 6, Article ID e2324, 2008.

[60] G. Brugerolle, "*Cryptophagus subtilis*: a new parasite of cryptophytes affiliated with the Perkinsozoa lineage," *European Journal of Protistology*, vol. 37, no. 4, pp. 379–390, 2002.

[61] B. W. Ibelings, A. De Bruin, M. Kagami, M. Rijkeboer, M. Brehm, and E. Van Donk, "Host parasite interactions between freshwater phytoplankton and chytrid fungi (Chytridiomycota)," *Journal of Phycology*, vol. 40, no. 3, pp. 437–453, 2004.

[62] L. Mendoza, J. W. Taylor, and L. Ajello, "The class Mesomycetozoea: a heterogeneous group of microorganisms at the animal-fungal boundary," *Annual Review of Microbiology*, vol. 56, pp. 315–344, 2002.

[63] P. A. Rublee, D. L. Remington, E. F. Schaefer, and M. M. Marshall, "Detection of the dinozoans *Pfiesteria piscicida* and *P. shumwayae*: a review of detection methods and geographic distribution," *Journal of Eukaryotic Microbiology*, vol. 52, no. 2, pp. 83–89, 2005.

[64] T. Lindholm, P. Öhman, K. Kurki-Helasmo, B. Kincaid, and J. Meriluoto, "Toxic algae and fish mortality in a brackish-water lake in Aland, SW Finland," *Hydrobiologia*, vol. 397, pp. 109–120, 1999.

[65] K. A. Steidinger, J. M. Burkholder, H. B. Glasgow et al., "*Pfiesteria piscicida* gen. et sp. nov. (*Pfiesteriaceae* fam. nov.), a new

[66] L. L. Rhodes, J. E. Adamson, P. A. Rublee, and E. Schaefer, "Geographic distribution of *Pfiesteria* spp. (*Pfiesteriaceae*) in Tasman Bay and Canterbury, New Zealand (2002-03)," *New Zealand Journal of Marine and Freshwater Research*, vol. 40, no. 1, pp. 211–220, 2006.

[67] J. M. Burkholder and H. B. Glasgow, "*Pfiesteria piscicida* and other *Pfiesteria*-like dinoflagellates: behavior, impacts, and environmental controls," *Limnology and Oceanography*, vol. 42, no. 5, pp. 1052–1075, 1997.

[68] H. B. Glasgow, J. M. Burkholder, S. L. Morton, and J. Springer, "A second species of ichthyotoxic *Pfiesteria* (Dinamoebales, Dinophyceae)," *Phycologia*, vol. 40, no. 3, pp. 234–245, 2001.

[69] K. Rengefors and C. Legrand, "Toxicity in *Peridinium aciculiferum*—an adaptive strategy to outcompete other winter phytoplankton?" *Limnology and Oceanography*, vol. 46, no. 8, pp. 1990–1997, 2001.

[70] J. Huisman, H. C. P. Matthus, and P. M. Visser, *Harmful Cyanobacteria*, Springer, Dordrecht, The Netherlands, 2005.

[71] B. Díez, C. Pedrós-Alió, and R. Massana, "Study of genetic diversity of eukaryotic picoplankton in different oceanic regions by small-subunit rRNA gene cloning and sequencing," *Applied and Environmental Microbiology*, vol. 67, no. 7, pp. 2932–2941, 2001.

[72] R. Massana, J. Castresana, V. Balagué et al., "Phylogenetic and ecological analysis of novel marine stramenopiles," *Applied and Environmental Microbiology*, vol. 70, no. 6, pp. 3528–3534, 2004.

[73] B. J. Finlay, E. B. Monaghan, and S. C. Maberly, "Hypothesis: the rate and scale of dispersal of freshwater diatom species is a function of their global abundance," *Protist*, vol. 153, no. 3, pp. 261–273, 2002.

[74] F. Round, R. Crawford, and D. Mann, *The Diatoms Biology and Morphology of the Genera*, Cambridge University Press, Cambridge, UK, 1990.

[75] P. C. Silva and G. R. Hasle, "(1087) Proposal to conserve Thalassiosiraceae against Lauderiaceae and Planktoniellaceae (Algae)," *Taxon*, vol. 43, no. 2, pp. 287–289, 1994.

[76] D. Karentz and T. J. Smayda, "Temperature and seasonal occurrence patterns of 30 dominant phytoplankton species in Narragansett Bay over a 22-year period (1959–1980)," *Marine Ecology Progress Series*, vol. 18, pp. 277–293, 1984.

[77] J. E. Cloern, B. E. Cole, R. L. J. Wong, and A. E. Alpine, "Temporal dynamics of estuarine phytoplankton: a case study of San Francisco Bay," *Hydrobiologia*, vol. 129, no. 1, pp. 153–176, 1985.

[78] M. Ribera d'Alcalà, F. Conversano, F. Corato et al., "Seasonal patterns in plankton communities in pluriannual time series at a coastal Mediterranean site (Gulf of Naples): an attempt to discern recurrences and trends," *Scientia Marina*, vol. 68, no. 1, pp. 65–83, 2004.

[79] D. Sarno, W. H. Kooista, L. K. Medlin, I. Percopo, and A. Zingone, "Diversity in the genus *Skeletonema* (Bacillariophyceae). II. An assessment of the taxonomy of *S. costatum*-like species with the description of four new species," *Journal of Phycology*, vol. 41, no. 1, pp. 151–176, 2005.

[80] O. Beltrami, M. Escobar, and G. Collantes, "New record of *Prymnesium parvum* f. *patelliferum* (Green, Hibberd & Piennar) Larsen stat. nov. (Prymnesiophyceae) from Valparaíso Bay," *Investigaciones Marinas*, vol. 35, no. 1, pp. 97–104, 2007.

[81] B. S. Leander, O. N. Kuvardina, V. V. Aleshin, A. P. Mylnikov, and P. J. Keeling, "Molecular phylogeny and surface morphology of *Colpodella edax* (Alveolata): insights into the

phagotrophic ancestry of apicomplexans," *Journal of Eukaryotic Microbiology*, vol. 50, no. 5, pp. 334–340, 2003.

[82] S. Genitsaris, K. A. Kormas, and M. Moustaka-Gouni, "Airborne algae and cyanobacteria: occurrence and related health effects," *Frontiers in Bioscience*, vol. 3, pp. 772–787, 2011.

[83] A. Chrisostomou, M. Moustaka-Gouni, S. Sgardelis, and T. Lanaras, "Air-dispersed phytoplankton in a mediterranean river-reservoir system (aliakmon-polyphytos, Greece)," *Journal of Plankton Research*, vol. 31, no. 8, pp. 877–884, 2009.

Permissions

The contributors of this book come from diverse backgrounds, making this book a truly international effort. This book will bring forth new frontiers with its revolutionizing research information and detailed analysis of the nascent developments around the world.

We would like to thank all the contributing authors for lending their expertise to make the book truly unique. They have played a crucial role in the development of this book. Without their invaluable contributions this book wouldn't have been possible. They have made vital efforts to compile up to date information on the varied aspects of this subject to make this book a valuable addition to the collection of many professionals and students.

This book was conceptualized with the vision of imparting up-to-date information and advanced data in this field. To ensure the same, a matchless editorial board was set up. Every individual on the board went through rigorous rounds of assessment to prove their worth. After which they invested a large part of their time researching and compiling the most relevant data for our readers. Conferences and sessions were held from time to time between the editorial board and the contributing authors to present the data in the most comprehensible form. The editorial team has worked tirelessly to provide valuable and valid information to help people across the globe.

Every chapter published in this book has been scrutinized by our experts. Their significance has been extensively debated. The topics covered herein carry significant findings which will fuel the growth of the discipline. They may even be implemented as practical applications or may be referred to as a beginning point for another development. Chapters in this book were first published by Hindawi Publishing Corporation; hereby published with permission under the Creative Commons Attribution License or equivalent.

The editorial board has been involved in producing this book since its inception. They have spent rigorous hours researching and exploring the diverse topics which have resulted in the successful publishing of this book. They have passed on their knowledge of decades through this book. To expedite this challenging task, the publisher supported the team at every step. A small team of assistant editors was also appointed to further simplify the editing procedure and attain best results for the readers.

Our editorial team has been hand-picked from every corner of the world. Their multi-ethnicity adds dynamic inputs to the discussions which result in innovative outcomes. These outcomes are then further discussed with the researchers and contributors who give their valuable feedback and opinion regarding the same. The feedback is then collaborated with the researches and they are edited in a comprehensive manner to aid the understanding of the subject.

Apart from the editorial board, the designing team has also invested a significant amount of their time in understanding the subject and creating the most relevant covers. They scrutinized every image to scout for the most suitable representation of the subject and create an appropriate cover for the book.

The publishing team has been involved in this book since its early stages. They were actively engaged in every process, be it collecting the data, connecting with the contributors or procuring relevant information. The team has been an ardent support to the editorial, designing and production team. Their endless efforts to recruit the best for this project, has resulted in the accomplishment of this book. They are a veteran in the field of academics and their pool of knowledge is as vast as their experience in printing. Their expertise and guidance has proved useful at every step. Their uncompromising quality standards have made this book an exceptional effort. Their encouragement from time to time has been an inspiration for everyone.

The publisher and the editorial board hope that this book will prove to be a valuable piece of knowledge for researchers, students, practitioners and scholars across the globe.

List of Contributors

Dandan Han
Key Laboratory of Regenerative Biology, Guangzhou Institutes of Biomedicine and Health, Chinese Academy of Sciences, 510530 Guangzhou, China
Graduate University of Chinese Academy of Sciences, 100049 Beijing, China

Luan Wen
Key Laboratory of Regenerative Biology, Guangzhou Institutes of Biomedicine and Health, Chinese Academy of Sciences, 510530 Guangzhou, China
Graduate University of Chinese Academy of Sciences, 100049 Beijing, China
Section on Molecular Morphogenesis, Laboratory of Gene Regulation and Development, Program on Cell Regulation and Metabolism, Eunice Kennedy Shriver, National Institute Child Health and Human Development (NICHD), National Institutes of Health (NIH), Building 18T, Room 106, 18 Library DR MSC 5431, Bethesda, MD 20892-5431, USA

Yonglong Chen
Key Laboratory of Regenerative Biology, Guangzhou Institutes of Biomedicine and Health, Chinese Academy of Sciences, 510530 Guangzhou, China

Christian Michel-Cuello
Programa Multidisciplinario de Posgrado en Ciencias Ambientales, Universidad Autónoma de San Luis Potosí, Avenida Dr. Manuel Nava No. 6, Zona Universitaria, 78210 San Luis Potosí, SLP, Mexico

Marco Martín González-Chávez, Miguel Ruiz-Cabrera, Imelda Ortiz-Cerda, Lorena Moreno-Vilet, Alicia Grajales -Lagunes and Mario Moscosa-Santillán
Facultad de Ciencias Químicas, Universidad Autónoma de San Luis Potosí, Avenida Dr. Manuel Nava No. 6, Zona Universitaria, 78210 San Luis Potosí, SLP, Mexico

Johanne Bonnin
Institut de Chimie Organique et Analytique, Université dÓrléans, Rue d'Issoudun, BP 16729, 45067 Orléans Cedex 02, France

Ling Jun Zhan, Lin Lin Bao, Feng Di Li, Qi Lv, Li Li Xu and Chuan Qin
Key Laboratory of Human Diseases Comparative Medicine, Ministry of Health, Institute of Laboratory Animal Science, Chinese Academy of Medical Sciences (CAMS) & Comparative Medicine Centre, Peking Union Medical Collage (PUMC), Pan Jia Yuan Nan Li No. 5, Chao Yang District, Beijing 100021, China
Key Laboratory of Human Diseases Animal Model, State Administration of Traditional Chinese Medicine, Pan Jia Yuan Nan Li No. 5, Chao Yang District, Beijing 100021, China

Gholamreza Bahari and Ebrahim Eskandari-Nasab
Department of Clinical Biochemistry, School of Medicine, Zahedan University of Medical Sciences, Zahedan 98167-43463, Iran

Mohammad Hashemi
Department of Clinical Biochemistry, School of Medicine, Zahedan University of Medical Sciences, Zahedan 98167-43463, Iran
Cellular and Molecular Research Center, Zahedan University of Medical Sciences, Zahedan 98167-43463, Iran

Mohsen Taheri
Genetic of Non-Communicable Disease Research Center, Zahedan University of Medical Science, Zahedan 98167-43463, Iran

Mahdi Atabaki and Mohammad Naderi
Research Center for Infectious Diseases and Tropical Medicine, Zahedan University of Medical Sciences, Zahedan 98167-43463, Iran

Crystal A. Conway and Jose V. Lopez
Oceanographic Center, Nova Southeastern University, Dania Beach, FL 33004, USA

Nwadiuto Esiobu
Department of Biological Sciences, Florida Atlantic University, Davie, FL 33314, USA

F. Baruzzi, M. Cefola, A. Carito, S. Vanadia and N. Calabrese
Institute of Sciences of Food Production, National Research Council of Italy (CNR-ISPA), Via., G. Amendola 122/o, 70126 Bari, Italy

Shatha F. Dallo, Soonbae Hong and Anyu Tsai
Department of Biology, The University of Texas at San Antonio, One UTSA Circle, San Antonio, TX 78249, USA

James Denno
Department of Biology, The University of Texas at Austin, 1 University Station, Austin, TX 78712, USA

Bailin Zhang and Jing Yong Ye
Department of Biomedical Engineering, The University of Texas at San Antonio, One UTSA Circle, San Antonio, TX 78249, USA

Williams Haskins
Department of Biology, The University of Texas at San Antonio, One UTSA Circle, San Antonio, TX 78249, USA
Pediatric Biochemistry Laboratory, The University of Texas at San Antonio, San Antonio, TX 78249, USA
Department of Chemistry, The University of Texas at San Antonio, San Antonio, TX 78249, USA
RCMI Proteomics and Protein Biomar Feers Cores, The University of Texas at San Antonio, San Antonio, TX 78249, USA
Center for Research & Training in The Sciences, The University of Texas at San Antonio, San Antonio, TX 78249, USA
Division of Hematology/Oncology, Department of Medicine, Cancer Therapy & Research Center, The University of Texas Health Science Center at San Antonio, San Antonio, TX 78229, USA

Tao Weitao
Department of Biology, The University of Texas at San Antonio, One UTSA Circle, San Antonio, TX 78249, USA
Department of Biology, College of Science and Mathematics, Southwest Baptist University, 1600 University Avenue, Bolivar, MO 65613, USA

Sanjoy Banerjee and Helena Khatoon
Institute of Bioscience, Universiti Putra Malaysia, Selangor, 43400 Serdang, Malaysia

Mei Chen Ooi
Faculty of Veterinary Medicine, Universiti Putra Malaysia, Selangor, 43400 Serdang, Malaysia

Mohamed Shariff
Institute of Bioscience, Universiti Putra Malaysia, Selangor, 43400 Serdang, Malaysia
Faculty of Veterinary Medicine, Universiti Putra Malaysia, Selangor, 43400 Serdang, Malaysia

Lynne E. Murdoch, Michelle Maclean, Endarko Endarko, Scott J. MacGregor and John G. Anderson
The Robertson Trust Laboratory for Electronic Sterilisation Technologies, Department of Electronic and Electrical Engineering, University of Strathclyde-Glasgow, Glasgow G1, 1XW, UK

Ivan M. Petyaev and Yuriy K. Bashmakov
Lycotec Ltd. Granta Park Campus, Cambridge CB21 6GP, UK

Naylia A. Zigangirova, Natalie V. Kobets, Valery Tsibezov, Lydia N. Kapotina and Elena D. Fedina
Gamaleya Institute of Epidemiology and Microbiology, Ministry of Health, 18 Gamaleya Street, Moscow 123098, Russia

Expedito K. A. Camboim, Franklin Riet-Correa, and Marcia A. Melo
Unidade Acadêmica de Medicina Veterinária, Universidade Federal de Campina Grande, Avenida Universitária, s/n, Bairro Sta., Cecília, Patos, PB, CEP: 58700-970, Brazil

Michelle Z. Tadra-Sfeir, Emanuel M. de Souza and Fabio de O. Pedrosa
Laborat´orio de Fixação Biológica de Nitrogênio, Departamento de Bioquímica e Biologia Molecular, Universidade Federal do Paran´a, Curitiba, PR, CEP: 81531-980, Brazil

Paulo P. Andrade
Departamento de Gen´etica, Universidade Federal de Pernambuco, Recife, PE, CEP: 50670-901, Brazil

Chris S. McSweeney
CSIRO Livestock Industries, Queensland Bioscience Precinct, Carmody Road, 306, St Lucia, 4067, QLD, Australia

Tiago Gomes Fernandes, Amanda Rafaela Carneiro de Mesquita and Eulália Azevedo Ximenes
Laboratório de Fisiologia e Bioqu´ımica de Microorganismos, Departamento de Antibióticos, Centro de Ciˆencias Biol´ogicas, Universidade Federal de Pernambuco, 50670-901 Recife, PE, Brazil

Karina Perrelli Randau
Laboratório de Farmacognosia, Departamento de Farmácia Universidade Federal de Pernambuco, Centro de Ciˆencias da Sa´ude, 50670-901 Recife, PE, Brazil

Adelisa Alves Franchitti
Department of Biochemistry, Kansas State University, 141 Chalmers Hall, Manhattan, KS 66506, USA

Alapati Kavitha and Muvva Vijayalakshmi
Department of Botany and Microbiology, Acharya Nagarjuna University, Guntur 522 510, India

Bushra Uzair, Sobia Tabassum, Madiha Rasheed and Saima Firdous Rehman
Department of Bioinformatics and Biotechnology, International Islamic University Islamabad, Sector H-10, 44000 Islamabad, Pakistan

Michael Kube
Department of Crop and Animal Sciences, Humboldt-University of Berlin, Lentzeallee 55/57, 14195 Berlin, Germany
Max Planck Institute for Molecular Genetics, Ihnestr., 63, 14195 Berlin, Germany

Jelena Mitrovic and Bojan Duduk
Department of Plant Pathology, Institute of Pesticides and Environmental Protection, Banatska 31b, P.O. Box 163, 11080 Belgrade, Serbia

Ralf Rabus
Institute for Chemistry and Biology of the Marine Environment, Carl von Ossietzky University of Oldenburg, Carl-von-Ossietzky Straße 9-11, 26111 Oldenburg, Germany
Department for Microbiology, MaxPlanck Institute for Marine Microbiology, Celsiusstraße 1, 28359 Bremen, Germany

Erich Seemüller
Institute for Plant Protection in Fruit Crops and Viticulture, Federal Research Centre for Cultivated Plants, Schwabenheimer Straße 101, 69221 Dossenheim, Germany

Hilal Colak, Enver Baris Bingol and Omer Cetin
Department of Food Hygiene and Technology, Faculty of Veterinary Medicine, Istanbul University, Avcilar, 34320 Istanbul, Turkey

Hamparsun Hampikyan
The School of Vocational Studies, Beykent University, Buyukcekmece, 34500 Istanbul, Turkey

Meryem Akhan and Sumeyre Ipek Turgay
Academic Hygiene KGaA, Training, Audit and Consulting Services, Kuştepe Mahallesi, Tomurcuk Sokak, İzmen Sitesi, Sisli, 34387 Istanbul, Turkey

Hervé Meder, Anne Baumstummler, Renaud Chollet, Sophie Barrier, Monika Kukuczka, Frédéric Olivieri, Esther Welterlin, Vincent Beguin and Sébastien Ribault
Merck Millipore, Lab Solutions, BioMonitoring, Research & Development, Applications Group, 39, Route industrielle de la Hardt, 67120 Molsheim, France

Gurbuz Gunes, Neriman Yilmaz and Aylin Ozturk
Food Engineering Department, Istanbul Technical University, Maslak, 34469 Istanbul, Turkey

Kuan-Hsun Wu
Department of Pediatrics, Wan Fang Hospital, Taipei Medical University, Taipei 116, Taiwan

Ke-Chuan Wang
Graduate Institute of Medical Sciences, College of Medicine, Taipei Medical University, Taipei 110, Taiwan

Lin-Wen Lee
Department of Microbiology and Immunology, School of Medicine, College of Medicine, Taipei Medical University, Taipei 110, Taiwan

Yi-Ning Huang and Kuang-Sheng Yeh
Department of Veterinary Medicine, School of Veterinary Medicine, College of Bioresources and Agriculture, National Taiwan University, Taipei 106, Taiwan

Ahmad Reza Bahrami
Faculty of Veterinary Medicine, Islamic Azad University, Shahrekord Branch, Shahrekord, Iran

Ebrahim Rahimi
Department of Food Hygiene, Faculty of Veterinary Medicine, Islamic Azad University, Shahrekord Branch, Shahrekord, Iran

Hajieh Ghasemian Safaei
Department of Microbiology, Faculty of Medical Sciences, Isfahan University of Medical Sciences, Isfahan, Iran

Martin Krátký and Jarmila Vinšová
Department of Inorganic and Organic Chemistry, Faculty of Pharmacy, Charles University, Heyrovského 1203, 500 05 Hradec Králové, Czech Republic

Vladimír Buchta
Department of Clinical Microbiology, Faculty of Medicine and University Hospital, Charles University, Sokolská 581, 500 12 Hradec Králové, Czech Republic
Department of Biological and Medical Sciences, Faculty of Pharmacy, Charles University, Heyrovského 1203, 500 05 Hradec Králové, Czech Republic

Ebrahim Rahimi and Zienab Torki Baghbadorani
Department of Food Hygiene and Public Health, College of Veterinary Medicine, Islamic Azad University, Shahrekord Branch, P.O. Box 166, Shahrekord, Iran

Fahimeh Abdos
Department of Food Science and Technology, College of Agriculture, Islamic Azad University, Shahrekord Branch, Shahrekord, Iran

Hassan Momtaz
Department of Microbiology, College of Veterinary Medicine, Islamic Azad University, Shahrekord Branch, Shahrekord, Iran

Mohammad Jalali
Infectious Disease and Tropical Medicine Research Center and School of Food Science and Nutrition, Isfahan University of Medical Sciences, Isfahan, Iran

Shujie Wang, Chunyu Li and Xuehui Cai
National Key Laboratory of Veterinary Biotechnology, Harbin Veterinary Research Institute, Chinese Academy of Agricultural Science, Harbin 150001, China

Peng Liu, Yafang Tan, Dongsheng Zhou and Yongqiang Jiang
State Key Laboratory of Pathogen and Biosecurity, Beijing Institute of Microbiology and Epidemiology, Beijing 100071, China

Joana S. Teodósio, Manuel Simões, Luís F. Melo and Filipe J. Mergulhão
LEPAE-Department of Chemical Engineering, Faculty of Engineering, University of Porto, Rua Dr. Roberto Frias s/n, 4200-465 Porto, Portugal

Manuel A. Alves
CEFT-Department of Chemical Engineering, Faculty of Engineering, University of Porto, Rua Dr. Roberto Frias s/n, 4200-465 Porto, Portugal

Andreas Oikonomou and Konstantinos Ar. Kormas
Department of Ichthyology and Aquatic Environment, School of Agricultural Sciences, University of Thessaly, 384 46 Volos, Greece

Matina Katsiapi and Maria Moustaka-Gouni
Department of Botany, School of Biology, Aristotle University of Thessaloniki, 541 24 Thessaloniki, Greece

Hera Karayanni
Department of Biological Applications and Technology, University of Ioannina, 451 10 Ioannina, Greece

www.ingramcontent.com/pod-product-compliance
Lightning Source LLC
Chambersburg PA
CBHW070154240326
41458CB00126B/4703